NEUROMETHODS

Series Editor
Wolfgang Walz
University of Saskatchewan
Saskatoon, SK, Canada

For further volumes:
http://www.springer.com/series/7657

Clinical Trials in Neurology

Edited by

Felipe Fregni

Laboratory of Neuromodulation, Center of Clinical Research Learning,
Spaulding Rehabilitation Hospital, Harvard Medical School,
Boston, MA, USA

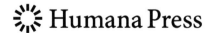 Humana Press

Editor
Felipe Fregni
Laboratory of Neuromodulation
Center of Clinical Research Learning
Spaulding Rehabilitation Hospital
Harvard Medical School
Boston, MA, USA

ISSN 0893-2336   ISSN 1940-6045 (electronic)
Neuromethods
ISBN 978-1-4939-9313-0   ISBN 978-1-4939-7880-9 (eBook)
https://doi.org/10.1007/978-1-4939-7880-9

Series Preface

Experimental life sciences have two basic foundations: concepts and tools. The *Neuromethods* series focuses on the tools and techniques unique to the investigation of the nervous system and excitable cells. It will not, however, shortchange the concept side of things as care has been taken to integrate these tools within the context of the concepts and questions under investigation. In this way, the series is unique in that it not only collects protocols but also includes theoretical background information and critiques which led to the methods and their development. Thus it gives the reader a better understanding of the origin of the techniques and their potential future development. The *Neuromethods* publishing program strikes a balance between recent and exciting developments like those concerning new animal models of disease, imaging, in vivo methods, and more established techniques, including, for example, immunocytochemistry and electrophysiological technologies. New trainees in neurosciences still need a sound footing in these older methods in order to apply a critical approach to their results.

Under the guidance of its founders, Alan Boulton and Glen Baker, the *Neuromethods* series has been a success since its first volume published through Humana Press in 1985. The series continues to flourish through many changes over the years. It is now published under the umbrella of Springer Protocols. While methods involving brain research have changed a lot since the series started, the publishing environment and technology have changed even more radically. Neuromethods has the distinct layout and style of the Springer Protocols program, designed specifically for readability and ease of reference in a laboratory setting.

The careful application of methods is potentially the most important step in the process of scientific inquiry. In the past, new methodologies led the way in developing new disciplines in the biological and medical sciences. For example, Physiology emerged out of Anatomy in the nineteenth century by harnessing new methods based on the newly discovered phenomenon of electricity. Nowadays, the relationships between disciplines and methods are more complex. Methods are now widely shared between disciplines and research areas. New developments in electronic publishing make it possible for scientists that encounter new methods to quickly find sources of information electronically. The design of individual volumes and chapters in this series takes this new access technology into account. Springer Protocols makes it possible to download single protocols separately. In addition, Springer makes its print-on-demand technology available globally. A print copy can therefore be acquired quickly and for a competitive price anywhere in the world.

Saskatoon, Canada   *Wolfgang Walz*

Preface

There has been a great development in biomedical research in the past few decades. This development has not been equally distributed among basic and clinical research. While there are a great number of basic research laboratories and scientists, there is a lack of its counterpart in the clinical research field. This contrast seems to be a bit more evident in the neurology field. Because of this imbalance, a number of discoveries in basic science have not been tested in clinical trials, thus hindering the translation of these discoveries. In addition, the increasingly positive efforts to use evidence-based medicine have increased the need for more clinical trials in neurology. Therefore, the expertise and ability to conduct high-quality trials for different conditions have become important issues to ultimately improve the quality of care in neurology.

This book aims to provide useful and important information to audiences in research, education, and clinical practice. It is intended for those involved in the medical field, in physical therapy, occupational therapy, nursing, public health as well as those concerned with health prospects and providing healthcare. Besides being an educational instrument for courses in clinical research and neurology, this book serves as an extensive reference for researchers and clinicians concerned with the development of critical thinking and innovative scientific evidence.

The chapters of this book were developed from a comprehensive literature search. In order to review and provide a useful guide to design clinical trials in neurology, we decided to systematically review the most cited articles in each reviewed field in neurology using the consort guidelines to discuss each aspect of these trials. With this review, we believe that we would then discuss the most important aspects of each field. Therefore, this book is not just a review of this topic, but also presents novel and important data in the field of clinical research in neurology. We reviewed the main aspects, using this framework, for 11 conditions in neurology.

We selected the 100 most cited articles using the following methodology: A review of the literature was conducted through the search of several electronic databases using specific keywords (as specified in each chapter). Based on specific inclusion and exclusion criteria, only randomized clinical trials published between 2005 and 2015 (except for the conditions where the number of clinical trials was not enough (in this case this period was increased)) were included. Each chapter reviews the methodology of the search, given there were some small differences between them.

During the writing of this book we actually learned that a good number of the most cited trials still have important limitations that we did not expect to find. In fact, this shows the real challenges of the field of clinical trials in neurology.

Hence, by gathering evidence on several neurological conditions as well as their primary challenges and future perspectives, we expect that researchers and clinicians will be able to address these aspects in new studies and apply this information on the pursuit of new knowledge. Finally, we expect that those involved with assessing evidence in clinical neurology will also find useful reviewing the methodology of clinical trial design in neurology.

Boston, MA, USA   *Felipe Fregni*

Contents

Contributors

CAMILA BONIN PINTO • *Laboratory of Neuromodulation, Center of Clinical Research Learning, Spaulding Rehabilitation Hospital, Harvard Medical School, Boston, MA, USA*

ALINE PATRÍCIA BRIET • *Laboratory of Neuromodulation, Center of Clinical Research Learning, Spaulding Rehabilitation Hospital, Harvard Medical School, Boston, MA, USA*

CELESTE R S DE CAMARGO • *Laboratory of Neuromodulation, Center of Clinical Research Learning, Spaulding Rehabilitation Hospital, Harvard Medical School, Boston, MA, USA*

LAURA CASTILLO-SAAVEDRA • *Laboratory of Neuromodulation, Center of Clinical Research Learning, Spaulding Rehabilitation Hospital, Harvard Medical School, Boston, MA, USA*

BEATRIZ TEIXEIRA COSTA • *Laboratory of Neuromodulation, Center of Clinical Research Learning, Spaulding Rehabilitation Hospital, Harvard Medical School, Boston, MA, USA*

RIVADÁVIO FERNANDES BATISTA DE AMORIN • *Laboratory of Neuromodulation, Center of Clinical Research Learning, Spaulding Rehabilitation Hospital, Harvard Medical School, Boston, MA, USA*

DANIEL DRESSER • *Laboratory of Neuromodulation, Center of Clinical Research Learning, Spaulding Rehabilitation Hospital, Harvard Medical School, Boston, MA, USA*

MIRRET M. EL-HAGRASSY • *Laboratory of Neuromodulation, Center of Clinical Research Learning, Spaulding Rehabilitation Hospital, Harvard Medical School, Boston, MA, USA*

ISADORA SANTOS FERREIRA • *Laboratory of Neuromodulation, Center of Clinical Research Learning, Spaulding Rehabilitation Hospital, Harvard Medical School, Boston, MA, USA*

FELIPE FREGNI • *Laboratory of Neuromodulation, Center of Clinical Research Learning, Spaulding Rehabilitation Hospital, Harvard Medical School, Boston, MA, USA; Spaulding Research Institute, Spaulding Rehabilitation Hospital, Harvard Medical School, Charlestown, MA, USA*

MELANIE FRENCH • *Laboratory of Neuromodulation, Center of Clinical Research Learning, Spaulding Rehabilitation Hospital, Harvard Medical School, Boston, MA, USA*

RODRIGO HUERTA-GUTIERREZ • *Laboratory of Neuromodulation, Center of Clinical Research Learning, Spaulding Rehabilitation Hospital, Harvard Medical School, Boston, MA, USA*

AURA M. HURTADO-PUERTO • *Laboratory of Neuromodulation, Center of Clinical Research Learning, Spaulding Rehabilitation Hospital, Harvard Medical School, Boston, MA, USA*

FELIPE JONES • *Laboratory of Neuromodulation, Center of Clinical Research Learning, Spaulding Rehabilitation Hospital, Harvard Medical School, Boston, MA, USA*

ALEJANDRA MALAVERA • *Laboratory of Neuromodulation, Center of Clinical Research Learning, Spaulding Rehabilitation Hospital, Harvard Medical School, Boston, MA, USA*

TAREK NAFEE • *Laboratory of Neuromodulation, Center of Clinical Research Learning, Spaulding Rehabilitation Hospital, Harvard Medical School, Boston, MA, USA*

ANTHONY O'BRIEN • *Laboratory of Neuromodulation, Center of Clinical Research Learning, Spaulding Rehabilitation Hospital, Harvard Medical School, Boston, MA, USA*

WELLINGSON SILVA PAIVA • *Department of Neurology, Hospital das Clinicas, Universidade de Sao Paulo - HC-FMUSP, Sao Paulo, Brazil*

CRISTINA RUSSO • *Laboratory of Neuromodulation, Center of Clinical Research Learning, Spaulding Rehabilitation Hospital, Harvard Medical School, Boston, MA, USA*

PRAKYAT SINGH • *Laboratory of Neuromodulation, Center of Clinical Research Learning, Spaulding Rehabilitation Hospital, Harvard Medical School, Boston, MA, USA*

RAFAELLY STAVALE • *Laboratory of Neuromodulation, Center of Clinical Research Learning, Spaulding Rehabilitation Hospital, Harvard Medical School, Boston, MA, USA*

ANA C. TEXEIRA-SANTOS • *Laboratory of Neuromodulation, Center of Clinical Research Learning, Spaulding Rehabilitation Hospital, Harvard Medical School, Boston, MA, USA*

AURORE THIBAUT • *Laboratory of Neuromodulation, Center of Clinical Research Learning, Spaulding Rehabilitation Hospital, Harvard Medical School, Boston, MA, USA*

FADDI GHASSAN SALEH VELEZ • *Laboratory of Neuromodulation, Center of Clinical Research Learning, Spaulding Rehabilitation Hospital, Harvard Medical School, Boston, MA, USA*

MARIA FERNANDA VILLARREAL • *Laboratory of Neuromodulation, Center of Clinical Research Learning, Spaulding Rehabilitation Hospital, Harvard Medical School, Boston, MA, USA*

RODRIGO WATANABE • *Laboratory of Neuromodulation, Center of Clinical Research Learning, Spaulding Rehabilitation Hospital, Harvard Medical School, Boston, MA, USA*

ANA LUIZA C. ZANINOTTO • *Laboratory of Neuromodulation, Center of Clinical Research Learning, Spaulding Rehabilitation Hospital, Harvard Medical School, Boston, MA, USA*

Chapter 1

Introduction

Isadora Santos Ferreira, Beatriz Teixeira Costa, and Felipe Fregni

Abstract

This book chapter discusses the main methodological and regulatory aspects of clinical research, guiding future investigators to conduct valuable, innovative, and appropriate trials. As clinical research has the main goal of generating knowledge to the medical field and advancing diagnostic/therapeutic methods, understanding essential concepts and foundations of clinical research is essential for improving the field. Therefore, important topics such as study population, trial outcomes, design, statistical analysis, and ethical aspects are contemplated in this chapter. As the neurology field is the focus of this book, practical examples on neurological disorders are also discussed.

Key words Clinical trial, Research, Ethics, Research methodology, Neurology

1 Overview

The concept of clinical research has evolved along with the development of new techniques and the constant need to improve the medical field. It intends to determine the safety and effectiveness of medications, devices, diagnostic instruments, and treatment regimens for individuals, by complying with systematic principles and regulations. To this end, clinical research has reached a broader meaning as its impact not only extends to individual patients by establishing alternative therapies but also to society by improving health care.

The will to understand or explain clinical events, as well as to explore new interventions, straightened the relationship between scientific evidence and clinical routine. Consequently, an evidence-based practice has been considered the optimal approach in medical management. As this method integrates clinical experience and patient values with the best available research information, it requires new skills from researchers and clinicians, including efficient literature-searching and critical thinking. By acquiring these skills, health professionals are able to acknowledge sources that contain high-quality material, apply the gathered information in

Felipe Fregni (ed.), *Clinical Trials in Neurology*, Neuromethods, vol. 138,
https://doi.org/10.1007/978-1-4939-7880-9_1, © Springer Science+Business Media, LLC, part of Springer Nature 2018

the process of clinical decision-making and finally generate useful knowledge by conducting research studies.

Evidence based medicine involves studies of different designs, which intend to provide answers to unsolved questions by evaluating the effectiveness and safety of therapies that are used to prevent, diagnose, or treat health conditions. However, clinical trials (especially randomized clinical trials) are known as the benchmark for comparing disease interventions [1] as they correspond to the gold standard of research. Although conducting clinical trials may seem straightforward, there are several scientific, ethical, and legal aspects involved. The rigorous methodology often makes investigators question themselves on which is the most appropriate study design for a specific disease, on how many groups a study should have, or even on which statistical test to use. Thus, it is essential for health care providers to understand the precepts of well-performed clinical trials, in order to pursue the safest and most effective therapies [2].

With a view to conduct a valuable and innovative clinical trial, several aspects must be considered. Most of them include methodological aspects, such as defining the research question and the study design, but some other aspects involve ethics, funding and risks involved. In addition, before addressing each of these aspects, it is important to have background information and basic knowledge regarding the mechanisms of the disease being investigated. As there is eventually a limited amount of information about a specific disease, it is crucial to counterbalance the risks and benefits for the patient before starting the trial. The risk analysis needs also to be followed by a financial cost-effectiveness analysis; in other words, what is the potential impact for society if results of a potential trial being planned are confirmed. Finally, after gathering the necessary resources, the clinical trial must comply with innumerous regulations as to obtain approval of an ethics committee. By following all these steps, the chances of achieving a feasible clinical trial highly increase, since the basic procedures and regulations are set in advance.

In neurology, as the mechanism of many diseases is not completely elucidated, designing a clinical trial is quite often a challenging task, given the lack of reliable markers for early development phases. Therefore, the investigators need to understand the specific challenges and methods to address them as to design and conduct successful trials. A final point worth to be highlighted: a successful trial is not the one that is positive, but the one that advances the field regardless being positive or negative.

In this book, we discuss the essential aspects involved in clinical trials of several neurological disorders as to guide clinicians and researches to develop future clinical trials. It is indeed important to divide by disorders given the differences across them. We provide an overview of the most relevant clinical trials in each neurological condition, describe their most common characteristics and correlate the findings with the basic principles of clinical research.

2 Trial Design, Site, and Phase

2.1 Trial Design

The study design is a methodological approach to eliminate systematic error (bias), minimize random error, estimate variability and ensure the external validity of study findings [3]. In general, as shown in Fig. 1, clinical research falls into two main categories: experimental and observational. The first one refers to investigations in which researchers manipulate the intervention, that is, they assign an exposure, for instance, an intervention (a drug, a behavioral therapy, etc.) to a given number of participants and observe the resultant variation between them and others that were not exposed. The main purpose of this study design is to investigate cause–effect relationships by comparing intervention groups in a "controlled environment" (ideally balancing the same conditions and characteristics in both groups). In observational designs, on the other hand, the exposure is observed rather than manipulated [3]. This type of study is useful especially when it is not ethical to assign an exposure for a subject (for instance, smoking) and provides usually large sample sizes for analysis; however, the risk of confounding and other biases increases. There are other types of designs, less used, such as pragmatic trials (that holds characteristics of both experimental and observational studies) and adaptive designs. In this book, we discuss only the experimental study design in neurology.

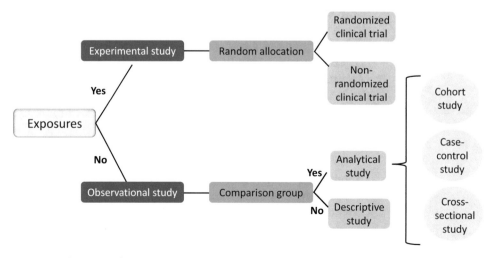

Fig. 1 Algorithm for types of clinical research. *Adapted from "The Lancet Handbook of Essential Concepts in Clinical Research"* [4]

As experimental studies aim to compare two or more conditions by controlling the effects of external factors, despite the existence of many alternative models, randomized clinical trials (RCTs) are considered the "gold standard" of experimental designs [4]. Although it may be costly, time consuming and involve several barriers such as recruitment and dropout, RCTs tend to be statistically efficient and are the best option to avoid selection bias, minimize randomness and quantify the uncertainty in a statistical procedure. Additionally, if properly conducted, they provide a higher likelihood of valid results, thus resulting in stronger evidence.

Clinical trials can be classified in several types, depending on the number of interventions and on how they are administered. In the first one, also known as parallel groups design, patients are divided in two groups (two arm design) or more than two groups (three arm, four arm, and so on) as to receive a treatment or a combination of treatments. In parallel studies, the experimental group, which receives the intervention being investigated, is compared to a control or placebo group. However, these studies can also involve standard therapies [5]. For instance, one group may receive a new drug plus standard therapy and the control group the standard drug plus placebo. This approach is commonly used for a given disease when there are ethical concerns involved with withdrawing a standard drug. Also, a RCT can have an active control, meaning that one group receives a new drug (ND) while the other receives a standard drug. Finally, both groups may differ regarding the amount of the drug or timing of administration (e.g., Low dose × High dose of ND; ND initially × ND delayed; ND intermittently vs. ND continuously).

Another type of design, which involves the testing of two or more treatments simultaneously for possible synergistic or antagonistic effects (combination treatments), is referred to as factorial design [5]. In a 2 × 2 form, subjects are allocated to one of four groups: A alone, B alone, both A and B, or Placebo (neither A nor B). This design is mainly applied to assess two independent interventions using only one sample, as to increase efficiency, or to test the interaction effects between both interventions. This is an efficient method as it allows simultaneous tests of two or more hypotheses in the same study: the effects of intervention A, the effects of intervention B and the effects of the interaction between A and B. However, this study design has a relatively small power to detect an interaction between groups (thus needing a large sample size). In addition, it assumes that efficacy of each regimen does not depend on the presence of the other regimen when a factorial design has the aim to assess each treatment separately.

In crossover designs, each patient is randomly assigned to a sequence of treatments, including at least two treatments of which one may be a standard drug or placebo. In other words, each

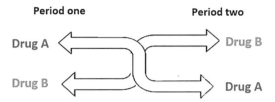

Period one   **Period two**

Drug A   Drug B

Drug B   Drug A

Fig. 2 Scheme of crossover designs

patient receives more than one treatment over time [5]. Therefore, those who start receiving a new drug will change to placebo for the active intervention and vice-versa (or two active drugs can also be used). A scheme of crossover designs is presented in Fig. 2. In addition, it is possible to have more than two interventions in crossover designs. This is a powerful design as each subject is able to be his/her own control, thus decreasing the biological variability. It also maximizes the number of subjects in each group, as all patients will receive both interventions. On the other hand, it is logistically more complicated to run as some factors may require extra attention: the carryover effect of the intervention and the "washout" period. The carryover effect is generally described as when an experimental treatment continues to affect the participant until it is time to switch treatments. This may lead to persistent physiological effects of drugs, which usually take weeks to be washed out from the body. Additionally, the first treatment or the interaction between both drugs might cure or permanently change the course of the disease. Therefore, the "washout" period between treatments should be long enough as to allow the complete reversibility of any treatment effect. Another potential issue is dropouts as dropping out due to one condition may lead to necessary exclusion of all the data for that subject.

Another design that is not frequently used but useful in few scenarios is the cluster-randomized trials. A cluster-randomized trial consists in randomizing a clinic, hospital, town or classroom to either two or more arms (instead of randomizing participants). This is useful when it is not possible to randomize subjects individually (i.e., when procedures are tested—for instance, a procedure in the emergency room). Accordingly, this study design reflects in a more complex statistical analysis.

Finally, another experimental method that has been used more often is the noninferiority design, which means the study is designed to assess whether one intervention is not noninferior to a standard treatment. In this case, subjects are randomized, for an example in a parallel design (usually); however, the question and the hypothesis of the study would differ. In addition, it is common that these studies need a large sample size given the statistical methods to show noninferiority (for further information, please read *Methodology of superiority vs. equivalence trials and non-*

inferiority trials [6], *A clinician's guide to the assessment and interpretation of noninferiority trials for novel therapies* [7], and *Understanding noninferiority trials* [8]).

2.2   Site

Clinical trial sites correspond to where patients are enrolled as to test therapies, medical devices and procedures, gathering accurate data on a certain investigation. Hence, a clinical trial can be conducted either in one site or even in two or more: single-center trials and multicenter trials, respectively.

Different aspects must be considered before choosing whether a trial is going to be single- or multicentered. As this criterion influences how the randomization process is conducted and may even have an important impact on authorship, it is fundamental that researchers counterbalance the pros and cons of each option. Single-center trials are usually conducted in hospitals, clinics, or laboratories and thus are less costly and logistically easier to be executed. In addition, a single-center study is often an essential starting point for the development of larger, multicenter studies as data collection is simplified and study population is less heterogeneous (diminishing possible confounders). Although there are several advantages involved in single-center trials, there are also some important shortcomings, such as their potentially limited external validity, since the subjects may not reflect the characteristics of a broader population, and the recruitment process, which is frequently a challenging. As a consequence, a single source of patients may result in an insufficient number of participants to run a feasible clinical trial [9].

Multicenter trials, on the contrary, are conducted at more than one hospital, clinic or medical facility, thus it corresponds to larger-scale studies. Phase III trials are particularly more prone to have this type of design as it allows a higher generalizability of the results, when compared to single-center trials. In this regard, this method increases the external validity of a study, especially if other centers are located in ethnically diverse geographic regions. Other advantages are the increased chance to achieve a larger sample size, thus providing sufficient power to detect smaller treatment effects, and the contribution of different research skills due to the diverse background from multiple investigators involved. Nonetheless, multicenter trials are considered more complex and expensive, as it requires more personnel and resources. Also, it is hard to ensure that the same procedures are followed in all the centers and that all centers have the same participation by playing a responsible role.

Therefore, it becomes evident that both single-center and multicenter trials have advantages and disadvantages. In fact, in some cases, it is only possible to do one or the other type of trial. The researcher needs then to understand whether it is feasible to design and run a multicenter trial.

2.3   Phase

The investigation of a new drug has well defined stages, also known as the phases of trials, before getting approved for clinical use. The preclinical stage corresponds to investigations that include only animals and evaluations of drug production and purity [10]. These studies with animal models explore the drug's safety, its mechanism of action (pharmacodynamics), how it is metabolized (pharmacokinetics) and, if the results look promising, the drug can be tested in human subjects.

The next step, phase I trials, starts with the administration of a new drug in healthy volunteers or groups with advanced disease. As this step is based on extensive preclinical safety evaluations, it intends to further investigate different aspects on safety, pharmacokinetics, and pharmacodynamics, as well as the interaction between subjects and the new treatment. In this interaction investigation, a dose escalation based on strict criteria is performed and subjects are closely followed up for evidence of drug toxicity. Also known as "human pharmacology" studies [11], phase I trials are usually performed in an unblinded way (open label) on a small number of healthy subjects (between 20 and 300). It usually takes a few months to be conducted and, after data collection on side effects, timing, and dosage are complete, investigators are able to run a phase II trial. There are some variations of phase I trials according to the disease and drug being tested. For instance, in oncology trials, because of drug toxicity, usually patients, rather than healthy subjects, are tested.

Phase II trials, referred to as "therapeutic exploratory," are usually placebo controlled and larger than phase I trials, besides being conducted in a small sample size [1]. The study sample, however, is not composed by healthy subjects, but by those who have the disease of interest. Phase II trials intend to test safety, pharmacokinetics, and pharmacodynamics. Yet they are also usually designed to explore fundamental aspects to the planning of a phase III trial, such as the determination of optimal doses, administration mechanisms and preliminary evidence of drug efficacy. Additionally, in these trials, patients often correspond to a homogenous sample (in terms of response to the intervention being tested) and the investigator tends to look for surrogate outcomes as clinical endpoints usually do not provide necessary power to test the study hypothesis. Also, several secondary outcomes are tested in phase II studies. Given the relative small sample size, it can be performed over a period of 2 years.

Finally, a phase III trial ("therapeutic confirmatory," "comparative efficacy," or "pivotal trial") is conducted in a larger and more diverse population to demonstrate drug's efficacy or confirm the preliminary evidence obtained from a phase II trial. Although the most common type of a phase III trial is comparative efficacy trials, which compares the new drug to either placebo or a standard drug, both equivalency and noninferiority trials are also often used.

Additionally, phase III trials usually have clinical outcomes as the main endpoint and rarely test markers or other laboratorial outcomes, as its main goal is to confirm effectiveness of the drug.

Once a drug is approved, some references may even mention the existence of "therapeutic use" or "post-marketing" studies, also known as phase IV trials. In this phase, although drug effectiveness in diseases and different populations may be evaluated, less common adverse events may be identified [12]. Furthermore, Phase IV trials include an ongoing investigation on therapeutic safety, possible drug interactions and contraindications, relationships and therapeutic delivery regimens, as well as risks, benefits, and optimal use of the drug. Therefore, the main objective of a phase IV trial is pharmacovigilance and postmarketing surveillance.

This model of preclinical and I–II–III–IV phase is not often used for medical devices investigation. Instead, clinical trials evaluating the effects of devices are usually classified into pilot, feasibility, pivotal, and postmarketing studies [13]. Furthermore, medical devices are classified as Class I, II, and III. Class I devices are those which provide a low risk and require the least regulatory controls, Class II are related to a higher risk when compared to Class I, thus requiring greater regulatory control and, finally, class III devices are usually the higher risk devices, requiring, as a result, the highest level of regulatory control [14].

3 Population

After defining which drug will be the aim of the investigation, researchers need to identify in which individuals the drug should be tested, considering the availability of resources and patients, the place that they come from (clinics, hospitals, medical facilities, and so on) and the main characteristics that will influence the study.

To that end, the understanding of some basic concepts is fundamental when it comes to planning a trial [15]. The target population or population of interest is the group of people to which researchers are interested in generalizing the conclusions. Usually, that is also the group that a researcher wants to draw conclusions or make inferences about. Accessible population, on the other hand, is a subset of the target population. This refers to those who would be available for the study. Finally, the study population or sample is a subset of the accessible population that corresponds to those who are included in the study. Please refer to Fig. 3 for a better understanding.

The execution of a trial, especially a robust one, requires the selection of an appropriate study population. The investigator needs to choose study population according to the mechanisms of the intervention being tested. For instance, if an intervention does not work in subjects taking a certain medication, then these sub-

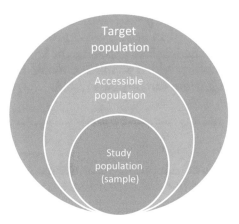

Fig. 3 Representation of target, accessible, and study population

jects need to be excluded. It is important therefore to know well the mechanisms of the intervention being tested.

Therefore, in order to choose an adequate sample, it is important to exclude conditions that might mimic the disease under study and do not respond to the therapeutic intervention; address patients with other comorbidities who might worsen because of these conditions, thus confounding the clinical trials results; And not overchoose patients by adding several inclusion or exclusion criteria items which may restrict the generalizability or external validity of the study.

The inclusion criteria describe the main characteristics of the target and accessible population that will qualify someone as a participant. To define the inclusion criteria, the investigator must account for different elements such as clinical findings, demographics and aspects that are important to the question being studied. The more restricted these criteria are, the more homogenous the population will be and, consequently, the more difficult to generalize the research findings. Exclusion criteria, on the contrary, correspond to the factors that would prevent an individual from participating in the trial, either due to safety reasons or factors that would generate potential confounders.

Regarding the external and internal validity of a trial, it is worth mentioning that they are not dependable. The first one refers to the extent to which the results of a study can be generalized to other situations and other people, and thus the observed results are unbiased estimates of the relationship between exposure and disease in the target population [16]. The internal validity, differently, reflects the extent to which a study minimizes systematic error (or "bias"). Although they do not depend on each other, if a study is not internally valid, it is not generalizable. However, the inverse is not necessarily true. Also, sometimes it is possible to reduce external validity to increase internal validity (by increasing the power of the study); but as aforementioned, internal validity can never be compromised.

Hence, before choosing the study population, it is important to consider which population would be responsive to a given intervention and can be safely tested.

4   Intervention

The intervention is one of the main aspects of a trial as it guides many of the other primary steps such as trial design, population, phase, and outcomes. Regardless if they are drugs, procedures, devices, or the combination of therapies, interventions have the intention to improve the condition of a group of individuals by relieving symptoms, improving quality of life or even promoting cure. According to the CONSORT 2010 guideline [17], researchers must not only describe the study interventions but also provide sufficient details as to allow replications. Consequently, there should be information regarding the type of intervention and its administration, particularly, how and when they were administered.

In addition, ethical aspects must be considered before defining the intervention in both experimental and control groups. In many conditions, especially in neurology, it is not ethical to keep the patient from receiving a drug that has been already proved efficient. So it is not always possible to run trials that involve only placebo in the control group. This concept is intimately correlated with the principle of equipoise, which states that there is a true uncertainty about which of the interventions are most likely to benefit the patient. A clinical trial is to follow the clinical equipoise in order to comply with ethical principles and ensure the patient's safety [18]. Hence, if there is an existing efficient drug, it is not commonly possible to use placebo alone, only when combined with a standard drug in comparison to the new drug plus the standard one.

Furthermore, there are a few factors that aid researchers to define a proper intervention for a trial [19], such as knowing the adequate dosage and mechanisms of the experimental therapy and gathering information about available treatments for the disease on investigation, By addressing these aspects and considering the ethics involved with a certain condition, the researcher is able to define a testable intervention as well as an adequate control. Finally, when the researcher does not know the optimal dosage of the intervention to be tested, he/she should run smaller phase II trials (or pilot studies) testing and comparing different dosages.

5   Outcomes

5.1   Types of variables

After defining which intervention is going to be investigated and the initial details of the study, it is time to specify what is going to be tested. Variables are characteristics or factors that can differentiate individuals of a group, thus being able to vary. By definition, variables are characteristics that have a quantity or quality that varies [19]. Research is then conducted to evaluate the relationship between two or more variables, in order to predict outcomes and to describe the role of those variables in a given scenario.

In experimental studies, such as randomized clinical trials, research variables are usually classified as independent and dependent according to how they are used. An independent variable or predictor variable, as the name suggests, is a condition or characteristic that will predict or influence a given outcome. Therefore, a clinical trial measures whether an independent variable has an effect on the dependent variable. On the other hand, the study outcome, also called as dependent variable, is a response or effect that is influenced by the independent variable [20]. In other words, the most commonly asked question is: does a variation in the independent variable causes or influences a change in the dependent variable? Sometimes a beginner investigator gets confused when there are different definitions for the variable names, especially in statistical packages (for instance, the dependent variable may be called as outcome, or response or output variable).

Besides being categorized into independent and dependent, variables can also be classified into one of two types: quantitative and qualitative. Quantitative variables, also referred to as numerical variables, are the ones represented by numbers. These variables can be further classified in discrete when the variables take on a countable number of values (e.g., 1, 2, 3, 4, …), and in continuous when the variable can assume any value in a specific range (e.g., 1.1; 1.2; 1.3; 2; …). So, in continuous variables, it is possible to obtain fractional numbers, while in discrete variables it is not. For instance, height corresponds to a continuous variable as an individual can have 163.5 cm, and number of children corresponds to a discrete variable, as an individual cannot have 1.5 children, only 1, 2, 3, 4, or more.

Qualitative variables, contrarily, are not represented by numbers but by categories, which might have an order or not. Nominal variables are the ones that do not have a natural ordering among categories, meaning that one category does not outrank the other (e.g., gender, ethnicity, or eye color). Ordinal variables, however, are identified when there is a hierarchical order among the categories, such as ranking scales or letter grades. For an example, toxicological stress can be categorized in without, mild, moderate, and severe, which can be represented by 0, 1, 2, and 3. Still,

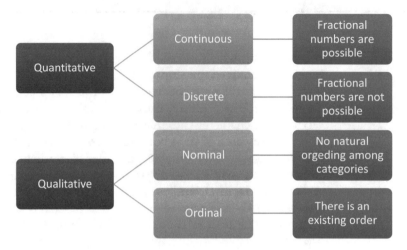

Fig. 4 Summary of the variables classification

it is important to notice that there is an order of severity (one category does outrank the other). Thus, toxicological stress is an ordinal variable. A summary of the variables classification is illustrated in Fig. 4.

One additional type worth mentioning is when a variable has only two possible values. These variables are said to be Binary or Dichotomous and they can usually be phased into a "yes or no" question or into gender (male and female). It is also important to notice that, despite this classification in qualitative and quantitative, a continuous variable can be transformed into categorical. For instance, the body mass index (BMI) is considered a continuous variable. However, it can be organized in categories: <18.5 = underweight; 18.5–25 = normal; 25–30 = Overweight and >30 = Obese. Researchers must identify which variables are more adequate for each study.

5.2 Primary and Secondary Outcomes

Outcomes can be classified according to their importance towards the unanswered questions of a study. The primary endpoint is the most important variable among the many dependent variables that are to be investigated in the study [21]. It measures the outcomes that will answer the primary and most important question of the trial. Therefore, it should be clearly defined during the study planning and protocol drafting. There are several reasons to why this should be a priority. Firstly, this prevents investigators from taking a glance at the significant results. If that was to happen, researchers might be tempted to present these results as the main findings of the trial. Secondly, the primary outcome is fundamental to the sample size calculation and the definition of the statistical tests, which should also be defined a priori.

Secondary outcomes also have an importance to the trial, but not as much as the primary one. For that reason, it is usually not the main focus of the trial. They include supportive or ancillary measures and their main purpose is to ask other relevant questions that are related to the independent variables [22]. In addition, some of the secondary outcomes can be analyzed post-hoc, after the study is completed. However, before the study starts, this aspect should also be stated in the study protocol.

The existence of a secondary outcome is not mandatory. It is quite the opposite. Ideally, a trial should not have too many outcomes, as this increases the complexity of the statistical analysis, often demanding more time and resources. Generally, studies in early phases of development have more secondary outcomes as there is more uncertainty and need to generate further hypotheses. However, larger trials usually have very few secondary outcomes. To that end, it is important to remember that the trial must have enough power as to analyze the primary outcome only.

Finally, when it comes to defining the primary and secondary endpoints, there is intensive discussion to whether they should be clinical or surrogate (Biomarkers). We discuss this topic in the following section.

5.3 Clinical Outcome vs. Surrogate Outcome

For the purpose of understanding how useful a drug really is, clinical outcomes are considered the most credible measure that can be assessed in a clinical trial. It corresponds to characteristics or variables that reflect how a patient feels, functions, or survives, thus being able to indicate the effects of a therapeutic intervention. An adequate clinical outcome should be objective, sensitive, unbiased, reproducible, easy to interpret, clinically relevant, and, most importantly, chosen a priori. Outcomes such as survival, decreased pain or absence of disease (cure) are some examples of clinical outcomes as these endpoints represent direct clinical benefits.

Contrarily, surrogate outcomes, also referred to as biomarkers, are indirect measurements of effect in circumstances where direct measurements of clinical response are not possible, either due to feasibility or practical reasons. They correspond to characteristics that are measured and analyzed as an indication of a normal biological process or pharmacologic responses to a chosen intervention. In addition, surrogate endpoints must be reliable predictors of clinical benefit, which might represent a challenge in some conditions [23]. On that account, it is important to remember that although all surrogates are biomarkers, not all biomarkers are proper surrogate endpoints.

Once an intervention alters the disease, the ideal biomarker should be the one capable of addressing those changes regardless of the biological level: molecular, cellular, or related to an organ itself. However, it is not always possible to observe and measure those changes caused by a certain intervention due to a lack of sensitivity and specificity of the biomarker. There are other useful biomarkers, which are not related to the clinical endpoint, but are affected in parallel by the intervention or by the disease. For instance, they can be good diagnostic markers but not good markers for progress and vice versa. Indeed, the same marker may have different applications in different contexts.

There are innumerous advantages regarding the use of surrogate outcomes. Besides being often cheaper and easier to measure, they can also be assessed more quickly and earlier. However, the use of biomarkers requires a thorough understanding about the pathophysiology of the disease and its mechanism of action. In neurology, differently than cardiology, there is a lack of reliable biomarkers and, in many disorders, only clinical outcomes can be measured. This scenario may change in a few years given the intensive research in this area. In this respective section, for each neurological disorder reviewed, you will be able to understand the most common biomarkers studied.

6   Statistical Analysis

6.1   Introduction

Clinical research attempts to generate and test hypotheses in order to make significant conclusions from a given amount of data. Before the launch of statistics, the interpretation of study results was mainly based on researcher's personal judgment, which led to the generation of unreliable results. By exclusively using clinical judgment and reasoning, researchers are not able to address the variability between individuals on investigation and are vulnerable to inaccurate findings. To date, statistics is applied to prevent an incorrect interpretation of research results and to generate reasonable inferences from a given data set. Indeed, statistical approaches allow researchers to identify fairly small differences between experimental groups and to potentially generalize these findings from the study sample to a larger population.

6.2   Descriptive Statistics

Descriptive statistics refers to the methods applied to arrange and summarize a given amount of information after data collection. Invariably, when dealing with quantitative data, researchers must organize a compilation of numbers with a view to transform it into useful measures [24]. Only after this step it is appropriate to conduct inferential statistics (in other words, using statistical tests to look for differences between experimental groups).

There is a wide variety of measures applied to summarize data. Depending on the goal of investigators, they may work with measures of central tendency and/or measures of dispersion. Central tendency values are applied to identify the central position of a given data set. In other words, they may be used if researchers desire to determine which value represents an entire distribution. The mostly used measures of central tendency in clinical trials are mean, mode and median. On the other hand, measures of dispersion, also called as measures of variability, are applied to precisely describe the spread in an amount of data. Accordingly, dispersion refers to how squeezed a given distribution is. Thus, by calculating these measures, researchers aim to identify the difference that exists among the numbers in a data set [24].

The most used measures of variability are range, quartiles and standard deviation. The application of each of these measures directly depends on the data distribution. Therefore, to fully characterize and summarize a given distribution of numbers, investigators must also describe the frequency distribution (table that displays the number of observations) and the shape of the distribution, along with a measure of central tendency and variability.

An important concept: the choice of these measurements will depend on variable characteristics. If the variable is ordinal or the continuous variable is not normally distributed, median and interquartile range should be used. Mean and standard deviation should be used when the variable is continuous and normally distributed.

6.3 Inferential Statistics

Differently from descriptive, inferential statistics refers to interpreting data distribution and making accurate conclusions from it. Using descriptive statistics alone does not always provide enough information in clinical research as it provides information only regarding the sample studies and, for this reason, making inferences is essential to draw powerful conclusions that can be applicable to the target population. Therefore, by using deduction and probability approaches researchers are able to infer whether differences between groups, for instance, would be likely seen in the target population. Inferential statistics is based on the estimation of population parameters and the test of hypotheses, which will be discussed in the following section.

6.3.1 Using Statistics to Compare Groups

When researchers decide to conduct statistical tests to compare interventions, there are two possible approaches to work with, both of which give complementary information. The first one is based on a probability theory and involves the calculation of a *p-value* [19]. This value is applied to determine statistical significance and it corresponds to the probability that a difference between experimental groups is caused by chance. In other words, investigators use *p-value* in clinical research as to assess whether an observed difference is due to a random sampling (meaning, for instance, choosing subjects by random chance that will be responsive in one group and not responsive in the other) or to a real difference between groups. Conventionally, a *p-value* lower than 0.05 refers to statistical significance.

The second approach used in statistical inference is based on the estimation of population parameters. Thus, researchers commonly use *interval estimates* for this purpose. These estimates represent a range of values in which a parameter of the population may lie. In clinical research, the most used estimated range is called *confidence interval,* which is based on the study sample and must contain the population parameter of interest. This range is commonly reported with a degree of confidence, which works as a probability percentage. Thus, researchers often work with a 95% confidence interval for the mean estimate. It indicates, for instance, that if you take 100 different samples, you would expect that the mean estimate would fall within that interval in 95% of the cases.

6.3.2 Hypothesis Testing

Conducting a successful clinical research includes collecting data, analyzing the sample by running statistical tests and testing hypotheses. This last step is crucial to draw conclusions about the results and clearly define if an adequate external validity was achieved. Before any clinical trial starts, researchers are required to generate hypotheses. The null hypothesis (Ho) states that there is no statistical significant difference between experimental groups, while the alternative hypothesis (Ha) states that the intervention is significantly different. Accordingly, by the end of statistical analysis, researchers must interpret the results and decide whether they should reject the null hypothesis, which would confirm that the Ha is correct, or not.

Furthermore, when researchers look for statistical significance during data analysis, they must be attentive to two possible errors. In the first place is the type I error, also called as alpha (α) which refers to the rejection of the null hypothesis when it is true and should not be rejected. Secondly, there is the type II error, also named as β, which corresponds to the failure to reject the null hypothesis, when it is false and should be rejected. It is expected that a study has a low probability of occurrence of both type I and II errors [19]. These are common potential errors involved in clinical research and their probability of occurrence must be well reported in any manuscript. Usually the clinician ignores this potential error when reading an article.

6.4 Statistical Tests

In order to define which statistical test is the most appropriate for a particular study, the primary step is to classify the chosen variables, both dependent and independent. As discussed in the Subheading 5 of this chapter, besides the classification in dependent and independent, variables can be subclassified in quantitative and qualitative. Among these, the most used types of variables are continuous, discrete, nominal, and ordinal. Additionally, regarding continuous variables, it is important to define whether or not they follow a normal distribution, which means that the collected data is symmetric, resembling a bell-shaped curve, and rely around the probability mean. If the investigator knows well the classification and distribution of their variables, it is not difficult to choose the appropriate statistical test. Then the next step is to understand what the results represent.

Computer programs are commonly used to perform statistical tests in clinical research. These tests have different applications and specific assumptions. The most used statistical tests will be discussed in the following sections.

6.4.1 Student's t-Test

The t-test is a hypothesis test formally applied with the goal of comparing two means. According to the study design, researchers may choose to apply an unpaired or a paired t-test. An unpaired t-test is performed when the compared samples are independent [25]. On the contrary, a paired t-test is used when samples are matched or correlated. For instance, if study subjects are used as their own control or if they are matched on relevant variables, the data is said to be paired, and therefore a paired t-test must be applied.

Classified as a parametric test, the t-test requires that the outcome variable follows a normal distribution. Other important assumptions required for t-test performance are the following: (1) the data collected must follow a continuous scale; (2) the study sample must be randomly selected from the population; and (3) the standard deviation of both samples is required to be approximately equal, which is also called as homogeneity of variance.

If some of the required assumptions are not met, there are alternative statistical tests to apply. For instance, research studies involving nonnormally distributed samples should not perform a t-test, but may use the Mann–Whitney U test, also called as Wilcoxon, which is analogous to a nonparametric t-test. Wilcoxon test has fewer assumptions than a standard t-test and, for this reason, can be easily assigned to data analysis. Although easier to conduct, as most nonparametric tests, Wilcoxon test may have a lower power when compared to t-tests (especially if the sample is normally distributed), depending on the study's sample size.

6.4.2 Analysis of Variance

Clinical investigators often choose to develop study designs with multiple experimental groups. In this case, when more than two conditions are being compared, an analysis of variance (ANOVA) must be performed. In fact, ANOVA can be seen as an extension for t-test. As the name says, ANOVA analyzes the variance among observations and it is mainly applied to determine whether a difference among a set of means is statistically significant [26]. Depending on the study design, researchers may choose to work with one of two different types of analysis of variance. The one-way ANOVA is used when a study involves only one intervention, but with multiple levels, such as different physical exercise movements. On the other hand, two-way ANOVA is attributed to studies that involve two or more interventions, such as pharmaceutical drugs and physical therapy. When there are many factors and some repeated measured, more complex ANOVA models can be used, but in this case, investigators need to understand how to use these models appropriately.

Similarly to the t-test, the analysis of variance is considered a parametric test, which means that a normal distribution of the study sample is required. Besides having extremely similar assumptions to that of t-test, if the assumptions for running an analysis of variance are not met, such as normality of sample distribution, there are analogous nonparametric tests to work with as well. The Kruskal–Wallis test is an alternative in this case as it is nonparametric and, in consequence, has fewer assumptions than the standard ANOVA.

In case the ANOVA is significant, the next step is to discover which means actually differ. To that end, pairwise comparisons are fundamental instruments, whether they are described a priori in the protocol or even after the main analysis has been completed (post hoc comparison). However, when repeated tests are used, the overall type I error increases. There are different methods to address this issue and consequently maintain a low overall type I error. One, in particular, is the Bonferroni correction [26] which divides the type I error by the number of comparisons made and provides a type I error rate for the pairwise comparison. Although this method is considered conservative, there are also other tests that could be used such as Tukey's test and Duncan's multiple-range test. Another alternative is the disclosure error method, in which the investigator discloses the error rate for the multiple comparisons (15).

6.4.3 Chi-Square and Fisher's Exact Tests

When both the independent and dependent variables of a study are categorical, there are specific statistical approaches to be used, as the abovementioned tests do not apply. Accordingly, in this case, researchers can work with the chi-square or the Fisher's exact test. For selecting one of these two tests, a few aspects of the research study must be taken into account. The chi-square is a simple test commonly performed when the study sample size is large enough to ensure its accuracy [19]. Another assumption for using this test is that the samples are independent. However, despite being easier to calculate by hand, this test provides an approximation of the p-value, which may not be ideal in certain studies.

Regarding Fisher's exact test, there are no requirements in terms of sample size and, as the name says, this test does not provide an approximation but an exact p-value [19]. This advantage may explain why it is applied more often in clinical research. Although it may be more precise, it may not be possible to run this test in some situations, especially when the sample size is too large (because of computing power). In this case, it would be more appropriate to use the chi-square test instead.

6.4.4   Correlation

In statistical significance testing, instead of assessing whether there is a difference between experimental groups, researchers may also investigate the relationship between two measured variables. This statistical analysis is called correlation and it aims to identify the strength and direction of a relationship between variables, such as blood pressure and body mass index [27]. After conducting a correlation test, a coefficient is generated, which expresses if a correlation is positive or negative. The most common statistical tests used to assess linear correlation are the Pearson Product-Moment correlation, a parametric test, and the Spearman's Rank Order correlation, an analogous nonparametric test.

In order to perform Pearson correlation, researchers must ensure that all the following assumptions are met: (1) the observations must be normally distributed; (2) there must exist a linear correlation between the variables, as any other type of correlation will not be identified; (3) both variables must be in a continuous scale; and (4) outliers must not significantly influence the analysis.

The Spearman's Rank Order determines the strength and direction of the monotonic relationship between two ranked variables, instead of analyzing the linear relationship between two variables like Pearson correlation [28]. In a monotonic relationship, as the value of one variable increases so does the other, or, as the value of one variable increases, the other decreases. Unlike Pearson correlation, there is no requirement of normality. The only assumption is that the variables are ordinal, intervals, or ratios.

6.4.5   Regression Model

As mentioned in the previous section of this chapter, there are different statistical approaches to analyze correlation between study variables. Nonetheless, if researchers are planning to estimate a relationship between variables in a basis of prediction, they are most likely to choose working with a regression model [27]. There are various types of performing regression with statistical methods and each of them is usually applied for a specific purpose. For instance, while simple linear regression is commonly performed as to understand the type of relationship between one dependent and one independent variable; multiple linear regression is applied when more than one independent variable is involved. In both cases, the main goal is to estimate the type of correlation and develop predictor functions, which are based on the collected data. These functions mainly generate an equation that may predict future outcomes. Additionally, multiple linear regression is also applied when researchers are interested in one specific variable effect but need to adjust the data for other variables.

In any regression procedure, investigators must meet particular assumptions in order to perform the tests. For simple and multiple linear regression, the main assumptions include: (1) there must exist a linear relationship between the variables; (2) the data must be normality distributed; (3) all values of the independent variables must have the same variance (homoscedasticity) [19].

Building regression models is not an easy task and it requires some training (different than running a t-test). It goes beyond the scope of this book to explain model building; but we encourage investigators to learn about it if planning to run multivariate regression models. In addition, regression models are useful to adjust results for covariates (covariate adjustment). A commonly used covariate adjustment model is called ANCOVA. Also, the investigator needs to note that ANOVA is actually a special type of model (more restrict).

Figure 5 presents the most used statistical tests in clinical research and summarizes their main applications.

6.5 Survival Analysis

Survival analysis is often used in clinical trials that aim to measure the follow up time of patients from a starting point to the occurrence of an event [27]. When measuring survival times, investigators are interested in measuring a specific event, such as death or disease remission, over a given period of time. Thus, survival analysis measures time-to-event data and commonly provides meaningful information to clinical practice [29]. Here the investigator has some options: (1) using Kaplan–Meier curve as a method of descriptive analysis (similar to reporting mean and SD for continuous data); (2) log rank test as to compare two survival curves (similar to using a t-test for continuous data); and (3) Cox proportional hazard models for running multivariate analysis (similar to using linear regression model for continuous outcomes).

7 Sample Size Calculation

The sample size is one of the most important points to ensure the quality of a clinical trial. Regardless of the statistical procedure, sample size calculation usually takes into account some standard parameters, such as power of the study, significance level (α), variation measure of the primary outcome, and anticipated treatment/ exposure effect size. In other words, a high-quality RCT must have a specific and previously stated number of participants to test whether the investigated intervention is statistically significant and has an adequate power [19, 30].

Although this topic plays a major role in the final results and is one of the main aspects that should be defined in the beginning of the study, many authors fail to mention the procedures involving the sample size calculation, thus providing incomplete information. For readers, this may represent an important issue, as the lack of information makes it harder, or even impossible, to define if the study had conducted the sample size calculation properly or if the authors just failed to mention their approach.

Dependent Variable			

	Continuous data		Categorical / binary
	Normal distribution	*Non-normal*	
Compare two groups (independent variable - binary)	Unpaired and paired t-test	Mann-Whitney Wilcoxon	Chi-square Fisher' s exact
Compare three or more groups (independent variable - categorical)	ANOVA (one or n-way ANOVA)	Kruskal-Wallis or Friedman test	Chi-square Fisher' s exact
Association between two variables (independent variable - continuous)	Pearson correlation	Spearman correlation	
Association between three or more variables (independent variable - continuous)	Multiple linear regression		

Independent Variable

Fig. 5 Table of statistical tests according to variables classification

As the most challenging task here is to estimate the expected effect size, there are few methods to make this estimation: (1) conduct a pilot study; (2) look at studies that used other interventions in the same disease or same intervention in other diseases (the scientific knowledge of the investigator will determine whether this method is valid); or (3) calculate the minimally clinical difference. Any of these methods has disadvantages that need to be considered.

7.1 Interim Analysis and Stoppage Criteria

Although an adequate statistical analysis may provide evidence that the trial is headed toward the right direction, the monitoring process goals go beyond statistical warnings for stopping.

In general, some of the circumstances that could lead to stopping a trial are poor data quality, poor adherence, lack of resources, severe adverse effects, fraud, and the discovery of new information that might question the relevance or ethics of the study. The decision usually lies in an independent data monitoring committee, which uses prespecified statistical stopping criteria to make a decision. Besides of safety assessment, interim analysis may look into efficacy data as to stop the trial for futility or for early efficacy. However, when the investigator decides to conduct interim analysis to stop for efficacy, there is a p-value penalty [31].

The interim analysis consists in conducting a data analysis before data collection has been completed. If an investigator plans on doing an interim analysis, the statistical stopping criteria should be prespecified in the protocol and, preferably, an independent statistician whom is not involved in the trial should conduct it. However, a lack of information is commonly noticed in the trials.

8 Randomization

8.1 Importance

The generation of a random allocation sequence, mostly called as randomization, is one of the essential criteria to ensure validity in clinical trials. In order to compare interventions with control groups in a research study, the study groups should be similar (or not different). For instance, covariates such as gender, age, and ethnicity, should be as similar as possible in both intervention and control groups. Accordingly, researchers must ensure that the groups being compared are equal in regard to any potential covariate. If this assumption is guaranteed, any significant difference found by the end of the study will be exclusively due to the intervention. Therefore, randomly assigning participants to study groups is the best strategy to ensure a fair comparison between them. Though, as the name says, a random process may fail especially when study sample size is small.

Randomization ends up being more than a simple generation of random numbers. It is a precise technique used in most clinical trials to specifically avoid biased comparisons [1, 32]. Hence, in clinical research, every enrolled participant should have an equal chance of being assigned to any group, which means that researchers must not directly or indirectly choose the allocation of subjects. In fact, according to what will be discussed in the Sect. 9 of this chapter, they should not know to which group a certain patient will be (or was) assigned to. When investigators have any prior knowledge of group assignments, this may lead to selection bias.

There are different techniques of randomization commonly used in clinical research. Each of these types will be discussed in this section.

8.2 Methods of Randomization

8.2.1 Simple Randomization

The simplest method of randomization is based in a generation of a single sequence of random numbers. It is an extremely easy technique to randomly assign participants, which makes this method commonly used in clinical research [33]. Generating the sequence of random assignments is effortless, as it may be done by performing a simple coin toss or even by using accessible computer software. The simple flip of a coin is not frequently applied as a randomization method, in order to avoid any manipulation of group assignments.

Besides its easiness and simplicity, a simple randomization may generate imbalanced groups in studies with small sample sizes. For instance, in a study with 20 subjects, this type of randomization may assign 13 patients to group A and seven to group B, which may jeopardize the comparison between both groups. In this case, it is recommended to use other approaches. On the other hand, in large sample size studies, the performance of simple randomization is usually appropriate, as the large number of participants may overcome the risk of imbalanced groups.

8.2.2 Block Randomization

The randomization in blocks is done with the main goal of avoiding imbalance of group assignments, potentially caused by simple randomization. To perform this technique, researchers must previously determine the size of each block, which must be a multiple of the total number of groups [19, 32]. For instance, a research study comparing the effects of drug X with placebo, the participants may be assigned to one of the two groups. Thus, the block size could be any multiple of two. Commonly, the block size is small and there must have an equal number of participants assigned to each group.

Although the block randomization is often used in clinical trials, this method can make the allocation of subjects predictable, especially when every block has the same size and number of participants assigned to each group. For this reason, researchers may use the block randomization with random block sizes. For instance, blocks of 6 and 8 may be generated among the study. In addition, the block randomization does not avoid confounders within and between groups.

8.2.3 Stratified Randomization

The stratified method of randomization is applied when baseline characteristics of subjects may potentially influence the study outcomes. Accordingly, this technique helps to control for covariates and to avoid imbalance between groups in regard to specific characteristics of participants. In order to perform this type of randomization, researchers must clearly define what covariates may indeed affect the clinical trial conclusions if not taken into consideration. After defining them, every combination of covariates will correspond to a block, also called as *stratum* [33]. For instance, if gender and age are important covariates to consider, male participants under 18 years of age could correspond to one *stratum*. Hence, the goal with this process is to assign patients with similar baseline characteristics to the same *stratum*.

At the time that study participants are assigned to specific *strata*, simple randomization is performed within each of them. As the simple method of randomization will not ensure balanced sample size in each group of the study, researchers commonly use the block randomization technique within each *stratum*. By the end of this process, it is expected to have balanced group assignments and covariates. However, the investigator needs to be aware that this method may lead to imbalances if there are too many strata to a relatively small sample.

8.2.4 Covariate Adaptive Randomization	This method of randomization is intended to minimize imbalance between study groups in regard to sample size and covariates. As the name says, it is a technique that adapts over time. Thus, participants are allocated based on their baseline characteristics and previous assignments of subjects. For instance, if in a study it is defined that gender is an important variable and the first five male subjects are assigned to one of the groups, then the allocation probability of next male to be assigned to the other group increases. This method can also lead to imbalances if not well planned and it depends on special software.

8.3 Allocation Concealment

In clinical research, concealment of group allocation refers to avoiding investigators, research staff and subjects to know or predict group assignments. To date, several strategies are applied to prevent awareness of the next allocations, in order to ensure a valid comparison between study groups. If the randomization scheme is known and not concealed, there are a few potential drawbacks. Firstly, researchers may, consciously or not, influence the enrollment of specific participants, depending on their baseline characteristics, which is called selection bias. This situation often leads to an overestimation of the intervention effects, which generates invalid results. Secondly, subjects may choose not to participate in the study if they discover, for instance, that the next assignment is to the control group.

There are different methods of concealment commonly used in clinical trials, which mainly include sequentially numbered envelopes and telephone-based or computer-based systems. If researchers choose to work with envelopes, these must be opaque and well-sealed, as to prevent their violation by nonauthorized individuals. Due to the easiness to decipher the allocation by using envelopes, this is not the preferred method of clinical investigators. On the contrary, computer-based systems are very common in randomized clinical trials, as they are usually easy to work with and protected from undesirable manipulations.

9 Blinding

9.1 Importance

Blinding refers to the process intended to conceal the allocation of participants after randomization. It consists in preventing participants, health-care providers or study assessors to be aware of subject assignments [34]. Generally, researchers undertaking clinical trials use blinding to prevent the measurement and performance bias and thus to ensure that the conclusions made by the end of the study are valid and reliable. Although blinding is an essential component of clinical trials, not all of them need to or can be blinded, especially studies exclusively investigating safety aspects of an intervention, such as phase I trials. The various types of blinding are discussed in the following section of this chapter.

9.2 Types of Blinding

The type of blinding applied in each clinical trial directly depends on whom needs to be blinded, in order to conduct the study successfully. As there is a wide variety of studies being conducted, researchers must invariably report who in the study was indeed blinded. According to that, the study is said to be single-, double- or triple-blind [34]. For instance, if only the participant or the clinician is blinded to treatment allocations, the trial is said to be single-blind. In this type of study, the participants are the frequently blinded personnel. If neither the study center clinician nor the participant knows the treatment assignment, the study is called double-blind, which is the most common type of blinding seen in clinical trials. However, if clinicians, participants and outcome assessors or statisticians are blinded, the study is considered triple-blind.

As previously mentioned, there are a few clinical trials are not blinded, which are called open label studies. Commonly, they correspond to pilot studies, as their goal is to exclusively assess safety aspects of an intervention, such as maximum tolerated dose and typical adverse events. The nature and design of a study (including the intervention characteristics) is what truly determines whether it needs blinding and, if yes, what kind of blinding it demands. If blinding is not possible, the investigator needs to be aware of how it can affect the results of the trial.

9.3 Bias Associated with Lack of Blinding

Important sources of bias may be introduced in clinical trials due to blinding failure. Research studies with inappropriate or no blinding are vulnerable to intentional or unintentional biases. Certainly, unsuccessful blinding may distort study results and negatively affect its internal validity. There are a number of potential biases likely to occur given a lack of blinding, of which two will be subsequently discussed.

Measurement bias is likely to occur if blinding is not adequate. In general, this type of bias arises when an investigator consciously or unconsciously influences outcome measurements due to personal judgments after learning treatment allocation. For instance, if a researcher knows that a subject was assigned to the active treatment, he/she may unintentionally be more perceptive when assessing adverse events. Indeed, perception bias threatens the validity of a study, as the data collection is jeopardized and consequently the data analysis.

Additionally, performance bias is another important type of bias. It may occur when study participants are not blinded and consciously or unconsciously change their performance after realizing their treatment allocation. For instance, a subject who discovers that he/she is in the control group may become less interested and underreport important positive effects.

9.4 Blinding Assessment

As the maintenance of blinding is not always an easy task for investigators, there are specific strategies to assess blinding effectiveness in every research study. If no assessment is done and blinding has failed, the study is completely vulnerable to different sources of biases [35]. For this reason, questionnaires are usually applied to subjects at the end of study visits, which contains questions about study treatments. Thus, the participants need to give their best guess of which treatment they are receiving and grade how confident they are about this guess. Although useful results of this assessment may be associated with treatment effects, it may not be easy to detangle these effects.

10   Conclusion

Conducting clinical trials involves considerable challenges and demands careful planning. With the goal of generating valid results, investigators must invariably follow every step of the research process, from the development of a research question and selection of study design to data analysis and drawing conclusions. As these steps intend to ensure the relevance and feasibility of clinical trials, several studies, which fail to comply with all the requirements, usually present invalid results. Thus, it is crucial that researchers develop a critical sense as to learn how to recognize and design a strong clinical trial.

In the neurology field, the development of high-quality research studies has made significant contributions to the clinical practice. Innumerous conditions, those which did not have a specific treatment or neurophysiological mechanisms that were not completely understood, now have several advances in diagnostics and therapeutics. Due to the rapid advance of technology and the new possibilities that have emerged as a result, the role of investigators and researchers in the scientific field have become even more demanding. Therefore, their ultimate goal is to not only explore the available literature but also to advance the medical field by conducting innovative and strong clinical trials.

References

1. Umscheid CA, Margolis DJ, Grossman CE (2011) Key concepts of clinical trials: a narrative review. Postgrad Med 123:194–204
2. Röhrig B, du Prel J-B, Blettner M (2009) Study design in medical research: part 2 of a series on the evaluation of scientific publications. Dtsch Arztebl Int 106:184–189
3. Carlson MDA, Morrison RS (2009) Study design, precision, and validity in observational studies. J Palliat Med 12:77–82

4. Schulz KF, Grimes DA (2006) The Lancet handbook of essential concepts in clinical research. Elsevier, Amsterdam

5. Evans SR (2010) Clinical trial structures. J Exp Stroke Transl Med 3:8–18

6. Christensen E (2007) Methodology of superiority vs. equivalence trials and non-inferiority trials. J Hepatol 46:947–954

7. Oczkowski SJW (2014) A clinician's guide to the assessment and interpretation of noninferiority trials for novel therapies. Open Med 8:e67–e72

8. Hahn S (2012) Understanding noninferiority trials. Korean J Pediatr 55:403–407

9. Meinert CL, Tonascia S (1986) Clinical trials: design, conduct, and analysis. Oxford University Press, Oxford

10. Mann H, Djulbegovic B (2003) Choosing a control intervention for a randomised clinical trial. BMC Med Res Methodol 3:7

11. Toloi D d A, Jardim DLF, Hoff PMG, Riechelmann RSP (2015) Phase I trials of antitumour agents: fundamental concepts. Ecancermedicalscience 9:501

12. Suvarna V (2010) Phase IV of drug development. Perspect Clin Res 1:57–60

13. Center for Devices and Radiological Health, C. for B. E. and R (2013) Design considerations for pivotal clinical investigations for medical devices - guidance for industry. Clinical Investigators, Institutional Review Boards and Food and Drug Administration Staff, Silver Spring, MD

14. Comparing medical device and drug trials: how they differ - premier research. Available at: https://premier-research.com/perspectivesmedical-devices-vs-drug-trials/. Accessed: 8th November 2017

15. Feise RJ (2002) Do multiple outcome measures require p-value adjustment? BMC Med Res Methodol 2:8

16. Akobeng AK (2008) Assessing the validity of clinical trials. J Pediatr Gastroenterol Nutr 47:277–282

17. Moher D, Schulz KF, Altman DG (2001) CONSORT Group (Consolidated Standards of Reporting Trials). The CONSORT statement: revised recommendations for improving the quality of reports of parallel-group randomized trials. Ann Intern Med 134:657–662

18. Freedman B (1987) Equipoise and the ethics of clinical research. N Engl J Med 317:141–145

19. Portney LG, Watkins MP (2009) Foundations of clinical research: applications to practice. Pearson/Prentice Hall, Upper Saddle River, N.J.

20. Flannelly LT, Flannelly KJ, Jankowski KRB (2014) Independent, dependent, and other variables in healthcare and chaplaincy research. J Health Care Chaplain 20:161–170

21. Andrade C (2015) The primary outcome measure and its importance in clinical trials. J Clin Psychiatry 76:e1320–e1323

22. Freemantle N (2001) Interpreting the results of secondary end points and subgroup analyses in clinical trials: should we lock the crazy aunt in the attic? BMJ 322:989–991

23. Aronson JK (2005) Biomarkers and surrogate endpoints. Br J Clin Pharmacol 59:491–494

24. Krousel-Wood MA, Chambers RB, Muntner P (2006) Clinicians' guide to statistics for medical practice and research: part I. Ochsner J 6:68–83

25. Whitley E, Ball J (2002) Statistics review 5: comparison of means. Crit Care 6:424–428

26. Bewick V, Cheek L, Ball J (2004) Statistics review 9: one-way analysis of variance. Crit Care 8:130–136

27. Motulsky H (2010) Intuitive biostatistics: a nonmathematical guide to statistical thinking. Oxford University Press, Oxford

28. Bewick V, Cheek L, Ball J (2003) Statistics review 7: correlation and regression. Crit Care 7:451

29. Bewick V, Cheek L, Ball J (2004) Statistics review 12: survival analysis. Crit Care 8:389–394

30. Röhrig B, du Prel J-B, Wachtlin D, Kwiecien R, Blettner M (2010) Sample size calculation in clinical trials: part 13 of a series on evaluation of scientific publications. Dtsch Arztebl Int 107:552–556

31. Kumar A, Chakraborty B (2016) Interim analysis: a rational approach of decision making in clinical trial. J Adv Pharm Technol Res 7:118

32. Vickers AJ (2006) How to randomize. J Soc Integr Oncol 4:194–198

33. Kang M, Ragan BG, Park J-H (2008) Issues in outcomes research: an overview of randomization techniques for clinical trials. J Athl Train 43:215–221

34. Karanicolas PJ, Farrokhyar F, Bhandari M (2010) Blinding: who, what, when, why, how? Can J Surg 53:345–348

35. Kolahi J, Bang H, Park J (2009) Towards a proposal for assessment of blinding success in clinical trials: up-to-date review. Community Dent Oral Epidemiol 37:477–484

Chapter 2

Stroke

Camila Bonin Pinto, Faddi Ghassan Saleh Velez, and Felipe Fregni

Abstract

This book chapter aims to discuss the methodology and design of the 100 most highly cited clinical trials in Stroke. We analyzed the main topics of stroke clinical trials aiming to provide insights and valuable information that may improve the development of future trials. We systematically reviewed the 100 most highly cited clinical trials in stroke between the years 2010 and 2015 using the web engine "Web of Science." A systematic analysis of the 100 most cited clinical trials in stroke based on the CONSORT criteria is discussed; in addition, a summary of all the topics regarding the design of a trial in stroke is provided.

Key words Stroke, Methodology, Clinical trials, Prevention trials, Therapeutic trials, CONSORT guidelines

1 Introduction

Stroke is a cerebrovascular event defined by a clinical, radiological, and/or neuropathological evidence of ischemia or hemorrhage [1, 2]. The disruption or severe reduction of oxygen supply and nutrients due to diminished blood flow to the central nervous system leads to cell death and permanent injury. An ischemic stroke is defined as an episode of neurological dysfunction caused by infarction of the CNS [3, 4], while a hemorrhagic stroke is caused by a focal collection of blood within the brain parenchyma, or ventricular system that is not caused by trauma, which leads to neurological dysfunction, manifested in clinical symptoms such as aphasia or hemiparesis [5–7].

The most common symptoms and signs of a stroke include sudden numbness or weakness of the face, arm or leg contralateral to the location of the lesion in the brain. Other common symptoms include sudden confusion, dysarthria, aphasia, gait abnormalities, visual disturbances, dizziness, loss of coordination or balance, and headache without a clear etiology [8–14]. The disabilities following stroke are intrinsically dependent on the affected areas of the central nervous system [15–18].

Felipe Fregni (ed.), *Clinical Trials in Neurology*, Neuromethods, vol. 138,
https://doi.org/10.1007/978-1-4939-7880-9_2, © Springer Science+Business Media, LLC, part of Springer Nature 2018

Stroke is the second cause of death worldwide and the leading cause of death in upper-middle income countries; about 6.7 million people died from stroke in 2012 [19]. According to the center for disease control and prevention, stroke is the fifth leading cause of death in USA and causes around 130,000 fatalities a year being responsible for one of every 20 deaths from all the death causes. Every year 795,000 Americans have stroke, about 185,000 of those (23.2%) are a recurrent stroke [20–23]. For the survivals most of the time a complete motor recovery is not possible, with the majority of stroke patients being unable to perform professional duties or activities of daily living by 6 months after their stroke [24]. In USA, stroke is the leading cause of long-term disability and stroke survivors can experience permanent physical, psychological, and social disabilities such as hemiplegia, aphasia, sensory disturbances, depression, and cognitive and memory impairments [25].

After the onset of a stroke, acute treatment options are limited and need to take place in a short period of time. Nonetheless, researchers and global policies are concentrating efforts in stroke prevention especially among high-risk populations.

This book chapter aims to address the challenges in stroke research as well as to point out important aspects to be considered while developing new clinical trials. We reviewed the 100 most cited articles in stroke, and discussed the characteristics of the design, target population, data analysis, and results of stroke clinical trials.

2   Methods

In this search, we systematically reviewed the 100 most highly cited clinical trials in stroke from 2010 to 2015 using the Web of Science platform. The retrieved articles were clinical trials that had the MeSH term "stroke" in the title. This search retrieved 2019 articles that were sorted according to the number of citations. As we aimed to assess interventional trials, protocols and diagnosis trials were excluded. The CONSORT guideline topics were included in this analysis; most of the trials followed these guidelines. In the search process, we analyzed 131 articles; the principal reasons for exclusion were noninterventional trials, or nonstroke trials (Fig. 1).

3   General Characteristics of Cited Articles

Overall, the included trials reported most of the topics analyzed since the majority of the targeted journal suggested to follow the CONSORT guidelines (85 trials) (Table 1) [26]. In some of the analyzed trials the primary and secondary outcome were not clearly

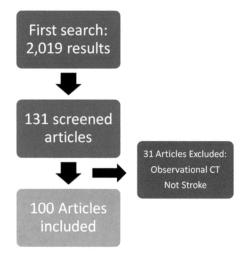

Fig. 1 Schematic Illustration of articles search and selection. In the first search we found 2019 clinical trials with the MeSH term "stroke" in the title. From these we screened 131 articles and excluded the observational and not stroke trials in order to finish with the 100 clinical trials included in this chapter

Table 1
General characteristics of cited articles

			# (Total N = 100)
Year		2010	40
		2011	21
		2012	19
		2013	12
		2014	5
		2015	3
Phase	Drugs	Phase I	2
		Phase I/II	1
		Phase II	6
		Phase II/III	4
		Phase III	35
	Devices and procedures	Exploratory-pilot	0
		Exploratory-feasibility	18
		Pivotal	34
		Post market	0
Trial design	RCT	2 parallel arms	80
		3 parallel arms	7
		4 or more parallel arms	4
		Factorial	0
		Crossover	1
		Non-RCT	8

(continued)

Table 1
(continued)

		# (Total N = 100)
Sample size	0–99	39
	100–499	25
	500–999	12
	1000–5000	10
	5001 and above	14
	Not applicable	0
Type of intervention	Drug	45
	Device	19
	Procedure	29
	Procedure and drug	4
	Device and drug	3
Placebo controlled	Yes	85
	No	15
Number of citations	31–40	8

Summary of the general characteristics analyzed in this book chapter. Overall idea of the clinical trials included in the search

identified and not properly organized; furthermore, the allocation concealment and data analysis of the secondary outcomes were missing for most of the studies. In Table 1 we present a summary of all 100 reviewed articles the findings. The majority of the articles were published in 2010; this might be due to the search methodology, since these articles had additional time to be cited than more recent articles. Regarding the phase, most were phase III trials, what is also expected once these are efficacy studies that are currently more relevant for the scientific and medical community. Additionally, the most common design was a two-arm parallel randomized clinical trial comparing drug vs. placebo. Regarding the sample size, most of the articles had an approximate sample size up to 99 subjects in total. As discussed in the following sections, preventive trials tend to have larger sample sizes when compared with therapeutic poststroke studies. Some of these characteristics will be further discussed in the following subsections.

3.1 Types of Study

Overall, most of the trials reviewed are related with poststroke interventions (70) and the most common topic is ischemic stroke (38), corresponding to 87% of the current ischemic stroke incidence and are usually related to acute stroke care. Moreover, 31 trials included participants with either ischemic or hemorrhagic stroke. The third most common topic is related to prevention of stroke (30) in a high-risk population (Fig. 2).

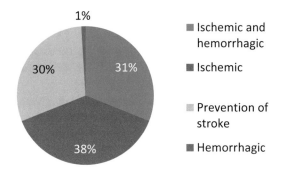

Fig. 2 Percentage of therapeutic poststroke trials divided by inclusion of ischemic or hemorrhage stroke and prevention of stroke clinical trials

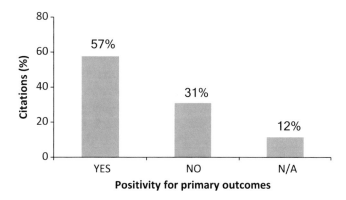

Fig. 3 Percentage of citations divided by positive or negative trials results for the primary outcome

3.2 Results of Primary Outcome

In this review, 66 clinical trials presented positive results while only 23 were negative for the primary outcome; in addition for 11 studies, the primary outcomes were unclear and thus difficult to define as positive or negative. Furthermore, it is important to point out that the number of citations is higher in studies with a positive outcome (Fig. 3) compared with the negative studies.

3.3 Journals and Impact Factor

Considering all included articles, only 15 clinical trials were not published in journals that endorse the CONSORT guidelines [26]. Nonetheless, all articles reviewed do not explicitly describe all items on the CONSORT checklist and some of the data was reported as "not specified/not applicable." The impact factor (IF) mean of journals which follow the CONSORT criteria was higher (21.13) than the ones (6.03) that are not included in the list. Moreover, only 21 trials were published in journals with IF of <5, and eight of these trials were published in journals that do not follow the guidelines. Overall, 33 different journals had trials published in the 100 most cited list. The journal with higher number of trials was STROKE journal (IF = 5.723) with 19 trials followed by New England Journal of Medicine (NEJM) (IF = 55.873) with 14 trials. Although IF is a controversial way of measure science quality, it can be used as an indirect assessments of scientific excellence.

3.4 Types of Interventions and Study Phases

From all the studies reviewed, the majority of them were drug trials (45) and were classified in phases I, II, III, or IV according to FDA definition of study phases. On the other hand, procedures (29) and devices (29) trials were classified as exploratory, pivotal, or confirmatory trials.

4 Trial Design, Study Settings and Phases

In this section, the goal is to discuss relevant aspects from previous positive or negative stroke trials and the importance of a carefully conceived trial design. Some characteristics need to be considered before designing phase I, II and III stroke trials for new drug therapies or exploratory and pivotal for devices and procedures. Among all the 100-analyzed studies, the great majority were phase III for trials in which the intervention was a drug and pivotal phase for those trials that involved procedures or devices as the intervention.

4.1 Phases of Clinical Trials

The development of new therapies is a long process that requires the investment of significant time and resources. In most of the stroke trials reported, the phase was not clearly defined and was common to find studies that combined trial phases. The success of a phase III trial is highly dependent of a carefully designed phase I and II clinical trials. The target population needs to be appropriate for the tested intervention, as well as the outcome that will be measured. The phase of the study is highly correlated with the study settings in stroke trials, with a few exceptions. The majority of the studies do not explicitly disclose the study phase, and the following criteria (Fig. 4) were used to establish the corresponding trial phase during the data analysis:

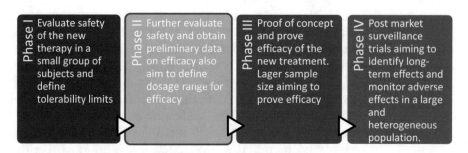

Fig. 4 Criteria to define study phase of drug trials according to FDA

In this review, there were two phase I and one phase I/II studies. These three trials tested cell transplantation (mesenchymal or bone marrow cells) in ischemic stroke patients; moreover two of them did it in acute phase and one in chronic stroke patients. All trials assessed safety and feasibility and there were no complications or adverse effects after the transplantation. In addition, all of these studies had a single-arm, with a nonrandomized design and were open label which is in agreement with the main goals of phase I clinical trials. It is important to point out that these phase I and pilot trials in stroke were not performed in healthy subjects, especially because of the intervention tested. These trials assessed the safety of stem cell by measuring serial cell blood counts, kidney function (urea, creatinine), biochemical blood analysis (i.e., electrolytes) as well as a neurological examination and complementary neuroimaging tests such as CT scan, MRI, radiography, and EEG until either the development of a novel stroke recurrence, cell-related adverse event during the procedure or during the follow-up period as well as death [27, 28].

We found six phase II trials in stroke. All were designed as randomized clinical trials (RCT); from these studies, three had two arms, two had four parallel arms, and one was a single arm study. In addition, five trials were multicenter while only one was single center. Among the six phase II trials, four trials had positive results. The goal of these phase II trials was to assess initial efficacy of the intervention as to gather data for the phase III trial design.

Late phase II trials, also called phase IIb or phase II/III can assess efficacy further in a larger sample size but smaller than phase III trials. We found four phase II/II RCTs. Only one of these had three arms, the majority of them were parallel RCTs with two arms. Moreover, only one study had negative results showing no difference between the studied groups, and was prematurely terminated.

In Phase III studies, the required sample sizes are larger and calculated based on the expected effects of the new intervention. Overall, 36 studies included in this review were phase III trials. Most of these are blinded, parallel, two-arm, multicenter RCT studies. Nonetheless one was a RCT with three groups and an additional one had four groups. Of these 36 trials, 12 had negative results and 15 had positive results while for other eight trials these categories were undetermined. The majority of these phase III trials tested drugs for prevention of stroke in a high-risk population.

In a phase IV trial the number of subjects required is even larger than that in a phase III trial, and the main goal is to evaluate potential long-term side effects. In this review, we found no phase IV trials.

We found 51 trials testing devices or procedures, among those; we found 18 exploratory-feasibility trials and 33 pivotal

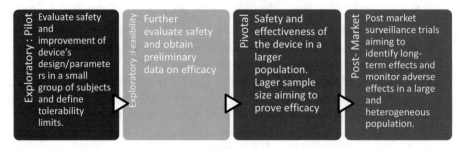

Fig. 5 Criteria to classify device and procedure trials

trials (Fig. 5). Device and procedure trials have some different characteristics compared with drug trials as they tend to have a relatively smaller sample size, major difficulties with blinding and also more diverse end points and aims to demonstrate safety. Interestingly, the number of stroke trials that test devices have been increasing over the recent years. According to the Web of Science between years 1990 and 2000, 87 trials on devices were found; by 2001–2010 this number increased to 304 and over the last 5 years there have been 371 device trials in stroke population.

Similar to phase I and II drug trials, exploratory-feasibility trials aim to evaluate safety and collect preliminary data regarding effectiveness in order to support a future pivotal/phase III trial. From the 18 exploratory-feasibility trials found in this search, eight trials tested a device as intervention while the other ten tested a medical procedure. Most of these studies were RCT, blinded, parallel, two arms, and single center studies. Of those 18 trials only two present negative results.

Moreover, pivotal trials' main goal is to show effectiveness in a larger sample size with the main goal to change clinical practice. In this review we found 33 pivotal trials, among those, 11 trials tested devices while 19 tested procedures as intervention. Most of these RCTs were multicenter, parallel two arms. Only one trial had a cross over design. In contrast to the drug trials, all pivotal trials analyzed were positive for the primary outcome, and most of them tested a device/procedure after stroke rather than for stroke prevention population.

When comparing drug vs. device or procedure phase III or pivotal trials; drug phase III trials had an average of 7681 subjects (min = 118 and max = 19,257), while devices pivotal trials had an average of 193 subjects (min = 48 and max = 707) and procedure pivotal trials had an average of 459 (min = 32 and max = 3120). Most of the drug trials were preventive trials (51%) while 89% of device trials and 93% of the procedure trials were therapeutic trials.

4.2 Study Settings   Most of the studies reviewed were multicenter clinical trials (76). Among these, 59 were phase III or pivotal studies. Two multicenter studies included in the analysis were phase I studies. Even

Table 2
Drug trial design, study settings and phases

Trial design		N = 49	Drugs trial phases							Study setting
			Pilot	I	I/II	II	II/III	III	IV	
RCT	2 parallel arms	40	0	0	0	1	1	1	0	Single center
			0	0	0	2	2	33	0	Multicenter
	3 parallel arms	2	0	0	0	0	0	0	0	Single center
			0	0	0	0	1	1	0	Multicenter
	4 or more parallel arms	3	0	0	0	0	0	0	0	Single center
			0	0	0	2	0	1	0	Multicenter
	Factorial	0	0	0	0	0	0	0	0	Single center
			0	0	0	0	0	0	0	Multicenter
	Cross over	0	0	0	0	0	0	0	0	Single center
			0	0	0	0	0	0	0	Multicenter
Non-RCT (Quasi-experimental)		4	0	0	1	0	0	0	0	Single center
			0	2	0	1	0	0	0	Multicenter

though the number of subjects was small the nature of the intervention—autologous bone marrow transplantation or autologous mesenchymal stem cells serum application—seems to be the reason why the studies were multicenter. Once this is an invasive procedure and the recruitment can be difficult due to the nature of the procedure itself in addition to the more specific inclusion and exclusion criteria required for this kind of clinical studies.

Due to the acute characteristics of a stroke and the requirement of a fast intervention in the nearest available center, most of the trials in acute stroke are conducted in multiple centers, what helps to address recruitment challenges given the limited enrollment windows. Tables 2 and 3 shows the study design of the drug and device trials, respectively.

5   Participants

In this section we present the inclusion and exclusion criteria for the reviewed stroke trials. Inclusion and exclusion criteria are essential to define the trial population. Population is chosen based on the "best chance to succeed" and also on safety. Therefore, participants who have lower chances of recovery such as older patients as well as participants with comorbidities that have high NIHSS scores or certain stroke subtypes are usually excluded.

Table 3
Device/procedure trial design, study settings and phases

Trial design		N = 51	Device and procedure trial phases				Study setting
			Exploratory-pilot	Exploratory-feasibility	Pivotal	Postmarketing	
RCT	2 parallel arms	40	0	8	6	0	Single center
			0	3	23	0	Multicenter
	3 parallel arms	5	0	1	2	0	Single center
			0	1	1	0	Multicenter
	4 or more parallel arms	1	0	0	0	0	Single center
			0	1	0	0	Multicenter
	Factorial	0	0	0	0	0	Single center
			0	0	0	0	Multicenter
	Crossover	1	0	0	1	0	Single center
			0	0	0	0	Multicenter
Non RCT (Quasi-experimental)		4	0	2	0	0	Single center
			0	2	0	0	Multicenter

Regarding study population in stroke trials, there are differences between prevention trials and poststroke trials. Considering this, we divided this section in two subsections: "Prevention Trials" and "Poststroke Trials." Most of the studies that we found for prevention trials used the CHAD score as the main inclusion criteria and for the poststroke group the main instruments used were the NIHSS and mRS scales.

5.1 Prevention Trials

The prevention studies can be classified as primary (no previous stroke) or secondary (previous stroke). The primary prevention trials aim to evaluate diverse interventions in patients who never had a previous stroke while the secondary trials assess a population in which the subject experienced at least one previous stroke or a transient ischemic attack (TIA). Considering all included prevention trials, most of the trials reviewed were secondary, including participants with prior stroke, TIA or atrial fibrillation (AF), and one or more additional risk factors.

Stroke prevention trials usually included participants with previous stroke or TIA; moreover, research groups are interested in individuals with diverse risk factors such as high blood pressure, diabetes mellitus, carotid or other artery disease, peripheral artery disease, atrial fibrillation (AF), and other cardiovascular diseases.

Table 4
Inclusion criteria for preventive stroke trials

Primary inclusion criteria	N
Atrial fibrillation and at least 1 other stroke risk factor	15
Stroke, TIA, or systemic embolism, or 2 or more	4
TIA or stroke	2
Cryptogenic ischemic stroke, and had a patent foramen ovale	1
Elevated Levels of C-Reactive Protein	1
Complete occlusion of an internal carotid artery and TIA or ischemic	1
SCA and previous stroke	1
2 diabetes mellitus, nondialysis CKD	1
Carotid artery stenosis	1
Non-cardioembolic cerebral infarction	1
TIA or stroke or 2 risk factors	1
Atrial fibrillation and CHADS2 score of 2	1

In this review, we found 30 preventive trials, among those, we found 25 trials using drugs as intervention; additionally, three using procedures and using two devices. Most of the trials were performed in nonvalvular atrial fibrillation adults who had at least one additional stroke risk factor (15) (Table 4).

Most of the prevention trials in this book chapter evaluated different anticoagulants for stroke prevention. An example of a prevention trial found in this review was the RE-LY trial, a multi-center noninferiority clinical trial that randomized 18,000 participants either to high (150 mg) or low dose (75 mg) of dabigatran or to warfarin for stroke prevention in patients with nonvalvular atrial fibrillation and at least one of the following risk factors for stroke: previous stroke or transient ischemic attack (duration of neurological symptoms <24 h); left ventricular ejection fraction of <40%; New York Heart Association heart failure symptoms of class 2 or greater within 6 months; at least 75 years of age; or at least 65 years of age with diabetes mellitus, hypertension, or coronary artery disease [29].

Here the investigator needs to weigh the number of risk factors to be used as inclusion criteria as more risk factors increase the likelihood of stroke and thus increasing the power of study as events are more likely but on the other hand decreases the size of accessible population for the study.

5.2 Poststroke Trials   Poststroke trials are studies aiming to reduce disability and mortality after stroke. Most of these include patients based on disability scores such as the NIH stroke scale and the modified ranking scale (mRS).

Two main stroke characteristics are usually considered: time since stroke (acute, sub acute or chronic) and type of stroke (ischemic or hemorrhage). We found 42 trials that included participants in the acute phase of a stroke (<90 days); most of these trials included ischemic stroke patients, few hours after the event onset with a detailed short range of NIH stroke score. Some trials are less restrictive including moderate and severe stroke. For example the trial conducted by Saver et al. [30] included participants with NIHSS scores from ≥8 up to ≤30. On the other hand some trials were more restrict, for instance Meng et al. [31] included participants with NIHSS 0-15.

When considering rehabilitation interventions, the investigator needs to consider the mechanisms of the intervention that is being tested and if there is any variation when applied in the acute or chronic phase. There is extensive literature discussing neuroplastic changes over time in stroke and the investigator is encouraged to take it into account [32–34]. Additionally, most of the rehabilitation trials chose to select patients in the acute stroke phase.

An important aspect to consider is the population enrollment potential. Kasner et al. performed a survey aiming to evaluate the decision making capacity of stroke patients or primary decision makers, showing that only 57% (95% confidence interval: 50–64%) of the eligible patients that answered the survey stated that they would participate in the proposed trial. Moreover, they showed that patients were more prone to participate than the proxy decision-maker group. This is an important aspect to be considered while designing an acute stroke trial, since the recruitment yield can be slow, jeopardizing overall enrollment.

From the 67 poststroke trials, only three studies included patients in acute and chronic phase, as shown in Table 5. Among

Table 5
Time since stroke onset versus type of stroke

	Total	Ischemic	Ischemic and hemorrhagic	Hemorrhage
Acute stroke	42	31	10	1
Acute and chronic	3	1	2	
Chronic stroke	15	4	12	
Subacute	7	2	5	

Acute <90 days
Subacute >90 days < 6 months
Chronic >6 months

these, two trials tested devices for stroke motor rehabilitation and one more evaluated the effects of botulinum toxin type A in spasticity. Moreover 15 trials were performed in chronic stroke population, all of them testing different types of rehabilitations techniques such as noninvasive brain stimulation, robotic-assisted therapies, botulinum toxin type A, and others. In these trials the main inclusion criteria was based on the participant's motor deficits (i.e., hemiparesis or spasticity). Likewise, seven trials include a sub acute stroke population and aimed to evaluate motor rehabilitation after stroke.

An additional important consideration is stroke localization. Only 16 (two of them NIBS) trials considered this criteria. Stroke localization may not be important for pharmacological therapies given in poststroke recovery; however, it may play an important role in targeted therapies where a lesion localization makes a difference such as in brain stimulation trials or studies with endovascular interventions [35–39].

6   Trial Duration and Follow-Ups

In this section, the importance of determining the proper duration for each type of intervention and the most appropriate time length for a follow-up and corresponding amount of follow-up visits are discussed.

As expected, the duration of intervention in stroke trials is a particular characteristic that requires to be previously analyzed due to the wide variety of interventions. There were a good number of single session procedures such as stent collocation, closure of patent foramen ovale as well as the application through an IV line of a thrombolytic agent or even long-term use of anticoagulants to prevent stroke in a high-risk population (Table 6).

Table 6
Trial duration for types of strokes

	Procedure duration	<1 month	1–3 months	>3 months– 1 year	>1 year
Prevention	0	0	2	11	17
Poststroke					
Hemorrhagic	0	0	1	0	0
Ischemic	0	4	21	10	3
Ischemic and hemorrhagic	1	9	13	7	1

6.1 Prevention Trials

In the case of prevention trials we found interventions without a predetermined time length because in such cases the outcome was time to event and thus subjects were followed for as long as possible (until they developed the event or trial was terminated) (right censoring). We considered the mean number of follow ups in these cases as the length of the trial.

As an example the ARISTOTLE study [40], a randomized, double-blind clinical trial compared the noninferiority of apixaban (at a dose of 5 mg twice daily) with warfarin (target international normalized ratio, 2.0–3.0) in 18,201 patients with AF and at least one additional risk factor for stroke. The enrollment started in 2006 and ended up in 2012, the median of follow-up was 1.8 years; additionally based on the study protocol the final duration of the trial was determined by the time required to acquire at least 448 primary efficacy events. The results showed that apixaban was superior to warfarin regarding stroke events or systemic embolism (annual incidence 1.27% vs. 1.60%). Moreover, apixaban was also associated with less major bleeding (annual incidence 2.13% vs. 3.09%) [40].

Overall the duration of the prevention trials was longer when compared with poststroke studies and most of them lasted for more than 1 year. The mean duration of preventive trials was two and a half years, being even frequent to follow patients up to 5 years for some studies [41, 42].

6.2 Poststroke Trials

In therapeutic poststroke trials the duration was heterogeneous varying from 1 day to more than 1 year. Besides that, most of the trials lasted about 3–12 months. As discussed in the previous section, most of the poststroke trials were performed in acute stroke participants and usually the last follow-up visit occurred after 90 days (22 from 42 acute stroke trials). Same pattern was observed in trials that enrolled chronic stroke participants (seven from 15 chronic stroke trials). Therefore, most trials in this phase that evaluated recovery outcomes only assessed patients for a relatively short duration. This is important to point out since most of these trials aimed to treat in some extent motor/sensory/cognitive impairments utilizing rehabilitation techniques.

7 Intervention: Drug/Device/Procedure

Here we reviewed the most common interventions used in stroke trials, the objective is to give an overall idea and describing the relationship between the study type, time since event and the chosen intervention. Most of the stroke trials use drugs as intervention, especially for treatment after an ischemic stroke (16 trials) or for stroke prevention (23 trials) (Table 7). Alternatively, studies including ischemic and hemorrhagic stroke population utilized procedures as intervention in the great majority of the trials analyzed.

Table 7
Types of interventions in therapeutic poststroke trials and in prevention trials

	Total	Ischemic	Ischemic and hemorrhagic	Hemorrhagic	Prevention
Device	19	9	8	0	2
Device and Drug	3	0	0	0	3
Drug	45	16	5	1	23
Procedure	29	9	18	0	2
Procedure and Drug	4	4	0	0	0

Table 8
Most common interventions in prevention stroke trials

Intervention	N	%
Oral anticoagulant	17	57%
Antiplatelet therapy alone or combined	3	10%
Closure alone or combined	3	10%
Others	3	10%
Surgical	2	7%
β-blockers and calcium-channel blockers, angiotensin inhibitor	1	3%
Statins	1	3%

7.1 Prevention Trials
Prevention trials aimed to evaluate the effects of an intervention in order to decrease the probability of a stroke event for the first time or in some cases a recurrent stroke event. As discussed before most of the studies utilized pharmacological therapy such as oral anticoagulants specially to reduce the probability of a thromboembolic event in subjects with AF (57%); furthermore, some studies evaluated antiplatelet therapies as the intervention. In addition to anticoagulant/antiaggregant therapies, we also found interventions in which the desired target of the intervention was to decrease blood pressure levels and in fact decrease the risk of a hemorrhagic stroke; in order to asses this, the research groups tested diverse antihypertensive groups of medications such as b-blockers (i.e., Pindolol), calcium channels blockers (i.e., Amlodipine) and angiotensin converting enzyme inhibitors (i.e., Perindopril) (Table 8). In particular we found interesting the study "the Switch trial" which in fact is the only study that included children and adolescents (5–18 years of age) with diagnosis of sickle cell anemia disease and the research group applied as intervention the drug hydroxyurea in conjunction with phlebotomy [43].

Table 9
American Heart Association Guidelines for anticoagulants evidence

Drug	Class	Level	Dosing	Target
Warfarin	I	A	INR 2–3	Vitamin K-dependent factors
Dabigatran	I	B	Twice a day	Thrombin
Rivaroxaban	I	B	Once a day	Factor Xa
Apixaban	I	B	Twice a day	Factor Xa

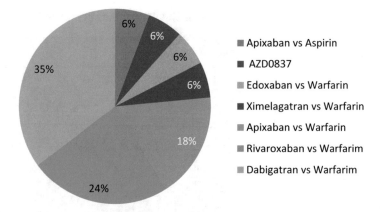

Fig. 6 Percentage of trials using specific types of anticoagulants

Finally, the last pharmacological treatment encountered in our search was the application of statins to decrease the risk of vascular events that can cause ischemic stroke. Besides pharmacological therapy, three clinical trials implemented surgical procedures such as carotid endarterectomy or arterial anastomosis to prevent ischemic stroke; moreover, we found three clinical trials that investigated the effects of closure of the left atrial appendage in order to decrease the risk of novel stroke events.

Due to the increased number of ischemic strokes encountered in subjects with cardiac arrhythmias, the topic of anticoagulation therapy is highly relevant during the development of stroke trials and also due to the wide variety of pharmacological agents available in the market (Table 9). Please refer to Fig. 6 to check the percentage of trials using specific types of anticoagulants.

An investigator in this area needs then to consider therefore this landscape of prevention trials before embarking in designing one. Currently, there are several drugs and procedures approved for stroke prevention; indicating that novel trials would certainly need to consider the comparison with standard treatment (such as the trials of comparing rivaroxaban vs. warfarin; mobilization protocol

(VEM) VS. standard stroke unit care (SC); alternative treatment (hydroxyurea/phlebotomy) to standard treatment (transfusions/chelation) for reduction of secondary stroke and improved management of iron overload or mechanical embolectomy (Merci Retriever or Penumbra System) VS. standard care among others) [41, 43–45]. Given the amount of time and large sample size required for prevention trials (especially considering the small effect sizes between active treatments), the cost benefit of trials such as these ones is very low. Trials such as this should try to find better outcomes and also have optimized plans of interim analysis and also the investigator needs to have clear evidence that new intervention is better or safer than current standard treatments.

7.2 Ischemic and Hemorrhagic Stroke Trials

In poststroke trials, the intervention can be directed for (1) the acute care of patients aiming to control and prevent medical and neurological complications due to the event, especially to revert the neurological effects of the stroke and in some cases avoid a new event or the hemorrhagic transformation and (2) for poststroke deficits, as an example motor rehabilitation, treatment for spasticity or aphasia.

7.3 Acute Stroke Care

In acute care, intravenous t-PA is the only current available approved pharmacological therapy that target reperfusion of brain parenchyma in acute ischemic stroke. For acute stroke trials (42 trials) most of the studies tested anti thrombolytic agents (13), among these, we found six pharmacological agents (rtPA or tenecteplase) and seven with nonpharmacological thrombolytic therapies (stents or combinations). Among the most highly cited studies, we encountered the IMS III trial [46] that compares IV-rtPA alone versus rtPA combined with endovascular therapy in acute ischemic stroke patients. This study had 481 citations and was published in a high impact journal, the New England Journal of Medicine. It was a randomized, phase III open label clinical trial and did not show any significant differences between the rtPA alone and the combined therapy; additionally, the study was stopped early due futility in the interim analysis. One of the primarily critics of this trial was the use of older recanalization devices, that could potentially influence the lack of results once the recanalization rates were lower than the ones reported by other trials using newer devices [46]. Most of the trials including hemorrhagic stroke patients aimed to improve poststroke rehabilitation; moreover, only two studies included hemorrhagic patients into the acute phase, one of them tested the angiotensin receptor blocker effects in order to decrease blood pressure (in ischemic and hemorrhagic participants) and the additional one tested pro coagulant therapy after a hemorrhagic stroke (Table 10).

7.4 Rehabilitation Trials

Most of the trials in post stroke population focus on improving recovery and reducing poststroke disabilities. As an estimate of disability, in 2010 ischemic stroke was the cause of the loss of

Table 10
Most common interventions in poststroke trials

Intervention	N	Stroke onset
Rehabilitation therapy	15	Acute, chronic, and subacute
NIBS	12	Acute, chronic, and subacute
Others	14	Acute, chronic, and subacute
Pharmacological thrombolytic	6	Acute
Stent retriever	5	Acute
Cell transplantations	4	Acute
Antiplatelet therapy	3	Acute
Botox	3	Acute, chronic, and subacute
Antihypertensive	2	Acute
SSRI	2	Acute
Stent retriever + pharmacological thrombolytic	2	Acute

39.4 million disability adjusted life years (DALYs) and hemorrhagic stroke lead to the loss of 62.8 million DALYs in low-income and middle-income countries [19]. Among these patients up to 50% are not able to achieve a level of recovery adequately enough to perform professional duties or activities of daily living [24]. Motor recovery is directly correlated to the achievement of independence in such activities and thus, is one of the main objectives in rehabilitation. Therefore, 47% of the poststroke trials aimed to evaluate techniques that may improve overall rehabilitation, with great emphasis in motor recovery as well as speech rehabilitation. Of these, 15 clinical trials looked at the effect of several poststroke rehabilitation techniques such as motor imagery, robot-assisted therapy, video game therapy, and walking protocols on a treadmill. Other common rehabilitation therapy tested was noninvasive brain stimulation techniques such as TMS (7) and tDCS (5). As an example, in 2014, low frequency rTMS of the contralesional motor cortex was recommended as Level B (probable efficacy) for stroke rehabilitation in an evidence-based guideline performed by Lefaucheur and collaborators [47].

8   Outcomes

One of the most important aspects while developing a clinical trial protocol is to have a clear definition of the research question and a definition of the expected outcomes. For this section we also divide

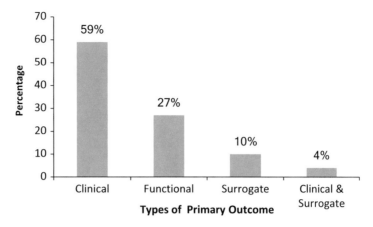

Fig. 7 Types of primary outcome

the outcomes in clinical, functional, or/and surrogate outcomes. Most of the clinical trials, as shown in Fig. 7, used a clinical outcome that focuses on assessment of how the participants feel and behave. Some of the trials preferred to use functional outcomes that measure changes in certain activities of daily life such as walking speed. Moreover, few clinical trials in stroke applied surrogate outcomes to indirectly evaluate the improvement of the participant's clinical status as the primary outcome.

In this section, we describe and discuss the most common endpoints for stroke trials. In order to guide and organize this discussion we divided the trials in prevention and after stroke trials.

8.1   Prevention Trials   Most of the prevention trials have as an end-point the occurrence of a new stroke and stroke-related mortality. These trials aim to obtain information in order to define new stroke occurrence and stroke-related mortality in a population with specific risk factors. These types of prevention trials have longer duration and higher overall costs. Besides this, the sample size also tends to be larger. Most of the prevention trials (30 trials) evaluated time to event considering a stroke, TIA or systemic embolism and/or death as the primary outcome (Table 11). As discussed before this prevention trials aims to decrease events rates using usually anticoagulant therapies. In this context, a clear definition to define the event is important, if the endpoint will be nonfatal stroke, fatal stroke, TIA, or a combination. For example, the authors are required to describe the following: (1) how the event was defined and confirmed; (2) if all the study centers used the same strategy; (3) if it was necessary to have an imaging confirmation; or (4) if the diagnoses were based in initial focal or global neurological physical exams. Besides that, how the classification of stroke types (i.e., hemorrhagic, thromboembolic, or other) was performed (by site/local physicians) and if there was a trial committee responsible to reevaluate and confirm the end points.

Table 11
Most common outcomes for prevention and therapeutic poststroke trials

	Stroke/TIA and/or systemic embolism and/or death	Safety/ adverse effects	mRS	FMMS	Rehabilitation	Imaging studies	Others
Prevention	24	2	0	0	0	0	4
Poststroke							
Hemorrhagic	0	0	1	0	0	0	0
Ischemic	3	6	10	3	1	5	10
Ischemic and hemorrhagic	3	1	0	4	15	0	6

Only two prevention trials had as a primary aim to evaluate the safety and adverse effects of a medication. The first one evaluated safety of endoxaban compared with warfarin to prevent stroke (ref put again 39). The primary outcome measures were related to safety and included bleeding and abnormalities in hepatic function and showed that the rate of bleeding was similar for both drugs [48]. The second one evaluated safety of the oral direct thrombin inhibitor AZD0837 in the prevention of stroke and systemic embolism in patients with atrial fibrillation [49] and showed an overall tolerability profile of immediate-release of the AZD0837.

8.2 Poststroke Trials

On the other side, ischemic and hemorrhagic stroke trials used poststroke recovery as the outcome measure rather than mortality or stroke event. Therefore, common single measurement tools such as the modified Rankin scale (mRS) and rehabilitation scales (i.e., for aphasia, walking distance) are applied. There is no consensus on the ideal outcome or on the amount of difference required in order to provide clinical significance. For this reason is clear that no single outcome will be able to measure the wide variety of aspects involved in stroke recovery. Therefore, is very common that these trials utilize more than one scale to measure stroke rehabilitation; consequently, some trials have multiple primary outcomes or measure the same aspect of recovery with different scales. Nevertheless, most of the trials focus in a single aspect of stroke rehabilitation and just a few studies have a more generalized approach trying to measure overall stroke rehabilitation. Consequently, a subject could participate in a clinical trial and improve its motor abilities obtaining a high score on a motor scale, but remain with severe aphasia, as indicated by the NIHSS score or a specific aphasia scale. The main objective of the trial based on the mechanisms of the drug or intervention need to be carefully considered when choosing this outcome. There are detailed articles discussing the differences of the outcome measures of neurological deficits in stroke [50].

From the 42 trials in ischemic stroke, ten used modified ranking scale as a primary outcome and most of these studies additionally analyzed NIHSS as secondary outcome and were pivotal or phase 3 trials. Moreover, for trials including ischemic and hemorrhagic stroke the most frequent primary outcomes were different scales to measure rehabilitation, such as aphasia scales, spasticity scales, and other motor scales (i.e., Jebsen–Taylor Hand Function).

9 Sample Size

Sample size calculation is an important issue to ensure trial validity. In order to apply sample size calculation methods in clinical trials with stroke participants is necessary to consider the wide variety of possible outcomes as well as of the usual parameters required to develop an estimate of sample size.

Most of the studies (64 studies) reviewed reported some parameters used to perform the sample size calculation, 41 trials (64%) out of the 64 reported in some detail aspects of the methods used to support the sample size calculations (Table 12). Prevention trials reported 80% (24 trials) of the times some method/parameters for the sample size calculation while therapeutic poststroke trials reported just 59% (41 trials) of the times. Moreover, most of the stroke trials based their calculations on results of previous trials (32).

As expected, prevention trials tend to have larger sample size when compared with the poststroke trial group. Some studies used a one-tailed test (8) instead of a two tailed test (20). If the

Table 12
Sample size calculation

		Number of trials total	One-tailed test	Two-tailed test	Not reported
Alpha	0.5	14	0	10	4
	0.025	7	4	0	3
Power	80%	13	1	7	5
	90%	10	2	2	6
	95%	3	1	1	1
	Others	4	0	0	4
Total		51	8	20	23
Method					
	Previous study	32			
	Pilot data	4			
	Detect minimally clinical differences	2			
	Others	4			

investigator decides to use a one-tail test this needs to be very well justified. When an investigator chooses for a one-tail test, it does mean that the result in the unexpected result cannot be considered; therefore this cost of disregarding a result in the opposite direction (for instance the new drug is worse than placebo) needs to be well justified and also it is always risky that reviewers may question and ask for additional evidence that this choice of one-tailed test was done a priori.

In this review, eight trials reported the use of a one tailed test and 20 a two tailed test. Regarding power and alpha level (chance of calling an ineffective treatment effective) 0.05 it is most commonly used by the reviewed trials. In fact 14 studies reported alpha level of 0.05, however if multiple groups are being compared, the alpha level may be set to a smaller value, an example of that was found in our search in which seven studies used 0.025 alpha level. Additionally, most of the trials (13) used 80% (i.e., a 20% chance of missing the hypothesized treatment effect) of power.

Most trials based the expected difference on previous trials, while only two based it on the minimally clinical difference. These two trials evaluated upper or lower limb motor function rehabilitation after stroke. For the sample size calculation it was considered the mean difference between the main groups of 5 points in the Fugl-Meyer score. Although still controversial, 5 points in this scale is considered an acceptable minimal clinical difference [51, 52].

10   Interim Analysis and Stoppage Criteria

Interim analysis is an important factor in the development of clinical trials. In this section, we discuss some criteria and aspects of early terminated stroke trials. In general clinical trials can be stopped early either due to positive or negative results found when usually 50% of the study participants completed half of the intervention [53–55]. In this review, 26 trials reported decision parameters for stopping the trial and the criteria for interim analysis. In most of the trials this section was not described.

Safety of participants is considered as one of the mayor aspects while developing a proper clinical trial for stroke population, consequently conducting interim analysis is an aspect that gains mayor importance during this process. Most of the times an independent Data Safety Monitoring Committee (DSMB) is required to monitor for safety and can suggest the early termination of the study. In this review, two trials had to stop due safety issues as recommended by the DSMB. The first one of these [48] had to prematurely terminate one of the four treatment arms (endoxaban 60 mg bid) because of an excess of bleeding in this specific group. The second one was stopped because an interim analysis showed significantly higher mortality in one of the groups [56].

A clinical trial can also be stopped because of overwhelming evidence of efficacy of one of the arms of the intervention. This way it would help to avoid that subjects in the other arms receive ineffective treatments for longer periods, therefore utilizing this positive result to develop a novel therapy that may be applied faster and in a larger population. Overall among those 33 trials that describe that performed an interim analysis, eight studies needed to be stopped early, among those, three of them were in favor of one of the interventions. However there is a cost for doing preliminary analysis for efficacy. The EXTEND-IA [57] and the ESCAPE [58] trials were stopped due an interim analysis confirming that tPA plus coadjuvant therapy with an intra-arterial intervention (i.e., thrombectomy) early after the onset of an ischemic stroke improves neurological outcomes and do not increase the incidence of intra cranial hemorrhage. Both interim analyses were not per protocol, but they were performed after the announcement of the results of the MR CLEAN trial, published in 2015 showing that endovascular therapy is highly beneficial when compared with intra venous tPA alone [59]. This is an exception case of not per protocol interim analysis; but acceptable if approved by a DSMB or the ethics committee.

On the other hand, stopping a trial due to negative results is a helpful measure to prevent subjects to be exposed to an unnecessary intervention that can be harmful in short and long term and in the same way decrease the inadequate use of resources. We found two trials that stopped by futility due to lack of difference between the intervention groups. As mentioned before, one of these is the IMS trial the most cited study of our search (481 citations). Notably, this trial published in 2013 showed no difference between IA therapies as a co adjuvant treatment for IV tPA. As discussed before, future trials such as MR CLEAN, EXTEND-IA and ESCAPE were able to show difference for this same intervention between the groups [57–59]. According to the MR CLEAN investigators, the failure of previous trials such as the IMS III was primarily due to lack of consistent neuroimaging documentation and the use of older stent retrievers such as the MERCI device compared to newer devices (used in 82% of MR CLEAN patients) [60].

11   Randomization

In this section, we discuss the most frequently randomization methods applied in the studies analyzed in our search. In the same way we analyze specifics aspects such as type of randomization, allocation concealment, and randomization sequence generation.

11.1   Types of Randomization

Randomization is the method utilized to allocate each subject to a specific arm of intervention in a RCT. There are different methods of randomization and its implementation varies depending on the amount of participants enrolled and how precise is the balance of the sample [61, 62].

Simple randomization is helpful while allocating participants into two groups (treatment or control), in the same way as flipping a coin, the research group would not be able to know in which group each participant will be allocated. Block randomization is utilized to assure an equal number of participants in each group. Finally stratified randomization helps to ensure that covariates are balances across groups of treatment [61].

Even though 79 trials were randomized, only 43 articles gave detailed information to understand the randomization method utilized. Overall, one trial utilized simple randomization with a sample size of 362 subjects to guarantee the adequate balance of groups. Interestingly, checking the Table 1 of this article, we found that actually groups were well balanced; however, a difference was found in the percentage of patients with baseline atrial fibrillation between groups. This factor is relevant since this trial compared acute poststroke treatment with endovascular therapy vs. IV-tPA and patients with atrial fibrillation may be receiving baseline anticoagulant therapy therefore being more prone to develop hemorrhagic transformations after intervention. Despite that, there was no difference in the percentage of fatal or nonfatal intracranial hemorrhage, serious adverse events or fatality rate between groups. In addition, 7 trials utilized block randomization, 13 more utilized block stratified randomization, and 21 utilized stratified randomization (Fig. 8). Additionally, all nonrandomized studies were poststroke therapeutic trials.

Among the 21 studies that utilized stratified randomization, 14 of them used just one stratum, while two trials used two strata and three studies used more than two strata to randomize. It is important to point out the importance of the study center as a factor to stratify, 11 of the clinical trials used it to randomize their sample size; in addition, if we compare with the studies that used block stratified randomization nine of those used also center as the criteria to stratify. Moreover, factors such as age, gender, NIHSS scale were frequently utilized. More uncommon was the use of the occlusion severity or the affected side of the brain as factors to stratify the groups in two clinical trials. However please note that trials that used stratification had a large sample size and on average randomized 2920 patients per intervention group [40, 44, 57, 58, 63–66].

11.2 Randomization Sequence Generators

Most of the studies that reported the randomization sequence generator method used a computerized system to generate their corresponding randomization list; among those computer programs, the most commonly applied were www.random.org. and SAS (version 8.2; www.randomizer.at; random number generator and the Moses–Oakford assignment algorithm; Microsoft Excel).

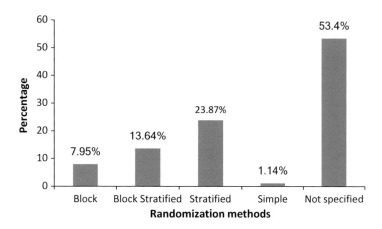

Fig. 8 Types of randomization

11.3 Allocation Concealment

The allocation concealment was reported in as few as 20 studies of our initial search; among them the most common methods utilized were opaque sealed envelopes (7/20) as well as an independent person or study assistant in charge of performing this task. Additionally, other different methods such as prepacked numbered drugs packs, randomization code, and phone telemedicine software were implemented.

11.4 Implementation of Randomization

The implementation of randomization was reported in 14 clinical trials, for most of the cases (5) an independent person such as a study coordinator, biostatistician, or study assistant took charge of this task. Additionally, we found three studies in which the pharmacist implemented the randomization; moreover, for the remaining studies some methods such as automated telephone system, a central telephone service, or even a research therapist were used in the proper development of this important stage of a research trial.

12 Blinding

The blinding section focus in analyzing and understanding the blinding methods that were more commonly found in the 100 reviewed stroke trials.

The type of blinding is highly dependent of the trial design and the outcome measure selected. The importance of adequate blinding strategies is the reduction of bias while assessing the clinical trial outcomes. Although for some more objective outcomes measures such as death, it is more difficult to introduce bias. On the other hand blinding is crucial in stroke trials in a more subjective outcome measure as for example recovery scales. In stroke trials is common the use of a blinded rater to perform the outcome evaluation. In this case the assessors are always blinded to the intervention, even when the subject is not. We found 44 open label studies most of them were post-ischemic-stroke phase III or pivotal trials. Furthermore, in single blinded studies either the investigator or the subject is blinded to the intervention throughout the entire study. We found 15 single blinded trials; most of them were poststroke therapeutic trials.

Meanwhile, in double blinded studies neither the investigator nor the participants have any knowledge regarding allocation of the intervention. Double blind studies are a difficult task in stroke trials, especially those requiring surgical procedures as an intervention. Similarly occurs in trials in which the intervention requires any type of rehabilitation technique especially because the participant is directly exposed to the intervention. In the opposite way, stroke studies utilize a great amount of pharmacological treatments that can be easily blinded to the participant and the investigator

We found 39 double blinded clinical trials, as shown in Table 13; most of them aimed to use drugs in order to treat an acute ischemic stroke. Besides that, approximately half of the prevention trials were double blinded. However, it is also important for the investigator to consider the long duration of the prevention trials and potential for unblinding given the different adverse effects of drugs (or drugs vs. placebos). This was an issue in a trial in which around 20% of participants of each group (warfarin vs. rivaroxaban) stopped the intervention therapy before the endpoint event or study termination date [67].

13   Adherence

Generally speaking, the adherence of a clinical trial can be improved just by simplifying the researcher's demands on participants and specially simplifying the procedures that subjects are required to undergo; in the same way, adherence increases if the subject comprehend deeply the study and gather a great amount of information in regard to the study.

Table 13
Common blinding methods for prevention and poststroke trials

	Total	Ischemic	Ischemic and hemorrhagic	Hemorrhagic	Prevention
Open label	44	21	10	0	13
Single blinded	15	4	9	0	2
Double blinded	39	12	12	1	14
Not described	1	1	0	0	1

Run-in phase is also a strategy to decrease nonadherence by having a period in which subjects receive only placebo intervention and nonadherents are excluded. For stroke trials during acute phase, the possibility of performing a run-in phase may be in some case extremely difficult due to the acute setting of presentation of a stroke event. However this may be a good strategy in prevention trials. On the other hand prevention trials are usually very large trials thus making more difficult and costly to have a run-in phase. In fact only one trial reported a run in phase [68]. For this trial, a 4-week prerandomization run-in period was developed in order to select patients that were capable to provide long-term (average of 1.9 years) protocol compliance to the study.

Furthermore, it is important to point out that stroke trials that involve one time dose interventions in which a great majority of participants are in an inpatient setting and the follow-ups are provided in short term periods will have a decrease rate of dropouts, therefore high adherence values. Contrary to this, prevention trials and studies with long term interventions usually utilize long-term follow-ups, in which the most valuable data is collected at the end of the study [69].

In the 100 trials analyzed, 75 reported adherences values from these 61 were poststroke trials and 14 were prevention trials. In the Fig. 9, we report the mean of adherence from prevention and poststroke trial, nevertheless only half of the prevention trials describe it.

Despite the fact of the major difficulties and challenges that can appear during recruitment process for stroke participants, our search reported that a great majority of studies had adherence rates equal or above 80%. However, the great majority of clinical trials (47%) did not specify these values. Tables 14 and 15 summarize adherence rates for drugs and device/procedure clinical trials. In this context considering at least a dropout rate of 10% when calculating the sample size of stroke trials is advised.

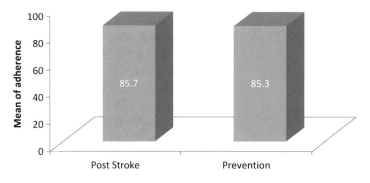

Fig. 9 Means of adherence of clinical trials that reported it

Table 14
Adherence of drug trials

Adherence	I	I/II	II	II/III	III
0–50%	0	0	0	0	0
50–69%	0	0	0	0	2
70–89%	0	0	4	2	6
90–100%	2	0	1	2	11
N/S	0	1	1	0	17

Table 15
Adherence of procedure and device trials

Adherence	Exploratory-pilot	Exploratory-feasibility	Pivotal
0–50%	0	0	0
50–69%	0	0	3
70–89%	0	6	8
90–100%	0	10	18
N/S	0	2	4

14   Statistical Analysis

Given the differences between prevention and poststroke trials, we divided this section in subsections.

14.1   Prevention Trials

For the prevention studies, most of the clinical trials presented in this review used survival analysis to evaluate time-to-event results. This analysis reflects the time until a participant presents a pre-specified event, in this case a stroke, TIA or death. Among the 30 prevention trials, 19 of them applied survival analysis to evaluate time to event. As discussed in Sect. 8 of this chapter most of prevention clinical trials evaluate the risk of developing a stroke, systemic embolism, TIA or death.

During a survival analysis, an important aspect is the estimation of the survival function or survival curve. The Kaplan–Meier estimator is a nonparametric statistic test that can be used to estimate the survival function. Seven of the 19 clinical trials that utilized survival analysis reported to use the Kaplan–Meier estimator. Additionally, in order to compare two group's survival curves the researchers can use a log rank test or use a Cox proportional hazard model as to adjust analysis for other important covariates. In this analysis, three articles reported to use log-rank test to compare the differences between survival times in the proposed groups and 16 articles reported the use Cox proportional hazard models. The covariates often chosen for the Cox proportional hazard models were age, gender, history of stroke or TIA, CHADS2, and comorbidities such as diabetes and hypertension [40, 63, 64, 66, 70–78]. In addition, some prevention articles used Fisher exact test, Bayesian Poisson models, or Wilcoxon rank test to analyze their outcomes.

Furthermore, in the 100 most cited articles we found subgroup analyses, and the majority of them utilized regression analysis to build prediction models in order to understand the role of some characteristics that may increase the risk of having a stroke.

14.2 Poststroke Trials

The type of analysis is more heterogeneous for therapeutic poststroke trials when compared with prevention trials as the outcomes are more variable (i.e., Walking speed test, mRS, measurements of cortical excitability through motor-evoked potentials). Nevertheless, logistic regression analysis was the most common statistical test used in nine clinical trials. These studies tested devices or procedures aiming to compare preintervention and postintervention measurements controlling for individual aspects such as time after stroke, lesion size, location, and demographic characteristics such as age and gender.

The second most used statistical tests were Student's t test and survival analysis, with seven trials for each. Five studies used independent t-test to compare the differences between groups postintervention and two trials to compare the differences between preintervention and postintervention. Most of these studies tested noninvasive brain stimulation techniques in motor recovery after stroke. Additionally, seven therapeutic trials used time-to-event analysis with Kaplan–Meier survival curves as well as with Cox proportional hazard model or log rank in order to compare the differences between groups. All of these studies were acute stroke trials aiming to reduce death events after an ischemic or hemorrhagic stroke. Please refer to Table 16 to check the trials that used survival analysis.

15 Limitations

In this section we discuss the most commonly found limitations throughout the 100 analyzed studies.

During our search, we found that 58 out of the 100 analyzed trials reported study limitations. The remaining 42 clinical trials did not specify any study limitations encountered during the trial; however, some of them suggested potential future considerations for next phases of the corresponding clinical studies. Among the 58 clinical trials that reported limitations, the most frequent drawback was small sample size, as shown in Table 17, which in fact

Table 16
Survival analysis and methods using comparing group's survival curves

	Total	Kaplan–Meier estimator
Cox proportional hazard regression	16	4
Log-rank	3	2

Table 17
Overall limitation in the 100 most cited stroke clinical trials

Limitation	N
Small sample size	18
Limited external validity	13
Not enough statistical power	9
Inadequate assessment scales	3
Imbalanced randomization	4
Problems with blinding	6
Open label—not randomized	3
Excessively strict inclusion and exclusion criteria	2
High rate of dropouts	2
Lack of experience of the research group	2
Sham method not adequate	1
Lack of control group	1
Duration of the study (extremely large—8 years)	1
No long-term followup	1
Measurement device unable to collect all the data	1

correlates with the increased difficulties in recruitment for this type of population. Approximately 32% of trials reported having a small sample size; this points out the importance of enhancing recruitment efforts during the whole process of the trial.

Additionally, limited external validity was a common limitation encountered. We found trials in which the population was predominantly male, limiting the external validity of the results. In order to avoid this limitation, future trials should design more inclusive criteria as well as a balanced randomization of participants.

Another important limitation is the assessment instruments specially for measuring neurological deficits in rehabilitation. Three studies reported limitations in the scales used to assess some of the outcomes, for example, if changes observed in NIHSS scale reflect an actual gain of function by the patients and if this may represent the overall patient recovery. Therefore, it is important to take in consideration that there are no scales that are able to measure global function after stroke. As a result it is important to select the adequate scales depending on the study question. As an example, some interventions specifically target motor rehabilitation (that might or not improve speech), while others target speech improvement (that might not improve motor function), and consequently each intervention will require a particular outcome assessment. Therefore, it is crucial to evaluate the available scales in order to select the most appropriate one [50, 79, 80].

References

1. Birschel P, Ellul J, Barer D (2004) Progressing stroke: towards an internationally agreed definition. Cerebrovasc Dis 17:242–252. https://doi.org/10.1159/000076161

2. Sacco RL, Kasner SE, Broderick JP et al (2013) An updated definition of stroke for the 21st century: a statement for healthcare professionals from the American Heart Association/American Stroke Association. Stroke 44:2064–2089. https://doi.org/10.1161/STR.0b013e318296aeca

3. Kunst MM, Schaefer PW (2011) Ischemic stroke. Radiol Clin North Am 49:1–26. https://doi.org/10.1016/j.rcl.2010.07.010

4. Mehndiratta P, Chapman Smith S, Worrall BB (2015) Etiologic stroke subtypes: updated definition and efficient workup strategies. Curr Treat Options Cardiovasc Med 17:357. https://doi.org/10.1007/s11936-014-0357-7

5. Smith EE, Rosand J, Greenberg SM (2005) Hemorrhagic stroke. Neuroimaging Clin N Am 15:259–272. https://doi.org/10.1016/j.nic.2005.05.003

6. Runchey S, McGee S (2010) Does this patient have a hemorrhagic stroke?: clinical findings distinguishing hemorrhagic stroke from ischemic stroke. JAMA 303:2280–2286. https://doi.org/10.1001/jama.2010.754

7. Smith SD, Eskey CJ (2011) Hemorrhagic stroke. Radiol Clin North Am 49:27–45. https://doi.org/10.1016/j.rcl.2010.07.011

8. Das K, Mondal GP, Dutta AK et al (2007) Awareness of warning symptoms and risk factors of stroke in the general population and in survivors stroke. J Clin Neurosci 14:12–16. https://doi.org/10.1016/j.jocn.2005.12.049

9. Judd SE, Kleindorfer DO, McClure LA et al (2013) Self-report of stroke, transient ischemic attack, or stroke symptoms and risk of future stroke in the reasons for geographic and racial differences in stroke (REGARDS) study. Stroke 44:55–60. https://doi.org/10.1161/STROKEAHA.112.675033

10. Kleindorfer D, Judd S, Howard VJ et al (2011) Self-reported stroke symptoms without a prior diagnosis of stroke or transient ischemic attack: a powerful new risk factor for stroke. Stroke 42:3122–3126. https://doi.org/10.1161/STROKEAHA.110.612937

11. Glymour MM, Maselko J, Gilman SE et al (2010) Depressive symptoms predict incident stroke independently of memory impairments. Neurology 75:2063–2070. https://doi.org/10.1212/WNL.0b013e318200d70e

12. Kothari R, Sauerbeck L, Jauch E et al (1997) Patients' awareness of stroke signs, symptoms, and risk factors. Stroke 28:1871–1875. https://doi.org/10.1161/01.STR.28.10.1871

13. Wall HK, Beagan BM, O'Neill J et al (2008) Addressing stroke signs and symptoms through public education: the Stroke Heroes Act FAST campaign. Prev Chronic Dis 5:A49

14. Lisabeth LD, Brown DL, Hughes R et al (2009) Acute stroke symptoms: comparing women and men. Stroke 40:2031–2036. https://doi.org/10.1161/STROKEAHA.109.546812

15. Howard VJ, Lackland DT, Lichtman JH et al (2008) Care seeking after stroke symptoms. Ann Neurol 63:466–472. https://doi.org/10.1002/ana.21357

16. Tyson SF, Hanley M, Chillala J et al (2006) Balance disability after stroke. Phys Ther 86:30–38. pii: 16386060

17. Adamson J, Beswick A, Ebrahim S (2004) Is stroke the most common cause of disability? J Stroke

Cerebrovasc Dis 13:171–177. https://doi.org/10.1016/j.jstrokecerebrovasdis.2004.06.003

18. Goljar N, Burger H, Vidmar G et al (2010) Functioning and disability in stroke. Disabil Rehabil 32(Suppl 1):S50–S58. https://doi.org/10.3109/09638288.2010.517598

19. Mozaffarian D, Benjamin EJ, Go AS et al (2015) Heart disease and stroke statistics--2015 update: a report from the American Heart Association. Circulation 131:e21. https://doi.org/10.1161/CIR.0000000000000152

20. Feigin VL, Lawes CMM, Bennett DA, Anderson CS (2003) Stroke epidemiology: a review of population-based studies of incidence, prevalence, and case-fatality in the late 20th century. Lancet Neurol 2:43–53. https://doi.org/10.1016/S1474-4422(03)00266-7

21. Mas JL, Zuber M (2005) Epidemiology of ischemic stroke. Cerebrovasc Dis 28:335–359. https://doi.org/10.1159/000108879

22. Wolf PA, Kannel WB (2011) Epidemiology of stroke. In: Stroke. Saunders, Philadelphia, PA, pp 198–218

23. Mukherjee D, Patil CG (2011) Epidemiology and the global burden of stroke. World Neurosurg 76:S85. https://doi.org/10.1016/j.wneu.2011.07.023

24. Hummel FC, Cohen LG (2006) Non-invasive brain stimulation: a new strategy to improve neurorehabilitation after stroke? Lancet Neurol 5:708–712

25. Di Carlo A (2009) Human and economic burden of stroke. Age Ageing 38:4–5. https://doi.org/10.1093/ageing/afn282

26. Schulz KF, Altman DG, Moher D et al (2010) CONSORT 2010 statement: updated guidelines for reporting parallel group randomised trials (Chinese version). J Chinese Integr Med 8:604–612. https://doi.org/10.3736/jcim20100702

27. Moniche F, Gonzalez A, Gonzalez-Marcos J-R et al (2012) Intra-arterial bone marrow mononuclear cells in ischemic stroke: a pilot clinical trial. Stroke 43:2242–2244. https://doi.org/10.1161/STROKEAHA.112.659409

28. Battistella V, De Freitas GR, Dias V et al (2011) Safety of autologous bone marrow mononuclear cell transplantation in patients with non-acute ischemic stroke. Regen Med 6:45–52

29. Hori M, Connolly SJ, Zhu J et al (2013) Dabigatran versus warfarin: effects on ischemic and hemorrhagic strokes and bleeding in Asians and non-Asians with atrial fibrillation. Stroke 44:1891–1896. https://doi.org/10.1161/STROKEAHA.113.000990

30. Saver JL, Jahan R, Levy EI et al (2012) Solitaire flow restoration device versus the Merci Retriever in patients with acute ischaemic stroke (SWIFT): a randomised, parallel-group, non-inferiority trial. Lancet 380:1241–1249. https://doi.org/10.1016/S0140-6736(12)61384-1

31. Meng R, Asmaro K, Meng L et al (2012) Upper limb ischemic preconditioning prevents recurrent stroke in intracranial arterial stenosis. Neurology 79:1853–1861. https://doi.org/10.1212/WNL.0b013e318271f76a

32. Dimyan MA, Cohen LG (2011) Neuroplasticity in the context of motor rehabilitation after stroke. Nat Rev Neurol 7:76–85. https://doi.org/10.1038/nrneurol.2010.200

33. Murphy TH, Corbett D (2009) Plasticity during stroke recovery: from synapse to behaviour. Nat Rev Neurosci 10:861–872. https://doi.org/10.1038/nrn2735

34. Dromerick AW, Edwardson MA, Edwards DF et al (2015) Critical periods after stroke study: translating animal stroke recovery experiments into a clinical trial. Front Hum Neurosci 9:231. https://doi.org/10.3389/fnhum.2015.00231

35. Barwood CHS, Murdoch BE, Whelan BM et al (2011) Improved language performance subsequent to low-frequency rTMS in patients with chronic non-fluent aphasia post-stroke. Eur J Neurol 18:935–943. https://doi.org/10.1111/j.1468-1331.2010.03284.x

36. Emara TH, Moustafa RR, Elnahas NM et al (2010) Repetitive transcranial magnetic stimulation at 1 Hz and 5 Hz produces sustained improvement in motor function and disability after ischaemic stroke. Eur J Neurol 17:1203–1209. https://doi.org/10.1111/j.1468-1331.2010.03000.x

37. Avenanti A, Coccia M, Ladavas E et al (2012) Low-frequency rTMS promotes use-dependent motor plasticity in chronic stroke: a randomized trial. Neurology 78:256–264. https://doi.org/10.1212/WNL.0b013e3182436558

38. Chang WH, Kim Y-H, Bang OY et al (2010) Long-term effects of rTMS on motor recovery in patients after subacute stroke. J Rehabil Med 42:758–764. https://doi.org/10.2340/16501977-0590

39. Kim D-Y, Lim J-Y, Kang EK et al (2010) Effect of transcranial direct current stimulation on motor recovery in patients with subacute stroke. Am J Phys Med Rehabil 89:879–886. https://doi.org/10.1097/PHM.0b013e3181f70aa7

40. Easton JD, Lopes RD, Bahit MC et al (2012) Apixaban compared with warfarin in patients with atrial fibrillation and previous stroke or transient ischaemic attack: a subgroup analysis of the ARISTOTLE trial. Lancet Neurol 11:503–511. https://doi.org/10.1016/S1474-4422(12)70092-3

41. Hankey GJ, Patel MR, Stevens SR et al (2012) Rivaroxaban compared with warfarin in patients with atrial fibrillation and previous stroke or transient ischaemic attack: a subgroup analysis of ROCKET AF. Lancet Neurol 11:315–322. https://doi.org/10.1016/S1474-4422(12)70042-X

42. Shinohara Y, Katayama Y, Uchiyama S et al (2010) Cilostazol for prevention of secondary stroke (CSPS 2): an aspirin-controlled, double-blind, randomised non-inferiority trial. Lancet Neurol 9:959–968. https://doi.org/10.1016/S1474-4422(10)70198-8

43. Ware RE, Helms RW, Investigators S (2015) Stroke with transfusions changing to hydroxy-urea. Blood 119:3925–3933. https://doi.org/10.1182/blood-2011-11-392340.There

44. Cumming TB, Thrift AG, Collier JM et al (2011) Very early mobilization after stroke fast-tracks return to walking: further results from the phase II AVERT randomized controlled trial. Stroke 42:153–158. https://doi.org/10.1161/STROKEAHA.110.594598

45. Kidwell CS, Jahan R, Gornbein J et al (2013) A trial of imaging selection and endovascular treatment for ischemic stroke. N Engl J Med 368:914–923. https://doi.org/10.1056/NEJMoa1212793

46. Khatri P, Yeatts SD, Mazighi M et al (2014) Time to angiographic reperfusion and clinical outcome after acute ischaemic stroke: an analysis of data from the Interventional Management of Stroke (IMS III) phase 3 trial. Lancet Neurol 13:567–574. https://doi.org/10.1016/S1474-4422(14)70066-3

47. Lefaucheur J-P, André-Obadia N, Antal A et al (2014) Evidence-based guidelines on the therapeutic use of repetitive transcranial magnetic stimulation (rTMS). Clin Neurophysiol 125:2150–2206. https://doi.org/10.1016/j.clinph.2014.05.021

48. Weitz JI, Connolly SJ, Patel I et al (2010) Randomised, parallel-group, multicentre, multinational phase 2 study comparing edoxaban, an oral factor Xa inhibitor, with warfarin for stroke prevention in patients with atrial fibrillation. Thromb Haemost 104:633–641. https://doi.org/10.1160/TH10-01-0066

49. Olsson SB, Rasmussen LH, Tveit A et al (2010) Safety and tolerability of an immediate-release formulation of theoral direct thrombin inhibitor AZD0837 in the prevention of stroke and systemic embolism in patients with atrial fibrillation. Thromb Haemost 103:604–612. https://doi.org/10.1160/TH09-07-0509

50. Baker K, Cano SJ, Playford ED (2011) Outcome measurement in stroke: a scale selection strategy. Stroke 42:1787–1794. https://doi.org/10.1161/STROKEAHA.110.608505

51. Gladstone DJ, Danells CJ, Black SE (2002) The Fugl-Meyer assessment of motor recovery after stroke: a critical review of its measurement properties. Neurorehabil Neural Repair 16:232–240. https://doi.org/10.1177/154596802401105171

52. Pandian S, Arya KN, Kumar D (2016) Minimal clinically important difference of the lower-extremity fugl-meyer assessment in chronic-stroke. Top Stroke Rehabil 23:233–239. https://doi.org/10.1179/1945511915Y.0000000003

53. Chan AW, Tetzlaff JM, Altman DG et al (2013) SPIRIT 2013 statement: defining standard protocol items for clinical trials. Ann Intern Med 158:200–207. https://doi.org/10.7507/1672-2531.20130256

54. Sankoh AJ (1999) Interim analyses: an update of an FDA reviewer's experience and perspective. Drug Inf J 33:165–176. https://doi.org/10.1177/009286159903300120

55. Geller NL, Pocock SJ (1987) Interim analyses in randomized clinical trials: ramifications and guidelines for practitioners. Biometrics 43:213–223. https://doi.org/10.2307/2531962

56. Rothwell PM, Howard SC, Dolan E et al (2010) Effects of β blockers and calcium-channel blockers on within-individual variability in blood pressure and risk of stroke. Lancet Neurol 9:469–480. https://doi.org/10.1016/S1474-4422(10)70066-1

57. Campbell BCV, Mitchell PJ, Kleinig TJ et al (2015) Endovascular therapy for ischemic stroke with perfusion-imaging selection. N Engl J Med 372:1009–1018. https://doi.org/10.1056/NEJMoa1414792

58. Goyal M, Demchuk AM, Menon BK et al (2015) Randomized assessment of rapid endovascular treatment of ischemic stroke. N Engl J Med 372:1019–1030. https://doi.org/10.1056/NEJMoa1414905

59. Fransen PS, Beumer D, Berkhemer OA, et al. MR CLEAN, a multicenter randomized clinical trial of endovasculartreatment for acute ischemic stroke in the Netherlands: study protocol for a randomized controlled trial. Trials 2014;15:343 doi: 10.1186/1745-6215-15-343

60. Berkhemer OA, Fransen PSS, Beumer D et al (2015) A randomized trial of intraarterial treatment for acute ischemic stroke. N Engl J Med 372:11–20. https://doi.org/10.1056/NEJMoa1411587

61. Vickers AJ (2008) How to randomize. J Soc Integr Oncol 4:194–198. https://doi.org/10.1016/j.bbi.2008.05.010

62. Schulz KF (2015) Subverting randomization in controlled trials. JAMA 274(18):1456–1458

63. Wang YY, Zhao X, Liu L et al (2013) Clopidogrel with aspirin in acute minor stroke or transient ischemic attack. N Engl J Med 369:11–19. https://doi.org/10.1056/NEJMoa1215340

64. Wallentin L, Lopes RD, Hanna M et al (2013) Efficacy and safety of apixaban compared with warfarin at different levels of predicted international normalized ratio control for Stroke prevention in atrial fibrillation. Circulation 127:2166–2176. https://doi.org/10.1161/CIRCULATIONAHA.112.142158

65. Duncan PW, Sullivan KJ, Behrman AL et al (2011) Body-weight-supported treadmill rehabilitation after stroke. N Engl J Med 364:2026–2036. https://doi.org/10.1056/NEJMoa1010790

66. Hijazi Z, Oldgren J, Andersson U et al (2012) Cardiac biomarkers are associated with an increased risk of stroke and death in patients with atrial fibrillation: a randomized evaluation of long-term anticoagulation therapy (RE-LY) substudy. Circulation 125:1605–1616. https://doi.org/10.1161/CIRCULATIONAHA.111.038729

67. Patel MR, Mahaffey KW, Garg J et al (2011) Rivaroxaban versus warfarin in nonvalvular atrial fibrillation. N Engl J Med 365:883–891. https://doi.org/10.1056/NEJMoa1009638

68. Everett BM, Glynn RJ, MacFadyen JG, Ridker PM (2010) Rosuvastatin in the prevention of stroke among men and women with elevated levels of C-reactive protein: justification for the use of statins in prevention: an intervention trial evaluating rosuvastatin (JUPITER). Circulation 121:143–150. https://doi.org/10.1161/CIRCULATIONAHA.109.874834

69. Tilley BC, Palesch YY (2011) Conduct of stroke-related clinical trials. In: Stroke. Saunders, Philadelphia, PA. https://doi.org/10.1016/B978-1-4160-5478-8.10060-0

70. Furlan AJ, Reisman M, Massaro J et al (2012) Closure or medical therapy for cryptogenic stroke with patent foramen ovale. N Engl J Med 366:991–999. https://doi.org/10.1056/NEJMoa1009639

71. Fox KAA, Piccini JP, Wojdyla D et al (2011) Prevention of stroke and systemic embolism with rivaroxaban compared with warfarin in patients with non-valvular atrial fibrillation and moderate renal impairment. Eur Heart J 32:2387–2394. https://doi.org/10.1093/eurheartj/ehr342

72. Sandset EC, Bath PMW, Boysen G et al (2011) The angiotensin-receptor blocker candesartan for treatment of acute stroke (SCAST): a randomised, placebo-controlled, double-blind trial. Lancet 377:741–750. https://doi.org/10.1016/S0140-6736(11)60104-9

73. Lip GYH, Frison L, Halperin JL, Lane DA (2010) Identifying patients at high risk for stroke despite anticoagulation: a comparison of contemporary stroke risk stratification schemes in an anticoagulated atrial fibrillation cohort. Stroke 41:2731–2738. https://doi.org/10.1161/STROKEAHA.110.590257

74. Piccini JP, Stevens SR, Chang Y et al (2013) Renal dysfunction as a predictor of stroke and systemic embolism in patients with nonvalvular atrial fibrillation: validation of the R2CHADS2 index in the ROCKET AF. Circulation 127:224–232. https://doi.org/10.1161/CIRCULATIONAHA.112.107128

75. Patel MR, Hellkamp AS, Lokhnygina Y et al (2013) Outcomes of discontinuing rivaroxaban compared with warfarin in patients with nonvalvular atrial fibrillation. J Am Coll Cardiol 61:651–658. https://doi.org/10.1016/j.jacc.2012.09.057

76. James SK, Storey RF, Khurmi NS et al (2012) Ticagrelor versus clopidogrel in patients with acute coronary syndromes and a history of stroke or transient ischemic attack. Circulation 125:2914–2921. https://doi.org/10.1161/CIRCULATIONAHA.111.082727

77. Wallentin L, Yusuf S, Ezekowitz MD et al (2010) Efficacy and safety of dabigatran compared with warfarin at different levels of international normalised ratio control for stroke prevention in atrial fibrillation: an analysis of the RE-LY trial. Lancet 376:975–983. https://doi.org/10.1016/S0140-6736(10)61194-4

78. Diener H-C, Eikelboom J, Connolly SJ et al (2012) Apixaban versus aspirin in patients with atrial fibrillation and previous stroke or transient ischaemic attack: a predefined subgroup analysis from AVERROES, a randomised trial. Lancet Neurol 11:225–231. https://doi.org/10.1016/S1474-4422(12)70017-0

79. Harrison JK, McArthur KS, Quinn TJ (2013) Assessment scales in stroke: clinimetric and clinical considerations. Clin Interv Aging 8:201–211. https://doi.org/10.2147/CIA.S32405

80. Barak S, Duncan PW (2006) Issues in selecting outcome measures to assess functional recovery after stroke. NeuroRx 3:505–524. https://doi.org/10.1016/j.nurx.2006.07.009

Chapter 3

Disorders of Consciousness

Anthony O'Brien, Prakyat Singh, Aurore Thibaut, and Felipe Fregni

Abstract

Disorders of consciousness (DOC) place numerous challenges to healthcare professionals and clinical investigators, as they involve a variety of pathophysiological mechanisms, which are not yet clearly understood. Research studies have been conducted for promoting a better comprehension on these neurological conditions and making progress on diagnostic and therapeutic approaches. This book chapter presents information collected in a review of the 70 most cited clinical trials in DOC. This review was conducted by using the online database "Web of Science" searching for articles published between 1995 and 2015. Essential methodology and design aspects of these reviewed studies, based on the CONSORT guidelines, are deeply discussed in this chapter with a view to improve the conduction of clinical trials on disorders of consciousness and provide recommendations for advancing the field.

Key words Disorders of consciousness, Review, Clinical trials, Study design

> *"It's exhilarating to be alive in a*
> *time of awakening consciousness;*
> *it can also be confusing,*
> *disorienting, and painful."*
>
> –Adrienne Rich

1 Introduction

Disorders of consciousness (DOC) are among the most challenging and poorly understood conditions of modern medical care. After severe brain injury (e.g., traumatic, anoxic, hemorrhagic) a patient can enter into a coma, a condition characterized by the lack of both awareness (i.e., no evidence of consciousness of his/her surrounding) and wakefulness (i.e., absence of eyes opening) [1]. A coma usually does not last longer than a few weeks, after which patients can progress to different clinical entities. In the worst case, patients decline into brain death (i.e., permanent loss of brainstem functions). Other patients may regain consciousness or develop a

Felipe Fregni (ed.), *Clinical Trials in Neurology*, Neuromethods, vol. 138,
https://doi.org/10.1007/978-1-4939-7880-9_3, © Springer Science+Business Media, LLC, part of Springer Nature 2018

vegetative state (VS). A VS is characterized by wakefulness, as evinced by opened eyes, without clinical evidence of awareness [2]. Patients in a VS can persist in this condition for months or even years, while others may develop a minimally conscious state (MCS), which is characterized by the recovery of some small signs of consciousness (e.g., visual pursuit, response to command or appropriate emotional response to external stimuli) [3]. Nevertheless, those patients remain unable to communicate with the environment.

The prevalence of DOC in the USA is ~315,000 cases, where VS represents around 40–168 cases per million people, and MCS is in the order of 45,000–250,000 cases [4–6]. Each year, 4200 new cases of VS are reported in the USA, whereas the incidence of MCS is still unclear [6]. One reason of this uncertainty could be a problem of diagnosis, as we know that about 40% of patients in VS are misdiagnosed and present some signs of consciousness [7]. DOC cost of care per patient ranges from $600,000 to $1,875,000 annually in the USA [8]. Lack of patient autonomy and interaction with their environment add to the complexity of DOC paradigm [9]. Until recently, the medical community has viewed patients in VS and MCS, especially chronic patients, with great pessimism regarding both prognosis and effective treatments. Unfortunately, this pessimism results in the neglect of patients in terms of health care as no improvement is expected. Nevertheless, in the past 10 years, a number of studies have reported that several treatments can enhance sign of consciousness [10–12].

1.1 Challenges in Clinical Trial Research

The public health challenges related to DOC also extend into the field of clinical research and its design. Currently, clinical examination is the gold standard to establish the diagnosis of patients with DOC. The clinical continuum of DOC leads to inconsistency and imprecision in its diagnosis and, as a result, it is challenging to select representative clinical trial participants. As neuroimaging techniques get more sophisticated, there will be more precision in diagnosing DOC; however, as of now, it is still limited. The development of these technologies will improve diagnostic accuracy as well as the understanding of patients' level of consciousness and may increase the interest and the number of studies performed in this challenging population. These technologies may help in reducing the uncertainty and human burden associated with caregiving and participation in clinical trials. In the meantime, it is important that standardized research protocols are developed to increase the number of high quality, uniform and comparable data.

This is a challenge, due to the multiple etiologies leading to DOC, the diverse pathophysiological occurrences, therapeutic approaches and unclear clinical outcomes. Another dilemma brought to light is the need to maintain long periods of follow-up in these populations, to gather a better appreciation on the evolution of DOC from the acute phase onward. This places added responsibility and expectations on researchers as it requires them to dedicate more time, resources, motivation and effort in maintaining long term relationships with caregivers, and fund-givers, for an already marginalized and misunderstood population [4].

In this chapter, a review of the literature was conducted to summarize the state of the art on clinical trial methodologies for DOC in the last 20 years with the intent of acting as a stepping-stone to guiding future clinical trials in DOC.

2 Methodology

The Web of Science database was searched for the most cited publications in "coma," "vegetative state," "unresponsive wakeful-ness syndrome," "minimally conscious state," and "conscious-ness." The search was restricted toward the most cited trials written in English and published between 2010 and 2015. A total of 29 articles were extracted, from which most were small, quasi-experimental trials with a low number of citations.

Consequently, in conjunction with the objectives of this book, the search strategy was modified to improve the quality and num-ber of publications on DOC. MEDLINE was searched using PubMed's therapeutic and narrow clinical query search function. The key terms explored were "coma," "vegetative state," "mini-mally conscious state," "comatose," "gcs < 9," and "cerebral malaria." We also expanded the range of years searched from 1995 to 2015. Altogether, 1316 results were attained from which there were 164 relevant titles. After reading the articles, 70 relevant clinical trial articles were eligible for review (Refer to Fig. 1).

3 Overview

Few trials focused primordially on VS and MCS specifically (13%). As a result, the search was centered on clinical trials that discussed acute DOC (e.g., coma). For example, cardiac arrest and other causes of secondary brain ischemia which may lead to disturbances in consciousness were retrieved. Severe traumatic brain injury (TBI) was the principal etiology discussed in association with DOC (67%), after which cerebral malaria (9%) and cardiac arrest (3%) were the most reviewed topics (Refer to Table 1). In conjunction with the general characteristics and the challenges presented, the lack of DOC oriented trials (VS and MCS) provides an avenue for improvement in many facets, including trial methodology.

3.1 Standardizing Trial Methodology

Transparency is essential to clinical trial reporting, and henceforth evidence based medical practice [13]. The adequate design of a clinical trial, not only ensures its reproducibility, a key component of the scientific method, but also reduces bias which can underestimate or artificially inflate an outcome [14]. Therefore,

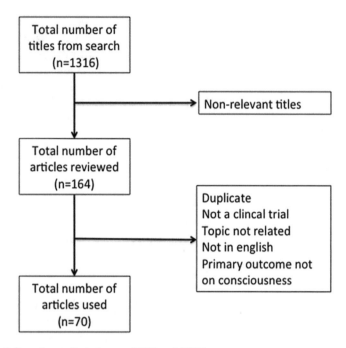

Fig. 1 Search results between 1995 and 2015

Table 1
Principle pathologies reviewed by articles

Pathology, *n*	1995–1999	2000–2004	2005–2009	2010–2015	Total
Acute ischemic stroke	0	0	0	1	1
Acute poisoning	0	0	0	1	1
Aneurysmal subarachnoid hemorrhage	0	0	0	1	1
Bacterial meningitis	0	0	0	1	1
Cardiac arrest	0	1	0	1	2
Cerebral malaria	1	1	2	2	6
Disorders of consciousness	0	0	2	7	9
Postanoxic encephalopathy from status epilepticus	0	0	0	1	1
Spontaneous supratentorial hemorrhage	1	0	0	0	1
Severe traumatic brain injury	5	7	17	18	47

validated guidelines like the CONSORT (Consolidated Standards of Reporting Trials) statement are essential tools to any researcher seeking to standardize the design, reporting and interpretation of their clinical trials [15].

3.2   Journal and Article Impact Metrics

Considering the importance of standardized design methodology, it is alarming that for DOC trials, the number of journals endorsing the CONSORT versus those which are not is almost the same (46% vs. 54%, respectively). Despite the expected growth in number of journal publications between 1995 and 2015 (see Graph 1), there is no discernable pattern that indicates a shift to DOC trial publications in more CONSORT journals over non-CONSORT journals [16].

The impact factor, a metric that evaluates the scholarly impact of a journal, was unequivocally higher in journals that followed CONSORT (Refer to Graph 2). This should be of interest to

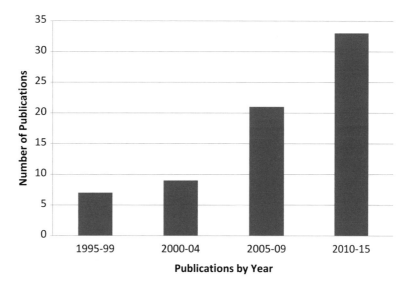

Graph 1 Number of DOC Publications by Year. Note for Graph 1: Please note that this includes various studies, nine that are exclusively on VS/MCS. The rest are principally written on coma

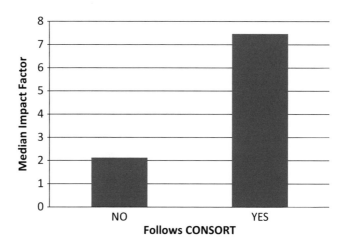

Graph 2 Median impact factor vs. if journal endorses CONSORT guidelines

editorial boards, readers and researchers alike in the DOC field. Comparatively, there were more journals that followed CONSORT with an impact factor ≥15 than non-Consort (20% vs. 1% respectively). Nonetheless, whether or not the journal supported CONSORT, the impact factor for journals publishing DOC trials was mainly 0–4.9 (67%), shadowed by journals with an impact factor ≥20 (19%).

Similarly, the number of citations reveals to some extent the academic impact of an article. The number of citations therefore can partially reflect quality in design and use in the scientific community. Most DOC articles (53%) were cited between 0 and 24 times. There were more articles published in CONSORT journals with the number of citations >100, in comparison to articles published in non-CONSRT journals (17% versus 6%). Overall, articles published in CONSORT journals had a higher number of citations than articles published in non-CONSORT journals (Refer to Graph 3).

3.3 Overview of Outcome of Journals and Articles

Concerning the outcome of the study, most articles reported positive results for their primary outcome (62%). In regard to the sample size, 59% of the trials had <100 participants, while 31% had 100–499 participants and 10% had more than 500. Please refer to Table 2 for more information. Briefly, the sample size relates to the phase of the trial, meaning that the reader can anticipate that the majority of DOC trials are either phase I, phase II, or exploratory in nature. Overall, one would expect that the interventions would most likely be looking at safety or efficacy, whether it is in drug, procedure, or medical devices.

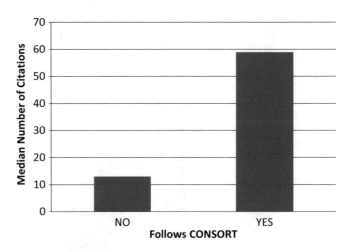

Graph 3 Median Number of Citations vs. if Journal Follows CONSORT Guidelines

Table 2
General characteristics of mined articles

	1995–1999	2000–2004	2005–2009	2010–2015	Total
Trial design, n					
Parallel (two arms)	7	8	18	24	57
Parallel (three arms)	0	0	1	3	4
Crossover	0	1	2	5	8
Adaptive	0	0	0	1	1
Factorial	0	0	0	0	0
Drug/device/procedure, n					
Device	0	0	0	3	3
Drug	5	6	12	22	45
Procedure	2	3	9	8	22
Primary outcome positive, n					
Yes	3	9	11	19	42
No	4	0	9	13	26
Not applicable	0	0	1	1	2
Journal follows consort guideline, n					
Yes	5	3	7	17	32
No	2	6	14	16	38
Sample size, n					
0–49	1	2	8	11	22
50–99	3	4	5	7	19
100–499	1	3	7	11	22
500–999	1	0	1	1	3
≥1000	1	0	0	3	4
Impact factor, n					
0–4.9	3	7	17	20	47
5–9.9	2	0	2	4	8
10–14.9	0	0	0	0	0
15–19.9	0	0	1	1	2
≥20	2	2	1	8	13
Number of citations, n					
0–24	1	1	8	27	37
25–49	0	1	5	2	8

4 Trial Design, Site and Phase

The phase objectives should be reflected in the clinical trial design. Knowing the design not only provides a broad outlook on the trial's aims, but also helps planning for future designs. In the studies listed in this chapter, most of the trials were single-center, phase II and used a two-arm parallel design (23%). After which single-center, exploratory, two-arm trials (14%) followed, and then multicenter, phase III, two-arm trials (10%) (Refer to Table 3).

Table 3
Trial design characteristics and related trial phase

Randomization	Trial design	Site	Phase								Total
			I	II	II/III	III	IV	Exploratory	Pivotal	Quasi-experimental	
Randomized	Parallel (2 arms)	Multi center	1	4	1	7	1	2	5	0	21
		Single center	5	16	0	1	0	10	3	0	35
	Parallel (3 arms)	Multi center	1	1	0	1	0	0	0	0	3
		Single center	0	0	0	0	0	1	0	0	1
	Crossover	Multi center	0	0	0	0	0	2	0	0	2
		Single center	2	3	0	0	0	1	0	0	6
	Adaptive	Multi center	0	0	0	0	0	0	0	0	0
		Single center	0	1	0	0	0	0	0	0	1
	Factorial	Multi center	0	0	0	0	0	0	0	0	0
		Single center	0	0	0	0	0	0	0	0	0
Non-randomized			0	0	0	0	0	0	0	1	1
Total			9	25	1	9	1	16	8	1	70

4.1 Trial Design in DOC

Before deciding which trial design best suits one's study, an adequate hypothesis and primary question should have been established a priori. Following this, researchers must ask themselves if the control variables and extraneous factors surrounding the trial are manageable (e.g., through randomization, group differences can become minimized). If the answer is yes, then the next step is to determine which trial design best suits the trial's features.

The most elementary design that can be used in a clinical trial is the one-way design which considers only one independent variable (also known as the intervention or control variable). Generally, the comparison is done before and after the intervention, and as such the researcher can look at the effect of the intervention between and within the selected groups. In this type of design, the researcher can compare the intervention against a control group, like a placebo/sham or an already established treatment. In the case of DOC, there are no established treatments as of yet to regain consciousness, as such it is most likely that a placebo/sham group would be the main comparator in strictly DOC trials.

An exception would be when different interventions or multiple doses are tested on two or more experimental groups. Another exception to this would be when considering the etiology of the DOC. For example, one can argue that a particular intervention for severe traumatic brain injury (TBI), like hypothermia, hypertonic solution or intracranial pressure monitoring, may lead to better outcomes in TBI patients and as such may also benefit, by extension, to patients with DOC following a TBI. In this sense, if a standard intervention already exists for a specific population or etiology, though it is not particularly designed for all DOC patients, it might be valuable for treating DOC patients as well. Nonetheless, as no orthodox treatment is yet established primarily for DOC it is best to design the trial with a control group in mind.

As mentioned at the beginning of this section and evidenced in Table 3, most of the clinical trials reviewed were one-way designed, two-arm trials. In terms of internal validity, this type of design is robust because it allows for testing the baseline similarities of the groups prior to intervention, and with randomization safeguards that they are going to be comparable. Indiscriminately attrition can be a main limitation in this design. Nevertheless, for trials conducted on DOC patients, attrition is not as prevalent due to the nature of the disease. Statistically speaking, they are also simple to evaluate; generally, a Student's t-test, chi test, or nonparametric equivalent like Wilcoxon rank sum test is adequate.

Expanding the aforementioned design, is the one-way, three-arm trial. Though not as commonly used in the articles reviewed in this chapter, it provides the same benefits as the two-arm trial while allowing for multiple experimental groups. On the other hand, as the number of experimental groups increases, as compare to the two-arm trial, a larger sample size is needed, which could be detrimental to the feasibility of this type of design in DOC. Also, the statistical analysis tends to involve more advanced techniques like analysis of variance, Kruskal–Wallis, and so forth.

In order to address the issue of sample size and variability, an alternative to the two-arm trial can be the crossover design. The advantage of the crossover design is that it uses the subjects as both the control group and the experimental group. This reduces inter-subject variability, and hence increases the internal validity of the trial comparatively. Another aspect to consider is the ethical component of the trial, which is even more important for this population. Though all trials are based on equipoise, in the crossover trial the subjects will always receive both the intervention and control, which can be ethically favorable, which may reduce dropout if patients regain consciousness, and/or which help attain and maintain informed consent. The main disadvantage to this design in DOC is that it should not be used when the treatment effect is slow. Therefore, it may be best for stable, yet early, cases of DOC, rather than long-term cases. In terms of statistical analysis, researches mostly rely on paired t-test, and/or Wilcoxon signed rank test.

4.2 Single-Center Versus Multicenter Trials in DOC

The decision of selecting a single-center trial, or a multicenter trial, can impact more on the success of the trial than just the obvious feasibility consequences. Single center trials, in general, are not only cheaper but tend to be easier to manage concerning logistics, negotiating protocols, data collection and population variability [17]. Therefore, for small trials that are focused on safety or efficacy, single-center trials may be better suited, than multicenter trials. Nonetheless, there are drawbacks that should be considered. The first is that single-center trials more frequently have larger intervention effects. Now, this may seem attractive, but it is important to understand the reason behind this observed phenomenon. A proposed reason is that single-center trials could be more prone to publication bias, though it is still uncertain [18–21]. An alternative explanation is that single-center trials are of lower methodological quality, although multicenter trials tend to undergo more missing data. Finally, the larger effect size may reflect different intervention protocols, a more homogenous population or more specialized personnel [18]. For the trials reviewed, 43 (62%) were single center trials; there was also an indication of more single center trials having a positive outcome (43%) than a negative one (17%). This reflects the nature of single center trials, in the sense that they are conducted on smaller more homogenous populations, which in the case of DOC is common since most of the trials are Phase II. Nevertheless, the use of multiple centers to cover recruitment numbers is a strategy that should not be undervalued.

4.3 Phases of Trial Design in DOC

Phase I trials and experimental trials, are generally smaller trials and conducted in healthy human subjects; <100 patients. However, due to the nature of DOC, the use of healthy subjects is unreasonable under most conditions. These trials are oriented toward safety, pharmacodynamics, pharmacokinetic dosing objectives, and/or performance appraisal. In the trials reviewed, nine (13%) were phase I and 15 (21%) were exploratory. Additionally, these trials may be used to help calculating the effect size for larger phase II and pivotal trials. Nonetheless, this suggests that in the last decade approximately one third of the DOC trials have only started being tested for drug, procedures and medical device safety.

Phase II trials and pivotal trials, are larger than phase I trials, involving ~50–300 subjects. They are designed to test the preliminary efficacy of an intervention, confirming dosing and looking at adverse events. In addition, they can be used as a means to power phase III trials. These trial phases have more homogenous samples, defined through strict inclusion/exclusion criteria, to narrow down the effect and control as many variables as possible. Additionally, the later aspect of inclusion/exclusion criteria is vital to contextualize, as it provides partial insight into the means by which the researcher attempts to control the noise-to-signal output for the phase's purpose, and prospective generalizability. These trials also tend to be randomized, single or double blind, with allocation concealment. Interestingly, while 11 (16%) of the phase II trials had between 100 and 499 participants, 14 (20%) phase II trials were between 0 and 99 participants. For DOC, this brings to light other aspects like sample size calculation, challenges with recruiting representative subjects, adequate and timely consenting, and difficulties maintaining long term follow-up.

Phase III trials are even larger than phase II trials; they can range from hundreds to thousands participants. They look at effectiveness of an intervention and therefore tend to have groups that are more heterogeneous. They are randomized and normally double or triple blind, comparing against a standard treatment or placebo. These trials influence heavily on FDA approval of an intervention. For DOC, 8/9 of the phase III trials were multi-center trials on severe acute TBI, while 1/9 was a single center trial on cerebral malaria. This also reflects the novelty of trial design in DOC, and the underrepresentation it has experienced in clinical trial research in the last 20 years. Note that we identified one phase II/III study.

Phase IV trials, also known as postmarket studies in drug trials or post-approval in drug/procedure trials, are designed to look at the long-term effects, risks, benefits, and best usage parameters once an intervention has been approved. Only one study was defined as being a phase IV. However, it did not match the aforementioned criteria and is more likely to be a phase III study.

5   Population

Most of the clinical trials (86%) were performed on acute (<1 week post insult) patients. Only nine studies focused on DOC subjects exclusively (subacute or chronic VS and MCS), two of which looked at children. The other studies mainly looked at acute (<28 days post insult) severe TBI subjects (Graph 4). The selection of the time point for the progression of DOC is of importance as it can have considerable impact on the observed outcome; literature review that acute stages tend to respond better to intervention than chronic stages [22, 23].

Out of the nine studies focusing on VS and MCS patients, eight were performed after 2010. As mentioned before, this highlights the novelty of the studies in this specific population of patients. The other 62 studies focused on coma patients (acute DOC), of which the majority ($n = 46$) included TBI patients only (71%), while 54 studies were conducted on adults (87%) and eight on children (13%).

5.1   Selection Criteria

As mentioned in the previous paragraph, the majority of the studies were performed on acute patients in "coma" without mentioning the current definition of a DOC (i.e., VS or MCS). It is difficult to delineate specific selection criteria for the DOC trials (VS or

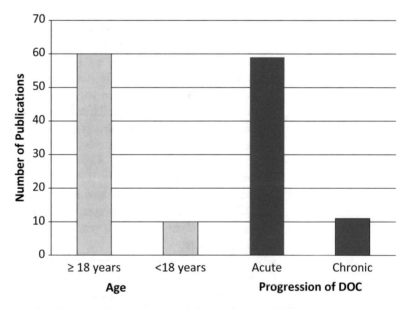

Graph 4 Number of Publications vs. Age or Stage of DOC. On the left (green): number of studies focusing on children (<18 years old) or adult (>18 years old). On the right (blue): number of studies performed on acute (<1 week post insult) or sub-acute (1 week to 3 months) or chronic (>3 months) patients

MCS) since only a small amount of studies focused particularly on this population.

For the majority of the studies, inclusion criteria were based on the GCS total score; patients with a score below 8 were diagnosed as being in coma [24] (see Table 4). Nevertheless, the level of consciousness cannot be entirely assessed using this scale and some patients may be incorrectly diagnosed (e.g., locked-in syndrome, patients that are unable to move or vocalize and have a GCS of 5). This is even more prone to occur in the acute stage since many patients can suffer from motor disabilities and cannot express that they are (partially) conscious. Therefore, using only the GCS, as observed in the majority of the trials (58%), may induce bias in the studies' results since the target population may be misdiagnose.

Most of the phase II and III studies used the GCS as a diagnosis and inclusion criteria (62% and 64%, respectively), which highlight even more the wide acceptance and use of this scale in the medical community. Nevertheless, as mentioned before, this scale is not sensitive enough to correctly diagnose patients with DOC. The scale does not assess the brainstem reflexes. To overcome this problem, in some studies, pupillary response to light was also evaluated.

Other scales were used to select patients, such as the Coma Recovery Scale-Revised (CRS-R) or the disability rating scale (DRS). These two scales are more complete and assess in more details the behavioral responses of patients. Another scale used in one study to select patients with DOC is the Ranchos Los Amigos scale. As highlighted in Table 4 only 16% of the reported studies used those scales. In comparison to the GCS, the CRS-R offers more sensitivity to disentangle patients in MCS from patients in VS [7, 25]. Nevertheless, it takes more time (30–45 min for the CRS-R compared to 5 min for the GCS) to be performed and requires specific training.

Another interesting observation is the age range chosen in the studies focusing on adults. Indeed, even if the majority of the studies included patients from 18 to 65, some studies included patients from 16 years old, or even 15. We observed the same difference for the upper limit. Even if the term adult is more related to the law, it is well known that cortical activity in young as well as old patients can be very different from each other, and this may influence the results of a trial; therefore more precision should be taken when defining the inclusion criteria.

CT-scan or MRI was often used to objectively evaluate a hemorrhage and its severity in TBI patients (15 studies out of 50–30%). Ten studies employed neuroimaging as an inclusion criteria were phase II and III (75%). While relying on objective measurements to enroll patients seems like a more reliable criterion than behavioral scales, in the present studies, different classification were used (i.e., Marshall's CT Classification, Classification system of the

Table 4
Selection criteria for mined DOC articles vs. study phase.

Eligibility criteria		N	Phases Quasi experimental	Exploratory	pivotal	I	II	II/III	III	IV
Behavioral scales	GCS	41	0	10	3	5	16	0	7	0
	CRS-R	5	1	0	0	0	3	1	0	0
	DRS	5	0	1	0	1	2	1	0	0
	Rancho Los Amigos	1	0	1	0	0	0	0	0	0
MRI-CT lesions		15	0	4	0	2	5	0	5	0
Pupil response		8	0	0	0	1	5	0	2	0
Stability of vital sign		7	0	2	0	1	2	0	2	0
Stable intracranial pressure		5	0	0	3	1	1	0	0	0
Seizure		3	0	1	0	0	2	0	0	0
No cardiac/vascular dysfunction		5	0	2	1	0	2	0	0	0
No multisystem injuries		3	0	0	2	0	1	0	0	0
Mechanical ventilation		4	0	0	0	2	2	0	0	0

GCS Glasgow coma scale, *CRS-R* coma recovery scale revised, *DRS* disabling rating scale

Traumatic Coma Data Bank, or own criteria), which reduces their reproducibility and comparability.

Even if most of the studies included acute patients <1 week post insult, only seven explicitly include patients in a "stable medical condition." In addition, only five studies on TBI out of 48, evaluated the ICP as an inclusion criteria. The stability of vital signs should be evaluated more carefully, since the efficacy of a treatment may be affected by the patients' state.

6   Intervention

6.1   Introduction to Interventions Used in DOC Trials

Disorders of consciousness are brought about by a variety of events; similarly, there is also a diversity of interventions to address these problems. Despite the variety of interventions tested, an effective intervention to treat DOC is yet to be established, resultantly most treatments are used on an off-label regimen based off of case reports and/or series [11]. The future of research is set to determining if these various treatments are effective and safe, while mapping out which patient type best responds to the interventions (Refer to Graph 5 for a review of the most used interventions vs. time).

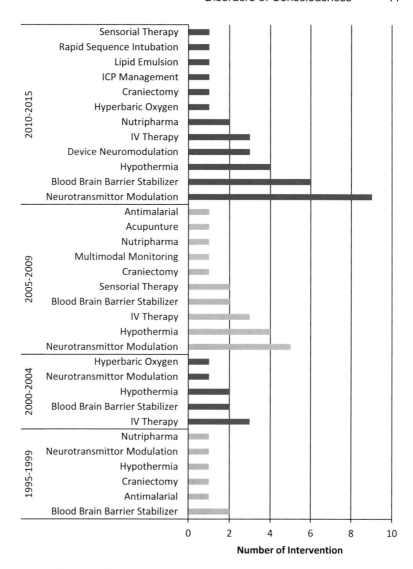

Graph 5 Number of Interventions Organized By Year

6.2 Blood–Brain Barrier Stabilization, Cytotoxic Reduction, and Neuro-degenerative Control

Severe TBI was the main etiology investigated for the DOC review; therefore, most interventions were rationalized around the mechanisms related to STBI. In STBI, at the time of injury, primary structural changes due to immediate cell death take place, then secondary injury ensues via various neurochemical cascades [26]. The maximum progression of edema occurs within 2–4 days of injury, and then decreases throughout the following 23 weeks. In fact, this corresponds approximately to a 6-month period, which ties in to the timeframe seen in several of the time points for evaluating the primary outcome and its follow-up. According to our review, most of these were phase I and phase III pharmacologic interventions. For example, trials involving drugs such as corticosteroids, phosphatidylcholine intermediates, methyl-xan-thine TNF-α synthase inhibitors, eicosanoids, NSAIDS, nonselec-tive bradykinin antagonist, vagal-nerve stimulation (VNS), and progestin are based on the hypothesis that they stabilize the blood–brain barrier or neurodegeneration, by reducing microvascular permeability, consequently reducing the effects of vasogenic edema, reactive oxygen species, cell death, and increased intracra-nial pressure [27–36]. Overall blood–brain barrier stabilizers (BBS) were the second most prevalent intervention in DOC trials.

6.3 Neurotransmitter Modulation

The shear forces that occur from STBI lead to disruption and dam-age of axon membranes, cellular skeleton and mitochondria of neurons. Extracellular glutamate levels rise to neurodegenerative levels, and NMDA receptors are activated to the point in which there is a large influx of intracellular calcium. This results in a dis-ruption of calcium balance and posterior cascade of calcium medi-ated biochemical events [37]. These events are connected with an increase in nuclear factor-kappa B (NF-kB), tumor necrosis factor-alpha (TNF-a), interleukin-1 beta (IL-1), and interleukin-6 (IL-6), which are associated with irregular excitotoxic neurotrans-mitter release [26, 38]. Traxprodil, an allosterically binding NMDA receptor antagonist, and similarly acting medications, are centered on the hypothesis that by preventing unstable NMDA activation, the posterior cytotoxic effects of glutaminergic release can be reduced [37]. GABA receptor and subtypes binding capacities are also altered. These changes all result in a maladaptive homeostatic neurotransmitter shift in which the brain enters into a state of dor-mancy known as DOC [39–41]. Trials with zolpidem are based on the hypothesis that GABA-v1 agonists may help return maladap-tive neurotransmitter release allosterically, via restoration of nor-mal metabolic activity in dormant neurons [39]. While trials with amantadine are based on its mechanism of action as an indirect dopamine agonist, which is associated with attentiveness, and arousal via nigrostriatal, mesolimbic, and frontostriatal circuits [11]. Parallel to this, methylphenidate is said to do the same [42]. Similarly, other than the anti-inflammatory effects mentioned early, VNS is said to activate areas like the locus coeruleus leading to norepinephrine release and increased activation of arousal net-works, while also reducing NMDA activation and GABAergic imbalance [27, 31–34]. Likewise, alternatives to benzodiazepines, like dexmedetomidine, propofol, and haloperidol, are being researched for mechanically ventilated patients since traditional sedatives like benzodiazepines act on GABAA receptors and as such may increase the risk for acute brain dysfunction, like coma [43, 44]. For this review, these were principally pharmacologic interventions, which were the focus of phase II trials DOC trials. By the end of 2015, drugs will account for the main interventions studied in DOC trials.

6.4 Hypothermia and Intracranial Pressure Stabilization

Hypothermia is utilized in severe TBI trials either as a mechanism to reduce raised intracranial pressure (ICP) or as a neuroprotective measure post-injury. When used to treat elevated ICP, hypothermia is generally administered within the first 24 h of injury and maintained during the zenith of ICP or until ICP normalizes. As a neuroprotector, hypothermia is initiated within the first 10 h post-event and upheld regardless of ICP changes for a preestablished time [45]. As reviewed by Clifton G. et al. there are enough quality clinical trials to support the use of hypothermia to reduce elevated ICP but not enough to support its use as a neuroprotectant for diffuse brain injury [45]. Comparably, craniectomy is generally used after the failure of first tier medication in controlling ICP, although evidence points to lower ICU stay, it is accompanied by increased adverse events [46]. Intravascular resuscitation with crystalloids or colloids, with varying osmolarities and compositions were also studied in several TBI trials. The general idea is that restoration to normal blood pressure can maintain cerebral blood pressure, while hypertonic solutions reduce brain edema by maintaining intravascular oncotic pressure, while balanced solutions prevent pH disturbances, therefore reducing maladaptive metabolic strain on the brain and its reserves [47]. Hypothermia was the most frequently reviewed technique in exploratory and pivotal trials, while IV therapy was the second most studied intervention for phase II and III trials.

6.5 Nutripharma, and Hyperbaric Oxygen

Nutripharma with branched chain amino acids is believed to enhance oxidative metabolism, synthesis of neurotransmitters, production of ATP, protein synthesis, insulin production and resultantly lead to improved brain ion homeostasis, neurotransmitter release, and metabolism [48]. Alike, the method of nutripharma administration is important because it can impact the rate of nutritional processing and hence energy spending, and catabolism; which is why intestinal mechanism of administration are being studied since they bypass the slow process of gastric emptying while reducing aspiration pneumonitis [49]. In this review Nutripharma was the fifth most assessed intervention. Grossly, hyperbaric oxygenation has been demonstrated to improve mitochondrial function, adenosine triphosphate (ATP) production, blood flow and to reduce intracranial hypertension, neuroinflammation, reactive oxygen species damage, and neurodegeneration in the hippocampus, around pericontusional and noncontusional brain parenchyma, all of which assist in functional recovery of the individual [50–55].

7 Outcomes

An outcome is an observed phenomenon that reflects a specific response in a subject toward a controlled intervention under precise conditions. It should also be reflected in the selection of an adequate and, if possible, valid outcome in order to tailor an optimal expected and observable response.

The outcome can encompass a plethora of direct patient responses, such as a repertoire of symptoms, the time to a particular event, a review of quality of life validated criterion, and behavioral/cognitive batteries. On the other hand, outcomes can also embody less patient-reported idiosyncrasies like clinical signs grouped through a variety of tools through behavioral scales, imaging techniques, biomarkers and physiological assessments. Consequently, the multidimensionality of outcomes can be narrowed down to clinical outcomes that represent a participant's survival, functionality or feelings, and then surrogate outcomes that substitute clinical outcomes for a more objective, yet indirect measurement of an effect [56] (Graph 6).

There are still no direct and objective measure of consciousness as of now its principal evaluation is accomplished via behavioral cues organized through a variety of scales. Other techniques include electrophysiological responses, or highlighted markers in neuroimaging; however, none of these methods has yet to be validated as a gold standard to observing and measuring consciousness.

In this review, the principal primary outcome for the majority of the studies was the Glasgow Coma Scale (GCS) and the Glasgow Outcome Scale (GOS) (see Table 5). The GCS is a clinical endpoint used to assess the level of consciousness after an injury and it can be applied over a short period of time to track changes. Nonetheless, despite its simplicity, the GCS has been criticized for its limitations when it comes to evaluating intubated/sedated

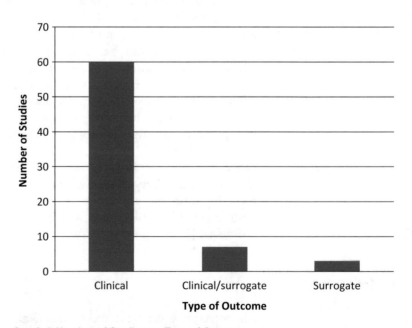

Graph 6 Number of Studies vs. Type of Outcome

Table 5
Main primary outcomes related to consciousness used in articles about Disorders of Consciousness

Description of primary outcomes related to consciousness		*n*
Behavioral scales	Glasgow outcome scale (GOS)	23
	Glasgow coma scale (GCS)	18
	Coma recovery scale-revised (CRSR)	7
	Disability rating scale (DRS)	5
	Glasgow outcome scale extended (GOSE)	4
	Barthel index score (BIS)	2
	Coma/near coma scale (CNCS)	2
	Pupillary response	2
Clinical	Time to regain consciousness	7
	Imaging (PET/CT/MRI)	3
	Time to sit unsupported	2
	Time to start oral feeds	2

individuals, brainstem reflexes, and subtle neurological changes [57]. Although GCS was the second most used scale in our revision, in an evidence-based review by the American Congress of Rehabilitation Medicine (ACMR), along with its modified version the Glasgow–Liege Scale, the GCS is not recommended for behavioral bedside assessment because of a lack of content validity, standardization, and/or reliability [25].

Unlike the GCS, the GOS is divided into a hierarchical, descriptive assessment of key aspects such as returning to work, personality disturbances, cognitive alterations, spontaneous eye closure and opening, etc. Overall, it is a broader scale with more subjective categories. Generally, it is not used on an individual bases, but rather in research during continuous or long-term assessment, and therefore graded to assess degree of return to autonomy in the daily life [58]. Its use is accepted when preestablished structured questions are employed; however, it is important to note that it was not included in the ACMR evidenced based recommendations for DOC appraisal [59, 60]. The Coma Near Coma Scale (CNCS) was developed as a revised version of the disability rating score (DRS). It is designed to notice subtle neurological changes in traumatic brain injury patients; the scale is recommended with major reservations by the ACMR, and backed by the Center for Outcome Measurement in Traumatic Brain Injury given that the evaluator is previously trained and that two independent researchers administer it at approximately the same time [25, 61]. Finally, according the ACMR the Western Neuro Sensory Stimulation Profile (WNSSP) should be used with moderate reservations, while the Coma Recovery Scale-Revised (CRS-R) has superior content validity, normalized administration, and scoring. Moreover, until now, it is the most sensitive scale to detect signs of consciousness in DOC [7, 25].

For selecting the outcome, studies should clearly understand the operational definition of the DOC being evaluated. Not only this, the evolution of the DOC, the subtle cues, and the time points to be evaluated should be clearly defined prior to starting the study. The tool used to measure consciousness should be reflective of this, and at minimum go beyond what is used most frequently, but rather be based on evidence-based recommendations. Studies that use unpublished scales have a higher probability of bias and exaggerating the effect size of an intervention [62, 63]. All trials reviewed in this chapter used a previously validated scales to evaluate consciousness. Nonetheless, most of the trials used the GCS, as a primary outcome, which, in itself, is not recommended by the ACMR. Additionally, considering the long term follow up occasionally required; it can be beneficial to have outcomes that not only can be administered by the bedside, but also via other mediums like the telephone. To this regard, it is understandable why the GOS or the extended version (GOSE) was frequently used in the studies reviewed. Moreover, aspects like the training prior to scale administration, the level of experience of the assessor, the quantification of the outcome, the time of day, and the masking of raters should be clearly disclosed.

Overall, we recommend using the CRS-R for DOC research under short-term scenarios (example: baseline and after intervention), when trained staff is available. If the scale's requirements are too demanding, we recommend the WNSSP or the CNCS as substitutes. For follow-up the GOSE is an adequate alternative, it is also the most frequently used scale for follow-up in the present review. Despite its simplicity, the use of GCS is not recommended as an outcome because of its lack content validity, standardization and reliability. Table 6 shows the time points of main primary outcomes in this topic.

8   Statistical Analysis

8.1 Statistical Analysis of Primary Outcome

Not only is the adequate selection of the outcome essential, but also the nature of their scoring and consequent statistical analysis should be clearly described. The broad idea is that any knowledgeable reader should be able to reproduce the statistical analysis if they had access to the original data [15]. For example, the GCS which goes from 3 to 15, is frequently divided into groups/categories (severe with GCS <8–9, moderate GCS 8 or 9–12, minor GCS ≥ 13); the categorical division of the GCS adds an additional avenue to its interpretation, in the sense that it becomes clinically simpler to appraise, but on the down side data is lost. Additionally, depending on the way the data is scored, the statistical approach

Table 6
Time points of main primary consciousness outcomes

Description of primary outcome		Baseline and/or at intervention	<1 month	1 month	3 months	6 months	>12 months
Behavioral scales	Glasgow outcome scale (GOS)	2	8	1	12	21	4
	Glasgow coma scale (GCS)	3	12	1	3	5	3
	Coma recovery scale-revised (CRSR)	0	1	0	1	1	3
	Disability rating scale (DRS)	1	5	0	0	0	0
	Glasgow outcome scale extended (GOSE)	0	0	1	0	4	1
	Barthel index score (BIS)	0	0	0	0	2	0
	Coma/near coma scale (CNCS)	0	1	0	0	0	0
	Pupillary response	0	2	1	1	2	0
	Blantyre coma score (BCS)	0	0	0	0	0	0

will change accordingly. For example, if the GCS is considered as an ordinal value, then the median, interquartile range and non-parametric approaches can be used for data analysis. However, if GCS is categorized then proportions, Chi2/fishers exact test, and/or logistic modeling may be more appropriate to analyze the data. Please refer to Table 7 for a more detailed description on how to statistically analyze frequently used behavioral scales.

In regard to the studies reviewed, the main method of evaluation for GOS was the Chi2 and Fischer test. For the GOS, the next preferred method of evaluation was logistic regression. When evaluating the relationship that an effect modifier and other covariates can have on the outcome, logistic regression is an invaluable approach, especially when considering the heterogeneity in the DOC population. A few studies utilized Wilcoxon Signed Rank Test (WSRT), and Mann–Whitney U (MWU) depending on if the data was paired or not. Other useful statistical approaches include the use of medians and interquartile range, rather than the mean and standard deviation, for nonparametric central tendency and dispersion analysis. Important to note is that a few studies did analyze these scales as continuous data using Student's t-test, which is not recommended. Congruent with the previous idea, the following neurobehavioral scales should be analyzed as categorical data: GCS, GOSE, CNC, RLCF, and CPC.

GCS was evaluated primarily as a continuous variable through Student's t-test. For the studies that did analyze it as categorical they used Mann–Whitney U, Chi2/Fischer, and generalized linear mixed models.

CRS-R was analyzed primarily through nonparametric approaches such as Wilcoxon Signed Rank Test and Mann–Whitney. These approaches are adequate considering that the scale is ordinal in nature. A few studies used multiple linear regressions; although this approach is possible, it is not the best option due to the lack of equivalent distant intervals inherent in the scale. Therefore, logistic regression may be a better approach, particularly if comparing neurological states like VS/MCS. Similarly, DRS was evaluated principally through logistic regression, Wilcoxon Rank-Sum test and mixed regression models. Finally, GOSE was reviewed via Chi2/Fischer, Mann–Whitney U, and logistic regression.

Not only it is enough to adequately and clearly state the nature of the variable and the statistical method that will be used to analyze the data, but also another crucial aspect to consider is the means by which the data will be managed all together. Randomization itself is essential to reducing bias in regard to the allocation of subjects to an intervention; as a result, in order to uphold the benefits that randomization has on the internal validity of a trial, an intention to treat (ITT) analysis is suggested. When ITT is conducted, the subjects data are analyzed in accordance to the intervention group they were assigned after being randomized,

Table 7

Primary consciousness outcomes reviewed and recommended statistical approach

Scale	Behavioral content	Item response set	Score range	Type of data	Recommended statistical test
Glasgow coma scale (GCS)	Visual, motor, verbal	Varies per item. Score ranging from 1 to 6	Severe, with GCS < 8–9 Moderate, GCS 8 or 9–12 (controversial) Minor, GCS ≥ 13 Lowest score is 3. Maximum is 15	Ordinal/ categorical/ time to event	Use data as categories or ordinal. When using data as categories chi or fisher exact test is recommended. However if transformed to a dichotomous variable then a logistic model may also be applied. When using as ordinal data apply nonparametric equivalents like Mann–Whitney U, Wilcoxon sum rank test and/or Kruskal–Wallis. Less frequently time to attain a certain score can be used, in these cases Kaplan–Meier scores, log-rank scores, and Cox- proportional hazard ratios can be used
Glasgow outcome scale (GOS)	Various categories of cognitive function	Descriptive categories of cognitive function from 1 to 5	1. Good recovery 2. Moderate disability 3. Severe disability 4. Persistent vegetative state 5. Death		
Glasgow outcome scale extended (GOSE)	Various categories of cognitive function	Descriptive categories of cognitive function from 1-8	1. Death 2. Vegetative state 3. Lower severe disability 4. Upper severe disability 5. Lower moderate disability 6. Upper moderate disability 7. Lower good recovery 8. Upper good recovery		
Coma near coma (CNC)	Visual, auditory, command, following, threat response, olfactory, tactile, pain, vocalization	"Occurs 2–3 times," "occurs 1–2 times," or "does not occur"	Average item score: 0.00–0.89 = no coma; 0.9–2.00 = near coma; 2.01–2.89 = moderate coma; 2.90-3.49 = marked coma; 3.50–4.00 = extreme coma		
Coma recovery scale-revised (CRS-R)	Auditory, visual, motor, verbal , communication, arousal	Subscales vary from 0-5	Total score = 0–23		
Disability rating scale (DSR)	Visual, communication, motor, activities of daily living, level of functioning, employability	Varies per item, 0–5	Score 0: None Score 1: Mild Score 2–3.5: Partial Score 4–6: Moderate Score 7–11: Moderately severe Score 12–16: Severe Score 17–21: Extremely severe Score 22–24: Vegetative state Score 25–29: Extreme vegetative state		
Barthel index score (BIS)	Activities of daily living, mobility	Varies per item, 0, 5, 10, 15	0–20 total dependency, 21–60 severe dependency, 61–90 moderate dependency, 91–99 indicates slight dependency.		

regardless if they dropout, withdraw, changed groups or had missing data. Table 8 shows the most applied methods of data analysis in DOC trials. Under the missing data scenario, strong assumptions regarding the method of data imputation must be made to complete the data analysis. Most of the articles reviewed here, did not state the precise method for imputing data, more so when they did they used last-observation carried forward, which despite its ease of use has been questioned as a reliable data imputation method [15]. Regardless of the analysis and the imputation method it is clear that in order for the scientific rigor of a trial to increase it is recommended that the methods be described rather than reduced to key words such as intention-to-treat. An example of this would be, "all subjects who were randomized to either placebo or intervention A were included in the analysis."

According to the consort guidelines, secondary outcomes are used to evaluate additional effects of the intervention. Considering the economic load of clinical trials, secondary outcomes also serve purposes such as (1) hypothesis generation for future trials, (2) reduction of costs by efficiently grouping other hypothesis into one trial instead of various trials, and (3) sustainability.

In the present sample of clinical controlled trials, 24% of the studies did not mention any secondary endpoints. The occurrence of adverse events during the following days or month after the inclusion in a study was measured in 19 (27%) of the reported studies. Most of them were phase II and III trials. This is a very relevant outcome to report and it is suggested to include this measure in all protocols that assess the effect of treatments in DOC. The rate of mortality (time range of data collection varies between 1 week and 6 months) was another widely used secondary outcome, mainly in studies focusing on acute setting. Note that the GOS/E was also often used and this scale encompasses both mortality and functional recovery. A certain amount of trials included cognitive and functional testing as a secondary outcome.

Table 8
Frequency and percentage of data analysis approach used in DOC articles reviewed

Method	Frequency, n	Percent (%)
Intention to treat	32	46
Complete case analysis	18	18
Unclear	20	20
Total	70	100

Other outcomes, such as time in intensive cares, intracranial pressure, brain lesions objectified with neuroimaging and blood tests were also observed in some studies.

For most of the studies, when the primary outcome was the GCS or GOS, secondary outcomes were physiological measures, intracranial pressure, blood sample tests or neuroimaging assessments. On the other hand, studies focusing on physiological measures chose GOS or other cognitive and functional scales as secondary outcome. GOS was assessed in almost all clinical trials as primary or secondary outcomes. This highlights the worldwide recognition of this scale but also its easy administration. It is recommended to assess the adverse effects of a treatment on every study, as well as the mortality and the GOS/E at follow-up as secondary outcome. This would give good feedback on the safety of a treatment and the functional outcome on the long-term.

9 Sample Size Calculation

The sample size is an essential component of designing a clinical trial; not only does it impact the probability of observing a statistically significant difference, but also ethically it determines the number of subjects whom will be exposed to an intervention and can undermine the feasibility of running a trial. It is critical that the following aspects be reviewed when designing a clinical trial: (1) the alpha (type I error), (2) the statistical power (type II error), (3) the primary outcome used to calculate the sample size, (4) the type of study from which the primary outcome was observed, (5) the expected result in the control group and the intervention group, (6) the direction of the effect and (7) the expected drop-out/attrition rate, with the adjusted sample size.

Consequently, only 28 of the DOC trials reviewed described at least one of the aforementioned sample size calculation aspects (Table 9). Contrastingly, 21 clinical trials reported the alpha level, power, and anticipated effect size; the three main elements to compute the sample size. The most commonly used alpha level was 0.05, as shown in Table 10, which is under a two-tailed hypothesis; when not being stringent or assuming directionality on hypothesis testing, these parameters are frequently used. Likewise, the most employed power was 80%, which in medical literature is considered sufficient powered to observe a statistically significant difference and prevent a type II error. Finally, the source and quantity for the expected effect size in both the control and active group should be stated. These were the least reported aspects in DOC sample size calculation, particularly the expected change in the control group (87%); while the remaining outcomes were based on observational data, previous clinical trials, pilot studies and expected minimally clinical difference. Tangentially from the six mentioned criteria, the design impacts the sample size; generally

Table 9
Sample size criteria reported in reviewed clinical trials

	Total
Alpha level	
0.05	
Two-sided	18
One-sided	0
0.025	
Two-sided	1
One-sided	1
0.01 Two-sided	1
Other	1
Not reported	48
Power	
80%	17
85%	2
90%	4
95%	1
Not reported	46
Control group expected effect	
Observational data	4
Previous clinical trial	5
Not reported	61
Active group expected effect	
Minimally clinical significance	13
Pilot studies	1
Previous clinical trial	7
Not reported	49
Attrition	
20%	0
10%	2
Other	3
Not reported	65

Table 10
Method used to generate the random allocation sequence

Sequence generation method	Frequency, n	Percentage (%)
Not clear	44	63
Computer generated	17	24
Random number tables	3	4
Odd and even numbers	3	4
Centralized web based	1	1
Automated telephone based	1	1
No sequence generation	1	1

paired variables lead to a reduced sample size, by reducing variability between the outcome measurement; which may be worthwhile to consider in DOC trials, for example once again the crossover design in chronic DOC cases shows its benefits.

9.1 Interim Analysis and Stoppage Criteria

Interim analyses are conducted to detect if a treatment is dangerous or beneficial, compared to the control group before the trial is completed. If such a difference is observed, the investigators are ethically obliged to stop the study earlier than planned.

For all the studies that reported interim analyses ($n = 11$), the purpose of the analyses was to evaluate if the expected effects of a treatment would be met or not. Nevertheless, only five clearly explained the analyses they performed. Out of the five studies, four used the O'Brien-Fleming method, with different alpha levels, depending on the study sample and time of the analysis. One study used the threshold of a p value inferior to 0.001 for stopping the study if there was evidence of significant benefit or harm in either arm of the study groups [64]. Another stopped earlier than expected after the second interim analysis due to a lack of effect [65].

Only one other trial [66] reported that the study stopped prematurely due to the request of the Safety and Monitoring Committee because of concern about the number of deaths and serious brain-related adverse events in two subjects. Nevertheless, when the authors compared the two groups (intervention and placebo) they did not observe any difference in outcome.

All studies that reported interim analyses or stoppage criteria had a sample size of more than 100 patients and were all conducted on acute patients. Two studies were pivotal, one was exploratory, while the other eight were phase II ($n = 2$), II/III ($n = 1$), III ($n = 4$) or IV ($n = 1$). It is suggested that all studies describe their interim analyses in their protocol. Moreover, as many studies on DOC are performed during the acute stage on patients with unstable medical condition and uncertain outcome, it is well thought-out to specify stoppage criteria before the start of the study.

10   Randomization

10.1   Importance

Randomization was first introduced in the early 1920s and then first implemented as part of a clinical trial, along with allocation concealment and triple-blinding, by 1946 [67–69]. The importance of randomization was thus set in the annals of history, from which the randomized clinical trial became the gold standard for conducting clinical trials. Succinctly, the randomization process, though diverse, essentially ensures that the sample groups, from which one is evaluating the effect of an intervention, are comparable and balanced in features. In other words, as put by Fischer R.A. it ensures that known and unknown confounders between groups are reduced to a minimum [70]. If properly done, randomization reduces the chance for performance and detection bias. Except for one trial, all the trials in this methodological review were randomized trials.

The term random itself must be used carefully; as the CONSORT 2010 states, the participant should have a known prospect of being assigned to each intervention, however the assignment itself is up to chance and therefore not foreseeable. Due to this, when odd and even number days, hospital numbers, or assigned physician names are used without there being a process of randomization, or unpredictability, the trial should not be considered randomized. Resultantly, as long as the allocation can be predicted the method should not be considered as a random allocation sequence [71]. In accordance with when the authors reported the method to generate the random allocation sequence, it was mainly computer generated (24%) or through random number tables (4%) (see Table 11). More than half of the articles reviewed did not report the method used for random allocation sequencing (64%). In addition, two trials reported using fixed days of the week as a method to randomly allocate the subjects, but as previously discussed, this is a misconception as it is not a true random process, and therefore they should not be considered as randomized clinical trials.

10.2   Types of Randomization

The method in which randomization is accomplished varies in accordance to the sample size, and interests of the researcher, overall though there should not be any direct influence on the random chance of the subject being allocated to an intervention.

One of the most commonly used randomization methods is simple randomization. In this procedure, the subject has an equal number of chances to be distributed to a group. A regularly used analogy is that of a coin-toss, in which on one single, independent event, there is a chance to get 50% heads or 50% tails; in this sense, there is a 1:1 allocation ratio. Despite the name and apparent simplicity, there are caveats to the procedure that every researcher

Table 11
Type of randomization utilized in reviewed articles

Type of randomization	Frequency, *n*	Percentage (%)
Simple	5	7
Block	15	21
Block/stratified	6	9
Stratified	10	14
Minimization	1	1
Unclear	32	46
None	1	1

should be aware of. The first thing is that simple randomization should be used in groups that are large, generally when the sample size is >100. The reasoning behind this is that even though the process is unbiased, it can lead to an irregular distribution of subjects in small sample sizes. For example, in a sample of four independent coin tosses it is possible to get four heads or four tails consecutively in a trial.

To address the issue of unbalanced groups, block randomization tends to be implemented. For example, for every group of ten subjects five will be allocated to one intervention, while the other five to the other. This can be done in many ways, fixed blocks of 2 can be set up so that in each block both interventions are evenly distributed, and by the time five sets of two blocks are completed ten subjects would have been proportionally distributed to each intervention. Unfortunately, this procedure comes with the expense of reducing the randomness afforded by simple blinding. To compensate for this random blocks, larger blocks can be used, but this does not eliminate the issue with sequence predictability altogether. In regard to the studies reviewed, 15 used block randomization, while six used block randomization mixed with stratification; block sizes of 4–10 were used most frequently. Interestingly, the majority of the clinical trials that used block randomization had between 100 and 499 subjects (60%).

Another aspect to contemplate is when idiosyncrasies can influence the outcome. In these situations, the subjects can be randomized in accordance with features that the researcher may consider impact the comparability between the groups. For example, in the present review on DOC, commonly used stratum was site of procedure (commonly used in multicenter trials), age, and severity of injury. For future trials, considering the heterogeneous aspects of the DOC population, various aspects such as the mechanism of trauma, the site of the lesion, the initial prognostic factors, the time since onset, the age of the individual, and a proper diagnosis (i.e., VS or MCS assessed using a sensitive scale), should be taken into account. A limitation to this method is that as the number of strata increase, so does the number of subjects required to adequately power the trial and hence reduce type 2 error; this was reflected in the DOC trials as 63% were larger than 100 subjects, and 75% of the stratified trials were multicenter. According to the CONSORT 2010, stratification is not effective without blocking; this is of interest as only 38% of the DOC trials combined stratification with blocking.

Minimization, in itself is not completely random. After the first individual is randomly assigned to a treatment group, minimization ensures that subsequent participants are allocated in such a way that the differences between the groups are weighted and reduced. Therefore, it can be extremely beneficial for small groups where stratification is not applicable. For DOC, this technique can be of particular use considering the difficulties related to recruiting large subacute and chronic DOC samples and the heterogeneity across individuals. Only one of the trials reviewed used this method of randomization, additionally it had a sample size larger than 500 subjects, which is somewhat counterintuitive to the reasoning behind minimization. Table 12 summarizes the type of randomization used in DOC trials.

10.3 Allocation Concealment, Generation and Implementation

The procedure by which the randomization sequence is obscured in the study is known as the allocation concealment. By protecting the assignment sequence, subjects cannot be purposively placed into a specific intervention (selection bias). Mechanisms to accomplish allocation concealment are: (1) Sealed, numbered, opaque

Table 12
Type of blinding, comparing if the trial described the individuals blinded adequately and if they used the correct terminology to describe the blinding

| Type blinding of | Description of individuals in text blinded | | Blinded individuals described in text not | | Total |
	Clear use of blinding terminology	Unclear use of blinding terminology	Clear use of blinding terminology	Unclear use of blinding terminology	
Open label	2	0	2	0	4
Single blind	2	17	0	0	19
Double blind	10	3	17	0	30
Triple blind	0	3	0	0	3
Unclear	0	0	0	14	14
Total	14	23	19	14	70

envelops, (2) Sealed, numbered, containers, (3) Automated systems, and (4) Third party assignment.

According to CONSORT, the range of omission on disclosing the method of allocation concealment ranges from 44% to 89% depending on the topic of interest; for DOC 63% of the trials did not clearly describe the mechanism of concealment [15]. Trials with inadequate concealment of the randomization sequence have hyperbolic effect size reporting, which should be of concern when designing, executing and reporting DOC clinical trials [15]. Additionally, it is also important to report the individual or system that is responsible for the generation of the allocation sequence. In DOC, the most frequent method of allocation sequence generation was through an automated system (10%), followed by another staff member or a third party center. The most secure method is through a third party center, followed by an automatic system, and then by an independent staff member. Depending on resources the researcher would have to decide which best suits their study; a good middle grounds method is through an automatic, interactive voice response system. Important to highlight is that the majority of the trials did not describe the method or individual involved in generating the allocation sequence (53%). Also, in the most ideal situation, the individual responsible for allocation generation should not be the same person responsible for implementation, as this can potentially lead to the random sequence being disclosed. This clearly occurred in four of the DOC trials reviewed, which involved third persons or pharmacist, and is of concern as most trials did not simultaneously report the individual responsible for allocation generation and/or implementation (69%).

11   Blinding

Blinding is a technique commonly implemented in clinical trial research to improve the signal-to-noise ratio between the intervention and the outcome (the cause and effect relationship, i.e., internal validity). For example, when the subject is not aware of the intervention, they are less likely to be influenced to change their behavior; that is, they will less likely use another intervention, or be less motivated and potentially dropout when knowledge of placebo is present. The same can be said about interventionist, as knowledge of the intervention can hinder their unbiased performance for caring or evaluating a subject.

Traditionally, blinding is divided into single blind, double blind and triple blind. Respectively, these terms traditionally mean that the subject is blinded, the subject and the interventionist are blinded, or the subject, interventionist and the evaluator are blinded. However, in a meta-analysis with over 200 RCT reviewing the agreement between the lexicon used and the description of the individuals blinded, there were over 20 different combinations to describe single blind, double blind and triple blind. For example, a single-blind study may have been considered as the subject blinded alone, the interventionist only blinded, or the evaluator only blinded. As in intention to treat analysis, the randomization procedure and other sections discussed in this chapter, effective communication of a concept is only achievable when there is consensus and consistency in the use of the terms used to encompass an idea. Therefore, it is recommended, for DOC trials, that the blinded individual should be clearly stated.

Regardless, academically speaking the term single blind is generally used for phase II trials when it is not possible to blind the intervention (for example in a surgical procedure). For DOC trials, the subject's awareness and vigilance is compromised to various extents; therefore, the family members or the caregivers should be blinded to the intervention, which brings up both ethical and methodological questions. Ethically, neuroimaging case reports have questioned the nature of VS and MCS subject awareness to external stimuli that can hinder if they are truly unaware to the intervention; how much this influences the outcome is uncertain [72]. In terms of methodology, should extra measures be taken around DOC subjects to not disclose the intervention? Are all DOC trials single-blinded? As seen in this review five trials were considered open label, which is questionable without description of how the researchers considered the subject's awareness. Also, if subject and caregiver were blinded would that constitute enough persons to term the trial double-blind?

Respectively, there were 19 single blind studies in the present review. Double blind and triple blind trials are more commonly used for phase II and phase III studies. Unfortunately, once again the term double blind has become more of a placeholder than an actual action in clinical trial research; nonetheless, when the terms are used correctly evidence suggest that the estimated effect sizes are less exaggerated [73]. In the studies included in this review, 55% of the double-blind trials did not clearly describe the blinded individuals. Considering the previously mentioned, this highlights a problem with how blinding is interpreted in DOC trials, how it is described when constructing the trial and how it is reported. This needs to be addressed, as it constitutes part of the construct validity of a trial.

Therefore, the authors of this chapter reiterate that it is of more use in designing a trial to clearly state the subjects that will be blinded than using traditional blinding terminology, including the DOC subject regardless of their state of consciousness.

12   Adherence and Informed Consent

Adherence is defined as the degree to which a participant follows the directives provided by the clinical researcher [74]. In DOC, this is less of a direct issue, and becomes a problem associated with a participants' proxy. Therefore, when designing a clinical trial for DOC, the researcher must be aware of the local regulatory laws governing clinical research in nonautonomous participants, as much as the institutions terms. Various aspects to consider would revolve around if emergency consent acceptable, if next of kin or a legal representative is required a priori (or a posteriori to emergency consent), the number of times a proxy should be contacted before using an emergency consent, and if the participant regains consciousness when and how is the best procedure to ask for consent. With the advent of technology, and its exponential progression, the concept of consent is almost certainly going to be adapted to include technology which bridges the gaps of communication seen in VS and/or MCS and should be of importance for researchers in the coming years [75].

13   Study Duration

In this section, the duration of the trial as well as the follow-up will be discussed. As many trials investigated the outcome or survival of DOC patients, follow-ups are of high importance for these studies.

13.1   Treatment Duration

Treatment duration was considerably dispersed, ranging from 1 day to 6 weeks of treatment (Refer to Graph 7). More than 70% of the trials were performed either using a single intervention (i.e., drug or surgery) or with an intervention that lasted 1 week. Only 14 out of 70 trials assessed the effects of longer interventions. This highlights the difficulty of long-term follow-up of these patients as well as the natural evolution of the disease. Indeed, studies including acute patients may face two problems: patient's death or natural recovery, which can bias the results of a treatment. Studies assessing the effect of treatment for several weeks were mainly performed on chronic patients with DOC. This circumvents the bias of natural recovery, since chronic patients with DOC do not express more than slight evolution of their condition.

Concerning the phases, phase III studies were performed using either a single or a short-term (i.e., 1 day to 1 week) intervention. This may be explained since phase III trials require a high number of patients and longer trials face a high rate of dropout, which makes conducting phase III trials very complicated. The same observation can be made for phase II studies, even if the number of required patients is less demanding. In this review, phase I and

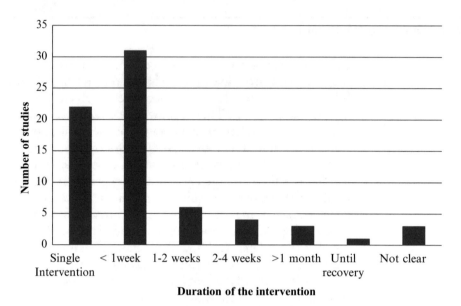

Graph 7 Number of Studies vs. Duration of the Intervention

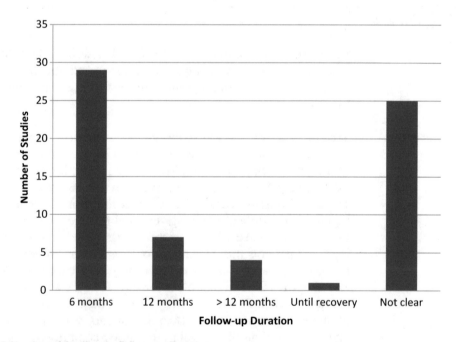

Graph 8 Number of Studies vs. Follow-up Duration

exploratory studies were of short-term duration which reflects the small sample size inherent to these trials.

13.2 Follow-Up Duration

Since most of the studies aimed to assess the recovery of patients in DOC, long-term follow-up is a crucial outcome. As shown in Graph 8, almost half of the trials reevaluated participants at 6 months.

Note that three crossover trials assessed the outcome of patients at 6 or 12 months. In these cases, the aim was not to compare the difference in outcome of the two treatments, but rather to evaluate the safety of the active treatment [27, 76].

Together with the overall short intervention duration previously discussed, 40% of trials did not perform any follow-up. This is of concern in chronic settings, since treatments can be applied for several months before there may be a discernible response.

14 Limitations

Interestingly, articles reporting limitations had the same amount of positive outcomes as compared to those that did not report any limitations. On the other hand, trials describing limitations were published in journals with higher impact factor (mean of 3.5, compared to 2). This observation highlights the importance of reporting limitations of a trial to publish it in a high impact factor journal. Indeed, they can help explain the nature of the result but also give insights and directions for future clinical trials. Note that 34 articles (48.5%) did not report any limitation in their discussion.

The most frequent limitation reported was small sample size ($n = 16$), nonhomogeneity in sample distribution ($n = 10$), lack of follow-up to assess the long term effect of the treatment ($n = 6$), absence or issue of blinding ($n = 6$) or study bias (i.e., no placebo group), acute patients included with potential natural improvement not related to the treatment, heterogeneity of the population or absence of direct assessment ($n = 5$). Other studies referred also to: lack of neuroimaging assessment, dropout, nonsensitive scales sued or statistic issues.

There was a high variability in the way authors explained the limitations. For half of them, limitations were only reported, while for the other half, limitations were explained and solutions proposed for future studies. As it is almost impossible to avoid all limitations, it is recommended that to clearly report all limitations and explore hypotheses and suggestions to circumvent them in future clinical trials.

15 Conclusion

The overarching goal of this review was to contextualize the methodological approach of the 100 most cited articles in DOC. However, the limited number of studies published exclusively on DOC hindered this approach. Consequently, the search was expanded to the last 20 years of research, and articles related to DOC or precisely on DOC were appraised. Only 70 relevant clinical trial articles were retrieved, of which only nine were about VS and/or MCS participants; underscoring the gap of clinical trial research in DOC, as confirmed by Fins et al. [4]. Despite the increase in publications over time in DOC, comparatively to other pathologies the growth is still scarce (Refer to Graph 3).

Noteworthy characteristics uncovered in this chapter include the following:

- Articles that were published in CONSORT endorsing journals were published and cited more than studies which did not; despite there being more journals which did not endorse CONSORT.

- Most trials were parallel, phase 2, single-center, drug trials. This reflects the contemporary novelty in DOC research, but also demonstrates the demand for larger phase III trials, which can reflect the heterogeneity of DOC better.

- Nonetheless, for negative phase II trials, it is worth reviewing aspects such as the idiosyncrasies of the sample population studied, the time of onset of the disease, the parameters of the intervention (e.g., frequency of doses, site of administration, etc.); to consider new hypothesis around the trials negative outcome in comparison to positive trials.

- Crossover design may be of particular use for chronic VS/MCS studies, while it seems less feasible if the population of interest is an acute setting.

- To evaluate consciousness on screening or at bedside, a large number of trials used GCS; this is not recommended. The CRS-R should be used, as it is the most sensitive scale to assess patients with DOC nowadays. If it cannot be used then the WNSSP, followed by the CNCS is recommended. Personnel should be trained accordingly, and also the scales should be applied at the same time, at different hours, by two independent researchers daily. Another alternative, is the FOUR (Full Outline of Unresponsiveness [77]). This scale, is easy and quick to perform, such as the GCS, but has the advantages to test the brainstem reflexes and, most importantly, it assesses the visual pursuit, which can detect a locked-in syndrome and is also one of the first sign of consciousness that patients with DOC recover [3, 78].

- Only 20% of the trials evaluated the effects of long-term interventions. For interventions that may be used in rehabilitation or chronic stages, protocols with a longer duration are needed.

- Follow-up is an essential part of long-term evaluation. This can be achieved using the same outcome for the direct effect or, using scales such as the GOS/E.

- To evaluate the long-term outcome of a treatment, 1, 3, and 6 months follow-up are recommended as minimum follow-up points.

- Most of the behavioral scales are ordinal in nature, and as such should be evaluated with nonparametric approaches. For effect modifiers, and additional covariates, logistic regression is an adequate alternative, though some studies did use generalized mixed models.

- Almost 70% of the studies did not report any sample size calculation. This is really problematic, as the number of patients included can highly influence the outcome of a study.

- The method of randomization was unclear in 63% of the trials. Regardless of the paper, all articles should explicitly explain the method of randomization used. Randomization by fixed days, months, or doctor, is NOT a method of randomization, and should not be used.

- Stratification can be a very useful technique to consider in larger phase III trials, due to the heterogeneity inherent in the DOC population. For example for the age group it can prove to be very useful, due to variation in cortical activity according to age.

- Blocking can be more useful for smaller phase II trials, since these trials are more homogenous.

- Almost 70% of the studies did not report who was responsible for allocation generation and/or implementation. This is a terrible mistake as it reduces the internal validity of the study, since it opens the possibility of the random sequence allocation being disclosed. For allocation generation and implementation, the use of an automatic, interactive voice response system is a good and feasible middle ground for future trials.

- Blinding terminology should not be a priority, instead ALL the personnel who were blinded must be described, for example: "[…] *the patient, family members, interventionist, evaluator and statistician were all blinded to the intervention. For this reason the trial was triple blind* […]"

- It is important to note that, even if patients with DOC are though unconscious, due to the high rate of misdiagnosis [7], they need to be blinded too.

- The method of concealment of the intervention should be clearly described also.

- An additional general recommendation is to include the precise description and delineation of the primary outcome, co-primary outcomes and secondary outcomes. Also the exact statistical tests should be described for these outcomes; it is not enough to say for example "[…] *for continuous variables t-test was used […]*", rather "*[…] the primary outcome CRS-R, an ordinal variable, was dichotomized and analyzed using Fisher's exact test […]*"

The authors of this chapter hope that the aforementioned take home points and the other aspects reviewed can contribute to the improvement of future clinical trial design in DOC, or at least provides a brief methodological panorama on DOC clinical trials, with the end purpose of highlighting that there is still much work to be done.

References

1. Jennett B, Plum F (1972) Persistent vegetative state after brain damage. A syndrome in search of a name. Lancet 1(7753):734–737

2. The Multi-Society Task Force on PVS (1994) Medical aspects of the persistent vegetative state. N Engl J Med 330(21):1499–1508. https://doi.org/10.1056/NEJM19940526 3302107

3. Giacino JT, Ashwal S, Childs N et al (2002) The minimally conscious state: definition and diagnostic criteria. Neurology 58(3):349–353

4. Fins JJ, Illes J, Bernat JL, Hirsch J, Laureys S, Murphy E (2008) Neuroimaging and disorders of consciousness: envisioning an ethical research agenda. Am J Bioeth 8(9):3–12. https://doi.org/10.1080/1526516080 2318113

5. Fins JJ, Master MG, Gerber LM, Giacino JT (2007) The minimally conscious state: a diagnosis in search of an epidemiology. Arch Neurol 64(10):1400–1405. https://doi.org/10.1001/archneur.64.10.1400

6. Beaumont JG, Kenealy PM (2005) Incidence and prevalence of the vegetative and minimally conscious states. Neuropsychol Rehabil 15(3-4):184–189. https://doi.org/10.1080/096 02010443000489

7. Schnakers C, Vanhaudenhuyse A, Giacino J et al (2009) Diagnostic accuracy of the vegetative and minimally conscious state: clinical consensus versus standardized neurobehavioral assessment. BMC Neurol 9:35. https://doi.org/10.1186/1471-2377-9-35

8. Whyte J (1998) The National Institutes of Health (NIH) Consensus Development Program: rehabilitation of persons with traumatic brain injury. NIH, Continuing Medical Education, Bethesda, MD

9. Laureys S, Boly M (2007) What is it like to be vegetative or minimally conscious? Curr Opin Neurol 20(6):609–613. https://doi.org/10.1097/WCO.0b013e3282f1d6dd

10. Schiff ND, Giacino JT, Kalmar K et al (2007) Behavioural improvements with thalamic stimulation after severe traumatic brain injury. Nature 448(7153):600–603. https://doi.org/10.1038/nature06041

11. Giacino JT, Whyte J, Bagiella E et al (2012) Placebo-controlled trial of amantadine for severe traumatic brain injury. N Engl J Med 366(9):819–826. https://doi.org/10.1056/NEJMoa1102609

12. Thonnard M, Gosseries O, Demertzi A et al (2013) Effect of zolpidem in chronic disorders of consciousness: a prospective open-label study. Funct Neurol 28(4):259–264 http://www.pubmedcentral.nih.gov/articlerender.fcgi?artid=3951253&tool=pmcentrez&rendertype=abstract. Accessed August 8, 2015

13. Rennie D (2001) CONSORT revisedDOUBLE-HYPHENimproving the reporting of randomized trials. JAMA 285(15):2006–2007 http://www.ncbi.nlm.nih.gov/pubmed/11308440. Accessed November 7, 2015

14. Wood L, Egger M, Gluud LL et al (2008) Empirical evidence of bias in treatment effect

estimates in controlled trials with different interventions and outcomes: meta-epidemiological study. BMJ 336(7644):601–605. https://doi.org/10.1136/bmj.39465.451748.AD

15. Moher D, Hopewell S, Schulz KF et al (2012) CONSORT 2010 explanation and elaboration: updated guidelines for reporting parallel group randomised trials. Int J Surg 10(8):28–55. https://doi.org/10.1016/j.ijsu.2011.10.001

16. Larsen PO, von Ins M (2010) The rate of growth in scientific publication and the decline in coverage provided by science citation index. Scientometrics 84(3):575–603. https://doi.org/10.1007/s11192-010-0202-z

17. Bellomo R, Warrillow SJ, Reade MC (2009) Why we should be wary of single-center trials. Crit Care Med 37(12):3114–3119. https://doi.org/10.1097/CCM.0b013e3181bc7bd5

18. Dechartres A, Boutron I, Trinquart L, Charles P, Ravaud P (2011) Single-center trials show larger treatment effects than multicenter trials: evidence from a meta-epidemiologic study. Ann Intern Med 155:39–52. https://doi.org/10.7326/0003-4819-155-1-201107050-00006

19. Dickersin KAY, Min Y (1993) Publication bias: the problem that won't go away. Ann N Y Acad Sci 703:135–148. https://doi.org/10.1111/j.1749-6632.1993.tb26343.x

20. Stern JM, Simes RJ (1997) Publication bias: evidence of delayed publication in a cohort study of clinical research projects. BMJ 315(7109):640–645. https://doi.org/10.1136/bmj.315.7109.640

21. von Elm E, Röllin A, Blümle A, Huwiler K, Witschi M, Egger M (2008) Publication and non-publication of clinical trials: longitudinal study of applications submitted to a research ethics committee. Swiss Med Wkly 138(13-14):197–203 2008/13/smw-12027

22. Pierce JP, Lyle DM, Quine S, Evans NJ, Morris J, Fearnside MR (1990) The effectiveness of coma arousal intervention. Brain Inj 4(2):191–197

23. Cope DN, Hall K (1982) Head injury rehabilitation: benefit of early intervention. Arch Phys Med Rehabil 63(9):433–437 http://www.ncbi.nlm.nih.gov/pubmed/7115044

24. Teasdale G, Jennett B (1974) Assessment of coma and impaired consciousness. A practical scale. Lancet 2(7872):81–84

25. Seel RT, Sherer M, Whyte J et al (2010) Assessment scales for disorders of consciousness: evidence-based recommendations for clinical practice and research. Arch Phys Med Rehabil 91(12):1795–1813. https://doi.org/10.1016/j.apmr.2010.07.218

26. Hatton J, Rosbolt B, Empey P, Kryscio R, Young B (2008) Dosing and safety of cyclosporine in patients with severe brain injury. J Neurosurg 109(4):699–707. https://doi.org/10.3171/JNS/2008/109/10/0699

27. Shi C, Flanagan SR, Samadani U (2013) Vagus nerve stimulation to augment recovery from severe traumatic brain injury impeding consciousness: a prospective pilot clinical trial. Neurol Res 35(3):263–276. https://doi.org/10.1179/1743132813Y.0000000167

28. Shakeri M, Boustani MR, Pak N et al (2013) Effect of progesterone administration on prognosis of patients with diffuse axonal injury due to severe head trauma. Clin Neurol Neurosurg 115(10):2019–2022. https://doi.org/10.1016/j.clineuro.2013.06.013

29. Shaikh AK, Mohammad QD, Ullah MA, Ahsan MM, Rahman A, Shakoor MA (2011) Effect of dexamethasone on brain oedema following acute ischemic stroke. Mymensingh Med J 20(3):450–458

30. Xiao G, Wei J, Yan W, Wang W, Lu Z (2008) Improved outcomes from the administration of progesterone for patients with acute severe traumatic brain injury: a randomized controlled trial. Crit Care 12(2):R61. https://doi.org/10.1186/cc6887

31. Roosevelt RW, Smith DC, Clough RW, Jensen RA, Browning RA (2006) Increased extracellular concentrations of norepinephrine in cortex and hippocampus following vagus nerve stimulation in the rat. Brain Res 1119(1):124–132. https://doi.org/10.1016/j.brainres.2006.08.048.

32. Marrosu F, Serra A, Maleci A, Puligheddu M, Biggio G, Piga M (2003) Correlation between GABAA receptor density and vagus nerve stimulation in individuals with drug-resistant partial epilepsy. Epilepsy Res 55(1-2):59–70. https://doi.org/10.1016/S0920-1211(03)00107-4

33. Lopez NE, Krzyzaniak MJ, Costantini TW et al (2012) Vagal nerve stimulation decreases blood-brain barrier disruption after traumatic brain injury. J Trauma Acute Care Surg 72(6):1562–1566. https://doi.org/10.1097/TA.0b013e3182569875

34. Borovikova LV, Ivanova S, Zhang M et al (2000) Vagus nerve stimulation attenuates the systemic inflammatory response to endotoxin. Nature 405(6785):458–462. https://doi.org/10.1038/35013070

35. Zafonte RD, Bagiella E, Ansel BM et al (2012) Effect of citicoline on functional and cognitive status among patients with traumatic brain injury: citicoline Brain Injury Treatment Trial (COBRIT). JAMA 308(19):1993–2000. https://doi.org/10.1001/jama.2012.13256

36. Das BK, Mishra S, Padhi PK et al (2003) Pentoxifylline adjunct improves prognosis of human cerebral malaria in adults. Trop Med Int Health 8(8):680–684. https://doi.org/10.1046/j.1365-3156.2003.01087.x

37. Yurkewicz L, Weaver J, Bullock MR, Marshall LF (2005) The effect of the selective NMDA receptor antagonist traxoprodil in the treatment of traumatic brain injury. J Neurotrauma 22(12):1428–1443. https://doi.org/10.1089/neu.2005.22.1428

38. Moein P, Abbasi Fard S, Asnaashari A et al (2013) The effect of Boswellia Serrata on neurorecovery following diffuse axonal injury. Brain Inj 27(12):1454–1460. https://doi.org/10.3109/02699052.2013.825009

39. Du B, Shan A, Zhang Y, Zhong X, Chen D, Cai K (2014) Zolpidem arouses patients in vegetative state after brain injury: quantitative evaluation and indications. Am J Med Sci 347(3):178–182. https://doi.org/10.1097/MAJ.0b013e318287c79c

40. Gerber BL, Ordoubadi FF, Wijns W et al (2001) Positron emission tomography using(18)F-fluoro-deoxyglucose and euglycaemic hyperinsulinaemic glucose clamp: optimal criteria for the prediction of recovery of postischaemic left ventricular dysfunction. Results from the European Community Concerted Action Multicenter study on use of(18) F-fluoro-deoxyglucose Positron Emission Tomography for the Detection of Myocardial Viability. Eur Heart J 22(18):1691–1701. https://doi.org/10.1053/euhj.2000.2585

41. Witte OW, Stoll G (1997) Delayed and remote effects of focal cortical infarctions: secondary damage and reactive plasticity. Adv Neurol 73:207–227 http://www.ncbi.nlm.nih.gov/entrez/query.fcgi?cmd=Retrieve&db=PubMed&dopt=Citation&list_uids=8959216

42. Moein H, Khalili HA, Keramatian K (2006) Effect of methylphenidate on ICU and hospital length of stay in patients with severe and moderate traumatic brain injury. Clin Neurol Neurosurg 108(6):539–542. https://doi.org/10.1016/j.clineuro.2005.09.003

43. Pandharipande PP, Pun BT, Herr DL et al (2007) Effect of sedation with dexmedetomidine vs lorazepam on acute brain dysfunction in mechanically ventilated patients: the MENDS randomized controlled trial. JAMA 298(22):2644–2653. https://doi.org/10.1001/jama.298.22.2644

44. Page VJ, Ely EW, Gates S et al (2013) Effect of intravenous haloperidol on the duration of delirium and coma in critically ill patients (Hope-ICU): a randomised, double-blind, placebo-controlled trial. Lancet Respir Med 1(7):515–523. https://doi.org/10.1016/S2213-2600(13)70166-8

45. Clifton GL, Valadka A, Zygun D et al (2011) Very early hypothermia induction in patients with severe brain injury (the National Acute Brain Injury Study: Hypothermia II): a randomised trial. Lancet Neurol 10(2):131–139. https://doi.org/10.1016/S1474-4422(10)70300-8

46. Cooper DJ, Rosenfeld JV, Murray L et al (2011) Decompressive craniectomy in diffuse traumatic brain injury. N Engl J Med 364(16):1493–1502. https://doi.org/10.1056/NEJMoa1102077

47. Myburgh J, Cooper DJ, Finfer S et al (2007) Saline or albumin for fluid resuscitation in patients with traumatic brain injury. N Engl J Med 357(9):874–884. https://doi.org/10.1056/NEJMoa067514

48. Aquilani R, Boselli M, Boschi F et al (2008) Branched-chain amino acids may improve recovery from a vegetative or minimally conscious state in patients with traumatic brain injury: a pilot study. Arch Phys Med Rehabil 89(9):1642–1647. https://doi.org/10.1016/j.apmr.2008.02.023

49. Taylor SJ, Fettes SB, Jewkes C, Nelson RJ (1999) Prospective, randomized, controlled trial to determine the effect of early enhanced enteral nutrition on clinical outcome in mechanically ventilated patients suffering head injury. Crit Care Med 27(11):2525–2531. https://doi.org/10.1097/00003246-199911000-00033

50. Rockswold SB, Rockswold GL, Zaun DA, Liu J (2013) A prospective, randomized Phase II clinical trial to evaluate the effect of combined hyperbaric and normobaric hyperoxia on cerebral metabolism, intracranial pressure, oxygen toxicity, and clinical outcome in severe traumatic brain injury. J Neurosurg 118(6):1317–1328. https://doi.org/10.3171/2013.2.JNS121468

51. Daugherty WP, Levasseur JE, Sun D, Rockswold GL, Bullock MR (2004) Effects of hyperbaric oxygen therapy on cerebral oxygenation and mitochondrial function following moderate lateral fluid-percussion injury in rats. J Neurosurg 101(3):499–504. https://doi.org/10.3171/jns.2004.101.3.0499

52. Matchett GA, Martin RD, Zhang JH (2009) Hyperbaric oxygen therapy and cerebral ischemia: neuroprotective mechanisms. Neurol Res 31(2):114–121. https://doi.org/10.1179/174313209X389857.

53. Rockswold SB, Rockswold GL, Vargo JM et al (2001) Effects of hyperbaric oxygenation therapy on cerebral metabolism and intracranial pressure in severely brain injured patients. J

Neurosurg   94:403–411.   https://doi.
org/10.3171/jns.2001.94.3.0403

54. Rockswold SB, Rockswold GL, Zaun DA et al
(2010) A prospective, randomized clinical trial
to compare the effect of hyperbaric to normobaric hyperoxia on cerebral metabolism, intracranial pressure, and oxygen toxicity in severe
traumatic   brain   injury.   J   Neurosurg
112(5):1080–1094. https://doi.org/10.3171
/2009.7.JNS09363

55. Wada K, Miyazawa T, Nomura N, Tsuzuki N,
Nawashiro H, Shima K (2001) Preferential conditions for and possible mechanisms of induction
of ischemic tolerance by repeated hyperbaric oxygenation in gerbil hippocampus. Neurosurgery
49(1):160–166 discussion 166–167. http://
www.ncbi.nlm.nih.gov/pubmed/11440438

56. Atkinson AJJ, Colburn WA, DeGruttola VG
et al (2001) Biomarkers and surrogate endpoints: preferred definitions and conceptual
framework. Clin Pharmacol Ther 69(3):89–95.
https://doi.org/10.1067/mcp.2001.
113989.

57. Jalali R, Rezaei M (2014) A comparison of the
Glasgow coma scale score with full outline of
unresponsiveness scale to predict patients' traumatic brain injury outcomes in intensive care
units. Crit Care Res Prac 2014:289803.
https://doi.org/10.1155/2014/289803

58. Hall KM, Bushnik T, Lakisic-Kazazic B, Wright
J, Cantagallo A (2001) Assessing traumatic
brain injury outcome measures for long-term
follow-up of community-based individuals.
Arch Phys Med Rehabil 82(3):367–374.
https://doi.org/10.1053/apmr.2001.21525

59. Lu J, Marmarou A, Lapane K, Turf E, Wilson
L (2010) A method for reducing misclassification in the extended Glasgow outcome score. J
Neurotrauma 27(5):843–852. https://doi.
org/10.1089/neu.2010.1293

60. Pettigrew LEL, Wilson JTL, Teasdale GM
(2003) Reliability of ratings on the Glasgow
outcome scales from in-person and telephone
structured interviews. Head Trauma Rehabil
18(3):252–258. https://doi.org/10.1097/
00001199-200305000-00003

61. Rappaport M (2000) The Coma/Near Coma
Scale. The Center for Outcome Measurement
in   Brain   Injury   http://www.tbims.org/
combi/cnc (accessed April 24, 2018

62. Mcdowell I, Newell C (2006) Measuring
health: a guide to rating scales, 3rd edn. Oxford
University Press, New York, NY

63. Streiner DL, Norman GR (2008) Health measurement scales: a practical guide to their
development and use, vol 14. Oxford
University   Press,   Oxford.   https://doi.
org/10.1378/chest.96.5.1161

64. Bernard SA, Nguyen V, Cameron P et al
(2010) Prehospital rapid sequence intubation
improves functional outcome for patients with
severe traumatic brain injury: a randomized
controlled   trial.   Ann   Surg   252(6):959–965.
https://doi.org/10.1097/SLA.0b013e
3181efc15f

65. Wright DW, Yeatts SD, Silbergleit R et al
(2014) Very early administration of progesterone for acute traumatic brain injury. N Engl J
Med   371(26):2457–2466.   https://doi.
org/10.1056/NEJMoa1404304

66. Morris GF, Bullock R, Marshall SB, Marmarou
A, Maas A, Marshall LF (1999) Failure of the
competitive N-methyl-D-aspartate antagonist
Selfotel (CGS 19755) in the treatment of
severe head injury: results of two phase III clinical   trials.   The   Selfotel   Investigators.   J
Neurosurg   91(5):737–743.   https://doi.
org/10.3171/jns.1999.91.5.0737

67. Bhatt A (2010) Evolution of clinical research: a
history before and beyond james lind. Perspect
Clin   Res   1(1):6–10.   https://doi.
org/10.4103/2229-3485.103599.

68. Streptomycin in Tuberculosis Trials Committee
(1948) Streptomycin treatment of pulmonary
tuberculosis: a medical research council investigation.   BMJ   2:769–782.   https://doi.
org/10.1136/bmj.2.4582.769

69. D'Arcy Hart P (1999) A change in scientific
approach: from alternation to randomised allocation in clinical trials in the 1940s. BMJ
319(7209):572–573.   https://doi.
org/10.1136/bmj.319.7209.572

70. Armitage P (2003) Fisher, Bradford Hill, and
randomization. Int J Epidemiol 32(6):925–
928. https://doi.org/10.1093/ije/dyg286

71. Hopewell S, Dutton S, Yu L-M, Chan A-W,
Altman DG (2010) The quality of reports of
randomised trials in 2000 and 2006: comparative study of articles indexed in PubMed.
BMJ   340(1):c723–c723.   https://doi.
org/10.1136/bmj.c723.

72. Monti MM, Vanhaudenhuyse A, Coleman MR
et al (2010) Willful modulation of brain activity in disorders of consciousness. N Engl J Med
362(7):579–589. https://doi.org/10.1056/
NEJMoa0905370

73. Devereaux PJ, Manns BJ, Ghali WA et al
(2001) Physician interpretations and text-
book definitions of blinding terminology in
randomized   controlled   trials.   JAMA
285(15):2000–2003.   https://doi.
org/10.1001/jama.285.15.2000

74. Robiner WN (2005) Enhancing adherence in
clinical research. Contemp Clin Trials 26:
59–77. https://doi.org/10.1016/j.cct.2004.
11.015

75. Bressman J, Reidler JS (2010) Willful modulation of brain activity in disorders of consciousness: legal and ethical ramifications. J Law Med AMP. Ethics 38(3):713–716

76. Meythaler JM, Brunner RC, Johnson A, Novack TA (2002) Amantadine to improve neurorecovery in traumatic brain injury-associated diffuse axonal injury: a pilot double-blind randomized trial. J Head Trauma Rehabil 17(4):300–313

77. Wijdicks EF, Bamlet WR, Maramattom BV, Manno EM, McClelland RL (2005) Validation of a new coma scale: the FOUR score. Ann Neurol 58(4):585–593. https://doi.org/10.1002/ana.20611

78. Bruno MA, Ledoux D, Lambermont B et al (2011) Comparison of the full outline of unresponsiveness and Glasgow Liege scale/Glasgow coma scale in an intensive care unit population. Neurocrit Care 15(3):447–453. https://doi.org/10.1007/s12028-011-9547-2

Chapter 4

Traumatic Brain Injury

Ana Luiza C. Zaninotto, Beatriz Teixeira Costa, Isadora Santos Ferreira, Melanie French, Wellingson Silva Paiva, and Felipe Fregni

Abstract

Traumatic brain injury (TBI) is one of the leading causes of disabilities and young adults' death. Considering all the challenges faced in research involving this field, this book chapter analyses the 100 most cited TBI trials in order to provide methodological recommendations for designing future trials in TBI. A literature review of 100 most cited clinical trials in TBI between 2010 and 2015 was conducted using the web engine: Web of Science. Based on the CONSORT guidelines, a comprehensive discussion of the main findings is provided as to guide and help investigators in this field to design TBI clinical trials.

Key words Traumatic brain injury, Research, Review, Trauma, TBI

1 Introduction

Traumatic injuries are responsible for more than five million deaths worldwide annually and traumatic brain injury (TBI) is one of the leading causes of disabilities and young adults' death. It is estimated that 1.7 million cases of TBI occur each year in the United States [1]. From those cases, according to the American Association of Neurological Surgeons, ~50,000 people die annually [2]. Even those with classified "mild" head injuries according to the Glasgow Outcome Scale suffer disabilities and impairments, 25% do not return to work and 84% have problems 1 year after the injury [3]. The Glasgow Outcome Scale (GOS) consists of five outcome categories: good recovery (able to work or go to school and live independently), moderate disability (able to live independently, but not able to return to work or school), severe disability (can follow commands, but not live independently), persistent vegetative state (in an unresponsive state and cannot interact with the environment), and death [4]. This global scale for functional outcome is commonly used to rate patient status after a traumatic brain injury.

TBI is characterized as an injury caused by external trauma. As a result, anatomical lesions of the skull, such as a fracture or tear in

Felipe Fregni (ed.), *Clinical Trials in Neurology*, Neuromethods, vol. 138,
https://doi.org/10.1007/978-1-4939-7880-9_4, © Springer Science+Business Media, LLC, part of Springer Nature 2018

the scalp, can lead to functional impairment of the meninges, brain, or their blood vessels. The level of impairment is divided based on its intensity: mild, moderate, and severe. This classification is usually defined by the Glasgow Coma Scale (GCS), which includes a patient's eye opening as well as verbal and motor responses [5]. Depending on the severity, brain disorders can cause either momentary or even permanent physical and neuropsychological (cognitive, behavioral, emotional) deficits [6]. Due to the burden that TBI can cause to families and patients, further research is still needed to develop better treatment methods for this disorder.

There are several challenges related to TBI clinical trials, such as protocols with a strict enrollment criteria, that hinders the generalizability of the results [7]; small sample sizes, which may affect the study outcomes [3, 7]; and heterogeneous TBI patient characteristics that make hard to delimitate an effective intervention for research studies [7, 8]. In addition, there are some logistical obstacles such as short study windows in several clinical trials for acute TBI studies, which make recruitment and implementation of the treatment difficult as certain types of treatments need to be implemented within a specific therapeutic window in order to be effective; lack of funding for TBI research, as compared to more well-known conditions, such as cancer and heart disease [3]; and the inaccuracy of Glasgow Coma Score (GCS) [3], which is often used to classify TBI patients in most clinical trials and does not capture all the possible injuries that could be noticed with the Glasgow Outcome Scale (GOS), for instance [7]. However, although GOS may provide more information regarding the patient, it still does not qualify all the multidimensional symptoms of TBI patients as even the extended version is not sensitive to differences in performance for those with high-scores: e.g., milder brain injury [7].

Considering these challenges, the main objective of this chapter is to contextualize and analyze the 100 most cited TBI trials, in order to provide methodological recommendations for designing future trials in TBI.

2   Methodology

A review of the literature was conducted through the Web of science database by searching the keyword "Traumatic brain injury" in the articles' title. Eligible studies were identified according to document type, "clinical trial," and year of publication (2005–2015). Unlike other chapters of this book, the search timespan for TBI was increased from 5–10 years since the initial idea did not yield an ideal number of articles that had a high quality and met the inclusion/exclusion criteria. In addition, this review included prospective observational studies and clinical trials. A few

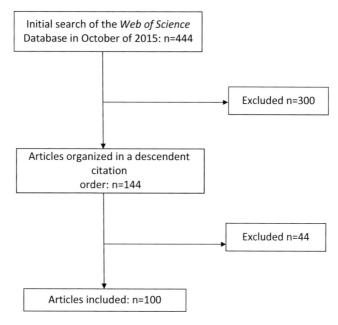

Fig. 1 Flowchart of the TBI clinical trials search

articles were excluded if they were not available in English or if they corresponded to a book chapter.

After the search completion (October 2015), a total of 444 articles was obtained, from which 300 were excluded. As to define the 100 most cited articles, they were sorted from a descendent citation order, highest to lowest. Consequently, from 144 articles, 44 were excluded. All the process is represented by Fig. 1, which summarizes the methodology strategy.

3 Overview

After defining the articles that were to be included in this review, data was collected and organized in a table based on the CONSORT 2010 criteria. The articles were then characterized according to different features: publication year, number of citations, impact factor and whether the journal complied with the CONSORT criteria or not (Table 1).

As outlined in the Table 1, the majority of the articles were published from 2005 to 2009 (77%) while the remaining 23 were published in a 5-year interval (2010–2015). This numbers may reflect a significant decrease in published articles related to TBI. The number of citations, on the other hand, had a wide distribution range from 21 to 307, with the highest concentration of articles (44%) between 20 and 30 citations. In regard to the impact factor, it could be noticed that only 11% of the articles were

Table 1

Number of publications related to clinical trials in TBI per variable selected

Year	2005–2007	41
	2008–2009	36
	2010–2015	23
# of Citations	20–30	44
	31–50	27
	51–75	11
	76–100	8
	100+	10
Impact factor	1–2	19
	>2–3	30
	>3–4	27
	>4–20	13
	>20	11
Consort criteria	Yes	44
	No	56

published in journals with an impact factor >20, while most of the other journals presented an impact factor between 2 and 4 (57%).

Overall, the trials assessed different interventions, including drugs, procedures and devices. Interestingly, there were a larger number of trials evaluating effects of procedures and devices, in comparison to drug trials. Differences were also observed in regard to the study design of each study. A two parallel-arm design corresponded to the most applied one, followed by three parallel-arm design. Lastly, the wide majority of the 100 most cited articles in TBI were randomized clinical trials (RCTs), accounting for 84%, while only 6% corresponded to open-label studies with no randomization. Please refer to Table 2 to see general characteristics of all included articles.

4   Trial Design

Experimental designs can be described according to different characteristics, especially when it comes to adopting an intervention as to manipulate a certain condition. An experimental trial intends to exert control over most threats to internal validity, providing the strongest evidence for causal relationships [9]. This section of the chapter discusses the use of different trial designs in TBI studies, including parallel-arm, crossover, and factorial designs, also considering the use of quasi-experimental design. Interestingly, RCTs were the most common type of study identified in the articles. This type of clinical trial allows investigators to directly compare an

Table 2

General trial characteristics of TBI articles search (*n* = 100)

Characteristic of the trial		Number of articles
Phase	Phase 1	9
	Phase 2	26
	Phase 3	5
	Exploratory	27
	Pivotal	33
Arms	Two parallel arms	72
	Three parallel arms	6
	Four parallel arms	2
	Factorial	1
	Crossover	9
Study design	RCT	83
	Observational	4
	Quasi-experimental	13
Sample size	0–25	28
	26–50	22
	51–100	24
	101–500	23
	>500	4
Type of intervention	Drug	40
	Device	3
	Procedure	57

intervention against either a control group or other interventions. As randomization avoids imbalances between groups, RCTs also allow researchers to draw conclusions about the cause–effect relationship between independent and dependent variables.

The majority of the RCTs had a parallel-arm design (*n* = 92), usually comparing subjects that received an intervention to a control group. Among these trials, 79.3% used a two-arm design, while only 18% chose a one-arm or three-arm design. As an example, Temkin et al. [10] assessed whether magnesium infusions have favorable effects in patients after TBI by using a two-arm parallel design, in which the intervention was compared to placebo. Furthermore, fewer clinical trials chose to use a crossover design, accounting for 7%. This type of design is extremely useful when testing interventions, as the participants are used as their own control, increasing the validity of results. Crossover trials in TBI mostly compared the effects of pharmaceutical drugs against placebo, while fewer studies evaluated effects of specific procedures. For instance, a study of Couillet et al. [11] assessed the effectiveness of a rehabilitation program for divided attention after severe TBI. Another example of crossover design includes a study of Jha et al. [12], which aimed to evaluate the efficacy of modafinil in treating fatigue after TBI in comparison to placebo.

Additionally, among the 100 articles, only one of them reported a factorial design study (2 × 2), which investigated the effects of erythropoietin and two different hemoglobin transfusion threshold (7 and 10 g/dL) on neurological recovery after TBI. In this trial, the comparisons were also made against placebo [13]. The lack of factorial design trials in TBI needs to be highlighted. Although this type of design allows the comparison of different interventions in a same study, it usually requires additional resources and can be more time demanding, which is possibly a justification for its small application in TBI studies.

Although RCTs are considered the gold standard in terms of scientific evidence, randomization is not always possible in numerous circumstances. Accordingly, 17% of the reviewed articles corresponded to non-RCT trials. Among these, quasi-experimental studies accounted for 2%.

4.1 Phases of TBI Trials

Clinical trials are designed and classified according to the type of intervention being tested and to the study goals. For trials evaluating the effects of drugs, the stages are referred to as phases I, II, III, and IV. On the contrary, for device and procedure trials, the stages are classified into feasibility, pivotal, and postmarketing studies. For this reason, the discussion of phases in TBI trials of different interventions will be conducted separately.

The phase of a study indicates in which stage the trial currently is and it is mainly determined by the sample size and the objective of the study (see below in the drug and procedure/device sections) [14]. The bigger the data on safety, efficacy, dose kinetics and side effects, the bigger the possibility to develop a more complex trial phase. For instance, phase IV trials are the ones conducted for assessing drugs that have already been approved by The Food and Drug Administration (FDA), thus being in a pharmacovigilance status (post marketing surveillance). The Food and Drug Administration (FDA) classification is used to determine the phase of drug [15] and device trials [16]. This system was also adapted to include procedure studies.

4.1.1 Drug Trials

Among the 100 most cited clinical trials, 41% evaluated the effects of drugs, based on the US Food and Drug Administration classification system [15] (Graph 1). The interventions related to the phase I trials mostly included cyclosporin A, progesterone, glycemic control, and methylphenidate. The reason for that is that most of the drugs tested in TBI have been tested in other conditions thus have been cleared in previous phase I trials. In addition, 89.2% of the drug trials corresponded to phase II, while only 12.2%

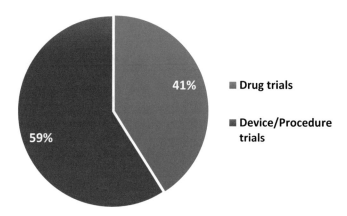

Graph 1 Frequency of trials according to the type of intervention

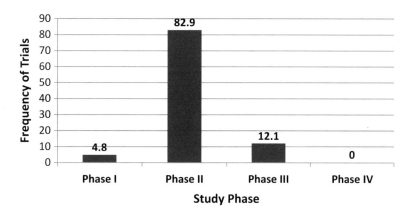

Graph 2 Frequency of the phases in TBI drug trials

corresponded to phase III studies. A few studies did not have the typical parameters of either a phase I or II. This can be explained by the fact that although they had a typical sample size or trial duration of a phase I trial, their primary aim was to evaluate efficacy rather than safety. In these cases, the studies were classified based on their aim instead of sample size or trial duration, as the goal to measure efficacy was also reflected in the trial's primary outcome. Lastly, there were no phase IV studies among the 100 most cited articles (Graph 2).

4.1.2 Device and Procedure Trials

Experimental studies of device and procedure accounted for 44.1% of the most cited trials in TBI. Among these, a number of 26 trials corresponded to feasibility studies, which usually have the goal of early evaluate devices/procedures for providing proof of principle and initial clinical data on safety [16]. As an example, a study of Cooper et al. [17] had the main goal of evaluating the feasibility of

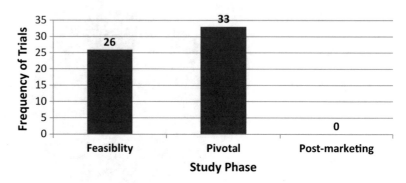

Graph 3 Frequency of the phases in TBI device/procedure trials

decompressive craniectomy surgery in patients with diffuse TBI and refractory intracranial hypertension. A study published in 2009, which aimed to understand the acute effects of packed red blood cell transfusion on cerebral oxygenation and metabolism after TBI, is also another example of feasibility trial [18]. This type of study design is extremely necessary for improving and making device modifications, also for refining and optimizing techniques. Considering that numerous devices and procedures under investigation may potentially change the prognostic of patients with TBI, the large number of feasibility trials is coherent and comprehensible (Graph 3).

Pivotal trials, which corresponded to 55.9% of device and procedure studies, are mainly designed to test effectiveness and measure adverse events involved with a procedure or device in larger sample sizes (over 50 subjects). Among these trials, two had 48 [19] and 49 subjects [20], but were classified as pivotal, as their sample size was close to the cut-off point. Sixteen out of the device/procedure trials (27.1%) were related to some type of rehabilitation program or behavioral intervention, such as problem solving training [21], attention training [22], or music therapy [23]. Also, other 16 studies reported medical procedures, such as paramedical intubation vs. hospital intubation [24], decompressive craniectomy [17, 25–27], and hypothermia [19, 25, 28, 29].

5   Population

The main eligibility criteria of the selected trials were age, severity of injury, injury duration, and admission status (in/out-patient). We reviewed the 100 most cited articles in which TBI was the main neurological condition on investigation; however, there were two atypical cases worth mentioning. For instance, the study of

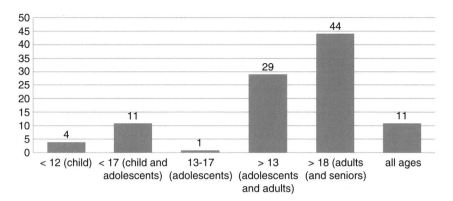

Graph 4 Distribution of the number of studies and enrolment age for the eligibility criteria

Katz-Leurer et al. assessed children with cerebral palsy and TBI [30], and O'Brien et al. assessed multiple sclerosis and TBI [31].

5.1 Age

In order to classify the articles according to age, the normalized aging classification in the USA was used: children (under 12), adolescents (13–17), adults (18–65), and seniors (over 65). As shown in Graph 4, most trials included adolescents and adults in their sample. The lack of studies involving a younger population might be related to ethical or logistic concerns.

5.2 Severity of the Trauma

The Glasgow Coma Scale was used in 71% of the studies to assess the severity of the trauma. Computed Tomography and time of Post Traumatic Amnesia were also used as a classification for trauma severity. Other types of classification were also reported: Injury Severity Score, Disability Rating Scale, Rancho de Los Amigos, Neurobehavioral Cognitive Status Examination, DoD Health Affairs criteria, Barthel Index, Rivermead Behavioral Memory Test severity scores, and classification of the American Congress of Rehabilitation Medicine-Head Injury Interdisciplinary Special Interest Group to "mild" trauma definition. Graph 5 represents the distribution of the studies and the criteria used to assess the trauma severity.

Interestingly, most of the trials (71%) included either severe or moderate to severe TBI patients. Also, 22% of the studies did not report TBI severity or the criteria applied for assessing it during enrollment. For the distribution of TBI severity in the studies, see Graph 6.

5.3 Stage of the Trauma

The majority of the studies (56%) enrolled patients during the post trauma acute phase, during hospitalization or while patients were in the Intensive Care Unit (ICU). Six studies assessed patients during the hospitalization stage while in a sub-acute phase, and the

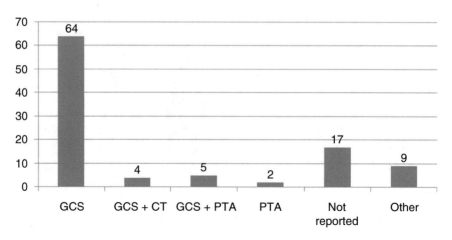

Graph 5 Distribution of the number of studies and the criteria used to assess the trauma severity. Subtitle: *GCS* Glasgow coma scale, *CT* computerized tomography, *PTA* post-traumatic amnesia

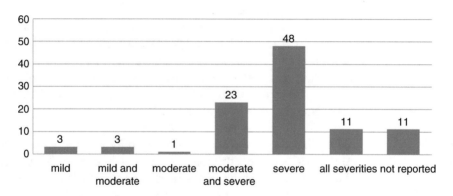

Graph 6 Distribution of TBI severity in the clinical trial

follow-ups were performed during an ambulatory setting. One third of the studies (35%) assessed patients in a chronic stage of the trauma. Graph 7 shows the distribution of the patients according to the stage of trauma during the randomization phase.

Thirty percent of the 100 most cited clinical trials in TBI reported the assessment of outpatients while 60% of the studies were related to inpatients. From all the outpatient assessments, 50% were related to behavioral or cognitive rehabilitation and the remaining ones were related to drug treatment, medical procedure, or device. On the contrary, studies that included inpatients were related to acute TBI and mostly assessed decompressive craniectomy, hypothermia, insulin therapy, among other pharmaceutical interventions. Graph 8 represents the admission status of the patient during the study setting.

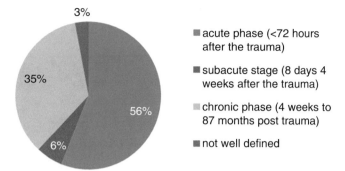

Graph 7 Distribution of the trials according to the stage of the trauma at the time of enrollment

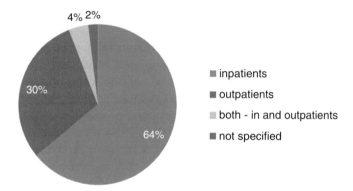

Graph 8 Distribution of the studies related to the admission status of the patient during the study setting.

6   Adherence and Recruitment

In this review, the large rate of adherence in several trials was related to the assessment of inpatients. Regarding the trials that presented an adherence rate larger than 80%, 44 of them assessed TBI during hospitalization, whereas only nine trials were related to outpatients during a chronic stage. Even in these trials that had many follow-up visits, the adherence rate was satisfactory.

On the other hand, some trials reported lower adherence of participants than expected. One of these studies, for instance, investigated a dopamine agonist in low-response children with severe TBI [32]. Of the 25 children recruited, only ten completed the trial. Likewise, another study reported a loss of 50% of their patients during the trial due to death [33]. Table 3 shows the adherence and recruitment rates of the 100 articles reviewed in this chapter.

Table 3
Rate of adherence and recruitment in the reviewed trials

Rate of adherence	N of trials	Rate of recruitment (enrollment/screened)	N of trials
<50%	1	0–20	11
50–70%	7	>20–40	12
>70–80%	10	>40–60	13
>80–90%	16	>60–80	6
>90–100%	38	>80	11
Not reported or N/A	28	Not reported or N/A	47
Total	100	Total	100

Ninety percent of the articles did not report any methods to improve adherence. However, five out of the ten remaining trials that reported this information mentioned the telephone or mail contact as to ensure patient's adherence to the study [17, 34–36]. For instance, one of the studies mentioned that if the subject did not complete all the assessments, the researchers would at least follow-up by telephone to collect the remaining necessary information [36]. Other techniques included changing the trial design to broaden the inclusion criteria [19]; using online training sessions; contacting multiple times for reminders/collecting data; opting for no randomization to increase adherence and benefit for patients; and having weekly meetings to monitor adherence. Although it is expected that open label trials increase adherence rate, this did not happen in all the cases. In this review, out of all the nine one-arm studies (where all participants/groups received study treatment), six of them had an 100% adherence rate. Despite that, three other trials reported a 77% [37], 67% [38] and 54% [39] adherence rate, respectively.

Based on the articles that did not directly report adherence methods (90%), there were still some noteworthy characteristics as many of the studies used ICU or inpatient care. Consequently, researchers were able to easily assess patients, as little to no effort was required from the patients to attend the appointments. In another study, a pilot trial designed to measure feasibility, although there were no reported methods used to improve recruitment in that study; these methods were employed in a later stage of the trial.

7   Interventions

The interventions of the included TBI trials were divided into two categories: Drug trials, which assessed the effects of pharmaceutical drugs; and procedure and/or device trials, related to behavioral or medical procedures, such as rehabilitation, cognitive training, hypothermia, and surgery. The results were analyzed according to the study phases: I, II, III, exploratory, or pivotal studies.

7.1 Drug Trials

As the treatment of TBI is still an ongoing debate, given the amount of possible therapies that are being tested, the chosen interventions in most clinical trials were also heterogeneous. However, after analyzing the 100 most cited articles in this field, it is possible to notice a pattern. From all drug trials, 32 compared the drug under investigation to placebo (78%). For instance, *Write et al.* [40] compared progesterone to placebo while *Temkin et al.* [10] administered magnesium to one group and placebo as a control.

A few studies also compared dosage. *Vespa et al.* [41], in their crossover trial, compared tight (80–110 mg/dL) vs. loose (120–150 mg/dL) glycemic control in patients with severe traumatic brain injury. In addition, *Ashman et al.* [42] combined dosage with placebo. They compared sertraline in doses starting at 25 mg and increasing to therapeutic levels (up to 200 mg) Vs. placebo. *Kraus et al.* [43], however, administered only one drug (Amantadine) in their open label trial. A few trials ($n = 4$) also compared two drugs. For example, Patrick et al. [32] compared pramipexole (Mirapex) to amantadine (Symmetrel) and Billota et al. [44] compared conventional insulin therapy vs. intensive insulin therapy.

Regarding each phase, phase I drug trials mostly assessed the effects of cyclosporin ($n = 4$), antibant ($n = 1$), methylphenidate ($n = 1$), glycemic control ($n = 1$), prostacyclin ($n = 1$), and progesterone ($n = 1$). Among all the phase II trials, the most used drugs included albumin fluid ($n = 1$), amantadine (vs. pramipexole or vs. placebo) ($n = 4$), bromocriptine ($n = 1$), creatine ($n = 1$), erythropoietin ($n = 1$), growth hormone (GH) replacement therapy ($n = 1$), hypertonic saline-dextran (vs. normal saline) ($n = 2$), intensive insulin therapy (vs. conventional insulin therapy) ($n = 3$), methylphenidate ($n = 2$), modafinil ($n = 2$), N-Acetyl Cysteine (NAC) ($n = 1$), progesterone ($n = 2$), rivastigmine ($n = 1$), rosuvastatin ($n = 1$), and sertraline ($n = 1$).

Finally, as mentioned before, phase III trials ($n = 5$) were mostly related to moderate and severe inpatients. All the interventions related to drugs were administered in a hospital setting and mostly consisted of saline-hypertonic [45], infusion of traxprodil [46], dexanabinol [47], magnesium [10], and citicoline [48].

7.2 Procedure and Device Trials

Relating to feasibility ($n = 27$) and pivotal studies ($n = 33$), three of them were related to devices, which included deep brain stimulation, discrete cerebral hypothermia system (the cooling cap) and hyperbaric oxygen. Also, all these studies were compared to sham condition. Medical procedures (26%), on the contrary, were related to those administered in a hospital setting while behavioral procedures (31%) referred to rehabilitation, cognitive training, and exercises in both inpatients and outpatients.

Among the trials that involved procedures, all of them ($n = 26$) were conducted with patients in a hospital setting. From these studies, nine were related to hypothermia, followed by those related to decompressive craniectomy ($n = 5$). Considering hypothermia, four studies assessed hypothermic treatments for the pediatric population and five chose to evaluate hypothermia in adults or elderly patients. Additionally, a few studies that assessed decompressive craniectomy compared this procedure with standard care ($n = 3$) or compared a limited type of craniectomy with standard treatment (STC) ($n = 2$).

Studies involving medical procedures ($n = 8$) also assessed monitoring intraparenchymal intracranial pressure (pressure-monitoring group) vs. imaging and clinical examination group ($n = 1$) [49]; hyperbaric oxygen vs. normobaric hyperoxia vs. standard care ($n = 1$) [50]; paramedic rapid sequence intubation vs. intubation in the hospital ($n = 1$) [24], red blood cell transfusion ($n = 1$) [18]; normobaric hyperoxia ($n = 1$) [33]; bone marrow-derived mesenchymal stromal cells ($n = 1$) [51]; transpyloric feeding vs. gastric feeding ($n = 1$) [52]; and intravenous administration of 6×10^6 autologous BMMNC/kg body weight (one arm study) ($n = 1$) [53].

The trials related to behavioral procedure ($n = 31$) presented more heterogeneous designs, thus the interventions varied accordingly. The most applied interventions were cognitive training and problem-solving program ($n = 14$), followed by interdisciplinary rehabilitation program ($n = 7$), psychotherapy ($n = 2$), family care intervention ($n = 2$), telephone support ($n = 2$), reminders on mobile phone (memory aid) ($n = 1$), paging system intervention ($n = 1$), music therapy ($n = 1$), and community intervention ($n = 1$).

8 Outcomes

8.1 Primary Outcomes

The primary outcome distribution was similar between both drug and procedure/device trials. On the other hand, the frequency of clinical primary outcomes was greater than surrogate endpoints. Besides that, a few studies worked with both clinical and surrogate outcomes. As an example, Zhao et al. [54] analyzed arterial blood glucose, lactic acid, and neurological deficits (indexed by the Glasgow Outcome Scale) as primary outcomes when comparing hypothermia with normothermia for the treatment of patients with TBI. Graph 9 shows the distribution of primary outcomes between the trials.

In numerous studies, the identification of the primary outcome was challenging, as it was not often explicitly stated. In these cases, an analysis of the variables listed in the results, study hypotheses, article title and abstract was performed in order to clearly

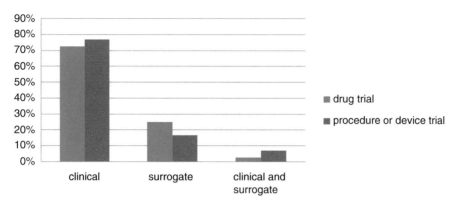

Graph 9 Distribution of primary outcomes between drug and device/procedure trials

determine the primary outcome. Yet if it was not possible to tell which outcome(s) were primary, they were labeled as "not clearly defined." Also, in two specific trials, the investigators used a composite measure [10, 55] that analyzed various types of assessments, so they were classified differently than the other studies. Table 4 summarizes the number of studies reporting the different categories of primary outcome according to study phases.

8.1.1   Drug Trials

As expected, most phase I studies used surrogate endpoints ($n = 8$) and only one reported a clinical outcome, which referred to length of stay in the ICU. Overall, the main surrogate endpoints included deoxy-D-glucose brain positron emission tomography; cerebral microdialysis and cerebral hemodynamic; cyclosporin A blood concentration, indexed by HPLC ultraviolet detection; lactate pyruvate ratio; safety, tolerability and pharmacokinetics of Antiban and intravenous infusion of CsA; and steady-state serum progesterone concentration. Four out of the nine phase I studies did not report nor evaluate a secondary outcome. However, four of the remaining trials did use surrogate endpoints to measure secondary outcomes, which included level of creatinine, platelets, hepatic function, white blood count, brain microdialysis, levels of bradykinin metabolite in plasma and cerebrospinal fluid, cerebral perfusion pressure, immunological evaluation and bacterial culture, and tissue histology.

As efficacy is the main goal of phase II trials ($n = 26$), most outcomes referred to clinical endpoints and only two studies used surrogate endpoints: rate of hypoglycemia [44] and intracranial pressure [56]. In trials that used mood end emotional symptoms as primary outcomes, researchers worked either with the Beck Depression Inventory [12], the Hamilton Depression Rating Scale [42], or both scales together [57]. Regarding cognitive assessment, it was performed with CANTAB [58], Hopkins Verbal Learn Test [59], working memory and attention [60], WISC-III,

Table 4

Number of primary outcomes in different study phases

Categories for primary outcome	Phase I	Phase II	Phase III	Exploratory	Pivotal	Total
Neuropsychological (cognitive and mood)	0	10	0	7	6	23
Biological	7	2	0	7	3	20
Physical performance or condition	0	0	0	3	0	3
Social or communication	0	0	0	0	0	0
Functional impairment/independence	0	8	3	3	9	23
Symptoms/adverse events	0	0	0	0	1	1
Satisfaction	0	0	0	0	0	0
Mortality	0	1	0	0	1	2
Status of employment/school/volunteer	0	0	0	1	1	2
Composite score	0	0	1	0	1	2
Sleep	0	2	0	1	0	3
Not clearly defined	1	1	0	2	1	6
Other	1	0	0	1	1	3
Mixed outcomes of categories above	0	2	1	2	9	12

Trail Making Test, COWAT, and Stroop in young TBI patients [61]. Also, social and behavioral assessments were applied by the Fatigue Severity Scale and Modified Fatigue Impact Scale, daily sleepiness (Epworth Sleepiness Scale), maintenance of wakefulness [12, 57, 60, 62], and Community Integration Questionnaire [59]. Finally, both clinical and functional outcomes were investigated through the Glasgow Outcome Scale [13, 34, 40, 63, 64]. Four of the studies, which used the Glasgow Outcome Scale studies, performed it in the pediatric population [32, 61, 65, 66].

From all the phase III studies (n = 5), four of them used Glasgow Outcome Scale Extended (GOSE) as the main outcome [10, 45–48] in moderate and severe adults or senior inpatients. Some of those studies used cognitive assessment and/or functional status as the outcome, as well as quality of life [10, 48].

8.1.2 Procedure Trials

To better understand how the authors selected their main outcomes, instead of dividing the studies according to their phases, they were divided according to the type of intervention: behavioral or procedures. Regarding behavioral trials (n = 15), only one used a surrogate endpoint as primary outcome, which corresponded to body composition index and cardio respiratory responses [39]. On the other hand, 14 studies used clinical measurements as primary outcomes, which included mood end emotional symptoms, cognitive assessment, social and behavioral assessments, functional outcomes (such as total wake time, sleep efficiency, and diagnostic criteria) and quality of life. For instance, studies of Tiersky LA et al., Thaut MH et al., and O'Brien A et al. performed cognitive assessments by using the Paced Auditory Serial Addition Test (PASAT) and Attention tasks [23, 31, 67];

Based on the total number of medical intervention trials ($n = 27$), one-third of the studies ($n = 9$) used surrogate measures as primary outcome for assessing severe or moderate-to-severe patients during hospitalization. These surrogate outcomes included brain tissue oxygen [18, 33, 50]; serial C Spinal fluid [68], temperature [69, 70], glutathione and antioxidant reserve [71], mesenchymal stromal cells/bone marrow; intracranial pressure ICP [28, 50, 72], Intracranial Pressure [19, 25, 28, 50, 72]; Brain tissue PO_2; Microdialysis. Nonetheless, most studies ($n = 18$) used clinical outcomes, such as cognitive assessment (mental status, working memory, information-processing speed, episodic memory and learning, verbal fluency, executive function, and motor dexterity [49]; and Galveston Orientation and Amnesia Test [49, 73]; functionality scales (Disability Rating Scale [49], Post-traumatic Disorder Checklist-Military Version [73], Glasgow Outcome Scale [17, 24–26, 49, 74, 75], and Immediate Post-Concussion Assessment [73]; other clinical measures, such as survival and pediatric logistic organ dysfunction [53]; Barthel Index [51]; Incidence of pneumonia [52]; Mortality [19, 49]; Arrhythmias [19]; Hemorrhage [19]; Coagulopathy [19]; and number of days in the ICU [49].

8.2 Secondary Outcomes

Several studies indeed reported at least one secondary outcome (77%). The most applied surrogate endpoints as secondary outcomes included intracranial pressure, hypoglycemia, plasma concentration, and cerebrospinal fluid. For clinical outcomes, investigators mostly used length in ICU, adverse events, Glasgow Outcome Scale, and Disability Ranting Scale. In these trials, the Beck Depression Inventory, Anxiety symptoms, and Apathy Scale were also used to assess patients' mood. Functional measurements and cognitive assessments were highly applied as well.

9 Sample Size

Sample size calculation is an analysis made a priori to collecting data in order to establish the required number of subjects to determine and detect a statistically significant difference for an intervention based on an expected effect size. The smaller the effect size, the larger the required sample [9]. Lack of power in clinical trials can increase the risk of Type II error (β), consequently increasing false negative results, or increasing the probability of failing to reject the null hypothesis. To avoid misinterpreting results and conclusions, it is suggested to be prudent and conservative in estimating the effect size. Usually, most trials tend to assume $\beta = 0.20$, which corresponds to a power of 80% and represents a reasonable protection against Type II error [9].

Regarding the review, there was a relative balance between the trials. Nonetheless, most of them (74%) did not have a sample size larger than 100, which elucidates that most studies of TBI have a small to moderate sample size. Although it is fundamental to properly describe how the final sample size was obtained, 73% of the studies did not report this information. For the studies that did report the sample size calculation, it was observed that 11 trials set up the power as 80%. Eight trials described the same approach while accounting for dropouts and loss to follow-up, thus adjusting their sample size calculation. For example, Hutchison et al. [76] reported their sample size calculation by adding a 10% rate of loss to follow-up. Yurkewicz et al. [46], on the other hand, stipulated a loss of 7% of dropout rate. Cooper et al. [17] calculated their sample size and additionally performed an interim analysis. There were also some variations on the methods of calculating the sample size. Bulger et al. [45] and Robertson et al. [13] used initially a one-sided interim analysis with a power of 62.2% and then they performed a second test for futility using a 91% power. Other studies have chosen to use a 90% [77] or 95% [10] power, for example, to test for adverse event. Interestingly, only one study opted to use a 66.8% power of estimation [66].

10   Interim Analysis and Stoppage Criteria

The aim and design of a given study are usually what determine its stoppage criteria. Clear stopping rules allow the search for interim findings, which may lead to the termination or change of a research study. For instance, if researchers realize that a number of participants have experienced serious adverse events due to the intervention, stopping the trial may be necessary. Considering that, stopping rules for phase II trials of serious diseases are extremely important, due to the high risk of complications.

In regard to the studies reviewed, ten of them reported the performance of interim analysis and stopped the trial earlier than expected. Interestingly, all those studies involved patients with severe or moderate-to-severe TBI during acute stages. Additionally, three trials reported an interim analysis for a Safety Monitoring Board Review, such as the study of Bulger et al. [45], which assessed whether out-of-hospital administration of hypertonic fluid improves neurological outcome after TBI [45, 46, 53]. Despite that, none of these studies had to be stopped earlier because of safety issues. Besides carrying about safety, two trials also conducted an interim analysis for futility with significant level $\alpha = 0.15$ and applied stoppage criteria for this purpose [13, 29]. Furthermore, in one of these two studies, investigators interrupted the trial if the interim analysis for futility showed less than a 20% chance of confirming the primary hypothesis [29].

Four studies designed an interim analysis using O'Brien-Fleming boundaries, of which three were related to severe and one to mild-to-moderate TBI patients [24, 40, 49, 77]. In addition, some of the reviewed studies ($n = 28$) were related to procedure interventions during the acute stage of the trauma, such as craniotomy, hypothermia, and red blood cell transfusion. However, a large number of trials did not report the type of safety data boarding monitoring used or did not conduct safety interim analysis. Nonetheless, as most of the patients with TBI are in a severe and unstable state during hospitalization, it is recommended that safety monitoring is considered and well reported.

11   Statistical Analysis

As previously mentioned in chapter one of this book, choosing the most appropriate statistical test for the analysis of the primary outcome (dependent variable) depends on several aspects such as data distribution (whether it has a normal or a skewed distribution), type of variable (continuous, categorical, ordinal), group comparison (between groups or within groups), and research question (comparison of the scores, correlation, prediction, etc.).

In this review, the primary outcome, when surrogate, was usually measured as a continuous variable, as intracranial pressure, blood glucose, and temperature. In these cases which the dependent variable was continuous and the comparison was between groups, the most used tests were t-test (paired or unpaired) and ANOVA, if the distribution of the dependent variable was normal. In case the data was not normally distributed (skewed), the Wilcoxon test or Kruskal–Wallis was performed. Also, it was observed that 26% of the studies used at least one of those tests to analyze the results. For multiple comparisons correction and post hoc analysis, the Bonferroni test and Tukey test were applied, respectively.

As a few authors dichotomized the outcome according to the score (>4 or ≤ 4) [45] or the level of recovery ("good or moderate" versus "all other levels of recovery" [13, 40, 78], the statistical test also varied. In these cases, the chosen test was Fisher exact test or chi-square. Moreover, for studies that used GOS scores (continuous), the commonly used tests included Mann–Whitney or t-test [26, 45, 47, 63] and logistic regression for dichotomous outcomes [13].

Only four studies reported the use of regression analysis, correlation or survey analysis to investigate the primary aim of the study. In those trials, more than one primary outcome was analyzed. For instance, Rockswold et al. [50] investigated the effect of hyperbaric to normobaric hyperoxia on cerebral metabolism, ICP and oxygen toxicity in severe TBI. They used a mixed-effects linear model with fixed effects for treatment, time, day, and interaction between treatment and time. In terms of sensitivity analysis, only one study reported this element [76]. Graph 10 shows the frequency of the statistical tests used to analyze the primary outcome of the studies.

Regarding the secondary outcomes, which frequencies are described in Graph 11, an exploratory analysis through descriptive statistics was also used. Interestingly, mortality was described as a secondary outcome in eight studies. In these cases, survival or prevention were chosen to describe the intervention's effect. This type of analysis is important to understand the prognosis and treatment effectiveness. In this review, four trials analyzed survival by Kaplan–Meier Curves, two studies used Cochran–Mantel–Haenszel test and three used Cox proportional hazards model. Also, in some studies, mortality of patients was analyzed by chi-square test [76] or by Mann–Whitney U test [65].

The intention-to-treat (ITT) is a principle whereby data are analyzed according to the group assignments, regardless if subjects actually completed the study or not [9]. ITT analysis was used in 28% of the trials included in this review and the remaining studies did not report the type of analysis. Furthermore, some studies used both ITT and per-protocol analysis. Per-protocol analysis is restricted to the participants who completed the treatment protocol where they were originally allocated. If done alone, however, this analysis may lead to bias.

11.1 Subgroup Analysis

Subgroup analysis is a statistical approach that involves splitting the participant data into subgroups in order to conduct specific comparisons between them. In this review, 12 studies used subgroup analysis to assess differences between groups, which reflects how common this type of analysis is in RCTs. According to Brookes et al. [79], subgroup analysis, especially exploratory subgroup analysis, can lead to potentially erroneous identification of subgroup differences and affect conclusions if not conducted properly. However, the most common approach used to identify differences between groups was the analysis of covariance (ANCOVA). This statistical procedure is used to compare two or more treatment groups while controlling for the effect of one or more covariates or effect modifiers, based on the F statistic.

Brown et al. [80], for example, used ANCOVA to analyze differences between the experimental and control groups. They considered six covariates: (1) pretest values on the outcome measures; (2) severity of ataxia; (3) level of functional independence; (4) gait

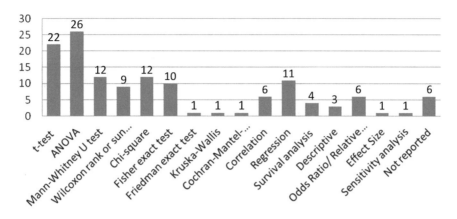

Graph 10 Frequency of the different statistical tests used to assess primary outcomes

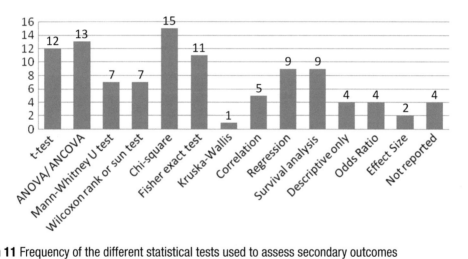

Graph 11 Frequency of the different statistical tests used to assess secondary outcomes

symmetry; (5) degree of trunk control; and (6) assistive device used. Moreover, Pearson correlation coefficients determined whether interrater reliability was acceptable or not.

12   Randomization

The randomization process is essential for controlling differences that may exist between the groups being compared. Ensuring a random assignment means that each subject has an equal chance of being designated to any group regardless of personal judgment [9]. It is considered one of the crucial characteristics of clinical trials, as only randomization can truly attempt to determine cause–effect relationships between the independent and dependent variables. In this review, 83% of the studies included sample

Table 5

Distribution of randomization type, sequence and implementation method in TBI trials

Randomization sequence	Phase I (9/9)	Phase II (24/26)	Phase III (5/5)	Exploratory (18/27)	Pivotal (27/33)	Total
Telephone	0	0	0	0	2	2
Computer	2	7	3	3	10	25
Not reported	4	15	0	10	10	39
Not clearly specified	0	1	2	1	0	4
Random number method/ number table	2	1	0	2	3	8
Not true randomization	1	0	0	2	1	4
Computer or phone	0	0	0	0	1	1
Type of randomization						
Simple	0	3	0	0	1	4
Blocked (not stratified)	5	8	2	3	7	25
Stratified	0	8	3	2	11	24
Other	0	0	0	0	2	2
Not reported	4	5	0	13	6	28
Implementation of randomization						
Not reported	8	18	0	14	17	57
Study staff	0	1	3	2	3	9
Pharmacy	1	4	1	0	0	6
Non-study staff/independent party/statistician	0	1	1	2	7	11
Concealment						
Sealed envelopes	0	4	0	2	4	10
Identical appearance	2	5	3	0	0	10
Restricted assess	0	0	1	0	3	4
Not reported or specified	7	15	1	16	20	59

randomization. The remaining 17% of the trials were nonrandomized and corresponded to quasi-experimental or observational trials. Out of the 40 drug trials, only two were nonrandomized open-label studies [37, 43]. Please refer to Table 5 for more details of randomization and blinding.

12.1 Randomization Sequence Generation

The randomization process should be conducted through a system that does now allow the prediction of the results by investigators, thus all patients would be assigned to an experimental group by chance. In this review, numerous trials (47%) did not report the method used for generating the randomization sequence. On the contrary, 26 trials used computer-generated systems to perform the randomization and three reported the use of a random number method or table/list [60, 81, 82]. In addition, one multicenter trial used either a computer or telephone depending on the site location [74], while two trials used a telephone based randomization sequence [17, 76].

A few other studies also referred to different randomization approaches such as a randomization system by a company called ALMAC [48] and a "randomly generated numeric code" [45]. An exploratory trial allocated the participants based on their admission sequence [83] and one pivotal trial determined the allocation sequence according to the day of the month (odd numbers received treatment and even control) [25]. Two studies reported the use of a randomized number table, but did not specify how the tables were generated [67, 84].

Finally, there were a few randomization methods described that, in fact, do not qualify as a true random sequence generation, as most of them allowed researchers to guess the assignment of the patients. For instance, one of the drug trials used the hospital file number of subjects for assigning each of them, meaning that patients with odd file numbers were assigned to one group, and the ones with even numbers were assigned to another [85].

12.2 Types of Randomization

Another critical aspect to consider when conducting randomized clinical trials is the type of randomization to be used. In regard to the trials reviewed, only four of them used simple randomization and their sample size varied from 40 to 81 subjects, which still allows the generation of imbalanced groups. On the other hand, block randomization was the most used type, accounting for 25 studies, followed by stratified randomization (24 studies). Although a few studies mentioned using a block randomization approach, they also mentioned the reassignment of a couple of participants after they were randomized [52] and technical difficulties that lead to an alteration in the randomization process [50]. Therefore, they were classified as "other."

12.3 Implementation of Randomization

Most drug trials (65%) failed to report or did not report clearly the person or group of people responsible for implementing the randomization. Among the studies that did report this information, six trials referred to study staffs, which corresponded to either the pharmacy or an independent researcher. Generally, when the intervention was to be allocated by someone from the study staff, the article also specified how involved the specific individual was with the study. For instance, a research assistant who was not involved in the clinical management [36], or biostatistician [69].

12.4 Allocation Concealment

After the randomization is assigned, it is also critical to have a system in place to make sure the intervention is concealed to staff, subjects and those related to the trial as allocation concealment prevents selection bias. Concealment and blinding can be maintained by using similar interventions, as identical capsules [62, 86, 87], identical infusion injection [13, 46, 88], active vs. sham device [89], or double-dummy procedures [12]. Due to its importance, 42% of the drug trials specified how the allocation concealment was maintained.

However, regarding the studies that were not randomized, only two reported blinding of evaluators to aspects related to the treatment or procedures [75, 90]. In one of the articles, which discussed about a rehabilitation program, the investigators reported that their evaluators were blinded to their previous history of rehabilitation [90]. In another case-control study, the authors mentioned that the treatment group was measured but the evaluators were blinded to the results [75]. Additionally, Ouellet M.-C. et al. reported that there was a randomization at baseline, but not at any other time points [91] during the study.

13   Blinding

Blinding involves innumerous techniques to reduce experimental bias by keeping the subject and/or investigators unaware of group assignments and research hypothesis. Also, blinding is also important to control bias in reporting, analyzing, or interpreting collected data. In this review, if researchers were unaware of treatment allocation for at least the primary outcome (s), they were considered blinded (Table 6).

Considering single blinded studies, in which only the researcher or the participant are blinded but not both, there were 14 studies that reported this method. Some researchers may also consider severe TBI patients in the acute phase blinded as most patients are cognitively impaired to understand the study or their allocation after an acute severe TBI. In this review, there were several studies that fell into this category [18, 50, 52, 71, 92, 93]. However, as the trials did not report blinding for these populations, a conservative approach was admitted and these studies were allocated in the category of "no blinding reported."

In a double-blind design, as both the participant and the researcher are blinded, thus reducing even more the risks of bias, it is usually the most commonly used design in RCTs. Within the 31 drug studies that reported blinding, 27 were double-blinded. One of these studies reported that the outcome assessor was blinded, but did mention whether the participant was blinded or not [66]. Therefore, we identified this study as single-blinded. Finally, regarding triple-blind design, just one study reported this approach in which patient, assessors and physicians were blinded [94]. The lack of studies with a triple blind design may be due to the fact that this design is more costly and requires extra resources.

Table 6

Types of blinding used in TBI trials

Blinding	Phase I	Phase II	Phase III	Exploratory	Pivotal	Total
Single	0	1	0	3	10	14
Double	6	16	5	1	3	31
Other	0	3	0	2	2	7
None or not clearly reported	3	4	0	12	12	31
Total	9	24	5	18	27	83

A few articles mentioned atypical blinding procedures, thus being classified as "Other." For an example, this was considered when some of the primary or secondary outcomes were blinded and others were not [19, 83, 95]. Other cases included studies in which only some assessors were blinded and studies in which there was only blinding at baseline [30].

14 Study Duration

There was a large range in regard to trial duration from the analyzed studies. The average time of trial participation was 180 days, ranging from 4 h to 2 years. The following section describes how the clinical trials designed their follow-up periods and, for a better understanding, it was divided based on the trial phases.

In phase I drug trials ($n = 9$), the duration of the studies ranged from 4 days to 6 months. In five of those trials, the duration of the study was up to 8 days, in which the follow-ups consisted in biological measurements usually collected or measured repeatedly after 12 or 24 h [41, 92, 93, 96]. Moreover, four studies reported longer follow-ups, ranging between 1 and 6 months (Graph 12). For these trials, GOS and biological measures were used as endpoints [81, 82, 85, 86].

For phase II trials, however, the duration of the studies ranged between 4 h and 1 year. Among the shorter trials ($n = 4$), which the duration was up to 2 weeks, three of them were performed during the hospitalization period and/or in an acute stage of the trauma [65, 87, 97]. For the trials ranging from 2 weeks to 3 months ($n = 8$), 1–4 follow-ups were performed in total. Usually, the trials were designed to have a baseline and a follow-up assessment after the intervention and the main outcomes were surrogate for both inpatients and outpatients [32, 40, 57, 58, 61, 94]. In longer phase II trials, ranging from 3 to 6 months, many started during the hospitalization stage (acute stage of TBI) and followed

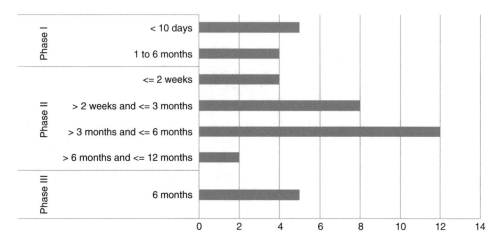

Graph 12 Number of studies reviewed in TBI trials per phase and duration of the trial

up the patient after the discharge. In those trials, the assessments were designed to be after the end of the intervention or the follow-ups were scheduled to be after 3 or 6 months. Interestingly, in phase III trials, all participants were older than 15 years old, and were classified as severe or moderate to severe. The duration of those five trials was 6 months, in which four trials were multicenter studies [45–48] and just one a single-center [10]. Those trials usually had two follow-ups after the intervention, frequently at 3 and 6 months. Graph 7 provides more details about the duration of phase I, II, and III trials.

The duration of exploratory studies (n = 27) was heterogeneous, ranging from 72 h to 2 years. Usually, trials involving behavioral procedures had a longer trial duration (from 3 months to 2 years), when compared to procedure or device intervention trials. These studies were mostly related to inpatients and the duration was shorter, ranging from 72 h to 6 months. For shorter trials, the follow-ups were usually surrogate endpoints related to biological measurements, whereas, for longer trials, follow-ups were frequently assessed by GOS.

Alike exploratory studies, the duration of pivotal studies was also heterogeneous (n = 33), ranging from 1 week to 1.5 years. However, for behavioral procedure studies (n = 16), a longer duration was seen, from 6 months to 1.5 years. Also, the total number of follow-ups ranged from 1 to 6. Another characteristic from behavioral procedure studies was the variability of the duration according to the type of intervention.

Finally, as to procedure and device interventions, there was a larger range of trial duration, from 2 days to 2 years. Furthermore, the number of follow-ups was related to the type of the primary outcome: a larger number of follow-ups were performed when they were related to biological measurements (1 or 2 follow-ups after 6 months usually using GOS). Graph 13 shows the duration of exploratory and pivotal trials.

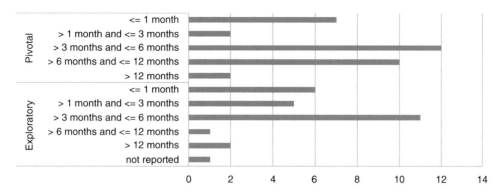

Graph 13 Number of studies reviewed in TBI trials and duration of the trials.

15 Limitations

Interestingly, one-third of the 100 most cited articles in TBI trials did not report their limitations. One of the main limitations in these trials is the small sample size, resulting in a lack of power in statistical analysis [33, 67]. Indeed, the lack of sample size calculation increases the risk of type I and II errors, which may lead to erroneous conclusions. Other limitations reported were short trial duration [18], lack of randomization [90], lack of blinding [80], and lack of covariate adjustment. According to Andelic et al. [78], a quasi-experimental study design cannot eliminate the possibility of confounding bias caused by unmeasured confounders.

Besides the small number of limitations reported, we could observe that just a few studies mentioned the method of missing data imputation. Another potential bias is related to the lack of multiple comparison adjustments. Moreover, the most common limitation of the studies was the description of the methodology used in these trials. A significant number of these studies did not report the primary outcome, the endpoints, duration of the follow-up, and the type of analysis. As a result, the low quality of the methods description leads to an unreliable statistical analysis and consequently doubtful results.

16 Conclusion

After reviewing the most cited clinical trials of TBI, interesting patterns related to study design, population, outcomes, and statistical analysis could be identified. Accordingly, the majority of TBI studies corresponded to phase II trials, which shows that safety aspects and efficacy of interventions are a major concern of investigators in this field. Also, it could be noticed that numerous trials often involved young adults and applied clinical assessments as primary outcomes, such as neuropsychological evaluations. Even though most articles provided important information, a significant number of studies presented methodological limitations and, in consequence, doubtful conclusions.

Another important aspect worth mentioning after this review is the difficulty that most investigators found regarding patient's adherence to the studies. Especially for trials recruiting participants from outpatient facilities, the dropout rate was higher than expected, possibly due to the level of TBI severity of these patients. Furthermore, there was a large number of procedure/device trials in a pivotal stage (n = 33), which mostly reported positive results for at least one primary outcome. Nonetheless, most of these trials (n = 20) had a small sample size, with fewer than 100 subjects, which elucidates the need of developing larger TBI clinical trials in the future.

In general, research studies of traumatic brain injury have been criticized for not generating significant results to the scientific community. This is possibly due to the high variability of patients with TBI, which may compromise the reliability of study findings or even their generalizability. Curiously, Maas and colleagues described a decrease in the amount of clinical trials of TBI over the past decade due to the lack of positive results in larger multicenter trials and increased costs for conducting these studies.

Although TBI is an important and meaningful research topic for clinical trials, developing these studies is often a challenging task. Thus, future investigators must be encouraged to follow consort guidelines and give major attention to the principles of clinical research, in order to improve the quality of TBI studies and achieve the higher level of evidence. With that ensured, future trials will have the potential to advance patient's quality of life. To that end, it is expected that this review will provide useful and fundamental information for guiding healthcare providers in the development of well-designed clinical trials in TBI.

References

1. Faul M, Xu L, Wald MM, Coronado VG (2010) Traumatic brain injury in the United States: emergency department visits, hospitalizations, and deaths. Centers for Disease Control and Prevention, National Center for Injury Prevention and Control, Atlanta, GA, pp 891–904. Available from: http://www.ncbi.nlm.nih.gov/pubmed/23630120

2. Surgeons AAoN (2015) Traumatic brain injury. Available from: http://www.aans.org/patient information/conditions and treatments/traumatic brain injury.aspx

3. Wilson M, Zolfaghari P, Griffin C, Lockey D, Tolias C, Verma V (2014) The future of traumatic brain injury research. Scand J Trauma Resusc Emerg Med 22(Suppl 1):A7. Available from: http://www.sjtrem.com/content/22/S1/A7%5Cnpapers3://publication/doi/10.1186/1757-7241-22-S1-A7

4. Wright J (2011) Glasgow outcome scale. In: Kreutzer JS, DeLuca J, Caplan B (eds) Encyclopedia of clinical neuropsychology. Springer, New York, NY, pp 1150–1152. Available from: http://link.springer.com/10.1007/978-0-387-79948-3

5. Teasdale G, Murray G, Parker L, Jennett B (1979) Adding up the Glasgow coma score. Acta Neurochir Suppl (Wien) 28:13–16

6. Menon DK, Schwab K, Wright DW, Maas AI (2010) Position statement: definition of traumatic brain injury. Arch Phys Med Rehabil 91:1637–1640

7. Maas AI, Menon DK, Lingsma HF, Pineda JA, Sandel ME, Manley GT (2012) Re-orientation of clinical research in traumatic brain injury: report of an international workshop on comparative effectiveness research. J Neurotrauma 29(1):32–46. Available from: http://www.ncbi.nlm.nih.gov/pubmed/21545277%5Cn, http://online.liebertpub.com/doi/pdfplus/10.1089/neu.2010.1599

8. Maas AIR, Menon DK (2012) Traumatic brain injury: rethinking ideas and approaches. Lancet Neurol 11:12–13

9. Portney LG, Watkins MP (2009) Foundations of clinical research: applications to practice. Pearson Education, Inc., Upper Saddle River, NJ. 892 p

10. Temkin NR, Anderson GD, Winn HR, Ellenbogen RG, Britz GW, Schuster J et al (2007) Magnesium sulfate for neuroprotection after traumatic brain injury: a randomised controlled trial. Lancet Neurol 6(1):29–38

11. Couillet J, Soury S, Lebornec G, Asloun S, Joseph PA, Mazaux JM et al (2010) Rehabilitation of divided attention after severe traumatic brain injury: a randomised trial. Neuropsychol Rehabil 20(3):321–339

12. Jha A, Weintraub A, Allshouse A, Morey C, Cusick C, Kittelson J et al (2008) A randomized trial of modafinil for the treatment of fatigue and excessive daytime sleepiness in individuals with chronic traumatic brain injury. J Head Trauma Rehabil 23(1):52–63. Available from: http://content.wkhealth.com/linkback/openurl?sid=WKPTLP:landingpage&an=00001199-200801000-00009

13. Robertson CS, Hannay HJ, Yamal J-M, Gopinath S, Goodman JC, Tilley BC et al (2014) Effect of erythropoietin and transfusion threshold on neurological recovery after traumatic brain injury. JAMA 312(1):36. Available from: http://jama.jamanetwork.com/article.aspx?doi=10.1001/jama.2014.6490

14. Chittester B (2014) Medical device clinical trials – how do they compare with drug trials? Master Control, Inc., Granville, OH. Available from: http://cdn2.hubspot.net/hub/149400/file-646377456-pdf/docs/mc-n-med-dev-trials-compare-with-drug-trials.pdf

15. Administration USFaD (2015) Step 3: clinical research. [Internet]. [cited 2015 Oct 10]. Available from: http://www.fda.gov/ForPatients/Approvals/Drugs/ucm405622.htm

16. Food and Drug administration. What does it mean for FDA to "classify" a medical device? [Internet]. [cited .2017 Nov 30]. Available from: https://www.fda.gov/AboutFDA/Transparency/Basics/ucm194438.htm

17. Cooper DJ, Rosenfeld JV, Murray L, Arabi YM, Davies AR, D'Urso P et al (2011) Decompressive craniectomy in diffuse traumatic brain injury. N Engl J Med 364(16):1493–1502. Available from: http://www.nejm.org/doi/abs/10.1056/NEJMoa1102077

18. Zygun DA, Nortje J, Hutchinson PJ, Timofeev I, Menon DK, Gupta AK (2009) The effect of red blood cell transfusion on cerebral oxygenation and metabolism after severe traumatic brain injury*. Crit Care Med 37(3):1074–1078. Available from: http://content.wkhealth.com/linkback/openurl?sid=WKPTLP:landingpage&an=00003246-200903000-00040

19. Adelson PD, Ragheb J, Muizelaar JP, Kanev P, Brockmeyer D, Beers SR et al (2005) Phase II clinical trial of moderate hypothermia after severe traumatic brain injury in children. Neurosurgery 56(4):740–753

20. Bourgeois MS, Lenius K, Turkstra L, Camp C (2007) The effects of cognitive teletherapy on reported everyday memory behaviours of persons with chronic traumatic brain injury. Brain Inj 21(12):1245–1257. Available from: http://ovidsp.ovid.com/ovidweb.cgi?T=JS&CSC=Y&NEWS=N&PAGE=fulltext&D=emed8&AN=2008061172, http://sfxhosted.exlibrisgroup.com/cmc?sid=OVID:embase&id=pmid:&id=doi:10.1080%2F02699050701727452&issn=0269-9052&isbn=&volume=21&issue=12&spage=1245&pages=1245-1257&

21. Rivera PA, Elliott TR, Berry JW, Grant JS (2008) Problem-solving training for family caregivers of persons with traumatic brain injuries: a randomized controlled trial. Arch Phys Med Rehabil 89(5):931–941

22. Galbiati S, Recla M, Pastore V, Liscio M, Bardoni A, Castelli E et al (2009) Attention remediation following traumatic brain injury in childhood and adolescence. Neuropsychology 23(1):40–49. Available from: http://doi.apa.org/getdoi.cfm?doi=10.1037/a0013409

23. Thaut MH, Gardiner JC, Holmberg D, Horwitz J, Kent L, Andrews G et al (2009) Neurologic music therapy improves executive function and emotional adjustment in traumatic brain injury rehabilitation. Ann N Y Acad Sci 1169:406–416

24. Bernard SA, Nguyen V, Cameron P, Masci K, Fitzgerald M, Cooper DJ et al (2010) Prehospital rapid sequence intubation improves functional outcome for patients with severe traumatic brain injury: a randomized controlled trial. Ann Surg 252:959–965

25. Jiang JY, Xu W, Li WP, Gao GY, Bao YH, Liang YM et al (2006) Effect of long-term mild hypothermia or short-term mild hypothermia on outcome of patients with severe traumatic brain injury. J Cereb Blood Flow Metab 26(6):771–776. Available from: http://www.ncbi.nlm.nih.gov/entrez/query.fcgi?cmd=Retrieve&db=PubMed&dopt=Citation&list_uids=16306933

26. Jiang J-Y, Xu W, Li W-P, Xu W-H, Zhang J, Bao Y-H et al (2005) Efficacy of standard trauma craniectomy for refractory intracranial hypertension with severe traumatic brain injury: a multicenter, prospective, randomized controlled study. J Neurotrauma 22(6):623–628. Available from: http://www.liebertonline.com/doi/abs/10.1089/neu.2005.22.623

27. Qiu W, Guo C, Shen H, Chen K, Wen L, Huang H et al (2009) Effects of unilateral decompressive craniectomy on patients with unilateral acute post-traumatic brain swelling after severe traumatic brain injury. Crit Care 13(6):R185. Available from: http://ccforum.biomedcentral.com/articles/10.1186/cc8178

28. Qiu W, Zhang Y, Sheng H, Zhang J, Wang W, Liu W et al (2007) Effects of therapeutic mild hypothermia on patients with severe traumatic brain injury after craniotomy. J Crit Care 22(3):229–235

29. Adelson PD, Wisniewski SR, Beca J, Brown SD, Bell M, Muizelaar JP et al (2013) Comparison of hypothermia and normothermia after severe traumatic brain injury in children (Cool Kids): a phase 3, randomised controlled trial. Lancet Neurol 12(6):546–553

30. Katz-Leurer M, Rotem H, Keren O, Meyer S (2009) The effects of a 'home-based' task-oriented exercise programme on motor and balance performance in children with spastic cerebral palsy and severe traumatic brain injury. Clin Rehabil 23(8):714–724. Available from: http://journals.sagepub.com/doi/10.1177/0269215509335293

31. O'Brien A et al (2007) An investigation of the differential effect of self-generation to improve learning and memory in multiple sclerosis and traumatic brain injury. Neuropsychol Rehabil 17(3):273–282. Available from: http://ovidsp.ovid.com/ovidweb.cgi?T=JS&PAGE=reference&D=med5&NEWS=N&AN=17474057%5Cn, http://ovidsp.ovid.com/ovidweb.cgi?T=JS&PAGE=reference&D=emed8&NEWS=N&AN=2007218094

32. Patrick P, Blackman J, Mabry J, Buck M, Gurka M, Conaway M (2006) Dopamine agonist therapy in low-response children following traumatic brain injury. J Child Neurol 21:879–885. Available from: http://onlinelibrary.wiley.com/o/cochrane/clcentral/articles/476/CN-00572476/frame.html

33. Tisdall MM, Tachtsidis I, Leung TS, Elwell CE, Smith M (2008) Increase in cerebral aerobic metabolism by normobaric hyperoxia after traumatic brain injury. J Neurosurg 109(3):424–432. Available from: http://thejns.org/doi/10.3171/JNS/2008/109/9/0424

34. Coester A, Neumann CR, Schmidt MI (2010) Intensive insulin therapy in severe traumatic brain injury: a randomized trial. J Trauma 68:904. Available from: http://content.wkhealth.com/linkback/openurl?sid=WKPTLP:landingpage&an=00005373-900000000-99649

35. Vanderploeg RD, Schwab K, Walker WC, Fraser JA, Sigford BJ, Date ES et al (2008) Rehabilitation of traumatic brain injury in active duty military personnel and veterans: defense and veterans brain injury center randomized controlled trial of two rehabilitation approaches. Arch Phys Med Rehabil 89(12):2227–2238. Available from: http://linkinghub.elsevier.com/retrieve/pii/S0003999308014858

36. Zhu XL, Poon WS, Chan CCH, Chan SSH (2007) Does intensive rehabilitation improve the functional outcome of patients with traumatic brain injury (TBI)? A randomized controlled trial. Brain Inj 21(7):681–690. Available from: http://www.tandfonline.com/doi/full/10.1080/02699050701468941

37. Silver JM, Koumaras B, Meng X, Potkin SG, Reyes PF, Harvey PD et al (2009) Long-term effects of rivastigmine capsules in patients with traumatic brain injury. Brain Inj 23(2):123–132

38. High WM, Roebuck-Spencer T, Sander AM, Struchen MA, Sherer M (2006) Early versus later admission to postacute rehabilitation: impact on functional outcome after traumatic brain injury. Arch Phys Med Rehabil 87(3):334–342. Available from: http://linkinghub.elsevier.com/retrieve/pii/S0003999305014693

39. Bhambhani Y, Rowland G, Farag M (2005) Effects of circuit training on body composition and peak cardiorespiratory responses in patients with moderate to severe traumatic brain injury. Arch Phys Med Rehabil 86(2):268–276. Available from: http://linkinghub.elsevier.com/retrieve/pii/S0003999304004721

40. Wright DW, Kellermann AL, Hertzberg VS, Clark PL, Frankel M, Goldstein FC et al (2007)

ProTECT: a randomized clinical trial of progesterone for acute traumatic brain injury. Ann Emerg Med 49(4):391

41. Vespa P, McArthur DL, Stein N, Huang S-C, Shao W, Filippou M et al (2012) Tight glycemic control increases metabolic distress in traumatic brain injury. Crit Care Med 40(6):1923–1929. Available from: http://content.wkhealth.com/linkback/openurl?sid=WKPTLP:landingpage&an=00003246-201206000-00031

42. Ashman TA, Cantor JB, Gordon WA, Spielman L, Flanagan S, Ginsberg A et al (2009) A randomized controlled trial of sertraline for the treatment of depression in persons with traumatic brain injury. Arch Phys Med Rehabil 90(5):733–740. Available from: http://linkinghub.elsevier.com/retrieve/pii/S0003999309000835

43. Kraus MF, Smith GS, Butters M, Donnell AJ, Dixon E, Yilong C et al (2005) Effects of the dopaminergic agent and NMDA receptor antagonist amantadine on cognitive function, cerebral glucose metabolism and D2 receptor availability in chronic traumatic brain injury: a study using positron emission tomography (PET). Brain Inj 19(7):471–479. Available from: http://www.tandfonline.com/doi/full/10.1080/02699050400025059

44. Bilotta F, Caramia R, Cernak I, Paoloni FP, Doronzio A, Cuzzone V et al (2008) Intensive insulin therapy after severe traumatic brain injury: a randomized clinical trial. Neurocrit Care 9(2):159–166. Available from: http://link.springer.com/10.1007/s12028-008-9084-9

45. Bulger EM, May S, Brasel KJ, Schreiber M, Kerby JD, Tisherman SA et al (2010) Out-of-hospital hypertonic resuscitation following severe traumatic brain injury: a randomized controlled trial. JAMA 304(13):1455–1464

46. Yurkewicz L, Weaver J, Bullock MR, Marshall LF (2005) The effect of the selective NMDA receptor antagonist traxoprodil in the treatment of traumatic brain injury. J Neurotrauma 22(12):1428–1443. Available from: http://www.liebertonline.com/doi/abs/10.1089/neu.2005.22.1428

47. Maas AIR, Murray G, Henney H, Kassem N, Legrand V, Mangelus M et al (2006) Efficacy and safety of dexanabinol in severe traumatic brain injury: results of a phase III randomised, placebo-controlled, clinical trial. Lancet Neurol 5(1):38–45

48. Zafonte RD, Bagiella E, Ansel BM, Novack TA, Friedewald WT, Hesdorffer DC et al (2012) Effect of citicoline on functional and cognitive status among patients with traumatic brain injury: citicoline brain injury treatment trial (COBRIT). JAMA 308(19):1993–2000. Available from: http://jama.jamanetwork.com/data/Journals/JAMA/25500/joc120094_1993_2000.pdf

49. Chesnut RM, Temkin N, Carney N, Dikmen S, Rondina C, Videtta W et al (2012) A trial of intracranial-pressure monitoring in traumatic brain injury. N Engl J Med 367(26):2471–2481. Available from: http://www.nejm.org/doi/abs/10.1056/NEJMoa1207363

50. Rockswold SB, Rockswold GL, Zaun DA, Zhang X, Cerra CE, Bergman TA et al (2010) A prospective, randomized clinical trial to compare the effect of hyperbaric to normobaric hyperoxia on cerebral metabolism, intracranial pressure, and oxygen toxicity in severe traumatic brain injury. J Neurosurg 112(5):1080–1094. Available from: http://thejns.org/doi/10.3171/2009.7.JNS09363

51. Zhang Z-X, Guan L-X, Zhang K, Zhang Q, Dai L-J (2008) A combined procedure to deliver autologous mesenchymal stromal cells to patients with traumatic brain injury. Cytotherapy 10(2):134–139. Available from: http://linkinghub.elsevier.com/retrieve/pii/S146532490870166X

52. Acosta-Escribano J, Fernández-Vivas M, Grau Carmona T, Caturla-Such J, Garcia-Martinez M, Menendez-Mainer A et al (2010) Gastric versus transpyloric feeding in severe traumatic brain injury: a prospective, randomized trial. Intensive Care Med 36(9):1532–1539. Available from: http://link.springer.com/10.1007/s00134-010-1908-3

53. Cox CS, Baumgartner JE, Harting MT, Worth LL, Walker PA, Shah SK et al (2011) Autologous bone marrow mononuclear cell therapy for severe traumatic brain injury in children. Neurosurgery 68(3):588–600. Available from: https://academic.oup.com/neurosurgery/article-lookup/doi/10.1227/NEU.0b013e318207734c

54. Zhao Q-J, Zhang X-G, Wang L-X (2011) Mild hypothermia therapy reduces blood glucose and lactate and improves neurologic outcomes in patients with severe traumatic brain injury. J Crit Care 26(3):311–315. Available from: http://linkinghub.elsevier.com/retrieve/pii/S0883944110002418

55. Bell KR, Temkin NR, Esselman PC, Doctor JN, Bombardier CH, Fraser RT et al (2005) The effect of a scheduled telephone intervention on outcome after moderate to severe traumatic brain injury: a randomized trial. Arch Phys Med Rehabil 86(5):851–856. Available

from: http://linkinghub.elsevier.com/retrieve/pii/S0003999304013930

56. Cooper DJ, Myburgh J, Heritier S, Finfer S, Bellomo R, Billot L et al (2013) Albumin resuscitation for traumatic brain injury: is intracranial hypertension the cause of increased mortality? J Neurotrauma 30(7):512–518. Available from: http://online.liebertpub.com/doi/abs/10.1089/neu.2012.2573

57. Lee H, Kim SW, Shin IS, Yang SJ, Yoon JS (2005) Comparing effects of methylphenidate, sertraline and placebo on neuropsychiatric sequelae in patients with traumatic brain injury. Hum Psychopharmacol 20(2):97–104

58. Silver JM, Koumaras B, Chen M, Mirski D, Potkin SG, Reyes P et al (2006) Effects of rivastigmine on cognitive function in patients with traumatic brain injury. Neurology 67:748–755. Available from: http://ovidsp.ovid.com/ovidweb.cgi?T=JS&PAGE=reference&D=emed7&NEWS=N&AN=2006453041

59. High WM, Briones-Galang M, Clark JA, Gilkison C, Mossberg KA, Zgaljardic DJ et al (2010) Effect of growth hormone replacement therapy on cognition after traumatic brain injury. J Neurotrauma 27(9):1565–1575. Available from: http://www.liebertonline.com/doi/abs/10.1089/neu.2009.1253

60. Kim Y-H, Ko M-H, Na S-Y, Park S-H, Kim K-W (2006) Effects of single-dose methylphenidate on cognitive performance in patients with traumatic brain injury: a double-blind placebo-controlled study. Clin Rehabil 20(1):24–30. Available from: http://journals.sagepub.com/doi/10.1191/0269215506cr927oa

61. Beers SR, Skold A, Dixon CE, Adelson PD (2005) Neurobehavioral effects of amantadine after pediatric traumatic brain injury: a preliminary report. J Head Trauma Rehabil 20:450–463. Available from: http://onlinelibrary.wiley.com/o/cochrane/clcentral/articles/349/CN-00530349/frame.html

62. Kaiser PR, Valko PO, Werth E, Thomann J, Meier J, Stocker R et al (2010) Modafinil ameliorates excessive daytime sleepiness after traumatic brain injury. Neurology 75(20):1780–1785. Available from: http://www.neurology.org/cgi/doi/10.1212/WNL.0b013e3181fd62a2

63. Ponsford JL, Myles PS, Cooper DJ, Mcdermott FT, Murray LJ, Laidlaw J et al (2008) Gender differences in outcome in patients with hypotension and severe traumatic brain injury. Injury 39(1):67–76. Available from: http://linkinghub.elsevier.com/retrieve/pii/S0020138307003555

64. Xiao G, Wei J, Yan W, Wang W, Lu Z (2008) Improved outcomes from the administration of progesterone for patients with acute severe traumatic brain injury: a randomized controlled trial. Crit Care 12(2):R61. Available from: http://ccforum.biomedcentral.com/articles/10.1186/cc6887

65. Baker AJ, Rhind SG, Morrison LJ, Black S, Crnko NT, Shek PN et al (2009) Resuscitation with hypertonic saline–dextran reduces serum biomarker levels and correlates with outcome in severe traumatic brain injury patients. J Neurotrauma 26(8):1227–1240. Available from: http://www.liebertonline.com/doi/abs/10.1089/neu.2008.0868

66. Sakellaris G, Kotsiou M, Tamiolaki M, Kalostos G, Tsapaki E, Spanaki M et al (2006) Prevention of complications related to traumatic brain injury in children and adolescents with creatine administration: an open label randomized pilot study. J Trauma Inj Infect Crit Care 61(2):322–329. Available from: http://content.wkhealth.com/linkback/openurl?sid=WKPTLP:landingpage&an=00005373-200608000-00011

67. Tiersky LA, Anselmi V, Johnston MV, Kurtyka J, Roosen E, Schwartz T et al (2005) A trial of neuropsychologic rehabilitation in mild-spectrum traumatic brain injury. Arch Phys Med Rehabil 86(8):1565–1574

68. Buttram SDW, Wisniewski SR, Jackson EK, Adelson PD, Feldman K, Bayir H et al (2007) Multiplex assessment of cytokine and chemokine levels in cerebrospinal fluid following severe pediatric traumatic brain injury: effects of moderate hypothermia. J Neurotrauma 24(November):1707–1717

69. Harris OA, Muh CR, Surles MC, Pan Y, Rozycki G, Macleod J et al (2009) Discrete cerebral hypothermia in the management of traumatic brain injury: a randomized controlled trial. J Neurosurg 110:1256–1264. Available from: http://onlinelibrary.wiley.com/o/cochrane/clcentral/articles/002/CN-00707002/frame.html

70. Puccio AM, Fischer MR, Jankowitz BT, Yonas H, Darby JM, Okonkwo DO (2009) Induced normothermia attenuates intracranial hypertension and reduces fever burden after severe traumatic brain injury. Neurocrit Care 11(1):82–87

71. Bayir H, Adelson PD, Wisniewski SR, Shore P, Lai Y, Brown D et al (2009) Therapeutic hypothermia preserves antioxidant defenses after severe traumatic brain injury in infants and children. [Erratum appears in Crit Care Med. 2009;37(4):1536]. Crit Care Med 37(2):689–695

72. Liu WG, Qiu WS, Zhang Y, Wang WM, Lu F, Yang XF (2006) Effects of selective brain cooling in patients with severe traumatic brain injury: a preliminary study. J Int Med Res

34(1):58–64. Available from: http://imr.sage-pub.com/content/34/1/58.short

73. Wolf G, Cifu D, Baugh L, Carne W, Profenna L (2012) The effect of hyperbaric oxygen on symptoms after mild traumatic brain injury. J Neurotrauma 29:2606–2612

74. Andrews PJD, Sinclair HL, Battison CG, Polderman KH, Citerio G, Mascia L et al (2011) European society of intensive care medicine study of therapeutic hypothermia (32-35 °C) for intracranial pressure reduction after traumatic brain injury (the Eurotherm3235Trial). Trials 12(1):8. Available from: http://www.pubmedcentral.nih.gov/articlerender.fcgi?artid=3027122&tool=pmcentrez&rendertype=abstract

75. Adamides AA, Cooper DJ, Rosenfeldt FL, Bailey MJ, Pratt N, Tippett N et al (2009) Focal cerebral oxygenation and neurological outcome with or without brain tissue oxygen-guided therapy in patients with traumatic brain injury. Acta Neurochir 151(11):1399–1409. Available from: http://link.springer.com/10.1007/s00701-009-0398-y

76. Hutchison JS, Ward RE, Lacroix J, Hébert PC, Barnes MA, Bohn DJ et al (2008) Hypothermia therapy after traumatic brain injury in children. N Engl J Med 358(23):2447–2456. Available from: papers://72c6f5a3-fe57-4d09-a143-5358ae4043f5/Paper/p142

77. Giacino JT, Whyte J, Bagiella E, Kalmar K, Childs N, Khademi A et al (2012) Placebo-controlled trial of amantadine for severe traumatic brain injury. N Engl J Med 366(9): 819–826. Available from: http://www.nejm.org/doi/abs/10.1056/NEJMoa1102609

78. Andelic N, Bautz-Holter E, Ronning P, Olafsen K, Sigurdardottir S, Schanke A-K et al (2012) Does an early onset and continuous chain of rehabilitation improve the long-term functional outcome of patients with severe traumatic brain injury? J Neurotrauma 29(1): 66–74. Available from: http://www.liebertonline.com/doi/abs/10.1089/neu.2011.1811

79. Brookes ST, Whitely E, Egger M, Smith GD, Mulheran PA, Peters TJ (2004) Subgroup analyses in randomized trials: risks of subgroup-specific analyses; power and sample size for the interaction test. J Clin Epidemiol 57(3):229–236

80. Brown TH, Mount J, Rouland BL, Kautz KA, Barnes RM, Kim J (2005) Body weight-supported treadmill training versus conventional gait training for people with chronic traumatic brain injury. J Head Trauma Rehabil 20(5):402–415. Available from: http://content.wkhealth.com/linkback/openurl?sid=WKPTLP:landingpage&an=00001199-200509000-00002

81. Mazzeo AT, Brophy GM, Gilman CB, Alves OL, Robles JR, Hayes RL et al (2009) Safety and tolerability of cyclosporin a in severe traumatic brain injury patients: results from a prospective randomized trial. J Neurotrauma 26:2195–2206. Available from: http://onlinelibrary.wiley.com/o/cochrane/clcentral/articles/583/CN-00734583/frame.html

82. Olivecrona M, Rodling-Wahlstrom M, Naredi S, Koskinen LO (2009) Prostacyclin treatment in severe traumatic brain injury: a microdialysis and outcome study. J Neurotrauma 26(8):1251–1262

83. Cheng SKW, Man DWK (2006) Management of impaired self-awareness in persons with traumatic brain injury. Brain Inj 20(6):621–628

84. Wade SL, Michaud L, Brown TM (2006) Putting the pieces together: preliminary efficacy of a family problem-solving intervention for children with traumatic brain injury. J Head Trauma Rehabil 21:57–67. Available from: http://onlinelibrary.wiley.com/o/cochrane/clcentral/articles/924/CN-00554924/frame.html

85. Moein H, Khalili HA, Keramatian K (2006) Effect of methylphenidate on ICU and hospital length of stay in patients with severe and moderate traumatic brain injury. Clin Neurol Neurosurg 108(6):539–542

86. Marmarou A, Guy M, Murphey L, Roy F, Layani L, Combal J-P et al (2005) A single dose, three-arm, placebo-controlled, phase i study of the bradykinin B2 receptor antagonist anatibant (LF16-0687Ms) in patients with severe traumatic brain injury. J Neurotrauma 22(12):1444–1455. Available from: http://www.liebertonline.com/doi/abs/10.1089/neu.2005.22.1444

87. Hoffer ME, Balaban C, Slade MD, Tsao JW, Hoffer B (2013) Amelioration of acute sequelae of blast induced mild traumatic brain injury by n-acetyl cysteine: a double-blind, placebo controlled study. PLoS One 8(1): e54163

88. Mazzeo AT, Alves OL, Gilman CB, Hayes RL, Tolias C, Niki Kunene K et al (2008) Brain metabolic and hemodynamic effects of cyclosporin A after human severe traumatic brain injury: a microdialysis study. Acta Neurochir 150(10):1019–1031. discussion 1031. Available from: http://www.ncbi.nlm.nih.gov/pubmed/18781275

89. Schiff ND, Giacino JT, Kalmar K, Victor JD, Baker K, Gerber M et al (2007) Behavioural improvements with thalamic stimulation after severe traumatic brain injury. Nature 448(7153):600–603

90. Sarajuuri JM, Kaipio ML, Koskinen SK, Niemelä MR, Servo AR, Vilkki JS (2005) Outcome of a comprehensive neurorehabilitation program for patients with traumatic brain injury. Arch Phys Med Rehabil 86(12):2296–2302. Available from: http://linkinghub.elsevier.com/retrieve/pii/S0003999305009172

91. Ouellet M-C, Morin CM (2007) Efficacy of cognitive-behavioral therapy for insomnia associated with traumatic brain injury: a single-case experimental design. Arch Phys Med Rehabil 88(12):1581–1592. Available from: http://linkinghub.elsevier.com/retrieve/pii/S0003999307015560

92. Empey PE et al (2006) Cyclosporin A disposition following acute traumatic brain injury. J Neurotrauma 23:109–116. Available from: http://onlinelibrary.wiley.com/o/cochrane/clcentral/articles/206/CN-00562206/frame.html

93. Mazzeo AT, Kunene NK, Gilman CB, Hamm RJ, Hafez N, Bullock MR (2006) Severe human traumatic brain injury, but not cyclosporin a treatment, depresses activated T lymphocytes early after injury. J Neurotrauma 23(6):962–975

94. Tapia-Perez JH, Sanchez-Aguilar M, Torres-Corzo JG, Gordillo-Moscoso A, Martinez-Perez P, Madeville P et al (2008) Effect of rosuvastatin on amnesia and disorientation after traumatic brain injury (NCT00329758). J Neurotrauma 25(8):1011–1017. Available from: http://online.liebertpub.com/doi/10.1089/neu.2008.0554

95. Yang M, Guo Q, Zhang X, Sun S, Wang Y, Zhao L et al (2009) Intensive insulin therapy on infection rate, days in NICU, in-hospital mortality and neurological outcome in severe traumatic brain injury patients: a randomized controlled trial. Int J Nurs Stud 46(6):753–758. Available from: http://linkinghub.elsevier.com/retrieve/pii/S002074890900008X

96. Wright DW, Ritchie JC, Mullins RE, Kellermann AL, Denson DD (2005) Steady-state serum concentrations of progesterone following continuous intravenous infusion in patients with acute moderate to severe traumatic brain injury. J Clin Pharmacol 45(6):640–648. Available from: http://www.ncbi.nlm.nih.gov/pubmed/15901745

97. Willmott C, Ponsford J (2009) Efficacy of methylphenidate in the rehabilitation of attention following traumatic brain injury: a randomised, crossover, double blind, placebo controlled inpatient trial. J Neurol Neurosurg Psychiatry 80(5):552–557. Available from: http://jnnp.bmj.com/cgi/doi/10.1136/jnnp.2008.159632

Chapter 5

Parkinson's Disease

Maria Fernanda Villarreal, Rodrigo Huerta-Gutierrez, and Felipe Fregni

Abstract

Parkinson's disease (PD) is recognized as one of the most common neurologic disorders, with an imperative need for research and new therapeutic interventions. Considering these characteristics, the aim of this book chapter is to analyze the 100 most cited clinical trials in PD, thus outlining methodological aspects, limitations, and challenges involved in this field. A systematic search of the literature between 2010 and 2015 was performed using the electronic database Web of Science. The main findings are reported and discussed as to contribute to the advance of clinical research in this area.

Key words Parkinson disease, Research, Clinical trials, Methodology, CONSORT guidelines

1 Introduction

Parkinson's disease (PD), also called "primary parkinsonism or idiopathic Parkinson's disease" is a neurodegenerative illness characterized by progressive motor dysfunction as a result of a dysfunctional dopaminergic system in the basal ganglia. The clinical diagnosis of Parkinson's disease lies on the presence of postural instability, tremor, bradykinesia, and rigidity. This condition normally appears in adults and elderly population, with an average age onset of around 65–70 years [1].

Parkinson's disease is recognized as one of the most common neurologic disorders, affecting approximately one million people in the USA with a prevalence of 0.3% in the general population. The incidence of PD is estimated at 4.5–16/100,000 persons/year and the number of patients with Parkinson's disease in the world's most populated nations is projected to exponentially increase and be over eight million by 2030 [2]. Severe disability and death is expected in 25% of the patients within 5 years, 65% within 10 years and 80% of PD patients with 15 years of onset [3]. The costs associated with this disease are around $14 billion per year in the USA alone [4]. Therefore, an imperative need for research on new therapeutic interventions is required.

Felipe Fregni (ed.), *Clinical Trials in Neurology*, Neuromethods, vol. 138,
https://doi.org/10.1007/978-1-4939-7880-9_5, © Springer Science+Business Media, LLC, part of Springer Nature 2018

Generally by the time clinical symptoms first manifest about 60% of nigrostriatal dopaminergic terminals are lost. For this reason, most of PD research is to identify individuals at risk on prodromal phase and find diagnostic biomarkers [5]. The development of new treatment strategies seems promising and remarkable progress is being made, however no definitive cure has been provided in the extensive research of Parkinson's disease. An increasing number of therapeutical approaches are still in the preclinical stage. In fact, among one of the most exciting advances in PD research is the presentation of novel approaches targeting toxic α-synuclein to overtake the primary pathological mechanism in PD [5]. Therapies for Parkinson's disease are rarely restricted to just pharmacological compounds. Thus, the majority of current trials involve pharmacological therapies and additional interventions, such as: surgery, gene therapy, neuroprotective therapies, and physical or behavioral therapies. Delaying the progression of neurodegenerative diseases is one of the topics of major interest on the field. Nonetheless, translation to the clinical practice will still be contingent of well-designed clinical trials. Researchers willing to advance the knowledge of Parkinson's disease treatment must be aware of the challenges that it implies. Clinical trials have become one of the major standards for evaluating new therapies and interventions in Parkinson's disease.

The major challenges for clinical trials in Parkinson's disease include: lack of objective outcomes, recruitment strategies, progressive and irreversible nature of the disease, heterogeneous response to treatment, unspecified pathophysiology, high degree of comorbidity and the concomitant treatment complications. The purpose of this chapter is to outline proper scientific methodology to improve future development of research in Parkinson's disease.

2   Methodology

A systematic search was performed on published literature on Parkinson's disease clinical trials (July–September 2015) using the electronic database of Web of Science[1], by the title words "Parkinson's disease" and document type "Clinical trial." Initially, our total finding of articles was 596; we screened 124 clinical trials to include 100 clinical trials in this chapter (Fig. 1). These articles were assessed for accuracy by the authors (MF/RH) as interventional clinical trials and also arranged by number of citations (from higher to lower). All others results that did not include this criteria

---

[1] Access to Web of Science was provided by Countway Library—Harvard Medical School.

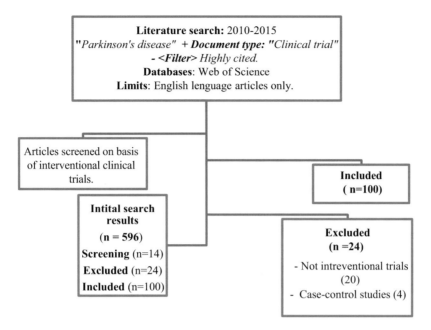

Fig. 1 Schematic illustration of search results and exclusion criteria

(noninterventional trials, observational studies, case control, etc.) were excluded from this search.

As a standardized search methodology[2] along with other chapters from this book, all results selected were articles published in English between the years 2010 and 2015. In addition, as a tool for validity in the experimental design we decided to use the CONSORT (Consolidated Standards of Reporting Trials) Guidelines to screen the articles accordingly on a 22-item checklist of clinical trial quality and description to report consistency in our findings. These guidelines are used for helping authors to determine how was the study conducted and analyzed. In the following subheadings of this chapter we will be discussing detailed information about Parkinson's disease clinical trial design, phases, eligibility criteria, and statistical analysis.

2.1 Initial Results

Overall, the percentage of clinical trials of Parkinson's disease in Neurology is expanding. According to clinicaltrials.gov, there were ~1642 registered in this website in 2015, of which more than 70% were active for recruitment. Arising from our findings, we found out that 70% of the trials studies resulted positive for primary outcome. Furthermore, we were able to determine the impact factor of the journals involved (see Graph 1). The journals presented in

---

[2] Some chapters in this book used an extended search for 10 years.

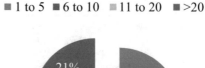

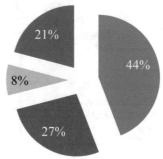

Graph 1 Impact factor distribution

our review, were the following; Movement disorders, followed by Annals of Neurology, Lancet, NEJM, Transitional Research, etc.

2.2   Impact Factor

Impact factor is a common measure to validate the importance of a journal and show the mean of citations within a year. In this review, the majority of the studies included an impact factor within 1–5. Nevertheless, this was followed by a 29% that had an impact factor above 11.

2.3   Consort Guidelines

The consort guidelines give a minimum set of recommendations for reporting randomized trials. Several studies suggest that use of the CONSORT checklist is associated with improved reporting of randomized trials. For this chapter, alongside with other chapters from this book we consider the consort guidelines a standard for reporting and we consequently presume that trials that accomplish most of these requirements carry a proper methodology for clinical trial design. In our review, 61% of the journals involved where following the consort guidelines, whereas the remaining 31% did not.

3   Trial Design–Phase–Settings

3.1   Trial Design

In our review, Table 1 describes the general characteristics of all trials analyzed. The most common study trial design was two parallel arms with a total of 47% of all trials included for this chapter, followed by nonrandomized clinical trial (non-RCT) design with 24%; that mainly corresponded to open-label studies. In fact, the two-arm parallel trial is still the optimal design and should always be considered.

Table 1
Review search results

		N
Year		
	2010	50
	2011	19
	2012	14
	2013	13
	2014	4
	2015	0
Phase		
	Not described	4
	Phase I	7
	Phase I/II	5
	Phase II	38
	Phase II/III	7
	Phase III	38
	Phase IV	1
Trial design		
	Two parallel arms	48
RCT	Three parallel arms	10
	Four or more parallel arms	3
	Factorial	5
	Crossover	9
	Non-RCT	25
Sample size		
	0–20	22
	20–45	26
	45–200	28
	>200	24
Type of intervention		
	Drug	54
	Device	19
	Procedure	27
# of Citations		
	0–20	9
	21–25	27
	26–30	13
	31–35	10
	36–40	7
	41–60	15
	60–80	6
	80–100	2
	More than 100	11

(continued)

Table 1
(continued)

		N
Journal impact factor		
	1–5	44
	5–10	28
	10–20	7
	More than 20	21
Study setting		
	Single-center	46
	Multicenter	54
Consort criteria		
	Yes	61
	Not endorsers	39

Based on the EMA guidelines [3], double blind, and placebo controlled studies of at least 6 months is recommended to establish efficacy maintenance and safety, due to the long and slow progression of the condition (time duration, time points and follow-up represent an important consideration for researchers, while selecting study design)

Among other type of design is the factorial design; useful for interaction effect between two independent variables, commonly substituted in PD trials by comparisons with placebo. Unfortunately, this design is not as widely used on PD trials as in other neurological conditions, representing 5% of all trial designs in our research. In the case of factorial designs, researchers can either assume that there is no interaction or calculate the synergetic or antagonistic effects of two interventions. Researchers can also opt to use a partial factorial design in which not all the combinations of two treatments are tested. For instance, Hauser [6] and Olanow [7] tested a levodopa-carbidopa gel against on immediate release oral intervention. This scenario made mandatory the use of a version of the partial factorial design, the double-dummy strategy, to produce effective blinding.

On the other hand, crossover designs are not used as frequently as other designs but have some advantages. One of its major advantages is the diminution of the intersubject variability that allows the recruitment of a lower number of subjects. However, an important drawback results from the unpredictable carryover effect or period by treatment interactions that can bias the estimated outcome. In our sample, we identified nine clinical trials that used this approach. For instance, Sawada [8] used a crossover design to investigate the effects of amantadine on dyskinesias. Contrasting to other papers on the topic, they chose a washout period of 15 days to limit carryover effects. Moreover, in order to control for period and carryover effects, they build a model that included these factors and included it on the statistical analysis.

There are several illustrative trials with this design that have shown the profitability of this design selection. For instance, the factorial and the crossover designs can be used simultaneously in a trial. Likewise, Ravina [9] studied the safety and efficacy of a cholinesterase inhibitor, donepezil hydrochloride, for the treatment of dementia in Parkinson's disease (PD). In this trial, the author used a crossover trial (AABB:BBAA) with four different periods analogous to the 2×2 design except that two consecutive periods of each treatment, comparing placebo vs. donepezil, were implemented. This strategy is referred as Latin square. With this respect, the Sequential Multiple Assignment Randomized Trials (SMART) will have a roll in the empirical selection of the tested sequence [10].

An effective method to save resources in clinical research is to use adaptive designs. In these trials, the protocol can change according to the information collected during the study. For instance, Simuni [11] planned to change the inclusion criteria to accept patients with hypertension in a trial of isradipine. Another method can be exemplified by the study of Schapira [12] in which sample size was calculated on the basis of an internal pilot study. Furthermore, futility designs allow to stop a trial in case that a prespecified margin of efficacy is not met. An example can be found in the trial of Kieburtz [13] that tested the efficacy of creatinine monohydrate and had to stop early due to futility. These designs can also be used to find the patients that are more responsive to an intervention or the more suitable outcomes to measure and endpoint.

A particular case of adaptive design is the delayed-start trial. These have been developed to provide evidence of an intervention to be neuroprotective. The design of these trials is divided on two phases: one placebo-controlled and the second one open-label. The rationale for this approach is that during the time between both phases the disease will continue progressing. Therefore, at the end of the open label phase there should exist a difference between the patients that received the active treatment since the beginning of the trial [14].

This kind of design requires an interim analysis at the end of phase I. Missing data should be imputed differently in each phase, as covered in the section below. In our review, five trials used this design and two were positive for primary outcome. Three statistical hypotheses must be tested: if there is a treatment difference at the end of phase I and II and if the difference in phase

I is maintained during the second phase. For the last purpose, a noninferiority procedure is recommended to test the slopes [15].

An alternative for this study design the one used by Barker et al. [16] to test the effects of exenatide in disease progression. The design of this trial consisted in comparing the participants of the trial with a matched cohort that did not receive any treatment. This trial was also an example of how drugs approved for other purposes can be used in degenerative diseases when biological plausibility for an effect exists. Also, a washout period could be incorporated at the end of the trial to guarantee that the effect of the treatment persists after the finalization of the trial. Another approach consists in incorporating a different washout period in each treatment groups performed by Devos et al. [17].

Although not used in our sample, innovative designs such as Multiphase Optimization Strategies (MOST) will allow identifying the components of a therapy that yield better results. This may have particular importance in complex interventions such as physical therapy and eHealth [10].

3.2   Phases

The phases in PD clinical trials offer information about the relation of dosage with treatment, safety and efficacy; and these are proportionally related with the number of subjects and the intended objective. More importantly PD trial phases can be frequently challenged by sample size; also, it is important to distinguish the difference among trials involving devices and drugs, as the phase changes dramatically in the number of subjects involved (see Fig. 2).

Table 1, as outlined before summarizes the different types of phases from all the reviewed trials in this chapter. The majority of the studies performed phase II trials with 41% followed by phase III trials (26%). An important challenge for PD trial phase selection is the intervention. Based on our findings, we can infer that most researchers consider the phase as a result to the number of subjects randomized and trial layout (objectives, endpoint, etc.) rather than the research question itself (Table 2). Thus, the selection of subjects (inclusion and exclusion criteria) is directly correlated with the challenges for power and study reliability.

3.2.1   Phase I

In order to limit the potential damages to patients in a clinical trial, the development of new interventional therapies should follow an order in which risks are balanced to benefits. In that sense, phase I trials are devoted to safety and therefore, usually recruit healthy subjects. Seven of the analyzed trials were compatible with the characteristics of phase I trials. The majority of these trials had small sample sizes, were concerned about adverse effects, were performed in a single-center, did not have a randomization and had no control group. Regarding the intervention involved, five included a drug, of which two were gene therapy, and two used procedures.

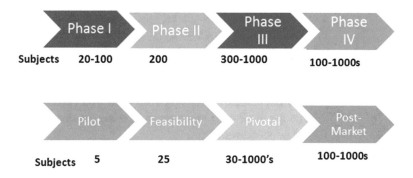

Fig. 2 Schematic illustration differences between drug and device phases

Table 2
Study design selection

Design according to objectives for PD trials

Treatment
1. Disease-modifying: Aimed to modify disease progression—Delayed start—trials
  • Delay disease progression
  • Modify late motor complications
  • Postpone late motor fluctuations

2. Symptomatic relief: Aimed to alleviate symptoms—Placebo controlled studies (2+ study arms)
  • Early stage PD before l-dopa
  • PD with l-dopa
    – Insufficient control of motor symptoms
    – Motor fluctuations
    – Rapid changing motor fluctuations
3. Substitution of neuronal loss

3.2.2 Phase II

The main objective of phase II trials is to select the optimal dose that maximizes the benefits and minimizes the risks. Since the sample sizes of these trials should be small, having homogeneous populations would help to find an effect if it exists. Therefore, eligibility criteria must be more stringent. Likewise, Devos et al. [17] reported a trial of an iron chelating compound. The authors used the UPDRS and iron deposition as outcomes illustrating a general principle of phase II trials: the use of surrogate measurements.

3.2.3 Phases I/II and II/III

For each new technology that gets FDA approval, other nine do not complete the long journey of clinical testing which makes the cost of the products very high. Therefore, several strategies to reduce these costs have been encouraged [18]. One of them consists in accomplishing the objectives of several phases of the development of a technology in one single study. For instance, the phase

I/II trials assess both safety and dose/response in opposition to classical literature that uses the highest dose that does not produce adverse effects of a phase I trial to conduct a phase II trial. Likewise, Palfi [19] performed a trial using several doses of ProSarvin which were escalated if safety was confirmed. Interestingly, besides adverse events, the UPDRS III was also used as a primary endpoint. This can be a solution to gene therapy trials since many of them report low dose as a limitation.

Another similar approach to limit the expenses of clinical research is to use a phase II/III trial. These are conducted in two stages: the first one serves to select the most appropriate intervention between similar candidates and the second is to test clinical efficacy. An important characteristic of these studies is that they can be stopped for futility or efficacy. An important example is the Long-term study one designed by the NET-PD group in which creatine was selected as a molecule for further testing. However, the trial was later stopped due to futility [13].

3.2.4 Comparative Effectiveness Research

Pragmatic Clinical Trials

In contrast to explanatory trials, pragmatic studies are designed to test the effectiveness of an intervention in a real-life situation. Therefore, they are more concerned with external validity which is reflected on the broad inclusion and exclusion criteria [20]. Likewise, the two pragmatic trials in this search [21, 22] selected the population by the presence of the disease and the independence or availability of a caregiver. However, the absence of psychiatric comorbidity was also considered, which make these studies less naturalistic according to the Pragmatic–Explanatory Continuum Indicator Summary (PRECIS). Interestingly, both of these trials tested a procedure as an intervention. However, only Munneke used a cluster randomization approach and took into account the analysis of cost–benefit as a secondary outcome. This last author has consolidated a network of health-care professionals to integrate the treatment of PD in the Netherlands. Although the network has improved patient satisfaction, quality of care and costs, there was not any effect on other health outcomes. The authors suggested that the negative results come from the initial phases of the study which consequence was professionals with not enough expertise [23].

3.3 Setting

Study setting plays an important role in Parkinson's disease trials. There is an increasing trend for support group and home-based programs that make the recruitment easier and therefore facilitate the collaboration of more centers. Initial phases, tend to be single center due to the smaller number of patients required. On the other hand, multicenter studies are conformed by larger number of subjects and therefore require an increased generalizability among PD patients.

Table 3
Trial design, study setting, and phase

Trial design	N	Study setting	Phase							
			I	I/II	II	II/III	III	IV	ND	Total
			7	5	38	7	38	1	4	100
Two parallel arm	48	Single center	2	1	11	1	3	0	1	19
		Multicenter	0	0	11	2	15	0	1	29
										48
Three parallel arm	10	Single center	0	0	2	0	0	0	0	2
		Multicenter	0	0	2	0	6	0	0	8
										10
Four or more parallel arms	3	Single center	0	0	0	0	0	0	0	0
		Multicenter	0	0	0	0	0	0	0	1
										7
Factorial	5	Single center	1	0	1	0	0	0	0	2
		Multicenter	0	1	1	0	2	0	0	9
										2
Crossover	9	Single center	1	1	1	2	0	0	0	0
		Multicenter	0	0	1	1	0	0	0	2
										9
Other[b]	2	Single center	0	0	1	0	0	0	1	2
		Multicenter	0	0	0	0	0	0	0	0
										2
Non RCT[a]	23	Single center	3	1	6	0	4	0	1	15
		Multicenter	1	1	2	1	2	1	0	8
	100									23
										100

[a]Nonrandomized clinical trials, under this category are unblinded studies
[b]Other, this category belongs to studies where authors did not specify study design

The following table (Table 3) uses seven types of trial design to show the different settings and phases used in all included articles for this chapter. Overall, the frequency of multicenter trials is higher than single center trials (54%). More importantly, we can see the distribution of according to phase and center.

4   Study Population

4.1   Recruitment

Recruitment represents an important challenge for Parkinson's disease trials; this is mainly due to the low prevalence of PD in general population. It is highly recommended that the number of patients

Table 4
Selecting participant eligibility

Enrollment of participants PD clinical trials
Severity—stage
Early
• Untreated PD patients
• Early start of dopaminergic therapy
• Delayed initiation of dopaminergic therapy
Advanced
• Motor fluctuations—long-term progression of the illness.
Other
• L-dopa responder (+)
• L-dopa nonresponder (−)

recruited be increased to prevent misclassification (which usually occurs in early-stage PD) [3]. A poor recruitment strategy in Parkinson's trials threatens the power of studies and introduces bias; it is of utmost importance that researchers are aware and know how to address this problem. According to Harris et al. 68% of PD patients would be likely to participate in a clinical trial if one were available in their area [24].

Performing a study in all the members of a target population could be challenging. Hence, it is imperative to choose an accessible population and, through sampling, reach the study population. In order to do so, clinical researchers must establish inclusion and exclusion criteria. We dedicate the following section to analyze the factors that determine the selection of the eligibility criteria.

4.2 Selecting Participant Eligibility

The choice of the aforementioned criteria will depend on the objective of the trial, determined by the stage of development of the intervention. Studies in early phases would need to focus in increasing the safety of the participants by relying on stringent inclusion criteria (Table 4). Likewise, most of phase I trials in our sample excluded patients with comorbidities, including depression and dementia. For providing benefits while decreasing risk, interventions that involved cell or gene therapy included only advanced and untreatable stages of PD. Another important situation is shown by the use of exercise as an intervention. Pilot studies in physical therapy selected patients that were physically capable of performing the activities. This leads us to make clear that assuring the safety of the patients will allow an intervention to be further tested.

Following the previous statement, some trials limit their recruitment in accordance to demographic characteristics. The demographic selection is usually based on age and gender. Although the preferred age range consisted in patients over 30 years old, it varied among all studies. In order to limit confounding symptoms due to aging, many upper limits for age were used. However, for studies in patients with dementia, a higher lower limit was set. Lower limits were established to rule out degenerative diseases other than PD. All studies included both genders in their inclusion criteria. No study described inclusion of minority populations.

In terms of diagnosis, 44% of the studies used the United Kingdom PD Brain Bank criteria. Fifty-six percent of the studies only stated that the diagnosis was made by a neurologist or did not specify clearly. For some trials trying to investigate the effect of a treatment in early stages of a disease required that the diagnosis was recently made. These studies also excluded patients with motor complications that indicate disease progression and atypical motor symptoms, linked to an etiology other than PD.

As the heterogeneity of a population increases, the probability of detecting a true effect diminishes. Thus, trials that seek to proof clinical efficacy will need to reduce the noise in order to increase the signal. This can be done by directing the intervention to a study population that has a higher probability of success. In this regard, patients with a high Hoehn and Yahr stage are seen like more advanced in the natural history of the disease and less likely to respond to therapy. On the other hand, trials interested in evaluating subjects with complications of treatment measure the OFF time as an outcome and consequently establish a minimum amount of baseline OFF time to avoid ceiling effects. In the case of pivotal physical activity trials, the selection of subjects was based on the risk of falling assuming that it would determine the response to treatment. Another challenge to make homogeneous the effect of the intervention and its measurements is the daily variation of medication. Accordingly, 14 of the trials in our sample required the patients to be on stable doses of medication at least during 4 weeks before the start of the study.

In other trials, the hypothesized mechanism of action affected the inclusion criteria. For instance, some drugs with claimed neuroprotective effects would not be used in the late stages of the disease, when the dopamine producing cells can no longer be recovered. Hence, neuroimaging studies are sometimes required to demonstrate the dopaminergic activity of the substantia nigra. Another example is the case of trials in which a minimum response to L-Dopa is required. In our sample, three trials required an in-phase to certify the response to L-Dopa while other seven studies did not specify their approach in this respect but mentioned it as a prerequisite.

Many trials wouldn't be feasible without the full cooperation and involvement of the study participants. Likewise, studies that require the fulfillment of diaries exclude participants who are

Table 5
General eligibility criteria

Eligibility criteria		N
Characteristics	18	7
	20–39	29
	40–50	13
Age	50–70	5
	NA	46
	Range	35.45 ± 78.75
Gender	Both	100
Hoehn/Yahr (severity scale)	Yes	41
	No	58
Response to L-dopa	Yes	14
	No	86
Years from initial diagnosis	1–1.5	2
	4–5	10
	NA	88
On/Off	OFF	32
	ON	68

unable of doing so. However, this approach precludes generalization to a more extent population. Moreover, caregivers are often required (especially on trials recruiting for severe stage of PD) which highly relates to adherence and subject follow-up.

While choosing a narrow population has the advantage of increasing the signal of the effect, it comes with the drawback of limited external generalizability. Thus, clinical trials in PD that attempt to demonstrate effectiveness must recruit a sample with broader characteristics instead of choosing a population based on age or other baseline characteristics (Table 5).

Patient selection in Parkinson's disease is complex mainly due to underlying conditions that need to be evaluated along with primary endpoint. Even though most of the evaluations represent the motor manifestations of the disease, are a wide variety of studies assess non-motor symptoms (NMS) or even surrogate endpoints like sleep disturbance, psychosis, dementia, and pain. Even if a set of diagnostic criteria may apply for certain patients, they may not be the same as those ideal for people with PD. Furthermore, comorbidity adds a level of complexity to the endeavors of clinical trials in these patients. Hence, defining more accurate diagnostic criteria for NMS in PD is required to better select a study population.

Table 6
Eligibility criteria

Category	Number of studies
Demographics ($N = 54$)	
Age	See Table 1
Gender	100
Diagnosis ($N = 100$)	
UKPDBB	46
Not specified	56
Treatment ($N = 49$)	
Response to L-Dopa	14
Years treated with L-Dopa	2
Without recent changes	32
No treatment	3
Disability and complications ($N = 26$)	
Disability symptoms off time	3
UPDRS	6
L-Dopa induced complications	4
Inadequate medical response	9
Not described	4
Severity ($N = 48$)	
Hoehn and Yahr scale I–IV	8
Hoehn and Yahr scale II–IV	6
Hoehn and Yahr scale II–III	3
Not described	31
Physical capacity ($N = 15$)	
Ability to stand and walk	11
Independent	4
Care ($N = 3$)	
Caregiver required	3
Non-motor symptoms ($N = 9$)	
Depression	2
MMSE 10–24	3
Visual or auditory hallucinations	4
Symptoms ($N = 7$)	
Gait disorders	3
Symptom asymmetry	2
Recent fall or risk of falling	2

4.3   Inclusion Criteria   The inclusion criterion for PD is directly correlated to the aim of the study. Most clinical trials in Parkinson's disease use the Hoehn–Yahr scale to categorize severity (Table 6). Nevertheless, trials should also specify severity of according to a particular endpoint, for instance functional impairment or motor fluctuations. The researcher should consider that the more homogenous a sample is the less likely that these results can be generalized to a heterogeneous population.

4.4 Exclusion Criteria

Exclusion criteria, usually indicates those factors that would preclude participants from being part of the study. These factors are usually potential confounders to the results; therefore, the proper identification of these is required. Commonly reported exclusion criteria in our results was: Mini-mental State Examination (MMSE-25) score lower than 24 (usually indicates, cognitive impairment) alongside with debilitating conditions (such as, important physical impairment or serious medical conditions); concomitant neurological conditions; severe dyskinesia "on-off" period and motor fluctuations. Depending on the type of intervention, the exclusion criteria may be modified. We realized that most device trials particularly DBS, exclude subjects that do not complement fitness for surgery.

5 Interventions

5.1 Drugs

Most of the existing therapy for PD is designed to engage the dopaminergic network. However, with the decrease in the population of these cells, dopamine receptor agonists become less effective. On the other side, the dopaminergic system is not the only one that undergoes degeneration. For those reasons, new investigational drugs have focused on nondopaminergic agents. The motor symptoms observed in PD patients are the result of the aforementioned degeneration. Nonetheless, recent studies have shown great interest in how to treat NMS. Seven of the trials studied where designed for this purpose. The clinical conditions of interest were cognitive impairment, hallucinations, pain, and depression.

The objective of some drug trials in our search was to address the motor complications of levodopa therapy as to compare across different drugs (Table 7).

During the early phases of drug development, pharmacologic studies must be conducted before any efficacy measure is perused. In PD, the effects of a drug are highly dependent on its biodisponibility. To address this issue, several trials have tested innovative drug delivery methods. For example, three trials looked for the effectiveness of the intestinal delivery of a carbidopa gel. In these cases, the challenges in the design include the procedure that is needed in order to reach the intestine. Pharmacokinetics must not be the only concern of early stage PD trials. Some of our studies assessed the mechanism of dopaminergic action in the central nervous system.

Table 7
Interventions

Category	Number of studies
Drug ($N = 51$)	
Motor symptoms	36
Non motor symptoms	8
Devices ($N = 20$)	
Deep brain stimulation (DBS)	15
Non-invasive brain stimulation (NIBS)	5
Procedure ($N = 22$)	
Exercise/training	16
Telemedicine	2
Virtual reality	2
Cognitive behavioral training	2
Cell therapy ($N = 3$)	
Dopamine producing cells	1
Bone marrow cells	1
Retinal	1
Gene therapy ($N = 4$)	
GAD	1
Prosarvin	1
L-aminoaciddescarboxilase	1
Nertuin	1

5.2 Devices

Device trials are an emerging trend in the medical field, over the last few decades deep brain stimulation has become the most notable device intervention in Parkinson's disease patients. In fact, the advent effect of classic deep brain stimulation still remains a mainstay for patients who remain unresponsive to medical treatment and present a high rate of complications. Moreover, other devices such as noninvasive deep brain stimulation also represent of simplistic benefit on motor symptoms in patients [25].

A total of 21 studies used a device as an intervention (Graph 2). Moreover, 15 were related to deep brain stimulation (DBS), and five to noninvasive brain stimulation (NIBS). The preferred site of stimulation in 12 of the DBS trials was the subthalamic nucleus (STN). However, the pedunculopontine nucleus and the pallidal nucleus were also stimulated locations. It is worth it to mention that ten studies were aimed to manage certain aspect of the disease, whereas the remaining ten were designed to investigate the physiological consequences. In the following section, we discuss the challenges of designing trials for the management of PD when devices are involved.

Complications and medical treatment failure convey towards surgical approaches. Therefore, deep brain stimulation surgery has

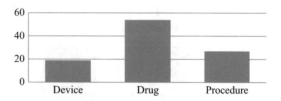

Graph 2 Type of intervention

become a standard procedure among PD patients. However, patients may benefit from an early intervention, this can be exemplified by the EARLYSTIM-study [26] that demonstrates the multiple challenges that surgical interventions bring to clinical trials. Firstly, sham surgery involves ethical considerations and was only used in one of the analyzed trials. The others compared either two stimulation sites or stimulation versus the best medical treatment. The use of guidelines for defining the best medical treatment was one of the innovations of the EARLYSTIM trial. In future research, the caregiver can be taken into account for this definition.

The lack of standardization of surgical procedures is also a barrier for DBS development. For achieving large sample sizes, multicenter trails are often required. Under these circumstances, the use of imaging techniques for planning the surgery, the verification of the electrode placement, and the use of a specified anesthesia are some of the parameters that can be agreed upon between different centers. Although the precise placing of the electrodes remains debatable, the avoidance of relevant structures (sulcus, vasculature) is a common feature between these assays. A study by Witts [27] raised the concern of negative neuropsychological outcomes associated with the placement of leads that traversed the internal capsule. Moreover, the investigation of Daniels [28] also reports executive dysfunction after ST-D Another innovative feature of EARLYSTIM was its monitoring of suicide. These strategies for surveillance of negative neuropsychological outcomes and its relationship with the lead trajectory could impact the practice of DBS.

Delayed-start trial methodology can be incorporated in the research of DBS and other neurosurgical approaches in order to find out their effectiveness for slowing the progression of the disease.

5.3 Procedures

Physiotherapy is one of the most conventional additional therapies in Parkinson's disease (PD) [29]. Rehabilitation programs for postural stability are most effective if they incorporate balance practice and adjustment to environmental demands. A common problem in these types of trials is high attrition, since it requires active participation in participants with compromised independence in daily activities (Graphs 3 and 4).

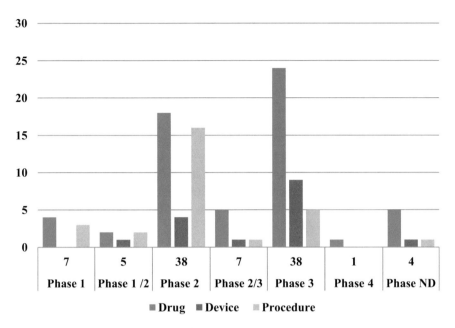

Graph 3 Type of intervention related to study phase

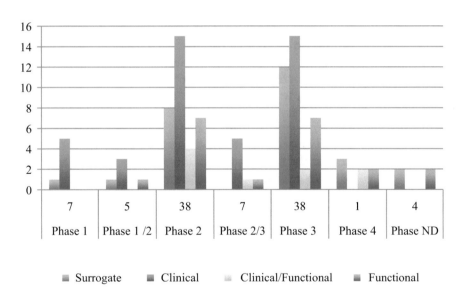

Graph 4 Type of outcome related to study phase

In our review, common procedures included: balance training programs, speech recovery and surgeries along with device implantation, physical training, cognitive and behavioral procedures, cuing, virtual reality, telemedicine, and educational programs. A total of 21 studies used a procedure, as the intervention.

Similarly to trials that involve surgery, trials of procedures face a big challenge regarding the delivery of standardize procedures across subjects. In order to allow for reliability and reproducibility, researchers should ensure that the procedure is delivered in the same way across participants if the trial is of pivotal nature.

In trials that use physical therapy, the preferred outcome scale was to use balance measures as primary outcomes, such as; Berg balance test, limits of stability test, and risk of falls. Despite the fact that most studies were of exploratory nature they used clinical outcomes to evaluate such endpoints. In addition, exploratory clinical trials in PD may beneficiate from the use of surrogate outcomes. Moreover, studies could be powered to evaluate nonmotor symptom response. The use of other functional outcomes like cortical excitability, brain connectivity, and physical fitness could give insights about the mechanisms of exercise action [30].

Selecting an appropriate comparator is a challenge in trials that evaluate exercise. In our search, patients in the control group receive one training program that is not similar to the active one. For example, stretching exercises instead of aerobic exercise.

Other limitations for trial conduct are separating the effects of exercise from other properties that the intervention may have. For example, physical activity usually requires social interaction, two of the included trials used a community-level approach and therefore the effects in the individuals can be mediated by the strengthening of the community. In this last case cluster randomization may be a powerful approach since it allows for controlling for factors that the members of a group of people have in common.

The preferred design for exercise trials was parallel two-arm study. Interestingly, three out of ten studies included three intervention arms. This illustrates the need of several comparators to dissect the source of the effect. No trial used a crossover design despite the fact that it would be appropriate to ease the variability of this intervention's effect between the subjects. A meta-analysis by [31] concluded that although physical training can have significant positive effects on motor outcomes of PD, clinical trials must be planned to have longer durations. Yet only two out of the ten exercise trials had a duration longer than 6 months.

Although the short time duration of trials that test exercise allows evaluating its immediate effects, the long-term consequences of the intervention remain to be clarified as well as its precise mechanism of action.

6   Study Duration

6.1   Study Duration

Studies must be long enough for the intervention to show an effect. However, trials that have a long duration may experience higher attrition rates. Therefore, clinical researchers must select the duration of the trial on the basis of the nature of the intervention and the outcome. For instance, trials which main objective is to

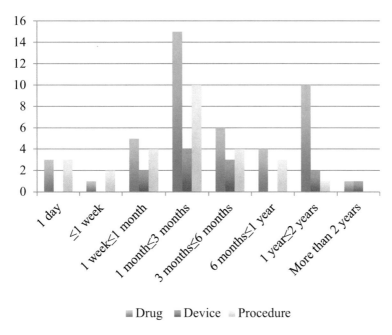

Drug   Device   Procedure

Graph 5 Participant duration

show a neuroprotective effect must last longer than those interested only in symptomatic relief. In our analysis, the duration of the clinical trials studied varied from 1 day to 7 years (Graph 5). The most common duration corresponded to the period of 1–3 months. The trials that engaged with long durations were interested in verifying the long-term effect of an intervention. For example, the trails that envisaged for cellular regeneration of substantia nigra searched for a dopamine marker using PET imaging to assure that living and metabolically active cells were still present across long periods of time. In the trails that lasted 1 day, the experimental condition was the states ON/OFF in relation to dopaminergic treatment or DBS. On the other hand, most of the trials that tested some form of procedure, including physical therapy, had lasted from 1–3 months.

6.2 Time-points

Another research issue is to define the number of time-points and the spacing between them. For instance, trials with more time-points will require more effort from the personnel involved. On the other hand, studies that consider fewer time-points are more likely to be less reliable. In our search, most of the studies analyzed used only 3–6 time-points (Graph 6). Overall, the study time points range was 2–72.

Many of the trials that chose only two time-points were performed during the experimental day and analyzed some aspect of the physiology of PD. In opposition, the trials with the highest number of time-points had a special focus on safety. Interestingly,

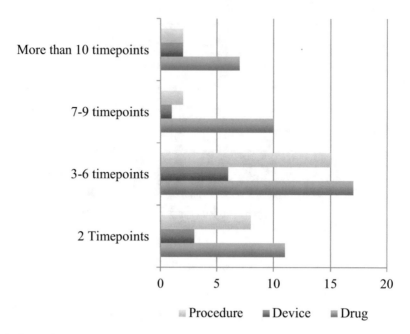

Graph 6 Study time points

20% of the trials with more than 10 time-points evaluated nonmotor symptoms like sleep and depression. For example, trials that include a titration phase like the one from Trenkwalder [32] have several time-points in a brief period.

This is also the case in early phases or drug development when the main interest is on pharmacokinetics. Another important feature for selecting the number of time-points is the outcome. For instance, Watts et al. [33] used the time-to-dyskinesia as an outcome. In these circumstances, the precision of the measurement would directly depend on the number of times that it is assessed. Some kinds of interventions require a higher number of time-points including the application of noninvasive brain stimulation and procedures like physical therapy to verify that the intervention is having an effect.

7   Outcomes

7.1   Primary Outcome

Investigators have to define the role of each endpoint and the primary intention that it plays in the clinical trial. In experimental medicine the dependent variable is represented by the outcome. The UPDRS has been considered to be the gold standard motor assessment for patients with Parkinson's disease, and it is the most widely used instrument for its clinical trials.

A researcher commonly faces the challenge of selecting one primary outcome out of a numerous amount of possibilities. A scientist should consider the stage of development of the intervention because this will determine the objective, the number of subjects tested and the design of the trials. To evaluate the possible outcome measurements, the researcher must bear in mind its validity, reliability, appropriateness and reproducibility. In our sample, we identified 47 different primary outcomes. These corresponded to the following endpoints: motor functions, nonmotor functions, quality of life, physiology, safety, feasibility, complications, and disability. We briefly revise each one of them.

Traditionally, the evaluation of PD has been based in motor symptoms. In this respect the United Parkinson's Disease Rating Scale was the primary outcome of 31 trials in our sample, making it the most used scale across all intervention modalities. This scale is divided on six parts that include motor function and quality of life. Nineteen trials relied only in the third part of this evaluation that corresponds to the physical examination. The use of this approach may avoid the subjective component of the evaluation. On the other hand, ten trials used its entire presentation. Furthermore, two trials selected the second and the fourth part of the UPDRS to evaluate quality of life or complications, respectively. Likewise, a large number of trials used the UPDRS V as an indicator of severity and used it as a covariate for the analysis or for the randomization. It is noteworthy that the UPDRS was not first developed using clinimetric methods and instead counted on a board of experts for its creation. Therefore, the Movement Disorder Society (MDS) revised it and in 2008 recommended several changes in order to make the scale more comprehensive [34]. Conversely, none of the trials analyzed used the MDS-UPDRS as a primary outcome.

Clinical trials that involve a procedure, including physical therapy, use other scales besides the UPDRS. For instance, to evaluate gait, several measurements can be utilized; gait speed, stand-up and walk test and walk time where some of the measurements that we found in our article selection. One advantage of using this kind of scales is that, unlike UPDRS, the result is less dependent on the rater. This makes it more objective and limits the influence of improper blinding. Furthermore, using an interval outcome such as distances or speeds allow a more diverse range of statistical analysis. Still, trying to resume complex traits like movements with a single value has its own limitations. For this reason, a trial preferred to perform kinematic analysis using accelerometers and gyroscopes to capture the complexity of motion. Another measurement that received attention from the three studies was the Berg Balance Scale.

The UPDRS and other scales can be measured while the patient is under the effects of therapy or when the effect of dopaminergic agents diminishes which is referred to as the OFF and ON periods. An investigator must decide which the most informative moment to perform a measurement is. Trials that select patients with recently diagnosed disease will not have a significant OFF period. Contrastingly, studies which objective is to decrease the degree of disability will be more interested on performing the assessments during the OFF period. Another very important aspect to take into account is to perform the measurements in the same period of the day in order to avoid the fluctuations of the therapeutical effects.

An intervention in an advanced phase of development may benefit from showing an increase in quality of life (QoL) outcomes. Five of the trials analyzed used QoL; three of them were procedures, while the other two were devices. The most frequently used outcome for analyzing QoL was the Parkinson's disease Questionnaire (PDQ-39) that was utilized in two trials with a device as an intervention. Other measurements applied when the interventions were procedures (e.g., physical therapy) were the HRQOL and the UPDRS-II.

Some trials are performed in the population with advanced features of PD like unresponsiveness to levodopa and abnormal movements. These trials are interested in other measures besides the UPDRS. Ten trials used the total OFF time as a primary endpoint. Nevertheless, other outcomes for measuring disability were performed.

Although a reliable marker for PD has not been found, contemporary clinical trials included in our sample use neuroimaging techniques to assess the potential benefits of a treatment. For instance Schapira [35] used SPECT to tract striatal dopamine. Another good example of the use of a surrogate outcome for neuroscientific investigations is the motor evoked potential (MEP) reported in the trials that applied noninvasive brain stimulation (NIBS). This has been proposed as a measurement of cortical plasticity and is used to better understand the mechanism of action of brain stimulation. Another application of this assessment is to unveil the pathogenesis of PD. For instance, Eggers [36] used theta burst stimulation to show and impaired motor plasticity in PD patients.

7.2 Secondary Outcomes

Secondary outcomes are useful to generate future hypotheses. Therefore, the proper identification and selection of secondary outcome will dictate the exploratory ends of a study. Additionally, secondary outcomes are supportive for primary outcomes in Parkinson's disease. Moreover, secondary outcomes underlie in the nonmotor symptoms associated with the clinical syndrome, and relate to quality of life improvement.

In the studies analyzed, an important proportion of trials used patient diaries to record the ON and OFF periods. Most of these trials used dyskinesia as a primary outcome. To make this measurement reliable, the researchers must ensure that the patients know about its appropriate usage. In fact, some researchers excluded from the study the patients that were incapable of successfully filling the diaries. Another purpose of the diaries was to record the quality and the amount of sleep. This is important in the studies that chose the OFF time as a primary outcome because the amount of hours spent in this period depends on the time that the patient is awake. Using patient diaries has also the advantage of limiting recall bias since the outcome is not only based in a single measure.

Before an intervention is translated to the clinical practice other aspects beside effectiveness should be accounted. These decisions are also supported by social and economic data that can be gathered through clinical research. For instance, trials that involved a procedure commonly described the additional benefits of an intervention. Examples of this outcomes are cost, caregiver burden and user satisfaction.

Even if the purpose of the trial is to establish efficacy or effectiveness, safety data should be collected and displayed. For instance, concerns about cognition in trials with DBS have made the measurements of this outcomes necessary practice.

Non-motor symptoms (NMS) of PD are beginning to be considered in the design of clinical trials. The NMS included in our search were: fatigue, insomnia, depression, sleep, dementia and psychosis. We recommend using NMS outcomes as secondary measures in trials designed to find a difference in motor outcomes and to keep QoL. A good example of this practice is the Movement Disorder Society-(UDPRS). To advance the understanding of PD, there is also a need to better characterize NMS. For this purpose, clinical trial could plan post-hoc analysis of clusters of patients exhibiting different NMS [24].

The results from the trials included for gene therapy (GT) delivery are negative for the primary outcome. Enrolling patients with recent diagnosis could be beneficial. Some of them used F-DOPA positron emission tomography (PET) as a biomarker for nigrostriatal dopamine pathways [37]. The development of cheaper and better biomarkers such as I-FP-CIT SPECT will also boost the field. Other improvement on the field will include: the use of higher doses and long term delivery. The possibility is also open for combining other treatment strategies with GT [38].

8   Sample Size

The sample size represents an essential part of the study and its importance relies on the ability to detect significant differences. In all randomized clinical trials the number of subjects should always

Table 8
Sample size calculation

Category	
Alpha	Proportion of articles
0.05	99%
Not described	1%
Power	
≤80%	53%
>80%	32%
Not described	15%
Methods for effect size estimation	
MCID	10%
Pilot	6%
Previous study	41%
Mixed	4%
Not described	39%
Estimated dropout rate	
<10%	6%
10–20%	18%
≥20%	12%
Unspecified	6%
Not taken into account	58%

be large enough to provide a reliable response to the research questions addressed [39]. Therefore, considering a small sample size should always be carefully planned, as the power tends to be low thus the desired effect may not be successful.

Sample size calculation represents a pivotal part of clinical trial design. Trials with a small sample size could be underpowered and therefore lead to failure to reject the null hypothesis. On the other side, studies with big sample sizes would expose more subjects to unnecessary risks. In order to replicate the calculation of a sample size of a given trial one needs to know the alpha, the power, and the expected treatment effect. While alpha and power are defined by the researcher, the mean intervention effect needs to be established by choosing a minimal clinically significant (MCS) or use information from previous pilot studies, meta-analysis of RCT. In addition, accounting the expected amount of dropout rate is also important.

From the articles included, only 56% reported any sample size calculation. Only 20% included all the elements that enable proper replication of sample size calculation (Table 8). The preferred method for estimating the minimal clinically important difference was the use of previous literature. In general, increasing the sample needed based on dropout rate for purposes of sample size calculation should be made.

In some trials, the selection of more than one primary outcome was considered. In such cases, the estimation of the treatment effect must include both measurements. The studies that required this correction, one used the Bonferroni method and two implemented a stepwise gate keeping procedures [40]. In the first case the adjustment for multiple comparisons leads to a loose of power while the second avoids this expense. In addition, another important factor to consider when performing a sample size calculation is to choose between a one tailed and a two-tailed test.

Just one trial used an adaptive design for a sample size recalculation. This means that a prespecified time-point the authors check the assumptions that were made for the initial sample size calculations and make adjustments based on this information; however, the process needs to be prespecified and only nuisance parameters should be checked. Another trial recalculated the sample size based in a power analysis but did it in an unspecified approach [41]. In the former case, the recalculation yielded a sample size much bigger than anticipated. No previous trial had used this primary outcome to estimate the effect size of the intervention tested.

9   Analysis and Stoppage Criteria

Recommendations on trial design changes or early termination, could be detrimental to study credibility if the stopping rule is for efficacy. Only four of the analyzed trials had to stop prematurely. There are few examples to illustrate stoppage criteria in our findings. For instance, in one case, the shelf life of the active substance reached its limit [42]. In the trial of Trenkwalder [32] the manufacturing process changed in a way that compromised blinding. Both of these studies were phase III which highlights the importance of taking into account factors associated with the properties of substances for the planning of a trial. In a phase 4 trial the sponsor decided to call one trial off because of safety concerns lifted by animal models [43]. The last trial was stopped [41] because of the results of sample size recalculation. The latter is discussed in the Sect. 8.

An interim analysis is a prespecified time-point in which the data is analyzed for matters of security and/or efficiency. If it the results demonstrate that the treatment is effective, then the trial may need to stop. However, the cost of performing an interim analysis is the decrease of power associated with a higher likelihood to reject the null hypothesis. Three studies of our sample conducted an interim analysis. They were all conducted in a period of time longer than 6 months. Two of them involved DBS and

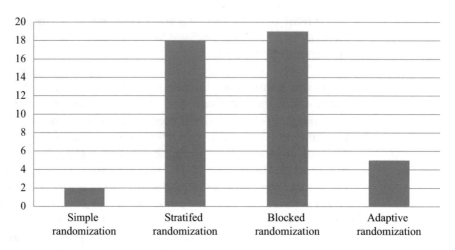

Graph 7 Type of randomization

therefore, safety was a main concern [44, 45]. The other one studied an intervention for which effect size is unclear and therefore was performed to verify the assumptions of sample size calculation [46]. Benninger [41] performed an interim analysis on the basis of subjective lack of improvement.

10   Randomization

10.1   General Concepts

The randomization process is an important highlight of randomized controlled clinical trials; it provides equality between the assigned intervention and control. Generally, inter-subject variability must be balanced throughout the randomization process. The randomization assignment of participants into different groups should be equal and completely free of personal judgment to avoid selection bias.

There are several methods for making random treatment assignments: simple randomization, permuted block randomization, stratified randomization and minimization/adaptive randomization. The selection of the randomization method should be related to the identification of relevant population characteristics (e.g., Hoehn–Yahr scores (severity), disease duration, age, and institution (multicenter trials)) the objective is to distribute the population into homogenous subsets and to avoid imbalances on baseline characteristics of treatment groups. Stratification results useful in small trials in which it can turn aside significant imbalances on prognostic factors. It will confer adequate balance (on the stratified factors) and probably slightly more statistical power and precision. On the other hand, an adaptive randomization design allows the randomization to be modified while the trial is ongoing.

When a trial has no further stratification in a blocked design, the randomization method selected should have a fairly large block size to reduce predictability.

In some of the reviewed studies, the sample size was too small to allow stratification by more than two or three variables. We were able to determine the factors that usually led to the stratification (center, severity scales, site, gender, concomitant drug use, wave of participant entry, etc.). The stratification of these factors is generally related to the prognosis of the primary outcome. The total studies that used stratified randomization was 18% (Graph 7), the majority of these trials where categorized by site/center (11%) followed by age and severity (Hoehn–Yahr/UPDRS scale) with (4%).

On the other hand, adaptive randomization (5%) consisted mainly of minimization technique. Minimization is a dynamic allocation method that aims to minimize imbalance among prognostic factors (14) the aim is to maximize efficiency in moderately sized trials containing 10–20 independent covariates. Thirty-nine percent of trials did not mentioned type of randomization.

10.2 Sequence Generation and Implementation

Randomization sequences are usually generated throughout centralized computer codes. A study needs to define who generated the allocation sequence, who enrolled participants, and who assigned participants. The personnel involved in generating the allocation mechanism should be different from the ones enrolling and assigning treatment.

10.3 Allocation Concealment

Allocation concealment prevents researchers from (unconsciously or otherwise) influencing which participants are assigned to a given intervention group [47]. Researchers should devote appropriate resources to doing the randomization generation properly and reporting their methods clearly.

11 Blinding

Blinding indicates that knowledge of the intervention assignment is hidden from participants, trial investigators or assessors [48]. Together with randomization, blinding is a key aspect to avoid bias in clinical research. Proper blinding prevents participants and assessors of knowing treatment allocation status. Therefore, researchers should devote interest in the methodological aspects of blinding.

According to the CONSORT guidelines; blinding should go further than just explaining single or double blind, but also detailing who was blinded and how. Investigators unaware of treatment assignment are more likely to remain impartial to the judgment of outcome measurement, thus leading to a greater credibility of results. In contrast, awareness of the treatment offered in any end (participant or investigator) leads to observer bias.

Appropriate blinding relies upon the nature of the treatment offered. In other words, due to the nature of certain interventions (such as, surgery or physical therapy) it may become unfeasible to blind. Under these circumstances, researchers should consider performing single-blind or nonblinded studies. Nonetheless, if blinding is not possible investigators can take other measures to reduce the risk of bias.

11.1 Blinding in PD Trials

Inferred from our findings, there is a moderate prevalence of open label trials in PD research (28%) most of which resulted as procedure or device trials (16%) (See Graph 1 of Chap. 4). More interestingly, some open label studies illustrate how sometimes mimicking an intervention is not possible. A good example for this is a study by Mueller et al. [49] where the main outcome was blood levels of homocysteine (surrogate outcome), after treating patients with levodopa/carbidopa intestinal gel. Complex procedures like this imply a series of challenges for the researcher to address properly, without unblinding in a masked setting.

Outcomes figure an important part of blinding, because they are prone to subjective interpretation and consequently more likely to have bias. Most outcomes in PD clinical trials are considered to be subjective outcomes (particularly, clinical outcomes, such as the UPDRS score) for that reason more blinding is needed for these types of outcomes. On the other hand, surrogate outcomes tend to be more objective and less likely to bias. Blinding becomes less important to reduce observer bias as the outcomes become more objective [48]. Another point from Mueller et al. is that knowledge of the intervention in this case does not affect measurement on the primary outcome—however this concept should be carried out with caution; as surrogate endpoint does not necessarily mean objectivity in the outcome.

The predominant trend of blinding in our search was double-blinded trials (49%). As expected, many of the drug trials where double-blinded studies (39%). Thus, the protection against bias is best achieved with this type of blinding. Consequently, placebo-controlled trials in our search were the only ones to report blinding assessment.

In contrast, single-blinded studies were less prevalent than double and unblinded studies in our search (23%). For the majority of these trials (Procedure 15%), participants and therapists were not blind given the nature of the treatment, but independent raters where. Participants were instructed not to discuss the nature of their procedure with the outcome assessors.

11.2 Blinding Techniques

Recent studies with DBS prove that many treatment arms in PD trials involve surgery; this trend has led to important discoveries in blinding techniques. Having said that, many researchers outsmart some of the challenges involved in surgery trials. A good example for this is a study from Follet et al. [50] where the authors compared bilateral deep brain stimulation to best medical therapy in patients with advanced Parkinson's disease; patients were required to wear caps to blind the evaluators to the presence or absence of brain surgery scars.

Another example is a study by Schuepbach, WMM et al. [51] where the author performed the blinded assessments based on preoperative and postoperative standardized video recordings obtained at baseline and at 24 months. Interestingly, other authors reported a similar blinding technique.

Furthermore, keeping blinding throughout clinical trials is challenging. Interventions that relate to a high rate of side effects (therefore, allow a possible detection of the intervention) tend to be more likely to unravel blinding; researchers can address these challenges by following a strict protocol (refer to section 12). Unblinding can occur due to several reasons, the most important are related to presence of side effects or an emergency situation. In our review, only two authors revealed the rate of unblinding [12, 52].

11.3 Blinding Assessment

As a result from our findings, the majority of the trials did not mention blinding assessment (73%). Nevertheless, authors should consider whether the participants where truly blinded and relate how successful the blinding was (especially, if it is a placebo-controlled trial). The following outlined box, shows a good example how a PD trial reported blinding assessment.

12 Adherence

12.1 Adherence in PD Trials

Adherence to study termination is crucial. Previous studies on Parkinson's disease clinical trials, suggest that PD participants are adherent to study medication and dropout rates are low [3]. Nevertheless, because of the natural evolution of this clinical syndrome, it is complicated for many patients with PD to remain untreated or remain on a stable dose of dopaminergic therapy for the duration of the trial follow-up.

The general percentage of adherence rate in the PD trials was of 86.92% (SD 9.85). Interestingly the trials with higher rate of adherence where those who had a lower rate on the follow-up visits; followed by trials that involved physical therapy as a primary endpoint. Thirty-eight percent of the trials did not report the adherence rate. Therefore, the estimation is based in trials that included the adherence rate (see Graph 8).

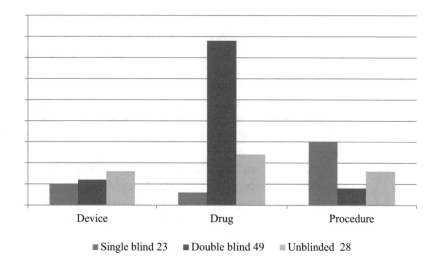

Device   Drug   Procedure

■ Single blind 23   ■ Double blind 49   ■ Unblinded 28

Graph 8 Blinding and type of outcome

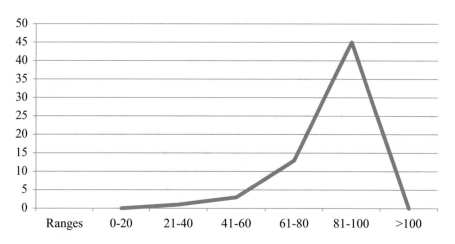

Ranges   0-20   21-40   41-60   61-80   81-100   >100

Graph 9 Adherence rate

***12.2 Methods
to Improve Adherence***

Lack of adherence in a clinical trial may jeopardize the internal validity of the study; therefore, ensuring participant adherence is imperative. In our review (see Graph 9), different methods where directed to promote participant adherence: dose titration (11%), dosage flexibility (3%), participant compensation (3%), and others 5% (home based programs, open label enrollment, compliance programs). The majority of the PD trials did not mention methods to improve adherence (72%) (Graph 10).

13   Statistical Analysis

13.1   ITT/
Per-Protocol

The intention-to-treat principle is defined as the analysis of all subjects randomized despite of the treatment received or their compliance to the study. This concept has a central role in RCTs because it protects the benefits of randomization. In other words, this principle normally safeguards against potential bias—helping to preserve the original balance of random assignment [54].

According to the CONSORT guidelines studies should state a clear description of exactly who was included in each analysis. More importantly, authors should also include a description of those who withdrew from the study before and after allocation and the reason why. This checkpoint is important mainly because it illustrates a flowchart that allows the reader to understand the breakdown of patient's compliance to the study. In our review, 64% of the trials used the ITT analysis, from which many performed a modified ITT analysis. The majority of these studies predefined their full analysis set as the patients that completed at least one treatment and one baseline measurement.

> … On the third day after surgery and at all subsequent visits, patients were questioned for opinions about treatment assignments. In the questionnaire about masking done 3 days after surgery, nine sham-treated patients thought their treatment was AAV2-GAD, seven stated they did not know, and five guessed correctly.—LeWitt, PA et al. [53]

For instance, during trials using cell and gene therapy as intervention, analyses included only the patients with an imaging study that showed the correct placement of the probe. Similarly, some authors decided to exclude patients from the primary analysis on the basis of their initial treatment response. Consequently, they tried to follow an attempt to find a study population in which the effect of a drug can be most readily demonstrated, also known as enrichment design. A different approach from Schapira et al. [35] was to include both ITT and modified ITT analysis depending on the pursued objective. If safety is the main concern, all the subjects that received the treatment should be analyzed as such.

However, if efficacy is the matter, then ITT is the gold standard. ITT and per-protocol analyses are advisable in noninferiority trials. This is substantial in PD trials, especially in delayed-start trials because authors need to test the hypothesis of noninferiority during the second stage [15].

Many researchers believe that ITT analysis may result in an underestimation of the treatment effect, therefore is commonly considered a conservative approach for data analysis. From this

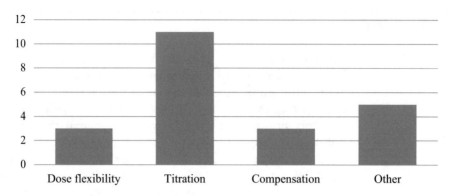

Graph 10 Methods to improve adherence

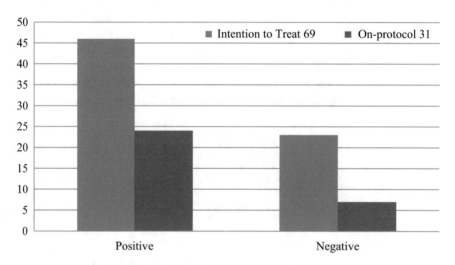

Graph 11 Negative and positive study correlation

viewpoint, it is important to mention that this approach has a greater potential to explore the robustness of the conclusions [39]. The ITT is regarded as a complete trial strategy for design, conduct, and analysis rather than analysis alone.

A key aspect for ITT analysis is to handle missing data (see Sect. 12.1) and perform a sensitivity analysis. Sensitivity analyses are useful to explore the effect of departures from the assumption made in the main analysis. On the other hand, the minority of the studies (36%) performed a completer analysis (on-protocol/on-treatment analysis). For instance, some trials that used physical therapy as an intervention replaced the patients that dropped out. In this case, using the data from both groups dropped-out subjects and replaced is necessary. Handling missing data in complete analysis should be managed cautiously, as failing to report properly may lead to overestimate the treatment effect. Having this said, we can also compare the number of ITT/per protocol trials with positive and negative outcomes (see Graph 11).

13.2 Hypothesis Testing

The goal of inferential statistics is to use measurements from a sample to make conclusions about a population. In order to do so, researchers need to perform a hypothesis testing. To select the best test, an investigator has to take into account the type of dependent and independent variable and the distribution of the data. A normally distributed sample can be analyzed with parametric tests whilst a not normally distributed sample requires the use of nonparametric methods or some degree of data modification. When the sample size is bigger than 100 subjects, the central limit theorem can apply. Surprisingly, 28 trials used parametric test despite a small sample. Many of them did not thoughtfully justify their choice, including the verification of the assumptions. On the other hand, 14 trials with a big sample size used nonparametric tests. Some of this trend can be explained because of the categorization of the outcomes into clinically meaningful ranges.

Because the UPDRS was the preferred primary outcome, we describe the methods that were used for analyzing it. As a continuous outcome, the UPDRS allows for a variety of statistical tests. For instance, 20 trials preferred to use parametrical tests as compared to only six that selected nonparametric analysis for hypothesis testing. Interestingly, phase I trials that did not had a comparator chose ANOVA and t-test for evaluating the outcome. This was specially the case of interventions related to gene therapy that had small sample sizes but long follow-up periods. The most commonly used parametric test was ANOVA ($n = 8$), followed by ANCOVA ($n = 6$), linear regression ($n = 4$) and t-test ($n = 3$). With respect to nonparametric tests, Kruskal–Wallis was used in two occasions while Mann–Whitney, Wilcoxon, Friedman, and permutation tests were used one time each one. In two cases, the statistical method for analyzing the UPDRS was not mentioned and in one only descriptive analysis was performed.

For statistical tests to yield robust results, a number of assumptions must be tested. For instance, trials with sample sizes of <100 should verify normality or transform the value of the outcome in the case that this requisite is not satisfied. Other assumptions like the homogeneity of variances must also be tested. Moreover, it is important to acknowledge that nonparametric tests must be applied if the assumptions necessary for parametric tests are not met.

Traditionally, researchers choose a priori a statistical method according to the nature of the outcome and the independent variable. If a trial evaluates only motor outcomes, its results can be misleading because other important endpoints can also be affected. This concern has motivated the development of other strategies for analyzing clinical trials in PD. Huang et al. used a multiple regression model termed global statistic test (GST) in order to "allow the assessment of clinically relevant outcomes rather than the artificial selection of a single primary outcome". In our article selection, only one trial by Murat et al. [55] chose not to select a primary outcome. In this case, authors used ANOVA for performing the analysis.

Another option for data analysis consists in determining what a clinical significant improvement means and then categorizing the data. This strategy provides the advantage of making the data easier to interpret from a reader's perspective and more meaningful from a clinical point of view. Recently, Schulman et al. [56] determined a cutoff for the clinical significant improvement in the UPDRS. The methods to assess clinical improvement include the anchor and the distribution-based approaches. Another strategy consists in using triangulation to determine a value that takes into account each method.

Two of the studied trials used only descriptive statistics. In one case, the population was too small to allow for any kind inferences whereas in the other scenario, the outcome was the result of a survey. Noticeably, two reports from trials did not include any description of statistical analysis. Predictive analytics is a field of statistics that uses previous information to anticipate events in the future. One of the trials used a simple prediction error learning model to try to prognosticate the influence of DBS in cognition [57]. These methods can be extended to other areas of PD research.

13.3 Missing Data

Missing data are values that could potentially change the results of a study if not accounted. Therefore, researchers should highlight the role of design and trial conduct to limit the effect of missing data on regulatory decisions [58]. There are many reported strategies to handle missing data; in this topic, we highlight the preferred methods for PD trials.

First of all, researchers should know how much of a challenge it is to deal with missing data. Authors should bear in mind that the strategies to handle missing data should be prespecified as a part of the research protocol, but more importantly to limit missing data in the first place (see suggested readings).[3] Another recommendation is to consider the use of more than one strategy to check if the inferred results vary.

In this review, last observation carried forward was preferred (31%). For instance, clinical trials using UPDRS score (or

---

[3] Little, R (2014) The Prevention and Treatment of Missing Data in Clinical Trials. N ENGL MED 367;14.

continuous outcomes scales) as primary outcome used this method. The National Research Council states in a report of 2014 that unless there is scientific justification to the assumptions made, this method should be used. Parkinson's disease is characterized by a slow and chronically progressive clinical stage, thus clinical rating scales are problematic in handling missing data because the assessments are time sensitive [59]. Therefore, the results cannot be retrospectively inserted.

An interesting assumption from this type of imputation; are disease-modifying trials in PD. The fact that a treatment is hypothesized to arrest the progression of the disease can make LOCF a powerful imputation method even if the patient becomes noncompliant, because the treatment may still have some long-term effect. Therefore, it is assumed to be more useful in this type of cases. Even though this simplistic method is widely applied in PD research the reliability should be handled cautiously as it often adds unrealistic results.

On the other hand, reported literature suggests [59] the use of prorated score since this implies substituting the mean of all values until half of the data is collected. Consequently, it proves to be a further trustworthy method. The reason why relies on the assumption that items share variance, thus measuring the same construct and subsequently using homogeneous values. An important inconvenience of this method is that it cannot be performed if more than 20% of the values are missing

Investigators should consider if they are willing to rely on unstable assumptions for missing data or reinforce avoiding missing data. Either way, focusing on the proper imputation should be more than just the type of outcome (clinical, surrogate or functional) but to dogmatically consider all the medical aspects of the condition studied.

13.4 Additional Analysis

The majority of subgroup analyses are exploratory in nature, and no significance testing should be performed unless an alpha level needed to achieve significance is attributed to the comparison of interest in advance [60]. According to the CONSORT guidelines subgroup analysis should be reported consistently as failing to do so may lead to bias. In other words, authors should do analysis only if they were previously prespecified in the protocol. Otherwise, authors should remain cautious to perform many subgroup analyses. In our review, 64% of the studies described secondary analysis for the interventions used, and 36% did not report additional analysis for secondary outcomes.

14   Limitations

The ultimate source of bias on randomized clinical trials is their restriction to interventions that are supposed to have a positive effect [61]. Moreover, this limitation is often related with the consistency of the research question, some clinical trials do not provide the answers to the questions initially proposed [61]. In fact, for many studies the perspective of the effects of the intervention in the quality of life or costs is not addressed, which is substantial for PD population. In addition, the CONSORT guidelines state that limitations should figure a part of the discussion and therefore clarify the general pros and cons of the study. For all potential sources of bias, it is important to consider the magnitude and the direction of the bias (2)

In our review, we summarized the frequent limitations that were often confronted in PD clinical trials, such as sample size restrictions, follow-up duration, study inclusion/exclusion criteria and blinding. Eighteen percent of the reviewed studies omitted to add limitations in their study.

The study sample size is amongst the most frequent limitations for this review (15%). Recent studies mention the challenges for recruitment in PD trials. The majority of trials involving PD have sorted patients accordingly to disease severity. This leads to a low recruitment yield attributed to high comorbidity, dependency and caregiver burden. Among other common limitations in PD are the study inclusion/exclusion criteria. Since it is possible to encounter difficulties to generalize the results because the studied population is variably different from the population treated in normal life. A good example for this is a study by Guidubaldi et al. Botulinum Toxin A Versus B in Sialorrhea, where the authors selected PD and ALS patients; in this case it is doubtful to rely in the results for PD population only.

Alternatively, extensive follow-up visits or lack of follow-up visits (20%) can lead to bias. For extensive follow-up visits—attrition bias is common; this term refers to systematic differences between groups in withdrawals from a study. Study withdrawals lead to incomplete outcome data; therefore, researchers should know how to address these challenges by improving compliance methods to the study and avoid dropouts. Similarly, patient exclusions refer to situations in which some participants are omitted from reports of analyses (treatment changes, motor complications, etc.), despite outcome data being available to the investigators.

On the other hand, due to the nature of some interventions, blinding could often become a limitation for some of the studies, particularly those involving a physical therapy intervention or surgical procedures, in which the study is normally was single blind

(23%) or open label, which leads to an awareness of the intervention. Thus, this bias leads to a less powerful conclusion. One example is the use of sham brain surgery in PD. This procedure is regarded as essential to control for placebo effects in surgical interventions for Parkinson's disease, particularly for the effects of DBS. The failure of multiple double-blind trials of surgical therapies for Parkinson's disease to confirm the results of open-label trials illustrates this point (3). Interestingly, some trials attributed open-label trials as unrandomized by default which also leads to less reliable results.

Other limitations (23%) included: technical constraints, for instance one study mentioned that the accelerometers where not able to produce valid information about spatial parameters of gait (such as, step length, walking distance, and speed). Unquestioned limitations, included: inadequate outcome measures, lack of control group, and uncontrolled analysis of data (multiple analysis, without performing alpha-spending). Researchers should be extremely aware of any source of limitations in the trial, and more importantly to report these in the manuscript. Reporting transparency leads to a higher credibility.

15 Conclusions

The concept of randomized clinical trials has been widely recognized over the years as a strong approach to improve clinical practice and treatment for patients with Parkinson's disease or any other conditions. Clinical trials are a powerful mainstay for clinical research; treatment improvement remains limited to modifying disease drugs and symptomatic relief research studies. An important aspect of Parkinson's disease trials is using the outcome assessment in order to determine the optimal treatment.

In this chapter, different studies outline the importance of a coherent and concise research question that goes followed by a proper methodology design and therefore, a better compliance to study protocol. Researchers should acknowledge that the proper selection of trial design is contingent to the research question, design, and the type of outcome selected.

In Parkinson's disease, the most common outcome measure of disease severity and treatment response is the motor outcome UPDRS score (before or after concomitant L-dopa use). Although the results of this study point towards an increasing number of future studies assessing novel biomarkers as surrogate endpoints. Further collection and categorization of clinical trials would be required to determine optimal methodology selection. The goal is to outline the importance of the research question and select the correct pathway to execute.

Throughout this revision we have verified that the nature of the intervention is critical for the design of a trial. Trials that test the disease-modifying potential of an intervention are fundamentally different that those involved in the management of complications or nonmotor symptoms. Likewise, nonpharmacologic interventions, including devices and procedures share challenges in some features in design including blinding and the choice of an appropriate comparator.

New clinical trials for Parkinson's disease should incorporate innovative designs, such as adaptive and delayed-start design. In addition, it can also be predicted that outcome-measuring tools will include Internet based applications. Clinical trials will benefit from the search of biomarkers including those from neuroimaging.

Acknowledgments

The authors would like to acknowledge Cristina Russo, Aura Hurtado, Felipe Jones, Anthony O'Brien, Beatriz Costa, Isadora Ferreira, and Felipe Fregni for the manuscript review.

References

1. Schapira AH, Jenner P (2011) Etiology and pathogenesis of Parkinson's disease. Mov Disord 26:1049–1055. https://doi.org/10.1002/mds.23732

2. de Lau LML, Breteler MMB (2006) Epidemiology of Parkinson's disease. Lancet Neurol 5:525–535. https://doi.org/10.1016/S1474-4422(06)70471-9

3. EMA (2012) Guideline on clinical investigation of medicinal products in the treatment of Parkinson's disease, vol 44. EMA, London, pp 1–16

4. Kowal SL, Dall TM, Chakrabarti R et al (2013) The current and projected economic burden of Parkinson's disease in the United States. Mov Disord 28:311–318. https://doi.org/10.1002/mds.25292

5. Jankovic J, Disease P, Clinic D et al (2015) The future of research in Parkinson disease. JAMA Neurol 71:2014–2015. https://doi.org/10.1001/jamaneurol.2014.1717.Conflict

6. Hauser RA, Hsu A, Kell S et al (2013) Extended-release carbidopa-levodopa (IPX066) compared with immediate-release carbidopa-levodopa in patients with Parkinson's disease and motor fluctuations: a phase 3 randomised, double-blind trial. Lancet Neurol 12:346–356. https://doi.org/10.1016/S1474-4422(13)70025-5

7. Olanow CW, Kieburtz K, Odin P et al (2014) Continuous intrajejunal infusion of levodopa-carbidopa intestinal gel for patients with advanced Parkinson's disease: a randomised, controlled, double-blind, double-dummy study. Lancet Neurol 13:141–149. https://doi.org/10.1016/S1474-4422(13)70293-X

8. Sawada H, Oeda T, Kuno S et al (2010) Amantadine for dyskinesias in parkinson's disease: a randomized controlled trial. PLoS One 5:6–12. https://doi.org/10.1371/journal.pone.0015298

9. Elm JJ, Goetz CG, Ravina B et al (2005) A responsive outcome for Parkinson's disease neuroprotection futility studies. Ann Neurol 57:197–203. https://doi.org/10.1002/ana.20361

10. Collins LM, Murphy SA, Strecher V (2007) The multiphase optimization strategy (MOST) and the sequential multiple assignment randomized trial (SMART): new methods for more potent eHealth interventions. Am J Prev Med 32:S112–S118. https://doi.org/10.1016/j.amepre.2007.01.022

11. Simuni T, Borushko E, Avram MJ et al (2010) Tolerability of isradipine in early Parkinson's

disease: a pilot dose escalation study. Mov Disord 25:2863–2866. https://doi.org/10.1002/mds.23308

12. Schapira AH, Stocchi F, Borgohain R et al (2013) Long-term efficacy and safety of safinamide as add-on therapy in early Parkinson's disease. Eur J Neurol 20:271–280. https://doi.org/10.1111/j.1468-1331.2012.03840.x

13. Kieburtz K, Tilley BC, Elm JJ et al (2015) Effect of creatine monohydrate on clinical progression in patients with Parkinson disease: a randomized clinical trial. JAMA 313:584–593. https://doi.org/10.1001/jama.2015.120

14. Athauda D, Foltynie T (2014) The ongoing pursuit of neuroprotective therapies in Parkinson disease. Nat Rev Neurol 11:25–40. https://doi.org/10.1038/nrneurol.2014.226

15. D'Agostino RB (2009) The delayed-start study design. N Engl J Med 361:1304–1306. https://doi.org/10.1056/NEJMsm0904209

16. Barker RA, Stacy M, Brundin P (2013) A new approach to disease-modifying drug trials in Parkinson's disease. J Clin Invest 123:2364–2365. https://doi.org/10.1172/JCI69690.2364

17. Devos D, Moreau C, Devedjian JC et al (2014) Targeting chelatable iron as a therapeutic modality in Parkinson's disease. Antioxid Redox Signal 21:195–210. https://doi.org/10.1089/ars.2013.5593

18. Dorsey ER, Venuto C, Venkataraman V et al (2015) Novel methods and technologies for 21st-century clinical trials: a review. JAMA Neurol 72:582–588. https://doi.org/10.1001/jamaneurol.2014.4524

19. Palfi S, Gurruchaga JM, Ralph GS et al (2014) Long-term safety and tolerability of ProSavin, a lentiviral vector-based gene therapy for Parkinson's disease: a dose escalation, open-label, phase 1/2 trial. Lancet 383:1138–1146. https://doi.org/10.1016/S0140-6736(13)61939-X

20. Williams HC, Burden-Teh E, Nunn AJ (2015) What is a pragmatic clinical trial? J Invest Dermatol 135:e33. https://doi.org/10.1038/jid.2015.134

21. Munneke M, Nijkrake MJ, Keus SH et al (2010) Efficacy of community-based physiotherapy networks for patients with Parkinson's disease: a cluster-randomised trial. Lancet Neurol 9:46–54. https://doi.org/10.1016/S1474-4422(09)70327-8

22. Goodwin VA, Richards SH, Henley W et al (2011) An exercise intervention to prevent falls in people with Parkinson's disease: a pragmatic randomised controlled trial. J Neurol Neurosurg Psychiatry 82:1232–1238. https://doi.org/10.1136/jnnp-2011-300919

23. Bloem BR, Munneke M (2014) Revolutionising management of chronic disease: the ParkinsonNet approach. BMJ 348:g1838–g1838. https://doi.org/10.1136/bmj.g1838

24. Schrag A, Sauerbier A, Chaudhuri KR (2015) New clinical trials for nonmotor manifestations of Parkinson's disease. Mov Disord 30:1490–1503. https://doi.org/10.1002/mds.26415

25. Martinez-Ramirez D, Hu W, Bona AR et al (2015) Update on deep brain stimulation in Parkinson's disease. Transl Neurodegener 4:12. https://doi.org/10.1186/s40035-015-0034-0

26. Deuschl G, Schüpbach M, Knudsen K et al (2013) Stimulation of the subthalamic nucleus at an earlier disease stage of Parkinson's disease: concept and standards of the EARLYSTIM-study. Parkinsonism Relat Disord 19:56–61. https://doi.org/10.1016/j.parkreldis.2012.07.004

27. Witt K, Granert O, Daniels C et al (2013) Relation of lead trajectory and electrode position to neuropsychological outcomes of subthalamic neurostimulation in Parkinson's disease: results from a randomized trial. Brain 136:2109–2119. https://doi.org/10.1093/brain/awt151

28. Daniels C, Krack P, Volkmann J et al (2010) Risk factors for executive dysfunction after subthalamic nucleus stimulation in Parkinson's disease. Mov Disord 25:1583–1589. https://doi.org/10.1002/mds.23078

29. Vivas J, Arias P, Cudeiro J (2011) Aquatic therapy versus conventional land-based therapy for Parkinson's disease: an open-label pilot study. Arch Phys Med Rehabil 92:1202–1210. https://doi.org/10.1016/j.apmr.2011.03.017

30. Earhart GM, Duncan RP, Huang JL et al (2015) Comparing interventions and exploring neural mechanisms of exercise in Parkinson disease: a study protocol for a randomized controlled trial. BMC Neurol 15:9. https://doi.org/10.1186/s12883-015-0261-0

31. Goodwin VA, Richards SH, Taylor RS et al (2008) The effectiveness of exercise interventions for people with Parkinson's disease: a systematic review and meta-analysis. Mov Disord 23:631–640. https://doi.org/10.1002/mds.21922

32. Trenkwalder C, Kies B, Rudzinska M et al (2011) Rotigotine effects on early morning motor function and sleep in Parkinson's disease: a double-blind, randomized, placebo-controlled study (RECOVER). Mov Disord

26:90–99. https://doi.org/10.1002/mds.23441

33. Watts RL, Lyons KE, Pahwa R et al (2010) Onset of dyskinesia with adjunct ropinirole prolonged-release or additional levodopa in early Parkinson's disease. Mov Disord 25:858–866. https://doi.org/10.1002/mds.22890

34. Goetz CG, Tilley BC, Shaftman SR et al (2008) Movement disorder society-sponsored revision of the unified Parkinson's disease rating scale (MDS-UPDRS): scale presentation and clinimetric testing results. Mov Disord 23:2129–2170. https://doi.org/10.1002/mds.22340

35. Schapira AH, McDermott MP, Barone P et al (2013) Pramipexole in patients with early Parkinson's disease (PROUD): a randomised delayed-start trial. Lancet Neurol 12:747–755. https://doi.org/10.1016/S1474-4422(13)70117-0

36. Eggers C, Fink GR, Nowak DA (2010) Theta burst stimulation over the primary motor cortex does not induce cortical plasticity in Parkinson's disease. J Neurol 257:1669–1674. https://doi.org/10.1007/s00415-010-5597-1

37. Van Der Brug MP, Singleton A, Gasser T, Lewis PA (2015) Parkinson's disease: from human genetics to clinical trials. Sci Transl Med 7(305):205ps20

38. Bartus RT, Weinberg MS, Samulski RJ (2014) Parkinson's disease gene therapy: success by design meets failure by efficacy. Mol Ther 22:487–497. https://doi.org/10.1038/mt.2013.281

39. Lewis JA (1999) Statistical principles for clinical trials (ICH E9): an introductory note on an international guideline. Stat Med 18:1903–1904. https://doi.org/10.1002/(SICI)1097-0258(19990815)18:15<1903::AID-SIM188>3.0.CO;2-F

40. Dmitrienko A, Tamhane AC, Wang X, Chen X (2006) Stepwise gatekeeping procedures in clinical trial applications. Biom J 48:984–991. https://doi.org/10.1002/bimj.200610274

41. Benninger DH, Lomarev M, Lopez G et al (2010) Transcranial direct current stimulation for the treatment of Parkinson's disease. J Neurol Neurosurg Psychiatry 81:1105–1111. https://doi.org/10.1136/jnnp.2009.202556

42. Moreau C, Delval A, Defebvre L et al (2012) Methylphenidate for gait hypokinesia and freezing in patients with Parkinson's disease undergoing subthalamic stimulation: a multi-centre, parallel, randomised, placebo-controlled trial. Lancet Neurol 11:589–596. https://doi.org/10.1016/S1474-4422(12)70106-0

43. Factor S, Mark MH, Watts R et al (2010) A long-term study of istradefylline in subjects with fluctuating Parkinson's disease. Parkinsonism Relat Disord 16:423–426. https://doi.org/10.1016/j.parkreldis.2010.02.014

44. Williams A, Gill S, Varma T et al (2010) Deep brain stimulation plus best medical therapy versus best medical therapy alone for advanced Parkinson's disease (PD SURG trial): a randomised, open-label, trial. Lancet Neurol 9:581–591. https://doi.org/10.1016/S1474-4422(10)70093-4

45. Marks WJ, Bartus RT, Siffert J et al (2010) Gene delivery of AAV2-neurturin for Parkinson's disease: a double-blind, randomised, controlled trial. Lancet Neurol 9:1164–1172. https://doi.org/10.1016/S1474-4422(10)70254-4

46. Li F, Harmer P, Fitzgerald K et al (2012) Tai chi and postural stability in patients with Parkinson's disease. N Engl J Med 366:511–519. https://doi.org/10.1056/NEJMoa1107911

47. Schulz KF, Altman DG, Moher D, for the CONSORT Group (2010) CONSORT 2010 Statement: updated guidelines for reporting parallel group randomized trials. Open Med 4(1):60–68

48. Schulz KF, Grimes DA (2002) Epidemiology series Generation of allocation sequences in randomised trials: chance, not choice Epidemiology series Generation of allocation sequences in randomised trials: chance, not choice. Lancet 359:515–519

49. Müller T, Kuhn W (2009) Homocysteine levels after acute levodopa intake in patients with Parkinson's disease. Mov Disord 24:1339–1343. https://doi.org/10.1002/mds.22607

50. Follett K, Weaver F, Stern M (2010) Pallidal versus subthalamic deep-brain stimulation for Parkinson's disease. N Engl J Med 362:2077–2091

51. Schuepbach WMM, Rau J, Knudsen K et al (2013) Neurostimulation for Parkinson's disease with early motor complications. N Engl J Med 368:610–622. https://doi.org/10.1056/NEJMoa1205158

52. Tickle-Degnen L, Ellis T, Saint-Hilaire MH et al (2010) Self-management rehabilitation and health-related quality of life in Parkinson's disease: a randomized controlled trial. Mov Disord 25:194–204. https://doi.org/10.1002/mds.22940

53. LeWitt PA, Rezai AR, Leehey MA et al (2011) AAV2-GAD gene therapy for advanced Parkinson's disease: a double-blind, sham-surgery controlled, randomised trial. Lancet

Neurol 10:309–319. https://doi.org/10.1016/S1474-4422(11)70039-4

54. Gupta SK (2011) Intention-to-treat concept: a review. Perspect Clin Res 2:109–112. https://doi.org/10.4103/2229-3485.83221

55. Emre M, Tsolaki M, Bonuccelli U et al (2010) Memantine for patients with Parkinson's disease dementia or dementia with Lewy bodies: a randomised, double-blind, placebo-controlled trial. Lancet Neurol 9:969–977

56. Shulman LM, Gruber-Baldini AL, Anderson KE et al (2010) The clinically important difference on the unified Parkinson's disease rating scale. Arch Neurol 67:64–70. https://doi.org/10.1001/archneurol.2009.295

57. Shiner T, Seymour B, Wunderlich K et al (2012) Dopamine and performance in a reinforcement learning task: evidence from

Parkinson's disease. Brain 135:1871–1883. https://doi.org/10.1093/brain/aws083

58. Little TD, Jorgensen TD, Lang KM, Moore EWG (2014) On the joys of missing data. J Pediatr Psychol 39:151–162. https://doi.org/10.1093/jpepsy/jst048

59. Goetz CG, Luo S, Wang L et al (2015) Handling missing values in the MDS-UPDRS. Mov Disord 00:1–7. https://doi.org/10.1002/mds.26153

60. Wang R, Lagakos SW, Ware JH et al (2007) Statistics in medicine — reporting of subgroup analyses in clinical trials. N Engl J Med 357:2189–2194. https://doi.org/10.1056/NEJMsr077003

61. Pannucci CJ, Wilkins EG (2010) Identifying and avoiding bias in research. Plast Reconstr Surg 126:619–625. https://doi.org/10.1097/PRS.0b013e3181de24bc

Chapter 6

Dystonia

Celeste R S De Camargo, Daniel Dresser, Laura Castillo-Saavedra, and Felipe Fregni

Abstract

Dystonia is a chronic neurological disorder that hinders the quality of life of affected individuals and causes serious disability. It is often an underdiagnosed or misdiagnosed condition due to the lack of clinical criteria and insufficient amount of research studies on the field. With the main goal of exploring methodological aspects of clinical trials in Dystonia, a review of the literature was conducted through the database engine "Web of Science," searching for the 100 most cited trials in Dystonia, published between 2005 and 2015. This chapter discusses the findings of this review and provides recommendations for augmenting future research studies in the field.

Key words Dystonia, Dystonic disorders, Neurology, Clinical trials

1 Introduction

The term "dystonia" was first used by Oppenheim in 1911 to describe a condition causing variable muscle tone and recurrent muscle spasms. Currently, dystonia is defined as a sustained muscle contraction, with repetitive twisting movements, and abnormal posture [1]. The cause of this disorder remains unknown; however, the most accepted theory is that it occurs as a consequence of damage to the basal ganglia or other brain regions that control movement; nonetheless, no abnormalities are found in diagnostic imaging. The etiology can be genetic, acquired or idiopathic. Some cases of dystonia have been attributed to medications, trauma, infections, vascular, and metabolic conditions, among others. Symptoms are diverse and can include uncontrolled simultaneous contraction of agonist and antagonist muscles,which can manifest as neck turning, uncontrollable eye blinking, dysphonic speech, among others. Dystonic movements can be both fast and slow, and tremor may be present or not [1].

It is difficult to assess the prevalence of dystonia, however, estimates show that around 300,000 people are affected in the USA

Felipe Fregni (ed.), *Clinical Trials in Neurology*, Neuromethods, vol. 138,
https://doi.org/10.1007/978-1-4939-7880-9_6, © Springer Science+Business Media, LLC, part of Springer Nature 2018

and Canada alone and primary dystonia occurs in 11.1 per 100,000 people [2]. It can affect men, women and children of all ages and backgrounds, and can be classified according to clinical characteristics, topographic distribution, age of onset and cause. It is frequently classified as generalized dystonia (affects most or all of the body), focal dystonia (restricted to a specific part of the body), multifocal dystonia (involves two or more unrelated body parts), segmental dystonia (affects two or more adjacent parts of the body), and finally hemidystonia (involves the arm and leg ipsilaterally). The most frequent types of dystonia include cervical, blepharospasm, hand, oromandibular, occupational, and laryngeal dystonia [3]. If patients develop symptoms before the age of 30, it is considered to be an early-onset or childhood-onset, and if this occurs afterward, it is referred to as late-onset or adult-onset. The classification according to age is extremely important because early onset cases usually evolve from a focal limb dystonia to the severe generalized form, while the late onset usually only affects craniocervical muscles, therefore, remaining local [1].

Dystonia is sometimes considered to be a rare condition, however, it is under or misdiagnosed; this might possibly be due to lack of specific clinical criteria. A recent meta-analysis on dystonia revealed that there are no existing guidelines on the topic [3]. Dystonia is a chronic and disabling condition [3] that can seriously hinder quality of life in patients who suffer from it, therefore there should be treatment alternatives that can be safely used in a long-term basis, effectively decrease symptomatology and improve quality of life. At the moment, there are no medications to prevent dystonia or slow its progression, all available treatments are directed toward minimizing symptoms, such as botulinum toxin, medications, and deep brain stimulation (DBS).

The aim of this chapter is to provide possible methodological recommendations that can guide researchers in the field of dystonia in designing clinical trials. We analyzed the characteristics of the 100 most cited Dystonia clinical trials since 2005. The topics covered are based on the CONSORT guidelines and include: trial designs and settings, phases of study, sample selection, interventions, choice of outcomes, sample size, interim analysis and monitoring, randomization, blinding, adherence, statistical analysis, assessment of biases, and confounding factors.

2 Methods

We conducted a review of the most cited articles on dystonia since 2005 using the Web of Science database. Our search was conducted using the term "Dystonia," which had to be included in all article titles. Results were then filtered so that only prospective clinical trials were selected. Initially, we limited our search to the

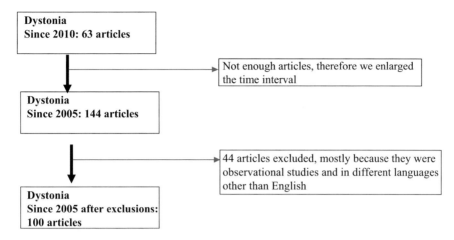

Fig. 1 Flowchart of search for clinical trials in Dystonia

last 5 years; however, this did not yield a sufficient number of articles (63). We subsequently expanded the time interval of the search to 10 years (2005–2015), this yielded 144 articles, of which 44 were excluded. The most prevalent exclusion criterion of the articles was an observational design. This is summarized in Fig. 1.

3   Initial Results

3.1   Publications Throughout Time

The articles analyzed were published between 2005 and 2015, with a relatively stable number of publications throughout the reviewed years. Yearly publications on dystonia ranged from a high of 15 in 2009 to a low of 6 articles in 2010 (excluding 2015) (Table 1).

3.2   Journals and Citations

From the 100 articles, only 37 followed CONSORT criteria, while 63 of them did not evidently follow this guideline. In these cases, in which the CONSORT items were not clear, we inferred the information, and if this was not possible, it was then classified as "not mentioned." Most articles were published in journals with impact factors lower than 6. Only 5% of them were published in top impact factor general journals: three in The Lancet Neurology and two in New England Journal of Medicine.

3.3   Primary Outcome: Positive or Negative and Journals

The great majority of the articles, 86% were positive for the primary outcome (Graph 1). This may indicate that articles with negative results were less published or received less citations. Publication of well-designed studies with negative results is essential, since learning that a certain drug/device/procedure is ineffective is as valid as learning the contrary. The ultimate goal of a researcher should always be advance scientific development.

Table 1
Summary of initial results

Summary of results	
Year	N
2005–2007	27
2008–2010	32
2011–2013	30
2014–2015	11
Phase	
Pilot	25
I	5
II	45
III	10
IV	2
Phase not clear	13
Trial design	
Not randomized	54
Randomized	46
Drug/device/procedure	
Drug	50
Device	4
Procedure	46
Primary outcome	
Positive	86
Negative	14
Journal follows CONSORT guideline	
Yes	37
No	63
Number of citations	
0–50	87
51–100	4
101–150	6
151–200	1
>200	2
Impact factor	
0–5	79
6–10	15
11–15	0
16–20	0
21–25	4

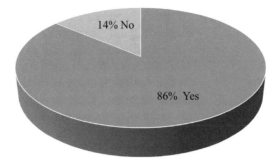

Graph 1 Description of positive vs. negative results for primary outcomes

4   Trial Design, Phases, and Settings

The choice of an appropriate trial design to fit the final purpose of the trial is crucial. In this analysis, 55% of the studies were randomized. Randomization is an important method to ensure internal validity and minimize imbalances between arms. Blinding both investigators and patients is necessary to avoid certain types of bias, such as performance and selection bias. Having a controlled study diminishes chances that results are due to placebo effect and thus attributes the findings to the actual intervention.

Fourteen percent of the trials we analyzed were crossover studies. In this type of design, patients receive both interventions at different times, and therefore it is possible to maximize the amount of information coming from patients as well as compare the effects of both treatments in the same subject, increasing the power of the study. This is even useful during recruitment, as patients are aware that they will all receive active drug/device/procedure, however, it is necessary to change from one treatment to the next without affecting blindness. A disadvantage of crossover designs is the carryover effect, which is when it is difficult to attribute the effect to a specific intervention, and therefore, the major problem is the inability to derive an unbiased estimate of the treatment effect when differences occur because of the different sequences in which treatments are applied [4]. Out of the 100 articles, 11 were crossover studies. Seven of them did not mention any issues regarding their study design and its possible consequences; however, Tassorelli et al. (article 29) stated that no carryover effect was observed. Bonanni et al. (Articles 32) justified the use of a cross over design to avoid the known placebo effect in the treatment of Parkinson's disease. Thomas et al. (article 56) recognized the possible problems with this study design; however, they believe that in the case of their study, the advantages outweighed the disadvantages. They diminished the inherit risk of bias by random delivery of both interventions at each visit, avoiding an order effect. Regarding any carryover effect, the authors assumed that they were negligible since subjects returned when symptomatic. Rieu et al. (article 96) adjusted for any issues regarding carryover effect in the statistical analysis by the use of mixed models with unstructured variance-covariance to compare the effect of epidural MCS on dystonia, spasticity, and pain intensity of patients including period, epidural MCS, sequence, and possible carryover as fixed effects.

Other factors that can invalidate the findings of a crossover trial include nonuniform pharmacologic and psychologic carryover effects, failure to return patients to their baseline state before the crossover, changes in the patients over time, and the use of time-dependent response measures [4]. In order to minimize these problems, wash-out periods should be implemented between one treatment and the other as to attribute each effect to it's respective intervention. All of the analyzed studies that made use of crossover designs mentioned the use of wash-out periods. The washout periods varied from 10 days (article 26) to 1 month (articles 84 and 90).

An open-label study is one in which neither the patient nor the investigator is blinded and therefore they are both aware of the intervention. This type of study is usually used for three main reasons: firstly to make the effective but not yet licensed drug available for subjects who have been initially randomized to the placebo arm (this may be due to ethical requirements or simply a means of enhancing recruitment) [5]; secondly, because there is the need for lengthier and more profound research on adverse effects that have not been thoroughly explored in the original RCT; and finally to demonstrate continued efficacy of the drug over a longer period of time or to show that participants randomized to receive the active treatment during the open-label phase achieved similar outcomes to those who received the drug from the beginning of the parent RCT [5]. A significant number of articles, 28%, were open-label trials. Hauser et al. (article 92) assessed Abobotulinumtoxin A's (Dysport) safety, tolerability and dose–response for cervical dystonia, and they chose to use an open label study so that their results could be comparable to other similar ongoing studies which also used this design. Tarsy et al. (article 77) preferred to use an open-label study design to test levetiracetam for the treatment of cervical dystonia since it was a pilot study.

Each study phase has a purpose and different characteristics regarding its objective, population size, and duration. Pilot studies are preliminary projects that access viability before a larger project is put into practice. Few trials (10%) were pilot studies. For example, Sanger et al. (article 36) chose to carry out a pilot study in order to obtain preliminary data on appropriate doses and side effects before designing a larger study, as they would first need this crucial information. Vidailhet et al. (article 7) also carried out a pilot study to assess the effects of bilateral pallidal stimulation on motor impairment, functional disability, and quality of life in patients with dystonic choreoathetosis. Indeed, the percentage of pilot studies is higher than one should expect and especially those studies receiving a large number of citations. The large number comes from the fact that it is challenging to run a large study in this condition and given the invasive therapies tested that usually are tested initially in a low number of subjects.

Phase I trials assess safety, side effects, timing, dosage, pharmacokinetics, and pharmacodynamics, and are usually carried out in healthy subjects, in sample sizes smaller than 200 subjects and generally take from 12 to 18 months. This was the case of 5% of the trials analyzed. For example, Chinnapongse et al. (article 86) made use of a phase I trial to investigate dose-escalation, safety, and tolerability of rimabotulinumtoxinb in in the treatment of cervical dystonia.

Phase II trials are used to analyze efficacy and safety, proof of concept, as well as dose adjustments. However, in this phase, safety is analyzed in a larger sample, usually ranging from 300 to 500 subjects, taking from 1.5 to 2.5 years to be completed. Out of all the articles we analyzed, 44% were phase II trials. Benecke et al. (article 6) for example, recruited 466 patients to prove the efficacy (proof of concept) of a botulinum toxin type A free of complexing proteins for the treatment of cervical dystonia.

Phase III trials are usually pivotal studies that evaluate effectiveness and safety in large sample sizes, around 2000–3000 subjects and take longer to be concluded (~3–4 years). Ten percent of the studies analyzed were phase III trials.

Phase IV trials are post marketing studies that test efficacy and safety in a heterogeneous population, analysis of long term effects, and are usually preformed as open labeled studies. This phase is vital in order to collect information of broad adverse effects the drug (procedure or device) may have as well as side effects that can be used to attribute more than one finality to the intervention. This was the case in 2% of the trials. Dressler et al. designed a phase IV trial to assess long-term efficacy and safety of incobotulinumtoxinA injections in patients with cervical dystonia.

Thirty-five percent of the trials were multicenter studies (Graph 2). This type of study has a good acceptance within the scientific community as it has many advantages: faster recruitment of patients, clearer results which are more convincing as the patient sample of multicenter trials representativeness [6]. More than one site taking part in the project guarantees a more heterogeneous study population and therefore more representative of the target population, increasing external validity and generalizability. However, multicenter trials require strong efforts for quality assurance concerning admission, treatment and follow-up, thus a highly

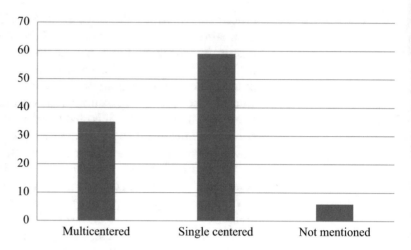

Graph 2 Study settings

developed coordinating center is needed to ensure and protocol compliance [6]. With the progression in phases, studies tend to a multicenter design, therefore phase I is usually single site trial, phase II is either single or multicenter trial, while phases III studies are almost all multicenter with phase IV studies occurring in the post marketing setting. Since most of the studies analyzes were still in the early phases (Pilot and phase II mainly), 59% of them were executed in a single center.

5   Population

Target population is the group of people in which the treatment is directed to, nonetheless, it is not possible to study all of these subjects. Patients who could feasibly be included in the study are the accessible population. Finally, the study population is composed of those subjects from the accessible population who fit the inclusion criteria and are therefore included in the study (Fig. 2). Similar to most clinical trials, convenience nonrandom sampling was the case in all the studies analyzed.

5.1   Eligibility Criteria

Eligibility criteria have to be very well determined and applied in order to maintain high validity, being that internal and external validity are inversely related. The stricter the inclusion and exclusion criteria the more homogeneous the study population becomes, which increases internal validity however decreases external validity since the target population is very heterogeneous. Nevertheless, broad eligibility criteria lead to a very heterogeneous population which would increase external validity and consequently generalizability, however, this may increase study variance and affect the likelihood of responding the research question adequately. For

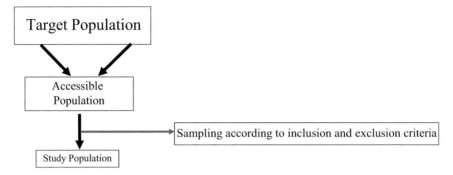

Fig. 2 Selection of study population

instance, if the diagnostic criteria is too inclusive; then it is possible that patients without the studied condition are included and thus affecting the internal validity of the study. Ideally, trials should have both high internal and external validity, therefore, a balance between them should be met.

Inclusion criteria from the articles analyzed included patients from both sexes, mostly between 18 and 65 years old (except for studies in which the target population was pediatric), diagnosed with primary generalized dystonia, and severe impairment in the ability to perform the activities of daily living despite optimal medical management, normal neurologic examination except for dystonia using 24 on the Mini-Mental State Examination (MMSE) as a cutoff.

A few trials, 22%, assessed generalized dystonia using the Burke–Fahn–Marsden Dystonia Rating Scale (BFMDRS). For trials assessing focal dystonia the Toronto Western Spasmodic Torticollis Scale (TWSTRS) was largely used to determine subject eligibility and is composed of rating scores in three main domains: Torticollis Severity, Disability, and Pain. TWSTRS scores for inclusion criteria used for severity were either 10 or 15, for disability and pain 3 and 1 respectively, and total scores were 20 or 30. This is summarized with percentages in Table 2.

6 Intervention

The most prevalent intervention was deep brain stimulation, in 41% of the trials, either pallidal (globus pallidus internus (GPi)) or thalamic. DBS allows for the performance of contemporaneous bilateral surgery with relatively low morbidity in these patients, being that the most beneficial results have been achieved in patients with primary genetic generalized and segmental dystonia, myoclonic dystonia, and complex cervical dystonia [7]. Pallidal DBS has been shown to be effective in complex cervical dystonia

Table 2
Cutoff point for the different TWSTRS subscales used to determine study eligibility and % of studies that used each cutoff value

Cutoff for eligibility	% Of studies
Severity	
15	75
10	25
Disability	
3	100
Pain	
1	100
Total	
30	50
20	50

Exclusion criteria used in the dystonia articles were mainly neurological diseases, pregnancy, breastfeeding, childbearing potential without using effective contraception for ethical and medical reasons, any secondary cause of dystonia and finally, contraindications for treatment (e.g., botulinum toxin)

yielding both symptomatic and functional benefit for up to 2.5 years of follow-up [6], and especially in dystonia conditions due to the DYT1 mutation [8] with improvements up to 90% [9]. Thalamic DBS for dystonia has proven to be a potent treatment in single cases with posttraumatic dystonia, post anoxic dystonia and paroxysmal nonkinesigenic dystonia; however, there is much less knowledge and experience on thalamic DBS for dystonia when compared to that of pallidal DBS [7]. There are many known advantages and beneficial effects of DBS for the treatment of Dystonia and serious complications from this therapy are rare [10]. There were 17 studies using this intervention, ten of them were Pilot studies, one was a phase 1 trial nine were phase 2 trials.

Botulin toxin was used as an intervention in 37%, mainly botulinum toxin type A, type B, incobotulinumtoxin A (Xeomin), abobotulinumtoxinA (Dysport), and onabotulinumtoxinA. This toxin has been studied since 1990 and today it is considered safe and effective for treatment of movement disorders and spasticity [11]. Some of the benefits that may derive from the application of botulinum toxin type A (BTX type A) include decreased muscle tone, improved range of motion, improvement of certain active function, improved hygiene, however, it is essential to balance doses according to effects and adverse effects that can possibly derive from this drug such as weakness and muscle stiffness [12]. The mechanism of action of BTX type A is via nerve terminal endocytosis and interaction with intracellular proteins that inhibit the vesicular release of the acetylcholine at the neuromuscular junction

[13, 14]. This results in a chemical denervation that paralyzes striated muscle, with optimum results at around 2 weeks after injections [15]; however, due to molecular turnover neuronal activity starts again after 3 months and returns to its complete original function at 6 months post injection [16].

Botulinum toxin type B became an FDA approved drug in 2000, and has proven to be a safe and effective treatment for cervical dystonia in patients who have responded to BTX-A and in those who developed resistance to BTX-A [11, 15, 17]. As with all the botulinum toxins, BTX-B acts at the neuromuscular junction inhibiting the release of Acetylcholine at the presynaptic membrane; however, in this case, it inactivates a different protein involved in the release of Ach [11].

Out of the 32 trials that used botulinum toxin as an intervention, seven of them were either phase 3 or 4 trials and 11 were phase 1 or 2 trials and another 12 were pilot studies. Some specific questions still remain unanswered such as the effect of botulinum toxin injection on lateral axial dystonia in patients affected by levodopa (L-dopa)-responsive parkinsonism, and this required a pilot study (article 32).

Other interventions included the use of drugs such as baclofen, levetiracetam, trihexyphenidyl, olanzapine, and risperidone, or even specific sequences of directed exercises and these were mostly phase 2 or pilot studies (Table 3).

7   Outcomes

The outcome is the end result the investigator is eager to explore, being that there may be a primary outcome which is the main outcome used for study design, and the presence of a secondary one is facultative but helpful to generate further hypothesis; however, the presence of a secondary outcome should not influence performance of first outcome. In the 100 articles analyzed, TWSTRS was most used as to determine primary outcome followed by the Burke–Fahn–Marsden dystonia scale for generalized dystonia (Table 4).

TWSTRS is an assessment scale (validated by Consky et al. in 1990 [18]) used to measure the impact of cervical dystonia on patients. The final TWSTRS score, which ranges from 0 to 87, is composed of three severity grades, disability (0–32), and pain. The Severity component varies from grade 0 to 35 and evaluates maximal excursion, duration factor, effect of sensory tricks, shoulder elevation/anterior displacement, range of motion, time (ability to maintain head within $10°$ of neutral position, up to 60 s). The disability scale evaluates the effect of cervical dystonia on ability to perform activities, including work activities of daily living (feeding, washing, dressing), driving, reading, watching television, activities

Table 3
Interventions performed in the Dystonia articles analyzed

Intervention	Number of trials
DBS	41
Botulinum toxin	37
Baclofen	2
Levetiracetam	2
Trihexyphenidyl	1
Olanzapine vs. risperidone	1
Others	16

Table 4
Frequently used primary outcome measures

Primary outcome	Number of articles
TWSTRS	29
Burke–Fahn–Marsden dystonia scale	15
Tsui scale	4
Oystonia discomfort scale (DDS),	2
Barry–Albright dystonia [20]	1
Others	49

outside the home and ranges from 0 to 32. Finally, the pain grade varies from 0 to 20 and analyses severity, duration and disability due to pain (21–23).

The Burke–Fahn–Marsden Dystonia Rating Scale is used to describe deep brain stimulation outcomes. It consists of provoking factors and severity factors, which vary from 0 to 4, each regarding many factors such as speech and swallowing, handwriting, hygiene, dressing, and walking among others, with a total score ranging from 0 to 120. However, this scale may provide little information about function and participation outcomes or changes in nonmotor areas [19].

There were yet other parameters used as primary outcomes in the trials, such as Tsui Scale, Dystonia Discomfort Scale (DDS), Barry–Albright Dystonia [20], Jankovic scores, Ashworth scale, and spatial discrimination threshold and temporal discrimination threshold (TDT) of tactile stimuli.

7.1 Type of Primary Outcome

There are three main types of outcomes: functional, clinical, or surrogate. Seventy-six percent of the trials analyzed chose to assess a functional outcome (Graph 3). Clinical outcomes analyze the subject's clinical conditions while functional ones assess improvements in their performances, such as disabilities and restricted movements. Surrogate endpoints measure the effect of a specific treatment using for instance laboratorial or imaging methods that may correlate with a real clinical endpoint. It is used when the effect sizes on a clinical outcome is small and/or the investigator needs to learn more about the mechanisms of the intervention. Only a small number of trials, 8%, used surrogate outcomes.

7.2 Secondary Outcome

73% of the trials reported secondary outcomes. Some of them were the modification of TWSTRS, Burke–Fahn–Marsden dystonia scale score, Barry–Albright Dystonia Scale, Ashworth Scale, and the Gross Motor Function Measure. More specifically, in trials where the intervention was the use of botulinum toxin, secondary endpoints were the effect of treatment upon the subsequent dose of BTX and the duration of the interval between the first and the second treatment. In some studies that used deep brain stimulation, the secondary outcome variables were the safety of bilateral GPi stimulation for the treatment of primary dystonia, incidence of permanent Adverse Events, Caregivers' burden Brief Psychiatric Rating Scale, and Clinical Global Impression—Severity.

Some more subjective secondary outcomes were improvement in cervical dystonia as determined by review of the video recordings and the patient's global impression of change, quality of life, anxiety and depression, and the effects of the intervention on patients' associated pain and mood.

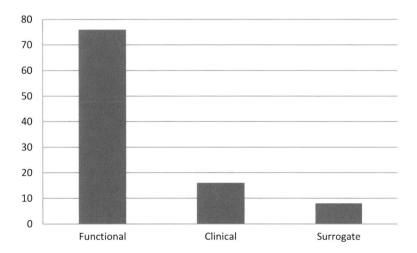

Graph 3 Types of primary outcomes

8 Blinding

Blinding refers to investigators, physicians, or subjects involved in the trial being unaware of treatment allocation, and it is one of the most important tools to avoid bias in clinical trials. If a subject is not blinded to its treatment allocation, there is an increased chance of performance bias, which occurs when a subject changes his or her behavior because they now know they are receiving either the active treatment or placebo. Another bias that can be avoided by blinding treatment raters is Detection bias.

Blinding is very important for a trial to succeed; however, it increases study costs and also makes the assessment of adverse effects more difficult. These are the reaons why some studies are not blinded: due to ethical concerns (for instance having to perform a sham surgery).

Studies can be single blinded, meaning that the investigator, physician, or study subject is unaware of the treatment. A double blinded study is when both the subject and the researcher/investigator are blinded; this minimizes different types of bias. However, as previously mentioned, it compromises the analysis of adverse effects as well as decreases the adherence of physicians as they may feel insecure about referring their patients for enrollment in a study if they are unsure of the actual treatment they may receive. Additionally, side effects may easily unblinded investigators. Single blind studies with a third blinded rater represent another option. It requires two investigators in contact with the patient; and is therefore much more complex, more expensive, demands more patient visits and consequently has a lower adherence.

In our analysis, 16% of the trials were single blinded. In 14% of the studies, it was not possible to blind neither the rater nor the patients nor the caregivers. Sanger (article 24) could not blind anyone involved since the population was especially vulnerable: children with cerebral palsy and secondary dystonia receiving trihexyphenidyl. Explanations for the lack of blinding involve assuring parents that they are aware of what their child is taking which consequently guarantees higher adherence, and also that in case of any complication or side effect, doctors could promptly intervene. Researchers do however recognize downsides to unblinded studies: increase placebo effect (like mentioned in article 24) and potential increase bias (like mentioned in article 72). Out of all the articles analyzed, 51% made use of devices or surgical procedures. Most of these, 43% (22 articles), were not blinded, and out of the 49 which had drugs as interventions, the majority, 39% (19 articles), was double-blinded while only eight were not blinded (Graph 4). This shows that researchers and doctors are more precautious with blinding when the intervention is a device or a procedure and prefer to keep the patients, doctor, and

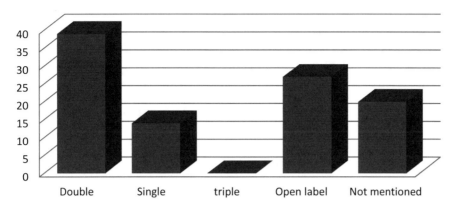

Graph 4 Types of blinding used in the trials analyzed

rater unblinded. This is not the case when drugs are used, probably because professionals feel safer about "hiding" the intervention since there is more knowledge about the medication and antidotes in the case of emergencies than when compared to what is known about procedures and devices.

A clinical trial is referred to as a triple-blind study when the treating or evaluating investigator, the physician, and the sponsor or the investigator responsible for data analysis are blinded. It is by far the safest option.

8.1 Similarity of Interventions

Both interventions, active and placebo treatment, should at least seem identical to patients and investigators in order to assure blinding. The trials that used deep brain stimulation had the same process carried out for both arms (basically stimulator was turned on and off), the time spent under the stimulator was the same, the instructions to patients were the same, making both interventions identical, except for the fact that in one case there was active stimulation and in the other, sham stimulation. Studies in which the intervention was medication, investigators made sure that the presentations were the equivalent. In clinical trials that used botulinum toxin as an intervention, injections were carefully prepared as to appear precisely the same. However, 37% of the trials did not mention the implementation of similarity of intervention and this was not applicable to 27% of the trials which were not blinded, meaning that only 36% of the studies mentioned what was done as to assure similarity of intervention.

8.2 Blinding Assessment

Only 3% of the studies reported the use of any kind of blinding assessment. In all of these, patients and investigators were asked to guess the allocations after the intervention. One of the reasons for this low number is that blinding assessment may also give biased results; because if the intervention is very successful, then participants in the active arm may guess active intervention not because of unblinding but because of therapeutic effects.

9   Recruitment

Recruitment is a topic that must be extremely well planned out and tested before the actual study begins, in order to avoid undesired interruptions. Sound recruitment plans decrease the time span in which subjects are introduced into the trial, saving both time and money. In single-site trials recruitment might be more complicated, while multicenter designs can facilitate it, as there is a larger accessible population.

There are three main recruitment strategies: broad based, target enrollment, and public awareness campaigns. Broad based methods consist of posting media advertisement, such as ads on television and radio channels. This allows recruiters to reach a large number of patients in a short period of time; however, it also attracts a large population of noneligible patients and therefore a second screening will be necessary, which might be time-consuming as well as expensive.

The target enrollment method is when the study recruits patients from a certain medical center and this therefore leads to a much more reliable accessible population and it is also a less costly method. However, this recruitment process depends on colleague referrals, and might consequently be subject to their different perceptions of the importance of the study. Public awareness campaigns are also advertisements however in this case in directed medias: medical local societies, pamphlets in hospitals, presentations in waiting rooms. This is a quite effective method as it incurs a low cost and achieves a relatively large reliable population.

10   Sample Size

Most of the studies analyzed for the purpose of this chapter included small sample sizes, which can be due mainly to challenges in recruitment and lack of funding sources for study development and conduction. Eighty-five percent of the trials enrolled 100 or fewer participants, and 19% of the included studies enrolled ten or fewer subjects.

10.1   Sample Size Calculation

In most conventional trial designs, determining the sample size is a critical element for study success. This determination is based on several parameters of the trial design, and should be clearly specified on the protocol. As mentioned elsewhere [21], four main parameters should be used to calculate the sample size, including (1) The significance level or alpha level, (2) the desired power of the intervention to obtain the hypothesized effect, (3) the estimated effect size, and an explanation on how this estimate was calculated; and (4) other variables that may affect the required sample, such as expected attrition.

Our literature review included mostly studies with small sample sizes, 77% of studies enrolled 50 or fewer participants (Table 5). Most of the trials included in this analysis do not report the method used for sample size calculation: of the included articles, only 11 described a method for calculating the size of their sample. Conversely, eight trials reported at least the three main parameters for sample size calculation, including significance level, desired power and effect size. The most frequently used power level was 80%, which is considered to be enough to encounter an expected effect in simple study designs (Table 6). For significance level, most trials reported using a 0.05 alpha level, and only one trial used a significance level of 0.025.

Table 5
Number of subjects enrolled in analyzed articles

Number of subjects enrolled	Frequency (#)
0–50	77
51–100	8
>100	15

Table 6
Parameters used for calculating the sample size and number of studies that applied them

Parameter	Frequency (#)
Alpha level	*10*
0.05	9
0.025	1
Power	*10*
70%	1
80%	8
90%	1
Effect size	*7*
Attrition	*5*
10%	2
20%	1
25%	1
30–40%	1

Only seven studies reported the estimated effect size. Bly et al. (article 42) determined this parameter based on previous studies also involving Learning Based Sensorimotor Training (LBSMT) and they reached the value of 0.9. McKenzie et al. (Article 66) expected an effect size equal to 0.5; however, after calculations, this value ranged from 0.4 to 0.7 for dependent variables.

10.2   Interim Analysis and Stoppage Criteria

An interim analysis refers to a preliminary review of available data from an ongoing clinical trial. Such evaluation can be conducted at different stages of the trial and it addresses the primary question of the study by focusing on different aspects of the study, such as efficacy and/or safety [22]. The methodology and objectives for an interim analysis should be clarified and established at the beginning of the trial, and they should be made in accordance with the main goal and specific phase of the study. As the goals of a trial vary according to the phase it belongs to, so should the objective of the interim analysis [21].

In our review, only one study conducted an interim analysis, but it was not a prespecified analysis, nor did it have defined specific stoppage criteria [23] In this case, investigators decided to conduct an interim analysis based on the report of severe adverse events, given that two study subjects presented worsening of their symptomatology till the completion of the trial. It is also important to note that investigators did not clarify the type of analysis they conducted during the interim review, but they did report that they did not find any association between the reported side effects and the intervention.

Only one study reported prespecified stoppage criteria, referring mainly to the possibility of "breaking the code" if severe adverse events occurred. Nonetheless, they did not clarify the definition of severe reactions. The remainder of the articles did not make any reference to stoppage criteria or prespecified interim analyses. Despite the general recommendation for reporting interim analysis in every trial protocol, most studies that include a small sample size or are conducted on patients with chronic, stable conditions do not establish predetermined preliminary reviews or criteria for protocol terminations. One of the reasons for the low number of interim analysis in these studies is the relatively small sample size studies.

11   Randomization

Randomization is a critical issue to ensure internal validity, as it minimizes imbalances in trial conduction, also avoiding selection bias. This is an essential issue that differentiates RCTs from observational studies, since randomization balances out characteristics among study arms. However, if the population is very small, separating arms in a totally random fashion may create disparities. In such cases there are certain types of randomization techniques that would be appropriate in order to minimize disproportions [21].

Randomization can be carried out in two main different ways, either by adaptive or by fixed allocation. In adaptive randomization, the allocation of subjects' changes as the trial progresses and can be based on group characteristics. Disadvantages of this method are that it may easily be unblinded and it demands a more complex statistical analysis. Fixed allocation randomization is when the probability of allocation does not change throughout the study.

Simple randomization is hard to unblind since the chance that the next patient will be randomized to either placebo or active treatment is not affected by the allocation of the previous patient, nevertheless it can lead to imbalances in small populations. In blocked randomization, there are fixed amounts of people in each group, being that the investigator usually decides the block size. This method reduces the chance of having different numbers of subjects in each group. Stratified randomization is used when population covariates need to be proportionally distributed among all the intervention groups.

Most of the studies included in this review did not include a randomized design; only 46 of the 100 most cited studies in dystonia had a randomized design. Of these, only 47% reported the type of randomization, and 65% described the method used for creating the randomization sequence. The most commonly reported type was simple randomization, and usually it was either a computer-generated or centralized sequence.

11.1 Randomization Sequence Generation

As previously mentioned, most of the reviewed randomized trials did not report the method used for sequence generation. In those were such method was described, the most common approaches used were computer-generated sequences and centralized randomization lists, with at least 14 of the 55 randomized trials using this method (Table 7). Nonetheless, other less frequently employed methods were also used for this purpose, such as the coin-toss method [24, 25].

11.2 Type of Randomization

The type of randomization can vary through studies and is usually determined by many factors, such as sample size, and goals of the trial. Regardless of the selected type, the chance of being allocated to a specific intervention should remain random. 32.6% of the articles analyzed used simple randomization (Table 8).

A reasonable number of trials, 14%, used simple randomization. Some of the methods used by these studies include centralized randomization list, random number tables, randomization table generated by the hospital's pharmacy, computer-generated,

Table 7
Methods used for randomization sequence generation, number and percentage of studies that applied them

Method for sequence generation	Frequency (#)	Percentage (%)
Centralized randomization list		
Web-generated	3	6.6
Other	4	8.6
Total	7	15.2
Computer-generated	7	15.2
Coin-toss	2	4.4
Random-number tables	1	2.2
Not specified	29	63

Table 8
Type of randomization employed, number and percentage of studies that applied them

Type of randomization	Frequency (#)	Percentage (%)
Simple	15	32.6
Blocked	11	24
Blocked-stratified	3	6.5
Pseudorandomization	1	2.2
Not specified	16	34.7

and finally coin method. Most of these trials (8) were phase 2 trials, four were pilot studies and two were phase 3 trials.

There was one study that used a pseudorandomization method (article 74) which appears to be random however the sequence is generated by a deterministic process. In this study, 20 patients with primary cervical dystonia were pseudo randomly assigned to either real or sham bilateral cerebellar cTBS (continuous theta burst stimulation) in order to test the hypothesis that long-term modulation of cerebellar–cortical interactions could induce clinical and neurophysiological changes in these patients.

11.3 Implementation of Randomization

In our review, we found that most studies did not specify the professional in charge of implementing randomization. Only eight articles reported how randomization was implemented (17%). Most of the time, investigators requested a member of the study staff to implement the randomization sequence. Pharmacy personnel was in charge in 8.6% of the trials, and in 4.3% of cases the randomization implementation was carried out by a third party or staff not involved in the study (Table 9). If study staff allocated the intervention, most articles specified the level of involvement of the researcher.

11.4 Allocation Concealment

As previously mentioned, allocation concealment refers to the method used to protect and obscure the random assignment sequence. Of the randomized studies that were included in our review, only six described how the randomization sequence was concealed (Table 10), and of these studies four reported who was directly responsible for this task.

Pappert et al. (article 15) described the method of allocation concealment they used: the principal investigator (PI) contacted an interactive voice response system for a subject randomization number and the treatment allocation for the randomization number was forwarded to the site pharmacist who prepared the study drug

Table 9
Method for implementing randomization, number and percentage of studies that applied them

Randomization implementation	Frequency (#)	Percentage (%)
Study staff	2	4.3
Pharmacy	4	8.6
Nonstudy staff/independent party/statistician	2	4.3
Not specified	38	82.6

Table 10
Method for implementation of allocation concealment, number and percentage of studies that applied them

Allocation concealment	Frequency (#)	Percentage (%)
Reported	6	13
Not reported	40	87

and had no contact with the subject or injector. Other trials were less descriptive about the method used and only mentioned that the randomization sequence was known only to the clinical trials coordination center (article 38) or kept only by the hospital's pharmacist (article 45) or by a single research nurse administering stimulation (article 50).

12 Adherence

Adherence is a critical component of protocol success, and refers to how much participants are willing and able to adhere to the recommendations given by the investigator and to the study rules. Many of the studies reviewed required multiple visits to the study center, either for intervention implementation or for measurement of outcomes; nonetheless, only three trials reported measurement of adherence levels (Table 11). One trial evaluated adherence using medication count, and the other two did not specify the method used.

In our review, 48 trials included pharmacological interventions that required multiple drug administrations throughout the study period. In these cases, adherence becomes an even more relevant issue, given that it will directly affect the intervention outcome and can lead to underestimation of its effects. Of these trials, 40 included administration of injectable botulinum toxin for prolonged periods of time, which required visits to the study center in the course of 3 or more months. This type of scenario poses a huge burden on both the investigator and subject because they require

Table 11

Methods implemented for improving adherence and frequency of studies that used them

Adherence improvement	Frequency (#)
Phone calls	2
Informative pamphlets and brochures	1
Supervision by study staff	3
Decreasing pain and discomfort related to intervention	1

The dropout rates in the articles analyzed ranged from 0.06 (article 30) to 0.55 (article 95). A possible interpretation for this discrepancy may be the intervention. Article 30 has a very low dropout rate and the intervention was the application of botulinum toxin, while article 95's intervention is much more invasive: DBS, and this could possibly explain its high dropout rate, for patients may feel more insecure about the procedure and consequently leave the study continuous communication, and constant interventions from the study site to ensure compliance with study visits. Unfortunately, most of these studies did not report the methods they used to guarantee adherence to the intervention or study visits.

Regarding the studies that did report a method intended to improve adherence, we found a high variability in techniques. Most of them were geared towards improving comfort and making participation as easy as possible. For example, some studies that included physical training as an intervention [26–28], provided supervised training sessions or constant and easy communication with physical therapists. They also offered extensive informative sessions and documents, to enable subjects to easily access the information they required. A study reviewing the effects of botulinum toxin A injection in children tried to increase adherence to the intervention by decreasing injection discomfort through offering the possibility to receive anesthesia and analgesia during the procedure. They also conducted frequent phone calls to evaluate possible side effects and encourage caregivers to continue the protocol regimen [20].

13   Study Duration and Follow-Ups

We found a large variability in study duration, which is a reflection of the great heterogeneity of study designs and tested interventions, which ranged from single-session studies, to clinical trials with large follow-up periods. This may be due to the fact that different interventions require different time intervals for follow-up; additionally, this may be a cost limitation issue.

13.1   Trial Duration

We identified a high heterogeneity in trial duration for the studies included in this review, ranging between 0.83 (single sessions) and 2190 days, with a mean duration of 399.38 ($\pm$516.9). In general, several studies required the administration of serial treatments or conduction of serial procedural interventions; as well as extended follow-up periods. Nonetheless, four of the 100 included studies were conducted on a single-session design (Graph 5). This introduced great variability to our sample. It is also noteworthy that 21 studies did not specify the total duration of patient participation.

13.2   Follow-Up Duration

We also encountered high variability in length of follow up in the reviewed articles (see Graph 6). The mean duration of the follow up period was 269.8 days ($\pm$513.980), ranging between 0.83 and 3600 days.

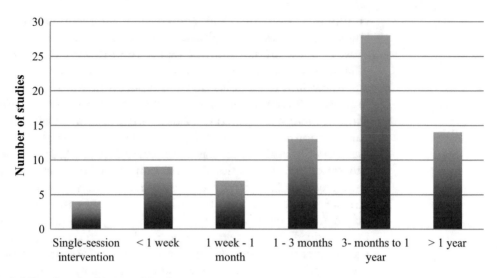

Graph 5 Duration of subject participation

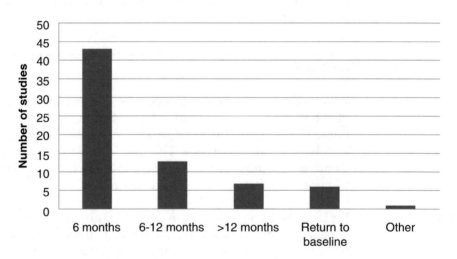

Graph 6 Length of subject follow-up

14 Statistical Analysis

Determining the best statistical approach for analyzing data collected in a clinical trial depends on several parameters. Amongst them, it is important to mention the variable type, its distribution in the study population, and how our research question intends to address the data. Given that the studies that were included in this review have highly heterogeneous designs and study populations, it is clear that the statistical methodology will also be diverse [21]. It is also important to note that most of the studies recruited a small and diverse sample, leading to high frequency of non-normally distributed outcomes, for which it is necessary to employ nonparametric statistics for analysis.

In our review, most of the analyzed primary outcomes were treated as continuous variables. In this case, the statistical tests employed included, but were not limited to, ANOVA, Student's *t*-test, Wilcoxon ranked, and Kruskall–Wallis test (when data is not normally distributed most probably due to small sample size studies), and linear and mixed regression models, among others. The two most commonly used statistical tests were ANOVA and Wilcoxon signed ranked test, which were employed in 28% and 32% of trials, respectively (Graph 7).

Five studies in this sample employed both linear and mixed regression models to analyze primary outcomes: used when the investigator feels the need to adjust for important covariates. It is also noteworthy to mention that two studies employed only descriptive statistics to analyze and present their results.

14.1 Subgroup Analysis

The main objective of subgroup analysis is to identify potential differences between subset of subjects with similar characteristics or a common denominator. This approach can aid in recognizing possible modifications to the intervention effect that were previously masked by the initial analysis; nonetheless, this technique can also introduce errors to the analysis if incorrectly used.

In our review, only ten out 100 studies reported a subgroup analysis. Of them, only one had clearly specified these groups a priori, and only six presented a specific account of how the analysis was conducted.

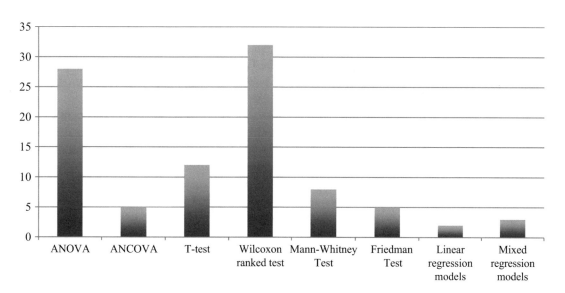

Graph 7 Type of statistical analyses used most commonly

The main characteristic for subgroup analyses was different VAS (Visual Analogue Scale) pain scores (articles 18 and 90). Troung et al. (article 18) separated the population in three groups: patient's change in VAS Pain scores (mm) for subgroups with VAS >40 at week 4, 8 and 12. There was a very small differential effect in the exploratory efficacy outcomes comparing the subgroups and which may possibly be attributed to the time differences between them: the group analyzed at 4 weeks had an efficacy outcome of 26.5 in the Dysport group and 10.8 in the placebo 1 while the subgroup analyzed at 12 weeks had an efficacy outcome of 18.4 in the Dysport group and 14.6 in the placebo 1.

14.2 Intention-to-Treat Analysis

Only 35% of the studies reported using an intention-to-treat analysis. This approach is employed as a mean to decrease effects of attrition bias that can arise from nonrandom and differential attrition from a clinical trial. None of the articles analyzed mentioned the method used for the intention-to-treat analysis; however, Vidailhet et al. (article 4) did state that ITT, in their study, underestimated the overall findings, which was the benefit of bilateral pallidal stimulation.

15   Limitations

Upon careful review of the selected literature, we found that 69% articles reported the limitations in their study design. Among others, we found that small sample sizes leading to low power were commonly cited (59%) (Table 12).

The methodology and design of each clinical trial lead to inherent limitations, and therefore hinder the ability of a protocol to encounter the hypothesized effect. Given that most of the articles included in this review did not include blinded outcome assessments, placebo-controlled or randomized designs, it is easy to infer that a great number of limitations will arise in this specific domain. For example, 28 studies included an open-label design, which intrinsically incurs in a series of limitations; for example, in introduces a high possibility for placebo effect that cannot be accounted for in the design. It also leads to expectancy bias, both from subject and observer-expectancy effect.

As previously discussed in this chapter, as well as in preceding chapters, the sample size and study design are critical determinants for hypothesis testing, as they will dictate the likelihood of finding the expected effect. The clinical nature and pathophysiology of dystonia highly restricts investigator's ability to design large-scale controlled trials, given that it can be challenging and possibly unethical to test new interventions against placebo. Also, access to this population can be challenging because of the severe effect dystonia can have on patient's ability to complete their daily activities,

Table 12
Type of limitations reported

Limitation reported	Frequency (#)
Study design	
Open-label study	7
Unblinded assessments	15
Sample	
Small sample size	41
Convenience sampling	5
Heterogeneity	
Pathology heterogeneity	5
Population heterogeneity	8
Others	
Sample not representative	4

rendering the patient partially or completely disabled. All of these effects can create limitation on accessing the population and creating study designs that can be feasible.

We also identified a high variability in the way limitations were addressed. A high proportion of authors only reported the existing limitations of their study, and only a small proportion of them offered further explanations on such limitations, and/or possible ways to overcome them in the present or future studies. Although one would expect that the vast majority of trials have some sort of limitation inherent to design and methodology, due to a relative underreporting in the majority of articles we reviewed, it is unclear what percentage of trials truly have a limitation that may introduce bias or confounding factors into their observed effects.

References

1. Tarsy D, Simon DK (2006) Dystonia. N Engl J Med 355(8):818–829. https://doi.org/10.1056/NEJMra055549
2. Defazio G, Abbruzzese G, Livrea P, Berardelli A (2004) Epidemiology of primary dystonia. Lancet Neurol 3(11):673–678. https://doi.org/10.1016/S1474-4422(04)00907-X
3. Albanese A, Barnes MP, Bhatia KP, Fernandez-Alvarez E, Filippini G, Gasser T, Krauss JK, Newton A, Rektor I, Savoiardo M, Valls-Sole J (2006) A systematic review on the diagnosis and treatment of primary (idiopathic) dystonia and dystonia plus syndromes: report of an EFNS/MDS-ES Task Force. Eur J Neurol 13(5):433–444. https://doi.org/10.1111/j.1468-1331.2006.01537.x
4. Woods JR, Williams JG, Tavel M (1989) The two-period crossover design in medical research. Ann Intern Med 110(7):560–566
5. Taylor WJ, Weatherall M (2006) What are open-label extension studies for? J Rheumatol 33(4):642–643
6. Messerer D, Porzsolt F, Hasford J, Neiss A (1987) Advantages and problems of multi-

center therapy studies exemplified by a study of the treatment of metastasizing renal cell carcinoma with recombinant interferon-alpha-2c. Onkologie 10(1):43–49

7. Krauss JK (2002) Deep brain stimulation for dystonia in adults. Overview and developments. Stereotact Funct Neurosurg 78(3-4):168–182. 68963

8. Coubes P, Vayssiere N, El Fertit H, Hemm S, Cif L, Kienlen J, Bonafe A, Frerebeau P (2002) Deep brain stimulation for dystonia. Surgical technique. Stereotact Funct Neurosurg 78(3-4):183–191. 68962

9. Kupsch A, Kuehn A, Klaffke S, Meissner W, Harnack D, Winter C, Haelbig TD, Kivi A, Arnold G, Einhaupl KM, Schneider GH, Trottenberg T (2003) Deep brain stimulation in dystonia. J Neurol 250(Suppl 1):I47–I52. https://doi.org/10.1007/s00415-003-1110-2

10. Ostrem JL, Starr PA (2008) Treatment of dystonia with deep brain stimulation. Neurotherapeutics 5(2):320–330. https://doi.org/10.1016/j.nurt.2008.01.002

11. Jankovic J, Hunter C, Dolimbek BZ, Dolimbek GS, Adler CH, Brashear A, Comella CL, Gordon M, Riley DE, Sethi K, Singer C, Stacy M, Tarsy D, Atassi MZ (2006) Clinicoimmunologic aspects of botulinum toxin type B treatment of cervical dystonia. Neurology 67(12):2233–2235. https://doi.org/10.1212/01.wnl.0000249308.66959.43

12. Francisco GE (2004) Botulinum toxin: dosing and dilution. Am J Phys Med Rehabil 83(10 Suppl):S30–S37

13. Rossetto O, Seveso M, Caccin P, Schiavo G, Montecucco C (2001) Tetanus and botulinum neurotoxins: turning bad guys into good by research. Toxicon 39(1):27–41

14. Simpson LL (1981) The origin, structure, and pharmacological activity of botulinum toxin. Pharmacol Rev 33(3):155–188

15. Pagan FL, Harrison A (2012) A guide to dosing in the treatment of cervical dystonia and blepharospasm with Xeomin(R): a new botulinum neurotoxin A. Parkinsonism Relat Disord 18(5):441–445. https://doi.org/10.1016/j.parkreldis.2012.02.008

16. Brin MF (1997) Botulinum toxin: chemistry, pharmacology, toxicity, and immunology. Muscle Nerve Suppl 6:S146–S168

17. Comella CL, Jankovic J, Shannon KM, Tsui J, Swenson M, Leurgans S, Fan W, Dystonia Study G (2005) Comparison of botulinum toxin serotypes A and B for the treatment of cervical dystonia. Neurology 65(9):1423–1429. https://doi.org/10.1212/01.wnl.0000183055.81056.5c

18. Consky E, Basinki A, Belle L, Ranawaya L, Lang A (1990) The Toronto Western Spasmodic Torticollis Rating Scale (TWSTRS): assessment of validity and inter-rater reliability. Neurology 40(suppl 1):445

19. Gimeno H, Tustin K, Selway R, Lin JP (2012) Beyond the Burke-Fahn-Marsden dystonia rating scale: deep brain stimulation in childhood secondary dystonia. Eur J Paediatr Neurol 16(5):501–508. https://doi.org/10.1016/j.ejpn.2011.12.014

20. Guettard E, Roze E, Abada G, Lemesle C, Vidailhet M, Laurent-Vannier A, Chevignard MP (2009) Management of spasticity and dystonia in children with acquired brain injury with rehabilitation and botulinum toxin A. Dev Neurorehabil 12(3):128–138. https://doi.org/10.1080/17518420902927994

21. Portney LG, Watkins MP (2009) Foundations of clinical research: applications to practice, 3rd edn. Pearson/Prentice Hall, Upper Saddle River, NJ

22. Stallard N, Whitehead J, Todd S, Whitehead A (2001) Stopping rules for phase II studies. Br J Clin Pharmacol 51(6):523–529

23. Volkmann J, Mueller J, Deuschl G, Kuhn AA, Krauss JK, Poewe W, Timmermann L, Falk D, Kupsch A, Kivi A, Schneider GH, Schnitzler A, Sudmeyer M, Voges J, Wolters A, Wittstock M, Muller JU, Hering S, Eisner W, Vesper J, Prokop T, Pinsker M, Schrader C, Kloss M, Kiening K, Boetzel K, Mehrkens J, Skogseid IM, Ramm-Pettersen J, Kemmler G, Bhatia KP, Vitek JL, Benecke R, dystonia DBSsgf (2014) Pallidal neurostimulation in patients with medication-refractory cervical dystonia: a randomised, sham-controlled trial. Lancet Neurol 13(9):875–884. https://doi.org/10.1016/S1474-4422(14)70143-7

24. Chan HY, Chang CJ, Chiang SC, Chen JJ, Chen CH, Sun HJ, Hwu HG, Lai MS (2010) A randomised controlled study of risperidone and olanzapine for schizophrenic patients with neuroleptic-induced acute dystonia or parkinsonism. J Psychopharmacol 24(1):91–98. https://doi.org/10.1177/0269881108096070

25. Levin J, Singh A, Feddersen B, Mehrkens JH, Botzel K (2014) Onset latency of segmental dystonia after deep brain stimulation cessation: a randomized, double-blind crossover trial. Mov Disord 29(7):944–949. https://doi.org/10.1002/mds.25780

26. Byl NN, Archer ES, McKenzie A (2009) Focal hand dystonia: effectiveness of a home program of fitness and learning-based sensorimotor and memory training. J Hand Ther 22(2):183–197; quiz 198. https://doi.org/10.1016/j.jht.2008.12.003

27. McKenzie AL, Goldman S, Barrango C, Shrime M, Wong T, Byl N (2009) Differences in physical characteristics and response to rehabilitation for patients with hand dystonia: musicians' cramp compared to writers' cramp. J Hand Ther 22(2):172–181; quiz 182. https://doi.org/10.1016/j.jht.2008.12.006

28. Rice J, Waugh MC (2009) Pilot study on trihexyphenidyl in the treatment of dystonia in children with cerebral palsy. J Child Neurol 24(2):176–182. https://doi.org/10.1177/0883073808322668

Chapter 7

Meningitis

Felipe Jones and Felipe Fregni

Abstract

Although the development of conjugate vaccines has decreased meningitis' incidence in developed countries, acute bacterial meningitis is still a life-threatening condition worldwide. Therefore, the aim of this chapter is to summarize and discuss methodological characteristics of the 100 most highly cited clinical trials in meningitis between 1995 and 2015. Based on the CONSORT guidelines, a review of the literature was conducted in order to discuss the main challenges involved in this field and guide researchers to develop future clinical trials.

Key words Meningitis, Research, Review, CONSORT guidelines, Methodology

1 Introduction

Meningitis is the inflammation of the leptomeninges, the membranes surrounding the central nervous system. The inflammatory process can result from infection, cancer, subarachnoid hemorrhage, or chemical irritation (i.e., drugs) [1]. Such myriad of causes have turned meningitis into one of the most important problems in medicine. The prevalence of its causative etiologies varies according to the population under study. Although its incidence in developed countries has declined due to the recent development of conjugate vaccines, acute bacterial meningitis is still a life-threatening condition [2]. Approximately 1.2 million cases occur each year worldwide [3] with a mortality rate of 20% in developed countries and 50% in low-income ones [2]. Moreover, around 21–28% of survivors develop neurological sequelae such as hearing loss, focal neurologic deficits, and neuropsychological impairment [4, 5].

The disease is a major burden in low-income countries and areas with high HIV prevalence. In the "African meningitis belt," around 400 million people are affected by bacterial meningitis epidemics, and more than 900,000 cases were reported in this area within the last 20 years [1]. In addition, among the 35 million individuals infected by the HIV worldwide, 625,000 die annually

Felipe Fregni (ed.), *Clinical Trials in Neurology*, Neuromethods, vol. 138,
https://doi.org/10.1007/978-1-4939-7880-9_7, © Springer Science+Business Media, LLC, part of Springer Nature 2018

due to cryptococcal meningitis (CM) [6, 7]. Although the incidence of CM infections has declined in patients who have access to antiretroviral therapy, CM is still a leading cause of death in developing countries with limited access to treatment [6, 7]. Additionally, tuberculous meningitis (TBM) is another highly prevalent condition in such areas, and results in death or neurological disability in about 50% of the cases [8].

The improvement of therapies for systemic cancer has resulted in increased incidence of leptomeningeal metastases [9, 10]. As a result, neoplastic meningitis (NM) has been a frequent complication in cancer patients. NM affects up to 5% of patients with cancer, and is associated with poor prognosis [9]. The median survival of such population varies between 6 and 8 weeks, without treatment, and 2–8 months with tumor-target therapies [11].

Besides its epidemiological impact, meningitis disorders present unique challenges for clinical researchers. Clinical trials on meningitis have struggled on finding new therapies with easy route of administration and that prove to be safe and cost-effective. Furthermore, the comparison of results between clinical trials may be difficult due to heterogeneity. There is a lack of standardized definitions of clinical cases and ambiguous consensus of response criteria for some conditions. Finally, the vulnerable profile of research subjects, most of them with severe systemic disease and poor prognosis, adds to the complex task of being a clinical researcher in the field.

This chapter is designed to review these challenges and guide researchers to planning future clinical trials in meningitis. In order to accomplish this goal, we will summarize and discuss relevant methodological characteristics of the 100 most highly cited clinical trials in meningitis since 1995. The topics covered herein are based on the CONSORT guidelines and include trial designs and settings, phases of study, sample selection, interventions, choice of outcomes, sample size, randomization, blinding, adherence, statistical analysis, assessment of biases, and confounding factors.

2 Methods

We conducted a search of the 100 most highly cited articles describing interventional clinical trials on the treatment of Meningitis using the Web of Science Database. To identify eligible articles, we searched the key term "Meningitis" in the title and restricted the findings only to document file "Clinical trial." In contrast to other chapters of this book, we expanded the search for the period 1995–2015, since the original strategy (2010–2015) did not yield sufficient number of papers that met our inclusion criteria. To achieve 100 papers, we screened 186 publications using a descending citation order. This process and the 86 exclusions are schematically represented in Fig. 1.

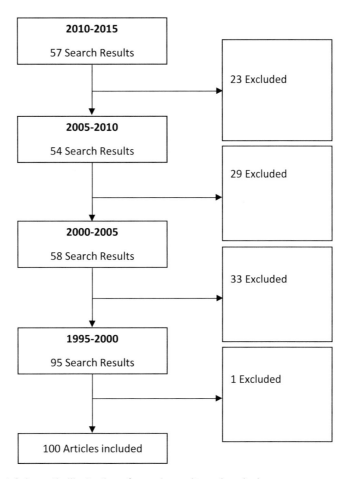

Fig. 1 Schematic Illustration of search results and exclusions

3   Overview

In the sample of articles reviewed in this chapter, only 54 were published in journals that endorse the CONSORT guidelines. The remaining 46 studies did not explicitly describe all items on the CONSORT checklist and data was either deduced by the reviewers or otherwise reported as "not specified." Moreover, journals that do not endorse the CONSORT guidelines tend to have lower impact factor (IF) (<5.0), except from the Clinical Infectious Diseases (IF of 9.20). These findings are indirect indicators that a significant percentage of trials in meningitis are not reported in high standards of clinical research, which raises concerns regarding potential biases related to the lack of transparent reporting; but also gives an opportunity to improve reporting in this area of research for future trials. It is therefore important that the investigator (especially beginners) follow guidelines of reporting (i.e., CONSORT) and guidelines to design a trial—such as the SPIRIT (http://www.spirit-statement.org/)—in order to improve the quality of research in this field.

Table 1 shows the distribution of these trials according to year, phase, trial design, type of intervention, positivity for primary outcome, sample size, to whether the journal follows the CONSORT guideline, number of citations, and impact factor. One interesting conclusion is that there is a uniform distribution of these trials according to some characteristics such as sample size, year of publication, number of citations, and positivity for primary outcome (yes/no). But also for trial design, for instance, more than half of the trials were RCT with parallel design; showing that the simplest approach may be the one that provides a more robust answer for a clinical question. Also for type of intervention, all included studies tested a new drug as the type of intervention, which indicate the importance of drugs (e.g., antimicrobials, corticoids, chemotherapy) as the main therapy in meningitis, but also may also highlight that the investigation of other types of intervention can have a great impact in improving current clinical practice. Another interesting result for the beginner clinical trialist is that the main outcome (positive or negative) does not seem to be related with number of citations (see Fig. 2). This finding underscores the importance of having a strong design to ensure validity of data.

Table 1
Overview of the 100 most cited articles in meningitis

		N
Year	1995–2000	16
	2000–2005	25
	2005–2010	25
	2010–2015	34
Phase	I	7
	II	26
	III	66
	IV	1
Trial design	RCT Parallel two arms	60
	RCT Parallel three arms	12
	RCT Parallel four arms	10
	RCT two-way factorial	5
	Single-arm nonrandomized	13
Drug/device/procedure	Device	0
	Drug	100
	Procedure	0
Primary outcome positive	Yes	54
	No	40
	Not applicable	06

(continued)

Table 1
(continued)

		N
Journal follows consort guideline	Yes	54
	Not in the endorsement list	46
Sample size	0–30	20
	30–60	19
	60–100	20
	100–300	21
	300–2000	20
Number of citations	0–15	26
	15–30	21
	30–60	25
	60–100	12
	>100	15
Impact factor	1–5	45
	5–10	36
	10–20	6
	>20	13

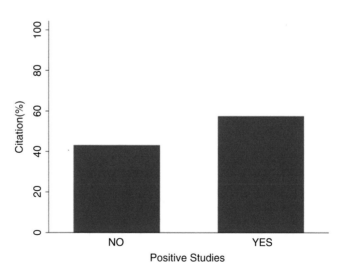

Fig. 2 Percentage of citation according to the studies with positive and negative results

4   Trial Design, Phases, and Settings

4.1   Importance of Phases on Choosing Trial Design

Before any new therapy being approved on the market, it has to undergo a long way under clinical research. Traditionally, clinical trials are conducted in a series of steps called phases. Each phase is designed to address specific research goals and, thus, the overall process of study planning and implementation is highly dependent on the study phase.

4.2   Phases and Trial Designs in Meningitis

Phase I clinical trials are designed to test a new therapy in a small group of subjects in order to evaluate its safety, determine a safe dosage range, and identify side effects [12]. In this sample of articles, there were seven phase I studies in meningitis. Five tested new drugs on neoplastic meningitis patients, whereas the other two investigated different dosages of preexisting drugs in children and neonates with tuberculous meningitis and bacterial meningitis, respectively. All of these studies were designed as a single-arm nonrandomized trial, which complies with the main goals of phase I trials.

One example for a phase I trial was a study testing the administration of intrathecal (IT) liposomal cytarabine in children with neoplastic meningitis [13]. In this study investigators tested three different doses of IT liposomal cytarabine (25, 30, and 50 mg), and assessed the appearance of cytarabine-related nonneurologic and neurologic adverse events using version 1 of the National Cancer Institute common toxicity criteria. Nonneurologic adverse events were dose-limiting if a grade 4 toxicity occurred in two patients of a cohort of six patients, or if a grade 3 toxicity was observed in four out of six patients. On the other hand, neurologic adverse events were dose-limiting whenever a grade 4 toxicity was observed in one patient of a cohort of three, or if two patients out of a six experienced a grade 3 toxicity. As a result, children receiving 50 mg of IT cytarabine had more dose-limiting adverse events compared to the other two groups, and authors concluded that the safest dose in the pediatric population is 35 mg [13].

Phase II clinical studies are designed to further evaluate safety of new therapies and provide preliminary data about efficacy in a group of patients [12]. We found 26 phase II studies in meningitis. Twelve were designed as randomized clinical trials (RCTs) with two arms, whereas three were RCT with three arms, four with four arms, two had factorial designs and five were single-arm nonrandomized trials. Therefore, as for phase II clinical trials, a simple and robust design such as a parallel two-arm trial is still the gold-standard for many reasons. First, this design allows for powerful statistical comparison, which is desirable in studies of this kind with relatively small sample sizes. Second, in the case of placebo-controlled studies, it allows better evaluation of adverse events, which is also important in this phase. Third, it is usually less costly and more feasible than trials with three and four arms, given their need to increase sample size. On the other hand, a relevant topic to deal in meningitis trials is the role of placebo-controlled studies. Because most causes of meningitis already have standard treatments with specific antimicrobials, direct comparison against placebo goes against the equipoise principle established by the Declaration of Helsinki [14]. Consequently, most studies compare the addition of a new drug to a standard therapy versus the standard therapy alone. Finally, three and four-arm phase II studies were mostly performed in CM—in which three or four different combinations of drugs were compared—and two phase II studies with factorial design compared different route of administration of interventions and different combination of antimicrobials in TBM and CM. Therefore, one additional challenge for the investigator looking for new drugs and designing a small trial is how to design a small phase II trial if the control group is not a placebo, meaning that the effect size against a standard drug may be small thus sample may need to be large. Few studies here addressed this question; such as the study of Jackson et al., which compared the addition of flucytosine to the standard therapy (fluconazole and amphotericin B) versus standard therapy alone in the treatment of cryptococcal meningitis [15]. In order to increase statistical power, investigators chose a simple two-parallel randomized design, and used a marker of infection clearance (i.e., early fungicidal activity) as a primary outcome, for surrogate outcomes are generally more powerful than clinical endpoints [15]. Another example of a small phase II study, not placebo controlled, is the trial conducted by Tansuphaswadikul et al., which compared a 1-week course of amphotericin B followed by fluconazole to the standard 2 weeks of amphotericin B followed by fluconazole [16]. Very similar to the previous example, this trial used a two-parallel randomized design with a surrogate marker of infection (CSF culture clearance) as primary outcome [16]. This way, investigators planning small phase II studies that lack placebo for ethical reasons should consider targeting a homogenous population, and designing a two-parallel randomized trial with a powerful surrogate outcome.

Phase III clinical trials are designed to confirm the efficacy of a new intervention in a larger study population [12]. In our sample of studies, the majority of articles (66) were phase III clinical trials. Among these studies, 48 were RCTs with two groups, nine were RCTs with three groups, and six with four arms and three had a factorial design. This distribution is compliant with the understanding that, as for phase III trials, the new intervention is usually compared to standard treatment in a heterogeneous, broader population. Therefore, in this phase, a two-arm design is the most used approach, because it is simpler and robust, as we mentioned above.

Other less employed design options, such as multiple-arm parallel design or factorial design, allow researchers to compare more than two drug schemes, route of administration or even durations of therapy. For instance, in three and four-arm RCTs the intervention is compared against two or three other treatments, respectively. These designs can simultaneously test whether the new drug is more effective than or at least as effective as the other treatments. It is especially useful for phase II studies in infectious diseases, where different drug combinations are usually utilized and there is a necessity for selecting optimal doses and treatment duration in order to avoid antimicrobial-resistance development. As an example, in one three-arm parallel RCT [17], Jarvis et al. compared three different therapies for the treatment of HIV-associated CM: intravenous amphotericin B and oral flucytosine (standard therapy); standard therapy plus two-doses of interferon-gamma 1b; and standard therapy plus six doses of interferon-gamma 1b. On the other hand, studies with a two-way factorial design, in which patients are randomized to four groups and two treatments are simultaneously tested, allow researchers to test the effect of two interventions independently and their interaction. This type of design is also an option for testing combination of drugs, which may be useful for clinical research in infectious meningitis. Scarborough et al. published a study in New England Journal of Medicine in which they randomized patients with BM to receive either dexamethasone or placebo and either intramuscular or intravenous ceftriaxone [18]. This is an example of a well conducted factorial-design trial, in which two different interventions are compared—the adjunctive dexamethasone treatment and the different routes of ceftriaxone administration. Nevertheless, though multiple-arm and factorial-design trials have advantages, they are less conducted than two-arm parallel studies because of some reasons. Firstly, trials with more than two-arms require more power and, therefore, increased sample size, as more comparisons are being made. Moreover, interventions with different route of administration provide blinding difficulties, and a double-dummy design may be required, increasing trial complexity, as we will discuss further in Subheading 10. Finally, the main drawback of factorial design is related to its main advantage. The interaction effect between both treatments might require even larger sample sizes if the combined intervention is just slightly more effective than each intervention separately. Again, the investigator has to make a careful risk-to-benefit assessment before adding more arms or using factorial designs in his or her research.

Phase IV clinical trials are conducted after drugs have been approved for marketing. They are designed to gather more information about drug efficacy in broader populations and to evaluate potential long-term side effects. In the context of meningitis, such trials can have a relevant role on periodic reassessment of drug efficacy, since antimicrobials are often susceptible to the development of microorganism resistance. Despite of this, we found only one phase IV trial that evaluated the efficacy and long-term safety of intrathecal chemotherapy for neoplastic meningitis using an open-label single-arm nonrandomized design. The lack of phase IV clinical trials on meningitis is not surprising and is a common finding in other medical fields, because of its relatively increased cost and timeframe compared to other phases. Anyway, the design of a phase IV trial is usually an open-label single-arm nonrandomized trial, because its main goal is to measure the long-term side effects of the approved therapy.

4.3 Study Settings

The choice of conducting studies in a single or multicenter setting must have a strong foundation. First of all, the number of centers involved in a study will have important implications in recruitment rate, costs, organizational aspects, and external validity of the results. Additionally, the phase of study, targeted sample size, aimed degree of generalizability, and specific characteristics of the disease under study are key factors for this decision. Conventionally, there is a trend of as phase progresses, more trials are multicenter than single-centered, mostly because of increased sample size and external validity requirements. In this sense, since phase I and II trials target more homogeneous population and a relatively small sample size, they are likely to occur in a single-center setting [12]. Meanwhile, phase III confirmatory trials are likely to occur in a multicenter setting with a broader population [12].

In spite of this traditional pattern, much of the meningitis research base in our sample is comprised of single-center trials (55%) (Table 2). This setting represented 71.4% of phase I studies, 73% of phase II, and 46.7% of phase III trials. A potential explanation for this pattern is because many studies involving CM, TBM and BM were conducted during a period of disease outbreak in a single location, thus, decreasing the necessity of involving multiple centers. On the other hand, as expected, 77.8% of multicenter studies occurred in phase III trials.

5 Population

For practical reasons, studying the entire population diagnosed with meningitis is not possible. Clinical investigators have to rely on specific selection criteria to target who is going to be studied. In order to accomplish this task, general concepts that guide well-planned eligibility criteria should be revisited. The eligibility criteria determine the balance between internal and external validity of a study. The more inclusion and exclusion items added to eligibility criteria, the more homogenous the study population will be. On one side, this enhances the ability to detect true results (internal validity), but on the other, it restricts its generalizability

Table 2
Trial design characteristics of manuscripts in meningitis

Trial design		N	Study setting	Phase I	II	III	IV
RCT	Two parallel arms	60	Single center	0	8	20	0
			Multicenter	0	4	28	0
	Three parallel arms	12	Single center	0	2	5	0
			Multicenter	0	1	4	0
	Four or more parallel arms	10	Single center	0	2	3	0
			Multicenter	0	2	3	0
	Factorial	5	Single center	0	2	3	0
			Multicenter	0	0	0	0
	Cross over	0	Single center	0	0	0	0
			Multicenter	0	0	0	0
Non-RCT (Quasi-experimental)		13	Single center	5	5	0	0
			Multicenter	2	0	0	1

(external validity). Therefore, researchers conducting phase I and II trials may target homogenous populations in order to optimize the proof of efficacy of an intervention. In contrast, phase III and IV studies are characterized for targeting broader populations in order to increase external validity.

Another important point to consider when planning the eligibility criteria for a study is the exclusion of conditions that mimic meningitis but might not respond to the study therapy. Similarly, comorbidities that will worsen the patient condition during the trial should also be excluded, since it can confound the results. Additionally, investigators have to ponder the feasibility of the selected eligibility criteria. In this sense, the research team have to be well-trained and able to apply the selected diagnostic criteria. Finally, the selected criteria will ultimately affect the recruitment of subjects, which is one of the biggest challenges for clinical investigators. We will address recruitment strategies on another section of this chapter.

Meningitis encompasses disorders that cause inflammation of the leptomeninges and is defined by an abnormal white blood cells (WBC) count in the cerebrospinal fluid (CSF) [5, 19]. The classic clinical triad is composed of fever, headache, and altered mental status, but varying presentations can be seen, especially in infants [19, 20]. Although clinical history, signs and symptoms are key pieces of information in establishing the syndromic diagnosis of meningitis, these findings alone are relatively unreliable for diagnostic purposes [19, 20], and do not establish the responsible etiology. Hence, much like the clinical practice setting, the diagnosis of meningitis in clinical trials is based on the combination of historical information, epidemiology and laboratory exams. However, in our sample, most studies do not specify the clinical features utilized for defining a case suspicion. Instead, the diagnosis relies on laboratory exams (Table 3), mainly those extracted from the cerebrospinal fluid (CSF). Thus, a unique feature of these trials is the necessity of performing lumbar puncture in order to get CSF laboratory exams. This can be challenging, since prompt diagnosis is crucial as many causes of meningitis are life-threatening, especially when diagnosis is delayed.

In this context, standard diagnostic testing for any cause of meningitis includes the following CSF lab exams: WBC count with differential, total protein, and CSF/blood glucose ratio or CSF glucose count itself. Nevertheless, the values of these tests overlap in many conditions and cannot be used alone for etiological diagnosis [19]. Therefore, different confirmatory exams are used in specific conditions. Since these populations have unique features, we will discuss the most frequently used diagnostic tests on clinical trials that target different causes separately.

5.1 Bacterial Meningitis

We found 32 clinical trials that investigated interventions for bacterial meningitis. Of these, ten targeted adult patients while 22 aimed children or infants. One possible reason for these numbers is the epidemiology of BM, in which patients <15 years old are mostly affected [19, 20]. Other explanation is the high rate of long-term morbidity and devastating effects in infants and children, driving researchers to study new adjunctive therapies to improve neurological outcomes.

In the inclusion criteria, all studies combined signs and symptoms of meningeal syndrome along with laboratory tests. However, as aforementioned, the majority did not specify which clinical features were indeed considered to define a case suspicion. Yet nine different laboratory tests were frequently reported (see Table 3 for more details). Standard testing for meningitis was composed by CSF cytology (96.9%), CSF–blood glucose ratio (12.5%), CSF protein (50%), and CSF glucose (50%). The most utilized confirmatory exams were presence of cloudy CSF (28.1%), CSF culture (59.4%), CSF Gram's stain (50%), latex agglutination test (25%), blood culture (25%), and polymerase chain reaction (3.1%).

Although CSF culture is a gold-standard diagnostic test of BM, the time to get results can be prolonged and its sensitivity and specificity reduces after antibiotic use [19]. In order to overcome these challenges, many studies have used the combination of less specific tests along with CSF culture and/or Gram's staining. In fact, Gram's staining is rapid, inexpensive, and helpful when antibiotics are administered prior to culture [19, 20]. Other confirmatory tool, the latex agglutination test was used in 25% of trials, but it requires lab infrastructure and may be of limited use in low-income countries [19]. In addition, CSF WBC count demonstrating increased WBC with predominance of neutrophils has

Table 3
Laboratory exams applied in meningitis clinical trials

	Total
Bacterial Meningitis	32
CSF WBC count	31
CSF culture	19
CSF Gram's stain	16
CSF protein concentration	16
CSF glucose	16
Cloudy CSF	9
CSF latex agglutination test	8
Blood culture	8
CSF–blood glucose ratio	4
Blood WBC count	4
C-Reactive protein	2
Polymerase chain reaction (PCR)	1
Micro-erythrocyte sedimentation	1
Cryptococcal Meningitis	31
CSF culture	16
India ink staining	16
Cryptococcal antigen (CRAG)	11
Blood culture	1
Tuberculous Meningitis	17
CSF lymphocyte count	14
CSF glucose	13
CSF Acid-fast bacilli	12
Chest radiography	12
CSF culture	9
Clinical evidence/history of extrapulmonary TB	9
CSF protein concentration	8
CT scan suggestive findings	8
Acid-fast bacilli in specimen other than CSF	7
Altered consciousness (GCS < 15)	7

(continued)

Table 3
(continued)

	Total
Duration of symptoms >5 days	7
Yellow CSF fluid	6
Mantoux skin test	5
Focal neurological signs	5
Gastric aspirate culture	3
Neoplastic Meningitis	**15**
CSF cytology	13
Histology	12
Karnofsky performance scale	11
Expected survival of 2 months	7
Imaging (MRI or CT)	5
Eosinophilic Meningitis	**3**
CSF cytology (eosinophil count >10%)	3
CSF culture	3
CSF latex agglutination test	3
CSF Gram staining	3
India ink staining	3
Cryptococcal antigen (CRAG)	3
Viral Meningitis	**2**
CSF cytology (pleocytosis)	2
CSF Polymerase Chain Reaction (PCR)	2
CSF viral culture	1
Viral serology	1

been the most used nonconfirmatory test, along with findings of increased protein and decreased glucose in CSF. Blood exams such as blood WBC count (12.5%), C-reactive protein (6.2%), and micro-erythrocyte sedimentation (3.1%) are less frequent and are usually used in trials of neonates, infants, or children.

5.2 Cryptococcal Meningitis

Cryptococcal meningitis (CM) had the second higher number of publications in this sample with 31 clinical trials. Since this is an AIDS-defining illness, 28 studies involved HIV-associated CM. However, the majority of trials did not report the selected diagnostic method for HIV. In addition, similarly to BM, diagnostic clinical features are underreported. On the other hand, studies on CM lack a standardized method for laboratorial case-definition. In sum, most trials (61.3%) combined two or more of the following exams: India Ink, Cryptococcal Meningitis antigen detection (CRAG), and CSF culture. Frequent employed combinations were positive India Ink and/or CRAG (22.6%); positive India Ink and/or CSF culture (6.4%); positive culture and/or CRAG (6.4%); and positivity for any one of these three methods (19.4%). Yet 16.1% of trials used CSF culture only, whereas 6.4% utilized India ink staining only.

Similarly to the case of other agents, results from CSF culture of *Cryptococcus neoformans* can take a long time to disclose the results (3–14 days) [19]. This is a main challenge for clinical trials testing new antimicrobials, because of the prompt need to treat these patients. In this context, the use of India Ink (time to result: 15 min) and CRAG (time to result: 1–48 h using ELISA/latex agglutination or 10 min using lateral flow assay) in RCTs is advantageous since they provide a faster diagnosis [19]. Overall, 16 trials (51.6%) in CM used India Ink alone or in combination with other exams (i.e., CRAG and CSF culture), and 11 studies (35.5%) employed CRAG as one of the laboratory exams to detect patients with CM. One example on how to employ these faster diagnostic exams can be found in the study conducted by Boulware et al., who tested the CRAG (CSF latex agglutination) in addition to CSF culture in patients with suspected meningitis [21]. In this trial, investigators screened all patients with suspected meningitis (regardless the etiology) for HIV infection status, CSF culture and the presence of CRAG in CSF at the time of hospital presentation. This approach reduced time-to-diagnosis and increased efficiency in recruiting patients for participating in the trial (recruitment yield: 177/389) [21].

Additionally, among the available tests that identify CRAG, it is worthy to mention a new test, the lateral flow assay (LFA). The LFA compares favorably to the traditional latex agglutination and ELISA immunoassay tests for CRAG (>99% sensitivity, >99% specificity), and constitutes a simple dipstick exam that only requires a single drop of blood or CSF, and is very quick (10 min) [19]. Moreover, it is a valuable alternative approach to provide diagnosis in low-resource settings, in which centralized laboratory facilities are limited, because it is very easy to perform, costs only $2, and can be stored at room temperature (even in warm weather) [19]. Although none of the trials included in this review specified which method was used to detect CRAG, clinical trialists should consider using this tool when screening subjects with suspected CM given its practical advantages.

5.3 Neoplastic Meningitis

Fifteen articles investigated drugs on neoplastic meningitis. The majority of them included histological examination along with the CSF cytology exam demonstrating increased cancerous cell count in the eligibility criteria. Because neoplastic meningitis presents in patients with cancer at a time when systemic disease has recurred or one or multiple chemotherapy regimens have already failed, most patients with NM have low life expectancy, and studied interventions are merely palliative. This way, based on what would be considered beneficial and ethical for the patient and his/her family, a minimal survival time of at least 8 weeks (2 months) was required in most studies. In this sense, researchers address the possibility of rapid clinical deterioration due to the disease's natural history. Likewise, most studies also request a minimum score of 50% in the Karnofsky Performance Status (KPS), with two other studies requiring a KPS of at least 60% or 70%, respectively. Additionally, articles that investigated subjects with solid tumors frequently applied neuroimaging studies such as magnetic resonance imaging (MRI) and cranial tomography (CT) as additional tools to identify the causative tumor.

5.4 Eosinophilic Meningitis

Three studies on eosinophilic meningitis were included in our sample. The selection criteria for all of them included the CSF eosinophil count >10% as an inclusion criterion. Following the principle of excluding conditions that may mimic the disorder under investigation, all studies excluded any other etiology of meningitis. Therefore, in order to identify such confounders, studies on EM uses CSF culture, CSF latex agglutination test, India ink painting, CSF Gram staining, and Cryptococcal antigen identification to identify and exclude patients affected by other etiologies.

5.5 Viral Meningitis

Two studies on viral meningitis were analyzed. One of them studied a population with herpes simplex virus, whereas the other investigated a population with enterovirus infection. Both studies included a finding of pleocytosis in CSF cytology and the identification of the specific pathogen using a polymerase chain reaction (PCR) test to detect the viral DNA/RNA.

5.6 Tuberculous Meningitis

Among the 17 studies on TBM, most of them used different case-definition criteria. Overall, the majority of papers classify study participants according to their likelihood of having TBM. Usually, the criteria are based on positive findings in the clinical examination, CSF laboratory data, chest radiography, and cranial tomography exams (Table 3), which are used to define three categories: possible, probable, or definite TBM.

On the other hand, despite the overall observed heterogeneity of diagnostic classifications, there was a tendency of standardization of such criteria between trials conducted after 2010. This results from an international effort for creating a uniform case definition

Table 4
Diagnostic criteria for meningitis established by Marais et al. [22]

Tuberculous Meningitis case definition
Definite
1. Clinical signs and symptoms plus one or more of the following: CSF acid-fast bacilli; CSF culture; or CSF positive nucleic acid amplification test. OR 2. Clinical signs and symptoms plus acid-fast bacilli seen in the context of histological changes consistent with tuberculosis in the brain or spinal cord and CSF biochemical changes, or visible meningitis on autopsy.
Probable
1. Clinical signs and symptoms plus a total diagnostic score of 10 or more points (when cerebral imaging is not available) or 12 or more points (when cerebral imaging is available) plus exclusion of alternative diagnoses. At least 2 points should either come from CSF or imaging criteria.
Possible
1. Clinical signs and symptoms plus a total diagnostic score of 6–9 points (when cerebral imaging is not available) or 6–11 points (when cerebral imaging is available) plus exclusion of alternative diagnoses. Possible tuberculosis cannot be diagnosed or excluded without doing a lumbar puncture or cerebral imaging
Not TBM
Alternative diagnosis established, without a definite diagnosis of TBM or other convincing signs of dual disease.

Table adapted from Marais et al. [22]. Diagnostic score is available in the mentioned reference.

in clinical research on TBM [22]. In 2009, an international tuberculous meningitis workshop took place in South Africa, and had its results published on LANCET in 2010 [22]. Through this document, a diagnostic score was established based on clinical, CSF, and radiological criteria, plus evidence of tuberculosis elsewhere, and exclusion of alternative diagnoses. The score ranges from 0 to 20, and it allocates participants in one of four categories presented on Table 4.

6   Interventions

The first step when planning a clinical trial is the development of a research question and hypothesis. Such hypothesis derives from previous knowledge about the disease itself and about the investigational therapy. In this sense, the investigator has to understand and hypothesize the potential effects of the intervention and to which population it is applicable. Moreover, the intervention's mechanism of action will influence the selection of an outcome that captures its effects, as well as the time length necessary to observe those effects.

All clinical trials included in our sample studied interventional drugs. The discussion of these drugs will be divided by etiology of meningitis. Interventions on infectious etiologies were classified as those directed to the specific pathogen (antimicrobials) or ancillary treatments applied to improve outcome (adjunctive therapy).

6.1 Bacterial Meningitis

Clinical trials in bacterial diseases have unique features. Antibiotics have been established as standard of treatment in a period before the randomized clinical trials era. Therefore, investigations of this class of drugs are usually not placebo-controlled, which would be at least questionable from an ethical perspective. Instead, many studies test either the addition of new antibiotics to the standard of care, or a direct comparison between new antibiotics and standard therapy. In addition, the development of resistant microorganisms is a major drive to conduct trials studying combined therapies, different dosages and even different durations of treatment.

Out of the 32 reviewed studies on bacterial meningitis, 11 investigated antibiotics. The most common design in these trials was randomized parallel with two groups of comparison—again, the simplest but still very robust design for confirmatory trials. Investigators were mainly interested in testing the efficacy of a new antibiotic against the standard treatment, or comparing different durations of therapy. In the latter case, a useful approach to comparing different durations of therapy can be exemplified in the study conducted by Singhi et al. They tested the efficacy of a shortened course of ceftriaxone (i.e., 7 days) against the standard duration of the same drug (i.e., 10 days) in children with bacterial meningitis. In order to minimize potential biases, all patients received the drug for the first 7 days and, in the seventh day of treatment, patients were randomized to two groups: continuation of antibiotic for more 3 days or matched placebo for the lasting 3 days. Therefore, both patients and investigators were appropriately blinded to the intervention.

On the other hand, investigation of adjunctive therapies to antibiotics has received a great amount of focus in this field. This is in agreement to the fact that bacterial meningitis is associated with high rates of neurological complications, such as hearing loss and neuropsychological disorders, which may be prevented by using adjunctive anti-inflammatory drugs. Hence, 19 out of 32 studies investigated ancillary therapies using dexamethasone and/or glycerol. Similar to antibiotic trials, the predominant study design for adjunctive therapies was randomized parallel with two comparison groups—in this case, adjunctive therapy and placebo. Additionally, in the context of adjunctive therapies, trialists had to choose whether to control for the concurrent antibiotic intervention, because this could be a source of variability to the trial results. In this sense, the majority of trials did not establish a standardized antibiotic therapy, whereas three clearly controlled which antibiotic patients received. For example, Wald et al. tested whether the addition of intravenous dexamethasone to ceftriaxone was more effective in preventing hearing loss in children with BM than ceftriaxone alone. Future investigations of ancillary treatments in this field should consider controlling the antibiotic therapy, for this may increase the internal validity of the trial.

Finally, it is worthy to mention that, in contrast to studies of antibiotics, a considerable amount of trials in adjunctive therapies are designed as randomized parallel with multiple-arms. These studies were mainly interested in investigating the individual effects of two ancillary treatments and their combination (e.g., dexamethasone vs. glycerol vs. combination of both vs. placebo). However, as previously mentioned, these are mainly phase II trials.

6.2 Cryptococcal Meningitis

Investigational therapies in cryptococcal meningitis usually address one of two characteristics of this condition: the high mortality rate and its relationship with AIDS. Mortality in CM is highly associated with rate of clearance of *Cryptococcus neoformans* from cerebrospinal fluid, thus, many trials in this condition test different treatment regimens with antifungals to achieve a rapid CSF clearance. On the other hand, other investigations target HIV activity, and explore antiretroviral treatment regimens.

Among the 31 clinical trials in cryptococcal meningitis, 23 studied different combinations and durations of antifungal therapy. Similar to antibiotic trials in BM, phase III trials of antifungals in CM were predominantly designed as randomized parallel with two groups: a new regimen vs. the standard therapy. Additionally, phase II trials were mostly randomized parallel with multiple arms, testing different doses, combination of drugs, and treatment durations. By the other side, all trials testing antiretroviral therapies were randomized parallel with two groups and tested the effect of timing of ART introduction in mortality of CSF clearance rate. Because the interventions in these trials were timing of ART introduction (e.g., immediately after CM diagnosis vs. delayed), all of them were open-label, since both patients and investigators knew the allocation group. However, because mortality and CSF clearance rate are objective outcomes—less associated with observer and experimenter biases—this design is a good alternative for comparing timing of ART therapy.

6.3 Tuberculous Meningitis

Clinical trials on TBM highly focused on the investigation of adjuvant therapies to modulate inflammatory response to the infection and reduce the occurrence of endarteritis and brain infarcts. Among the 17 trials on TBM, 11 investigated the role of corticoids, aspirin or thalidomide in improving neurological outcomes. These trials were mainly randomized parallel with two groups of comparison: adjuvant therapy vs. placebo. In addition, most trials ensured that all patients received the same antimycobacterial antibiotics, while testing the efficacy of these ancillary interventions. Finally, only one trial studied the role of ART and, similar to studies in cryptococcal meningitis, this was an open-label randomized parallel trial with two groups of comparison.

6.4 Neoplastic Meningitis

Interventional studies on neoplastic meningitis have investigated new chemotherapy regimens, comparing the combination of different drugs and treatment duration. Nine out of 15 articles on NM are nonrandomized single-arm trials that studied safety and pharmacology of such interventions, as mentioned in section 4. Other four studies tested new drugs against standard chemotherapies in a two-parallel randomized design. Furthermore, all tested interventions were administered intrathecally, which may explain why is difficult to have placebo-controlled studies in this condition.

6.5 Eosinophilic Meningitis

In our sample, studies on eosinophilic meningitis focused investigating an anthelmintic treatment (i.e., albendazole) and its association to an anti-inflammatory ancillary therapy (i.e., prednisone) in a two-parallel randomized design.

6.6 Viral Meningitis

Studies on viral meningitis investigated the efficacy of antivirals in a double-blinded, placebo-controlled, two-parallel design.

---

7 Outcomes

A critical step in the development of a sound research question relies on the determination of outcome assessments. Selection of primary and secondary outcomes is driven by trial phase, disease under study and intervention [23, 24]. Study phase determines the goal of a trial to the extent that early phases target safety and pharmacological measurements, whereas phase II and III trials aim to measure preliminary and definite efficacy, respectively. Moreover, outcomes vary according to pathophysiological mechanisms and clinical impact of the condition being investigated. In fact a thoughtful consideration of the study outcome is critical (read more in Chap. 1).

Meningitis, for instance, is an acute life-threatening condition associated with high mortality and morbidity rates. Thus, clinical outcomes in meningitis trials are predominantly associated to survival or neurological morbidity. Likewise, outcome assessments

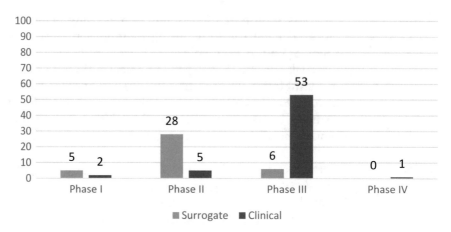

Fig. 3 Types of outcomes within study phases

are also settled on the mechanism of action of interventions that target either survival or morbidity improvement.

7.1 Types of Outcomes

Study outcomes can be classified as three different types: clinical, surrogate and functional. Clinical outcomes provide direct information on how patients feel or survive [24, 23]. Hence, they are robust measurements commonly employed to confirm the efficacy of a new intervention in phase III trials. In meningitis, the most frequently used clinical measurements evaluate either survival or neurological morbidity (e.g., hearing loss, neurological development delay in children). Additionally, because meningitis has an acute life-threatening profile, mortality becomes a key and suitable outcome measurement. On the other hand, assessing survival and morbidity can be unfeasible for phase II studies, given the characteristic smaller sample sizes of these trials and the necessity of higher statistical power to prove efficacy when using clinical outcomes [24, 26, 27]. Therefore, surrogates are useful alternatives as indirect measurements of clinical status. In this sense, the majority of phase II studies included in our samples assessed surrogate outcomes (84.4%). These measurements consisted on CSF biomarkers analysis, such as culture of etiological agents or cytology. Lastly, functional outcomes reflect patient's ability to perform daily activities. The analyzed trials in meningitis did not use functional assessments as primary outcomes.

The use of different types of outcomes within study phases is represented in Fig. 3.

7.1.1 Survival Outcomes

The ultimate goal of many therapies for meningitis is to improve overall survival. In accordance, 29 clinical trials measured survival as a primary outcome. To accomplish this, the researchers that conducted the studies included in our sample used one of three types of approaches. The most frequently used primary endpoint (14/29) for mortality was the measurement of time from randomization to death, also known as time-to-death. Alternatively, six studies primarily assessed the frequency of death in a specific time point, in other words: the mortality rate. Moreover, seven studies aimed to measure the effects of an intervention in either survival or neurological morbidity, whichever occurred until the completion of the study. The combination of two endpoints as a primary outcome, each equally important in determining efficacy, is defined as a co-primary outcome [28]. This way, researchers measured the frequency of death or neurological sequelae (mainly hearing loss) in five studies. The remaining two studies went even further and categorized the co-primary outcome. Finally, Nathan et al. defined the primary outcome as the percentage of treatment failure—a summary endpoint defined as death or worsening in many clinical parameters, such as Glasgow Coma Score, fever, and occurrence of convulsion, among others [29]. This latter approach is defined of a composite outcome, in which an event is defined as the first occurrence following an intervention of one of several predesignated outcomes.

Each one of these approaches has advantages and drawbacks. Mortality rate as a categorical outcome is the simplest primary outcome. It only requires a single measurement, which diminishes logistic issues. In addition, the statistical analysis of mortality proportions is relatively easy to perform and interpret [23]. However, this approach has two main drawbacks: (1) it does not capture the effect of an intervention over time, which is important, because the studied treatments can present survival rates that vary across time (therefore decreasing the amount of information and also analysis power); (2) it requires patients with similar follow-up periods, which may be unfeasible in some settings [30, 31]. An alternative to deal with these issues consist on using a time-to-death outcome. It enables researchers to study not only the occurrence but also the timing of death. Also, it allows the comparison of groups with different length of follow-up and, more important, patients who are lost to follow-up can still contribute to the study. A drawback of time-to-death outcomes is that it increases the number of endpoints and is more complex to analyze and interpret [30].

7.1.2 Neurological Morbidity Outcomes

We found nine trials that assessed neurological sequelae as a primary outcome. As previously mentioned, five of them combined a measure of neurological impairment with death in a co-primary outcome. Additionally, all trials assessed the frequency of neurological sequelae in their analysis. However, trials differ on how measured and categorized such outcomes. For example, hearing loss was evaluated in all nine trials, although the classification of hearing impairment varied across studies. Kanra et al. and Gijwani et al. categorized audiological impairment into mild (21–40 dB), mild to moderate (45–55 dB), moderate (56–70 dB), severe (71–90 dB), and profound (>90 dB) severity [32, 33]. The primary outcomes in those studies were the incidence of at least a moderate impairment. On the other hand, Sankar et al. used the following classification: mild (21–40 dB), moderate (41–60 dB), moderately severe (61–80 dB), severe (81–100 dB), profound (>100 dB) [34]. In this trial, the primary outcome consisted of frequency of abnormal audiological function. Lastly, Peltola et al. were only interested in the incidence of deafness, which was defined as no responses for sounds at ≥80 dB of intensity [35].

In addition, six trials assessed long-term neurological sequelae that were not hearing loss. The assessed outcomes were: blindness, hemiplegia, quadriplegia, psychomotor retardation, hydrocephalus requiring shunt, marked spasticity, seizure, and hypotonia. The majority of studies assessed a combination of such outcomes and classified them in a composite outcome defined as "severe neurologic sequelae." In this case, the primary outcome was frequency of severe sequelae. However, Sankar et al. divided neurologic impairments into two categories: major and minor [34]. The former relates to blindness, quadriparesis, hemiparesis, severe psychomotor retardations, and hydrocephalus, whereas the other was composed by monoparesis, moderate psychomotor retardation, and ataxia. In order to analyze these outcomes, the authors measured the frequency of both categories and compared across the groups.

7.1.3 Clinical Response Outcomes

Clinical response in meningitis trials refers to assessment of signs and symptoms of the disease other than neurological sequelae. Fourteen studies used clinical assessments as primary outcomes. Of these, only six used these results as a sole outcome, whereas four used clinical response as a co-primary outcome, and four as a composite outcome (See section 7.1.5).

Assessment of clinical response varied across clinical trials and etiology of meningitis. Some studies compared the frequency or number of patients with signs and symptoms across study groups. For example, three studies in EM compared the number of patients who still had headaches after 2 weeks of treatment. Additionally, Kapoor et al. assessed the frequency of fever, meningismus, dizziness, vertigo, diabetes insipidus, vomiting, and wound healing among patients with NM [36]. On the other hand, five studies categorized clinical success according to disappearance of signs and symptoms of meningitis, and compared the frequency of "cure/treatment success" across groups.

7.1.4 Surrogate Outcomes: The Role of CSF Analysis

We found 27 trials that used CSF examinations as primary outcomes. Sixteen of them evaluated these results in a single endpoint, whereas eight utilized them in a composite with other outcomes, and three assessed them with other co-primary outcomes separately. The CSF examinations varied across studies and etiology of meningitis (Table 5).

Table 5
Cerebrospinal fluid examination and the etiology of meningitis

	Total
Cryptococcal Meningitis	16
Mean-rate decrease in CFU in CSF	8
CSF culture conversion rate	7
Time to CSF culture conversion	1
Neoplastic Meningitis	6
CSF cytology clearance rate	6
Bacterial Meningitis	3
CSF culture conversion rate	3

Cryptococcal meningitis trials most frequently used CSF examinations endpoints, compared to other etiologies. Among the employed endpoints, the mean-rate decrease in cryptococcal Colony Forming Units (CFU) in CSF was observed in eight trials. This summary statistic is based on an average of repeated quantitative measures in the CSF of individual patients, and has been considered more powerful than the common comparison of CSF culture conversion rate in one single time point [23]. In addition, this measure is based on a log transformation of the number of CFU/ml of CSF per day, and is usually derived by the slope of the linear regression of log CFU against time for each patient [23]. On the other hand, the CSF culture conversion rate is also commonly applied, and has the advantage of being simpler and of easier interpretation. In contrast to the mean decrease in CFU, culture conversion rate is usually obtained after one single time point at 2 weeks of treatment for CM [23]. At last, only one trial in CM assessed the time to CSF culture conversion.

With regard to other etiologies, six studies on NM assessed CSF endpoints, and all consisted on CSF cytology clearance rate. In contrast, three studies on BM used of CSF culture conversion rate as their primary outcome.

7.1.5   Co-primary and Composite Endpoints

There are arguments for and against the use of co-primary and composite outcomes. Overall, these approaches ensure that a higher number of events will occur among the study population, which ultimately increases the statistical efficiency of a trial [28]. As a result, it may decrease the needed sample size and, consequently, the study costs and time length [28]. Nevertheless, composite and co-primary outcomes often make the sample size calculation a complex and imprecise process, since it is difficult to estimate the overall occurrence of the combined measures [28]. In addition, whenever assessing co-primary outcomes, researchers have to correct for multiple comparisons. Finally, the clinical interpretation of results might also be more difficult. Unless there is previous experience from pilot trials or other studies, the use of composite outcomes may not be the best approach for a confirmatory clinical trial.

8   Sample Size

A priori sample size calculation is one of the first steps when planning a study and it helps to frame the trial design. According to the CONSORT criteria, sample size estimations must be reported and justified in published articles [37]. In addition, whenever sample size is not calculated a priori, underestimation or overestimation of sample size might occur, and both situations have serious consequences. The first of them is increased type II error risk due to underestimation; which ultimately means that the intervention might be rejected when it could be beneficial. On the other side, overestimation increases trial costs and might overexpose subjects to a new and uncertain therapy, or equally expose them to unnecessary risk when randomized to a placebo arm [38]. Therefore, appropriate methodology of sample size calculation is of extreme importance for high-quality clinical research on meningitis.

In this context, we found 48 studies that reported sample size calculation (Table 10). Calculation of sample size is, however, conventionally executed with three main parameters: alpha level of significance (type I error), power, and expected effect size [38]. Hence, only 39 studies reported all three parameters—which is essential since it enables readers to recalculate sample size. It is therefore very important that investigators in this field report all three parameters used to calculate the sample sizes.

Among these parameters, the conventional designation of alpha 0.05 was the most employed level of significance (Table 9). This is expected since alpha levels below 0.05 are only appropriate in specific settings; such as an expensive drug associated with serious side effects, but that could be very beneficial. In these situations, researchers might want to be strict and minimize the risk of statistical error to 0.01 or even less [38]. Further, we found a predominance of studies powered at 80% level over 90%—which is in compliance with Cohen's suggestion that a power of 80% is sufficient to reasonably reduce the risk of type II error [39].

Additionally, finding the treatment estimate may be challenging. Although investigators often seek statistician's advice for sample size calculation, the appraisal of predicted effect size requires pertinent clinical reasoning. Thus, knowing sources from which is possible to make an "educated guess" is extremely valuable. Moreover, in order to predict effect size, a researcher has to make assumptions regarding both the control group response, and the

Table 6

Clinical trials in meningitis and the estimation of effect size

	Total
Alpha level	100
0.05	42
Two-sided	28
One-sided	7
Not specified	7
0.025 one-sided	1
Not reported	57
Power	100
80%	37
90%	9
71%	1
Not reported	63
Control group expected effect	100
Observational data	13
Previous clinical trials	8
Not specified	79
Active group expected effect	100
Minimally clinical significance	4
Pilot studies	0
Previous clinical trials	12
Not specified	86
Attrition	100
20%	5
10%	3
Other	5
Not considered	87

expected effect of the new intervention in the active group [38]. This way, most studies based the assumptions for control group on observational studies (13%) and previous clinical trials (8%) (Table 6). Though the majority of trials did not report the source from which they predicted the intervention response (86%), 12 studies relied on previous clinical trials results, while four employed the assumption of minimally clinical significance as a rationale to estimate the treatment effect.

9   Randomization

Random allocation qualifies RCTs as the gold standard in medical research. The main goal of randomization is to balance the distribution of known and unknown prognostic factors across groups [40, 41]. The allocation is accomplished under the light of probability theory—meaning that randomization is unsystematic and unpredictable [42]. Therefore, comparison groups differ only by their intervention, and any observed difference in outcome is likely to result from differences in treatment effect. This way, randomization minimizes the probability of selection and confounding biases.

Proper employment and clear reporting of randomization are central for scientific accuracy and credibility. The CONSORT criteria determine that researchers must report the following items in published articles: type of randomization, randomization sequence generation, implementation of randomization, and allocation concealment [37]. We will discuss each one of these topics herein.

9.1   Randomization Sequence Generation

A critical process for randomization in clinical trials is how to generate the allocation sequence. Most frequently used methods to create a randomization schedule include computerized random number generation and random-number tables. In our sample, we found 87 randomized trials, and only 45 reported how they generated the randomization sequence. The observed underreporting of this procedure (51.7%) is in accordance to a cohort of publications indexed on PubMed, in which only 36% of publications presented complete information [43]. Yet computerized random number generator was the most applied method among meningitis trials (86.7%). Random number tables came as the second preferred method (6.7%).

Interestingly, the remaining trials employed methods such as alternating days, and odd or even enrolment numbers to generate a randomization schedule. Although reported as randomization processes, these methodologies are not random since they are based on systematic occurrences [40, 41, 44]. Moreover, such schemes are prone to uncover allocation concealment, since it enables those who recruit subjects to know treatment assignment before it happens. This way, systematic methodologies for allocation are not truly random and are susceptible to introduce selection bias [44]. Therefore, it is recommended that investigators use computerized random number generators as a method for creating their randomization list.

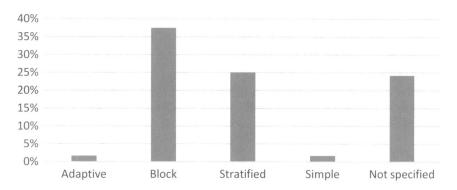

Fig. 4 Percentage of studies according to the type of randomization

9.2 Type of Randomization

There are different types of randomization, each one has advantages and disadvantages. Simple randomization is the most rudimentary method. Similarly to a toss of a coin, it gives each subject a 50% chance of receiving active or control treatments. The main drawback of such method is the risk of getting imbalanced groups, or groups with different sample sizes, when applying it to small sample sizes [41, 44]. In contrast, simple random allocation is unpredictable and an appropriate method to randomly balance groups in relatively big samples [41, 44]. In this context, we found only four trials that used simple randomization. Of these four, two used odds and even numbers to generate the allocation sequence and, therefore, are not truly randomized trials. The remaining two studies had sample sizes of 357 and 58, respectively. It is important to stress that the latter had a significant imbalance in baseline CSF glucose measurements, which highlights the limitation of using this method in relatively small samples.

Blocked randomization is an alternative when investigators want to ensure a balanced number of subjects across groups. Briefly, this method randomizes several subjects at time in blocks of fixed or permuted sizes [41, 44]. In this way, it is possible to guarantee that comparison groups will have equal sizes throughout the trial. This feature is very important when it comes to carrying out interim analysis or when researchers are working with small sample sizes [44]. On the other hand, this method does not abolish the risk of getting imbalanced groups for important prognostic factors. In addition, fixed blocks of small sizes (e.g., blocks of 4) increase the chance of uncovering the assignment schedule by those involved in enrolment tasks [41, 44].

In this context, we found 22 clinical trials that used block randomization (Fig. 4). Blocks of 10 were the most used size of blocking (5 out of 22), and were employed mainly in studies with more than 60 subjects. Blocks of 4 (4 out of 22) were the second most applied method, especially in trials with <100 subjects. Interestingly, two open-label trials used blocks of 16 in order to minimize the

Table 7
Applied factors for randomized stratification in each etiology of meningitis

	Total
Cryptococcal Meningitis	13
Center	8
Mental status defined by Glasgow Coma Score	7
Primary or recurrent infection	1
Prior therapy for acute infection	1
Severity of headache	1
CSF opening pressure	1
Tuberculous Meningitis	7
TBM severity	5
HIV status	2
Center	1
Bacterial Meningitis	6
Center	6
Neoplastic Meningitis	2
AIDS-related lymphoma	1
Primary tumor histology	1
Viral Meningitis	1
Primary or recurrent infection	1

chance of uncovering the allocation concealment, as discussed above. Moreover, only one study employed permuted block randomization, with block sizes varying between 2, 4, and 6. This way, although permuted block randomization is a robust method to ascertain unpredictability of random allocation, it has been less used than fixed block sizes.

Stratified randomization is intended to minimize imbalance for important covariates. This method creates separate randomization schedules for selected strata defined by important prognostic factors [41, 44]. However, stratification has important caveats. First of all, the statistical analysis of stratified trials has to address each variable that was used to stratify randomization. This creates complexity since increasing the number of covariates in a statistical model is not desirable. Secondly, stratification may not be useful in large trials because greater sample sizes lead to balanced groups with simpler methods (simple randomization) [41, 44]. One exception for this, however, is stratification by center in multicenter trials. Finally, stratification does not guarantee balanced groups if simple randomization is performed within each stratum; actually it can increase risk of imbalance since the total sample size is divided into each one of them. Therefore, stratification should be performed coupled with a type of restriction, usually blocking [44].

We found 30 studies that used stratified randomization and 20 of them used blocking as a restriction method. Eight papers did not specify the randomization method within each stratum, and the remaining two reported using an adaptive randomization technique. Moreover, the majority of trials (21) stratified for 1 factor only, while eight and one stratified for 2 and 3 factors, respectively. In addition, 15 studies stratified for center. In Table 7 we show the most utilized factors for randomized stratification in each etiology of meningitis.

9.3 Allocation Concealment

The second key process to achieve appropriate randomization is allocation concealment. Regardless of having a random allocation sequence, prevention of foreknowledge of treatment assignment avoids introduction of selection bias. Moreover, previous trials have shown that inappropriate allocation concealment undermines the randomization process, because those responsible for enrolment may assign patients for groups based on their beliefs [42].

Most studies in our sample did not report the employed allocation concealment method (46%). This is in conformity to reports that showed that authors omit this information in 45% of studies in general medical journals [42]. However, without proper reporting readers cannot be sure whether biases were introduced in the trial. Alternatively, among those trials that reported concealment methods, 32% used sealed envelopes, whereas 9% utilized sequentially numbered containers. These methodologies are considered appropriate, but reports have shown that investigators are able to subvert them in some occasions [42]. In this context, Schultz et al. described the minimum criteria for reporting allocation concealment schemes that include the use of one of the following mechanisms: sequentially numbered, opaque, sealed envelopes (SNOSE); sequentially numbered containers; pharmacy controlled; and central randomization [42]. In fact, with the currently available technological methods, it is fairly easy to implement concealment methods using a central computerized scheme. In addition to planning and describing the type of concealment utilized, it also important that investigators make use of tactics to prevent assignment subversion (i.e., cardboard inside the envelope, containers with equal appearance, and proper training of study staff involved central randomization office) [42].

9.4 Implementation of Randomization

Besides random sequence generation and proper allocation concealment methods, researchers have to plan how these processes are going to be implemented in order to avoid bias. This way, it is important to ensure that those who generate the allocation sequence and/or conceal the allocation schedule are not involved in enrolment or assignment of patients [42]. In our sample of trials, only 29% reported who generated the randomization list. It is therefore important that investigators comply with this requirement of trial planning and reporting.

10 Blinding

Study blinding, similarly to randomization, is critical to prevent the introduction of biases in clinical trials. Although, it is important to have clear in mind that blinding and randomization address distinct sources of bias. By keeping investigators unaware of group assignment, blinding reduces observer bias—which means that it reduces differential assessment of study outcomes by those assessors [45]. In addition, unblinded treating investigators may influence the behavior of subjects, or use different cointerventions, or make different adjustments to the intervention dose, or even take different decisions to withdraw participants or encourage them to discontinue the study. All those conscious or unconscious attitudes introduce what is called experimenter bias [45]. On the other way, knowledge of group assignment by participants can affect their response to the treatment because of changes in their lifestyle or through their expectation of improving (placebo effect) or worsening (nocebo effect) their condition. Further, unblinding subjects to their treatment allocation can affect their compliance to the protocol, and their adherence and retention to the study [45].

10.1 Level of Blinding

The levels of blinding are defined as: open-label, single-blinded, double-blinded or triple-blinded. Open-label studies are usually performed in phase I trials, in which patients undergo dose-escalating schemes and monitoring of adverse effects is central. Single-blinding refers to situations in which at least the investigator (assessor) or the participant is blinded, and is a useful approach whenever double-blinding is not possible. Double-blinding occurs when participants and both assessors and treating investigators are blinded. Finally, on triple-blinded studies, besides participants and investigators, the statistician, or the Data Safety Monitoring Board (DSMB), or the pharmacist is blinded.

In our sample of studies, the majority of trials were double-blinded (39), whereas 35 studies were open-label, eight were single-blinded, and 18 did not specify the type of blinding (Fig. 5). The high number of open-label studies is an interesting finding,

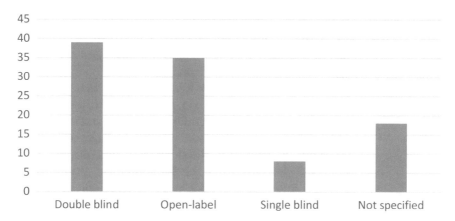

Fig. 5 Types of blinding applied in meningitis trials

since unblinded trials have high risk of bias. Indeed, the majority of these studies (30 out of 35) did not justify the reason for not establishing at least one level of blinding, whereas five were phase I trials. However, when analyzing for other trial characteristics, it is possible to establish that most of these trials compared combinations of different drugs, or different intervention doses, or different routes of administration. Another point to consider was that a good number of these subjects were assessed using surrogate, laboratorial exams (19/35 studies), which are considered objective measurements—less exposed to observer and experimenter biases. In addition, 11 of these trials studied interventions in patients with neoplastic meningitis, in which it is reasonable to affirm that there is an increased concern with safety and an open-label approach may be reasonable in order to monitor safety closely.

Anyways, recognizing the limitation of such inferences about open-label trials on meningitis, it is important to highlight similar studies that were successful on performing a single or double-blind condition. This way, seven of the eight single-blinded trials blinded the assessor who conducted clinical examinations or laboratory analysis, whereas one blinded only participants. By having blinded assessors, researchers diminish the risk of observer bias, but the experimenter bias is not addressed, and unblinded patients may not comply with the protocol. On the other hand, when only participants are blinded, the investigators may influence their behavior or introduce bias on outcome assessment.

10.2 Similarity of Interventions

The first step to maintain blinding during the trial relies on proper treatment manufacturing [45, 46]. In the context of meningitis, where all interventions consisted of drugs, it is crucial that pills or IV solutions are matched in appearance, taste, touch, weight, and smell in order to keep study participants and investigators blinded to group assignment. Additionally, the second step consists in ensuring that investigators do not know which intervention is in the treatment container [45, 46]. Therefore, drug containers must be labeled with only the patient number, and cannot present any treatment identifying information.

We found 29 double-blinded studies that clearly stated that interventions were matched in appearance. Some authors were even more careful with blinding reporting and stated that administration schemes were also similar. On the other hand, drug appearances could not be matched in four double-blinded trials, but researchers overcame this problem by ensuring that containers, syringes and infusion lines were opaque and undistinguishable, which is a valid approach. In addition, five studies that tested interventions with different route of administrations ensured proper blinding by using what is called double-dummy design.

10.3 Double-Dummy Design

Double-dummy design is a technique that ensures blinding when the study treatments cannot be made identical. In this technique, placebo conditions are manufactured for both interventions, and patients receive either only one active treatment or both placebos conditions (when there is a placebo group) [46]. For example, Singhi et al. were interested in studying the effects of glycerol as an adjuvant therapy for childhood bacterial meningitis [47]. Since dexamethasone was another adjuvant treatment already used by that time, the authors randomized patients to four groups: combination therapy, glycerol only (with placebo dexamethasone), dexamethasone only (with placebo glycerol), or double placebo. As glycerol is orally administered and dexamethasone has an intravenous route of administration, authors utilized an oral and an intravenous placebo that matched in appearance to glycerol and dexamethasone, respectively.

10.4 Blinding Assessment

The CONSORT statement recommends that researchers test for blinding success as a mechanism of evaluating the risk of bias [37]. However, in our sample, none of the trials reported any type of blinding assessment. Strategies for assessing blinding can help to assess whether bias was introduced in the trial.

11 Recruitment

Subject recruitment is one of the most challenging aspects of clinical research. Low recruitment rates can increase trial costs, time demanding, and determine the study to stop before reaching the calculated sample size, which affect the study power. Therefore, investigators have to be acquainted to recruitment strategies and, more importantly, know well their target population in order to effectively reach out to them.

The first step when planning recruitment of patients with meningitis is to define the target population. Researchers target their population when they plan the eligibility criteria. Therefore, as discussed in Subheading. 5, the inclusion criteria generally set the etiology of meningitis being studied, whereas the exclusion criteria is adjusted in order to eliminate factors that would decrease trial internal validity. Notwithstanding, these subtle changes on who is eligible to enter the study can have great impact on patient accrual. For example, Mfinanga et al. planned a trial that targeted patients with HIV-associated cryptococcal meningitis and a CD4 cell count <100 cells/μL [48]. In spite of conducting a trial in a high-prevalent HIV area (Zambia and Tanzania), the study had low recruitment rate, and the authors had to change the original criteria, then targeting patients with CD4 cell count <200 cells/μL.

The second step relates to a famous statement dubbed as Lasagna's Law. This principle states that researchers tend to overestimate the pool of available patients for a research study [49]. The popular rationale estimates that only 10% of the accessible population will be eligible for the study [49]. Therefore, investigators have to account for that error when making estimations of the eligible population.

Among the trials reviewed in this chapter, only 47 reported information regarding the number of subjects screened for participating in the trial. This information is extremely helpful as for providing an estimative of recruitment yield to researchers planning a clinical trial in meningitis. Recruitment yield is defined as the number of patients randomized divided by the number of patients screened, and is a good indicator of whether recruitment approaches were effective at reaching the targeted population of study. In this chapter, trials in different etiologies of meningitis had similar recruitment yield rates. In studies of bacterial meningitis, 68.9% of screened patients were indeed randomized, whereas in cryptococcal meningitis studies this rate was 63.25%, in tuberculous meningitis it was 62.11%, and neoplastic meningitis it was 58.21%. Therefore, clinical trialists should take these numbers in account when planning their study. These numbers are relatively high when compared to other chronic outpatient conditions such as chronic pain that yield rate can go as low as 1–5%.

12   Adherence

In the context of clinical trials, adherence refers to the extent to which subjects follow the treatment regimen and complete follow-up visits [50]. Poor adherence to intervention regimen increases outcome variance, and may mislead results. Likewise, lack of adherence to study appointments leads to underpowered testing, and invalid results because of missing data [50]. Moreover, similarly to

Table 8
Adherence rate of clinical trials in meningitis

	Adherence
Overall	89%
Tuberculous Meningitis	93%
Bacterial Meningitis	91%
Eosinophilic Meningitis	91%
Viral Meningitis	91%
Neoplastic Meningitis	88%
Cryptococcal Meningitis	86%

low recruitment rate, poor adherence increases trial costs and duration, and may determine the study to stop prematurely.

To avoid poor adherence, investigators have to understand its potential causes and address those issues when planning and implementing the study. In general, adherence is affected by factors associated both with patients, intervention regimen, and the disease under study [50, 51]. Patient factors vary from the reason for participating in the trial to their beliefs, psyche, comorbidities and relationship with the study staff. Therefore, as previously mentioned, it is important for researchers to understand their target population characteristics and needs in order to address potential problems. On the other hand, the type of treatment, its duration and associated side effects can be an issue for patients. Overall, trials with longer duration and negative side effects have lower retention than short term research with safer drugs. Additionally, the associated costs with transportation and time away from work are significant causes for low adherence, and are key factors to address during the study. Lastly, the natural course of disease and its severity also play an important role in this topic [50, 51].

In this context, the mean adherence rate of meningitis studies was 89% (Table 8), as defined as the percentage of participants who completed the protocol and study follow-up visits. Similarly, all studied etiologies had high adherence rates on average. Yet neoplastic and cryptococcal meningitis were associated with relatively smaller rates. This may be caused because CM and NM patients have unique characteristics that may be related with this finding. For example, CM patients are mostly co-infected with HIV, and NM patients commonly have systemic cancer progression. Therefore, participants with those conditions may drop out because of severe disease progression and necessity to get other drugs that may affect the trial results. In addition, studied interventions in CM and NM frequently have significant side effects, and toxicity is a common reason for which patients drop out the study.

Moreover, clinical researchers have developed strategies to address potential threats to participant adherence. In our sample, only five trials reported methods to improve adherence to study regimen and appointments. The two employed techniques were pill-counting and home visits. By counting pills, the study staff can track subject adherence to the treatment regimen and, thus, approach patients that are not following the protocol. On the other hand, home visits are especially useful for trials that have outpatient follow-ups, and may increase retention substantially. For example, Mfinanga et al. conducted a study in which trained lay workers were responsible to schedule and visit participants at their homes in order to deliver the study drug (Antiretroviral Therapy) and provide adherence support [48]. Similarly, after losing the first 16 patients to follow-up, Molyneux et al. improved the recording of patient's home addresses in order to trace them if they did not attend study appointments [52]. Not coincidently both studies had adherence rates >90%, which is an indicative of how methods to improve adherence must always be considered in clinical trials.

13   Study Duration and Time Points

While planning a clinical trial, investigators may wonder how long a study should last in order to provide useful data and respond their research question. In the case of meningitis, this decision will be based upon the studied etiology and study outcomes. In the following subsections we explore the trial duration for each etiology of meningitis.

13.1   Bacterial Meningitis

Most studies in BM followed patients through 6 or 12 months (Fig. 6). These trials generally aimed to measure death or neurological morbidity. Few studies had a shorter length, although

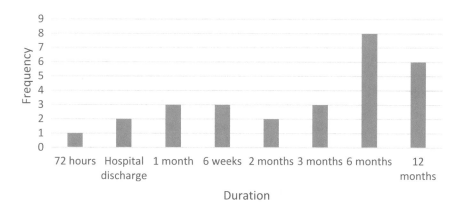

Fig. 6 Frequency of trials by study duration in Bacterial meningitis

they usually measured the same type of outcomes. In fact, trials that followed patients for 6 or more months have more power for measuring mortality and chronic neurological disability. In contrast to this pattern, a single study followed participants for 72 h during a meningitis outbreak in Niger, in which mortality rate and clinical signs and symptoms were measured in this short time frame. Nevertheless, because the number of patients and the mortality rate in that area is expressively high, the study had sufficient power to get robust results. Yet the short duration of trial limit generalizability of results for the long-term follow-up of patients, which is a limitation of that study.

13.2 Cryptococcal Meningitis

We found that the majority of studies on CM lasted 10 weeks (Fig. 7). The reason behind 10-week follow-up is the duration of therapy for CM. These trials typically tested a therapeutic scheme that consisted on 2 weeks of a set of drugs, followed by 8 weeks of maintenance therapy. Therefore, authors were mainly interested on measuring surrogate or clinical outcomes at 2 weeks and the follow-up at 10 weeks. On the other hand, a considerable number of trials lasted 6 and 12 months. Overall, these studies were phase III that studied long-term survival of HIV-infected patients.

13.3 Tuberculous Meningitis

For TBM, the majority of studies followed patients for either 6 or 9 months (Fig. 8). Similarly to CM trials, the reason behind these two study durations relies on the length of therapy for TBM. For these trials, researchers constrained their assessment to the time window of the therapy, which lasted 6 or 9 months among the included trials. In contrast, the studies that assessed 12 and 24 month outcomes were mainly interested on long survival analysis. The single trial that followed patients for only 1 month aimed to measure CSF biomarkers, thus, justifying the short length of the study.

13.4 Neoplastic Meningitis

The duration of studies in NM in our sample varied significantly according to trial phases and type of outcome. Phase 1 studies accounted for the trials that lasted 2 weeks and 1 month, and they evaluated CSF biomarkers and adverse events. On the other hand, the trials that lasted 9, 12, and 24 months measured long-term progression-free survival or quality of life survival (Fig. 9).

13.5 Eosinophilic Meningitis

The three studies on EM included in our sample varied considerably in their duration. However, these trials were homogeneous in their study outcomes and design as all of them aimed to measure symptomatic relief of EM, and researchers acquired the primary outcome data after 2 weeks of enrollment. Because the secondary outcome of these studies was the time until complete recovery of symptoms, the duration of trials varied depending on individual data.

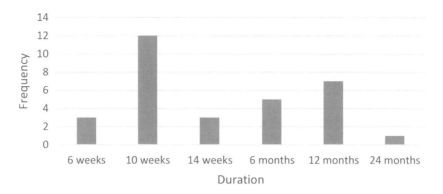

Fig. 7 Frequency of trials by study duration in Cryptococcal meningitis

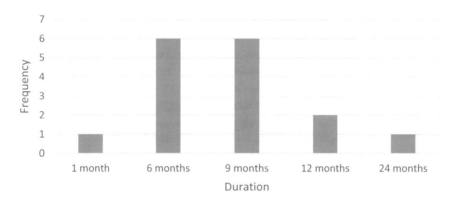

Fig. 8 Frequency of trials by study duration in Tuberculous meningitis

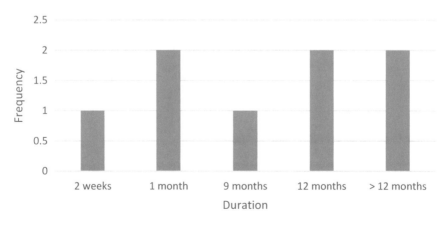

Fig. 9 Frequency of trials by study duration in Neoplastic meningitis

13.6   *Viral Meningitis*   The analysis of VM trials is limited to the single study included in this chapter. In this trial, researchers studied infection by herpes simplex virus 2 (HSV-2). Because this infection is chronic and the patients are subjected to recurrent episodes of meningitis, the researchers followed study participants for 2 years in order to measure the time to recurrence.

14   Statistical Analysis

Appropriate statistical analysis is critical for answering the tested research questions. Although there are many ways to analyze data, researchers have to follow basic assumptions when planning their statistical approach (see Chap. 1—Introduction). In addition, researchers have to be acquainted to the main advantages and disadvantages of each analysis method in order to extract the best of their data.

As discussed previously, the main outcomes on meningitis measure either survival or neurological morbidity or CSF outcomes or clinical improvement. Further, we showed that researchers have used these outcomes in different ways (e.g., time-to-death, mortality rate). In this context, study outcomes can be further classified in respect of their level of measurement (Consult Chap. 1—Introduction for more details). We will discuss the most frequently used statistical analysis for each type of data in order to guide clinical researchers when designing their trial.

14.1   Time-to-Event Data

We found 15 clinical trials that used time-to-event data as a primary outcome. The majority of these studies had time-to-death as their main outcome (Table 9). On the other hand, two studies had time-to-symptom disappearance as their main outcomes. Further, another study in CM had time-to-mycological response as a primary outcome, which represents the time to culture clearance. Finally, one study on neoplastic meningitis evaluated the progression-free survival, which is defined as the time-to-progression of the disorder.

Table 9
The use of time-to-event data in meningitis trials

	Frequency	Kaplan–Meier curves	Log-rank test	Cox-proportional hazards model
Overall	15	15	15	15
Time-to-death	11	11	11	11
Time-to-clinical response	2	2	2	2
Time-to-mycological response	1	1	1	1
Progression-free survival	1	1	1	1

Table 10
Primary outcomes applied in meningitis trials

	Frequency	Chi-square	Fisher's exact test	Logistic regression	Risk difference	Cochran–Mantel–Haenszel
Overall	43	21	8	9	3	2
Mortality rate	13	4	1	6	1	1
Mycological response (yes/no)	9	3	3	0	2	1
Clinical response (yes/no)	11	8	3	0	0	0
Audiological/neurological disability	6	3	1	3	0	0
Glasgow Outcome Scale	2	1	0	0	0	0
Meningitis recurrence rate	1	1	0	0	0	0
MRI outcome (yes/no)	1	1	0	0	0	0

All studies that analyzed time-to-event data performed Kaplan–Meier curves, log-rank test, and Cox-proportional hazards model to calculate median survival time, compare treatments and also adjust for other covariates, respectively (Consult Chap. 2—Introduction).

14.2 Categorical Data

In our sample, the majority of clinical trials had categorical data as their primary outcome. The primary outcomes can be seen in Table 10. Chi-square was the most employed statistical test for this type of data, mainly to analyze the clinical response outcome, which was categorized differently across trials (consult Sect. 7). Further, other eight trials analyzed categorical outcomes with Fisher's exact test. This test is an alternative to chi-square when researchers are dealing with small sample sizes, and the main assumption for this test is not fulfilled. Therefore, 29 studies on total analyzed their primary outcome with tests that compare observed and expected values in tables of contingence. These tests are relatively simple to perform, and easy to understand.

On the other hand, nine trials utilized logistic regression models to analyze their primary endpoints. Interestingly, researchers applied this approach to analyze mortality (6/9) or audiological and neurological disability outcomes. Logistic regression models allow researchers to adjust the statistical analysis for confounders and effect modifiers. For example, Peltola et al. opted to measure the effect of an adjuvant therapy on mortality rate in CM with a logistic model that included receipt of pretreatment antimicrobial

Table 11
Relation between the use of continuous primary endpoints and specific statistical tests

	Frequency	Linear regression	Student's T test	ANOVA	Mann Whitney U test
Overall	15	8	2	2	3
CSF biochemical data	3	0	0	2	1
CSF drug concentration	3	0	2	0	1
Mean rate decrease in the number of CFU in CSF	8	8	0	0	0
CSF opening pressure	1	0	0	0	1

drugs and timing of initiation of antimicrobial therapy as covariates [35]. This approach enabled them to identify small differences that could have remained undetected in direct comparisons between study groups.

14.3 Continuous Data

We found 15 trials that had continuous primary endpoints. Among these, eight studies analyzed the mean rate decrease in the number of Colony Forming Units (CFU) in the CSF of patients with CM (Table 11). All these studies employed linear regression in the analysis of this endpoint. This approach enables researchers to compare this mycological biomarker across groups controlling for its baseline values among subjects. In contrast, two studies analyzed other CSF biochemical data with repeated measure ANOVA models with treatment, time and interaction of treatment and time as independent variables. Both approaches are valid to measure changes in continuous data in meningitis trials, although the latter may suffice for most situations.

Moreover, two studies compared the mean of drug concentration in CSF across treatment groups with unpaired Student's *t* test, while three others used Mann–Whitney *U* test to analyze CSF opening pressure, CSF drug concentration or CSF biochemical endpoints. The option of using a nonparametric test is guided by the normality assumptions explained on Chap. 1—Introduction. Overall, parametric comparison of means has more power than nonparametric tests. Although, in the context of small samples that do not meet these assumptions, the use of nonparametric approaches is likely indicated for lack of normal distribution.

14.4 Missing Data

Missing data is a common feature of clinical trials. Despite all efforts to improve subject's adherence, and data collection and entry, it is uncommon to complete a study with no missing values. In fact, missing data can have serious consequences to trial results, since it reduces study power and may lead to biased results. This way, in addition to limit the likelihood of missing data, researchers have to know well the available methods for handling it and plan ahead the most suitable strategy for their trial.

Before presenting the most used strategies in meningitis trials, it is worthy to mention the types of missing data. Three missing data mechanisms exist: missing completely at random (MCAR), missing at random (MAR), and missing not at random (MNAR). The main idea behind this classification is whether the missing values are related to the study outcome and/or independent variables. Missing completely at random (MCAR) defines the case when missing values are independent of study outcomes and independent variables [53]. In contrast, missing at random (MAR) occurs whenever missing data depends on independent variables but do not relate to the outcome. Finally, in missing not at random (MNAR) the missing data relates to study outcomes [53].

To deal with the aforementioned types of missing data, researchers have different options depending on the context. The most simple but worst methodology consists in analyzing only the subjects who completed the study, which is defined as the Complete Case Analysis (CCA) or Per-protocol analysis. This approach is valid whenever the mechanism of missing data is MCAR. However, this approach leads to decreased study power, since it excludes subjects from the main analysis [53]. Therefore, the most recommended approach by the CONSORT criteria is to use a method of imputation of missing observations and analyze all participants as they were randomized, which is called intention-to-treat (ITT) analysis [37]. This approach allows researchers to preserve the benefit of randomization on statistical inferences, and can be used for MCAR and MAR mechanisms of missing data [37, 53]. Further, there are different methods for imputing missing data that comply with the ITT principle. Those methods range from simple imputation approaches (e.g., last observation carried forward, worst case scenario, mean/median imputation) to sophisticated techniques (e.g., multiple imputation). Finally, there is no efficient method to deal with MNAR data, since the missing observations are directly related to the study outcome.

In the context of meningitis trials, most trials in our sample followed the principle of intention-to-treat analysis (Table 12). However, only nine studies specified the strategy to deal with missing observations in those cases. Four of them used the last observation carried forward (LOCF), which consists on imputing missing data from the last known outcome value. The LOCF is simple and the most frequently used method of imputation in clinical trials [37]. Although, some authors have criticized this method, because its primary assumption consists in the lack of change in outcome values from previous observations, which is likely incorrect. Nevertheless, the Food and Drug Administration (FDA) has recommended this approach as a more conservative alternative to the

Table 12
Data analysis in meningitis clinical trials

	Adherence
ITT	72
Last observation carried forward	4
Multiple imputation	3
Worst case scenario	1
Not specified	64
Per-protocol	20
Not reported	8

per-protocol analysis [53]. On the other hand, three clinical trials in our sample employed multiple imputation models, which also are recommended by the FDA [53], and are a sophisticated but more complex approach to data imputation. Finally, one study considered the worst case scenario in order to impute the missing values, which consists on imputing the missing observation with the worst outcome. This approach is the most conservative of the three, but it may lead to robust results in case it demonstrates that missing data did not affected the study outcome.

14.5 Subgroup Analysis

A subgroup is defined by study participants that share a common characteristic at baseline (e.g., age, sex, race) [54]. The idea of analyzing subgroups come from a feature observed in clinical practice: patients with a specific condition differ greatly from each other, and respond differently to the therapy [54]. This way, subgroup analysis in the context of clinical trials serve the purpose of examining whether the treatment effects are different in patients that share a common characteristic. However, clinical researchers have to understand that subgroup analysis are merely exploratory, and the results generated from this type of analysis cannot be accepted as definitive. Therefore, subgroup analysis serve mainly to generate new hypothesis that can be tested in future studies.

In this sense, investigators may want to design a subgroup analysis for their trials to fully explore the data they have in hands. In order to achieve that, they have to follow some general concepts. Firstly, subgroups should be defined a priori, before the study starts. Secondly, the subgroup variable should be defined based on previous trials, or on the selected stratification variables in stratified randomized trials (e.g., site, age, severity of disease). Finally, the appropriate statistical tests for subgroup analysis usually consists in a regression model in which researchers test the interaction between treatment and the subgroup variable [54]. We will present how frequent subgroup analyses are utilized on meningitis trials, what variables are used to define subgroups, and what are the statistical tests used in this field.

In this review, the most frequently used variable for subgroup analysis is the causative pathogen. This was observed for BM trials, in which different pathogens are possible (e.g., Pneumococcus, Meningococcus, Haemophilus). The rationale behind this analysis is to address the potential differential prognostic effect of causative agents. Moreover, the second most used variable was likelihood of diagnosis, usually defined as probable and definite. This was also observed on some studies of BM trials, in which researchers opted to enroll all admitted patients with a suspected BM case. Since the confirmatory diagnosis is only available after cultures are analyzed, researchers preferred to analyze patients with regard of their likelihood of having BM. In contrast, HIV-infection status was used as a subgroup variable in both CM and TBM trials. Still in the context of TBM studies, TBM grades (e.g., British Medical Research Council Grade) were used to divide patients in subgroups according to severity of symptoms. Other variables used for subgroup analysis can be seen in Table 13 below.

14.6   Covariate Adjustment

Despite the randomization process being conducted appropriately, imbalances on some features can be observed in randomized trials [55]. When these characteristics are associated with the study outcome, they can increase their variability and result in bias. One method for controlling those imbalances was previously discussed on the randomization process, and consists on conducting stratified randomization. Another option is to use a statistical regression model in order to adjust the main analysis for important prognostic factors, which is defined as covariate adjustment [55]. This way, investigators are able to increase the efficacy of their analysis while reducing variability and bias risk.

In our sample, we found 29 studies that performed covariate adjustment in their analysis. Of these, most studies adjusted the analysis for the effect of baseline characteristics such as age, gender, altered mental status, and CSF white cells count. Characteristics such as age and gender are controlled because of their potential imbalance between groups, mainly when dealing with relative small sample sizes. Altered mental status is an important prognostic factor and was frequently used for TBM and CM trials. In contrast, CSF biochemical variables, such as white cells count, protein concentration, and CSF/blood glucose ratio were most used in BM trials. Other adjusted covariates are shown on Table 14.

Table 13
Variables applied in subgroup analysis in studies of meningitis

	Frequency
Overall	18 studies
Causative pathogen	7
Diagnosis (probable/definite)	3
HIV status positive (yes/no)	3
TBM grade	2
Time of death	1
CSF white cells count	1
Cryptococcal antigen positive (yes/no)	2
Glasgow coma score <15 (yes/no)	1
Prior antibiotic treatment (yes/no)	1
Indication for steroid use at baseline	1
Continent	1
Country	1
Age	1
Sex	1
Naïve to ARV (yes/no)	1
CD4 cell count	1
CSF culture results	1

Table 14
Adjusted covariates applied in studies of meningitis

	Frequency
Overall	29 studies
Age	8
Gender	7
Altered mental status	6
CSF white cells count	5
Blood laboratory tests	3
CD4 cell count	3
CSF total protein concentration	2

(continued)

Table 14
(continued)

	Frequency
CSF–blood glucose ratio	2
Concomitant rifampicin use	2
HIV viral load	2
CSF opening pressure	2
Causative pathogen	2
Duration of symptoms	2
Drug dose	2
Study site	2
CSF lactate concentration	1
Creatinine clearance	1
CSF fungal burden	1
Heart rate	1
CSF gamma interferon concentration	1
HIV status	1
Drug resistance	1
TBM grade	1
Race	1
History of *Pneumocystis carinii* pneumonia	1
Cryptococcal antigen group	1
Maintenance treatment (yes/no)	1
Use of flucytosine	1
Baseline organism load	1
Presence of hydrocephalus	1
Previous treatment with antibiotics	1
HIV status positive (yes/no)	1
Presence of seizures (yes/no)	1
Blood culture	1
Hypotension on admission	1

15   Limitations

Despite all efforts on planning and implementing a clinical trial, it is improbable that a study will not have limitations. Constraints to study implementation can raise from many sources and investigators should make a great effort to address most of them during trial planning. Yet some perceived flaws can be corrected during the trial, but always with a risk to affect study internal validity. For instance the use of different recruitment and adherence-enhancing techniques may be considered without a significant penalty to internal validity. In contrast, some limitations are directly related to the disorder or intervention under study, and it may be difficult for researchers to eliminate all potentials shortcomings in their trial. In this sense, the most frequently observed limitations in meningitis trials are related to small sample sizes, unblinded studies, and allocation concealment.

15.1   Small Sample Sizes

The most frequent limitation on meningitis trials was the presence of relative small sample sizes, which increases the risk of type II error. This issue can be mainly addressed on trial planning. As aforementioned, researchers have to know well their target population, and plan ahead efficient recruitment and adherence strategies in order to achieve and maintain an appropriate number of subjects. In case investigators foresee difficulties on patient accrual, the choice for multicenter trials might be a reasonable alternative when feasible. Another option is to adapt the trial design for a smaller population by reducing the possible variability in study outcomes. This is achieved by using a more homogeneous population, a robust outcome assessment (i.e., surrogate outcome), and having a strong rationale and background evidence for comparing the study groups. Notwithstanding, appropriate estimations of expected effect size and sample size calculations are critical on conducting trials with sufficient statistical power.

15.2   Unblinded Studies

As reported in Subheading 10, 35 out of 100 studies included in our sample were unblinded. Open-label studies have important caveats and can undermine the validity of trial results. In the context of meningitis—where most trials compared the combination of different drugs, or different routes of administration, or different duration of therapies—the use of matched placebos, or double-dummy designs are indispensable in order to reduce the introduction of biases. Moreover, researchers should not only put a great deal of effort on having similar interventions, but also on assessing whether the chosen blinding methodology was successful in their trials. It is noteworthy that none clinical trials in this chapter reported the use of blinding assessment.

15.3   Allocation Concealment

Another limitation highlighted in this chapter concerns to allocation concealment in meningitis trials. Most studies did not report the employed method for concealing the randomization schedule from study physicians and investigators. Moreover, due to lack of good reporting, it was not possible to evaluate whether the used methods were really efficacious. Anyway, it is important to highlight the central role of appropriate concealment in preventing introduction of selection and confounding biases to study results. Methods such as sequentially numbered, opaque, sealed envelopes (SNOSE); sequentially numbered containers; pharmacy controlled; and central randomization are considered the most adequate for clinical trials. In addition, researchers have to ensure that different persons are responsible for enrollment and generation of the randomization list in order to minimize potential biases.

References

1. World Health Organization (WHO) (2011) Health topics: meningitis. WHO, Geneva. https://doi.org/10.5860/CHOICE.41-4081

2. Kasanmoentalib ES, Brouwer MC, van de Beek D (2013) Update on bacterial meningitis: epidemiology, trials and genetic association studies. Curr Opin Neurol 26:282–288. https://doi.org/10.1097/WCO.0b013e328360415c

3. World Health Organization (WHO) (1997) Control of epidemic meningococcal disease. WHO practical guidelines. Int Organ 17:82. https://doi.org/10.1017/S0020818300002071

4. van de Beek D, de Gans J, Spanjaard L et al (2004) Clinical features and prognostic factors in adults with bacterial meningitis. N Engl J Med 351:1849–1859. https://doi.org/10.1056/NEJMoa040845

5. Scheld WM, Koedel U, Nathan B, Pfister HW (2002) Pathophysiology of bacterial meningitis: mechanism (s) of neuronal injury. J Infect Dis 186:225–233. https://doi.org/10.1086/344939

6. Desalermos A, Kourkoumpetis TK, Mylonakis E (2012) Update on the epidemiology and management of cryptococcal meningitis. Expert Opin Pharmacother 13:783–789. https://doi.org/10.1517/14656566.2012.658773

7. Park BJ, Wannemuehler K a, Marston BJ, et al (2009) Estimation of the current global burden of cryptococcal meningitis among persons living with HIV/AIDS. AIDS 2009 23:525–530. doi: https://doi.org/10.1097/QAD.0b013e328322ffac

8. Torok ME, Bang ND, Chau TTH et al (2011) Dexamethasone and long-term outcome of tuberculous meningitis in Vietnamese adults and adolescents. PLoS One 6:1–6. https://doi.org/10.1371/journal.pone.0027821

9. Gleissner B, Chamberlain MC (2006) Neoplastic meningitis. Lancet Neurol 5:443–452. https://doi.org/10.1016/S1474-4422(06)70443-4

10. Herrlinger U, Förschler H, Küker W et al (2004) Leptomeningeal metastasis: survival and prognostic factors in 155 patients. J Neurol Sci 223:167–178. https://doi.org/10.1016/j.jns.2004.05.008

11. Roth P, Weller M (2015) Management of neoplastic meningitis. Chinese Clin Oncol 4:1–8. https://doi.org/10.3978/j.issn.2304-3865.2015.05.02

12. Tamimi NA, Ellis P (2009) Drug development: from concept to marketing. Nephron Clin Pract 113:125–131. https://doi.org/10.1159/000232592

13. Bomgaars L, Geyer JR, Franklin J et al (2004) Phase I trial of intrathecal liposomal cytarabine in children with neoplastic meningitis. J Clin Oncol 22:3916–3921. https://doi.org/10.1200/JCO.2004.01.046

14. Organization WM (2013) World Medical Association declaration of Helsinki. JAMA 310:2191. https://doi.org/10.1001/jama.2013.281053

15. Jackson AT, Nussbaum JC, Phulusa J et al (2012) A phase II randomized controlled trial adding oral flucytosine to high-dose fluconazole, with short-course amphotericin B, for cryptococcal meningitis. AIDS 26:1363–1370. https://doi.org/10.1097/QAD.0b013e328354b419

16. Tansuphaswadikul S, Maek-a-Nantawat W, Phonrat B et al (2006) Comparison of one week with two week regimens of amphotericin B both followed by fluconazole in the treatment of cryptococcal meningitis among AIDS patients. J Med Assoc Thai 89:1677–1685

17. Jarvis JN, Meintjes G, Rebe K et al (2012) Adjunctive interferon-γ immunotherapy for the treatment of HIV-associated cryptococcal meningitis. AIDS 26:1105–1113. https://doi.org/10.1097/QAD.0b013e3283536a93

18. Whitty CJM, French N et al (2007) Corticosteroids for bacterial meningitis in adults in sub-Saharan Africa. N Engl J Med 357:2441–2450

19. Bahr NC, Boulware DR (2014) Methods of rapid diagnosis for the etiology of meningitis in adults. Biomark Med 8:1085–1103. https://doi.org/10.2217/bmm.14.67

20. Gray LD, Fedorko DP (1992) Laboratory diagnosis of bacterial meningitis. Clin Microbiol Rev 5:130–145. https://doi.org/10.1128/CMR.5.2.130

21. Boulware DR, Meya DB, Muzoora C et al (2014) Timing of antiretroviral therapy after diagnosis of cryptococcal meningitis. N Engl J Med 370:2487–2498. https://doi.org/10.1056/NEJMoa1312884

22. Marais S, Thwaites G, Schoeman JF et al (2010) Tuberculous meningitis: a uniform case definition for use in clinical research. Lancet Infect Dis 10:803–812. https://doi.org/10.1016/S1473-3099(10)70138-9

23. Montezuma-Rusca JM, Powers JH (2014) Outcome assessments in clinical trials of cryptococcal meningitis: considerations on use of early fungicidal activity as a potential surrogate endpoint for all-cause mortality. Curr Treat Options Infect Dis 6:326–336. https://doi.org/10.14440/jbm.2015.54.A

24. Ravina B, Cummings J, Mcdermott M, Poole RM (2012) Selecting outcome measures. In: Clinical trials in neurology: design, conduct, analysis, 1st edn. Cambridge University Press, Cambridge, pp 69–77

25. Atkinson AJJ, Colburn WA, DeGruttola VG et al (2001) Biomarkers and surrogate endpoints: preferred definitions and conceptual framework. Clin Pharmacol Ther 69:89–95. https://doi.org/10.1067/mcp.2001.113989

26. Walton MK, Iii JHP, Hobart J et al (2015) Clinical outcome assessments: conceptual foundation — report of the ISPOR clinical outcomes assessment – emerging good practices for outcomes research task force. Value Health 18:741–752

27. Ross S (2007) Composite outcomes in randomized clinical trials: arguments for and against. Am J Obstet Gynecol 196:119.e1–119.e6. https://doi.org/10.1016/j.ajog.2006.10.903

28. Nathan N, Borel T, Djibo A et al (2005) Ceftriaxone as effective as long-acting chloramphenicol in short-course treatment of meningococcal meningitis during epidemics: a randomised non-inferiority study. Lancet 366:308–313. https://doi.org/10.1016/S0140-6736(05)66792-X

29. Kim J (2012) Survival analysis. Pediatr Rev 33:621–643. https://doi.org/10.1016/B978-0-12-802387-7.00035-4

30. George B, Seals S, Aban I (2014) Survival analysis and regression models. J Nucl Cardiol 21:686–694. https://doi.org/10.1007/s12350-014-9908-2

31. Kanra GY, Ozen H, Secmeer G et al (1995) Beneficial effects of dexamethasone in children with pneumococcal meningitis. Pediatr Infect Dis J 14:490–494

32. Gijwani D, Kumhar MR, Singh VB et al (2002) Dexamethasone therapy for bacterial meningitis in adults: a double blind placebo control study. Neurol India 50:63

33. Sankar J, Singhi P, Bansal A, Ray P, Singhi S (2007) Role of dexamethasone and oral glycerol in reducing hearing and neurological sequelae in children with bacterial meningitis. Indian Pediatr 44:649–656. https://doi.org/10.1301/nr.2006.sept.385

34. Peltola H, Roine I, Ferna J et al (2007) Adjuvant glycerol and/or dexamethasone to improve the outcomes of childhood bacterial meningitis: a prospective, randomized, double-blind, placebo-controlled trial. Clin Infouctious Dis Am 45:1277–1286. https://doi.org/10.1086/522534

35. Kapoor R et al (2003) Blockers of sodium and calcium entry protect axons from nitric oxide-mediated degeneration. Ann Neurol 53(2):174

36. Schulz KF, Altman DG, Moher D, Group C (2010) Academia and clinic annals of internal medicine CONSORT 2010 statement: updated guidelines for reporting parallel group randomized trials OF TO. Ann Intern Med 1996:727–732. https://doi.org/10.7326/0003-4819-152-11-201006010-00232

37. Schulz KF, Grimes DA (2005) Sample size calculations in randomised trials: mandatory and mystical. Lancet 365:1348–1353. https://doi.org/10.1016/S0140-6736(05)61034-3

38. Cohen J (1988) Statistical power analysis for the behavioral sciences, 2nd edn. Lawrence Erlbaum Associates, Mahwah, NJ

39. Schulz KF (1995) Subverting randomization in controlled trials. JAMA 274:1456–1458

40. Vickers AJ (2008) How to randomize. J Soc Integr Oncol 4:194–198

41. Schulz KF, Grimes DA (2002) Allocation concealment in randomised trials: defending against deciphering. Lancet 359:614–618

42. Pildal J, Chan A-W, Hrobjartsson A et al (2005) Comparison of descriptions of allocation concealment in trial protocols and the published reports: cohort study. BMJ 330:1049–1040. https://doi.org/10.1136/bmj.38414.422650.8F

43. Schulz KF, Grimes DA (2002) Generation of allocation sequences in randomised trials: chance, not choice. Lancet 359:515–519

44. Schulz KF, Grimes DA (2002) Blinding in randomised trials: hiding who got what. Lancet 359:696–700

45. Schulz KF (2002) The landscape and lexicon of blinding in randomized trials. Ann Intern Med 136:254–259

46. Singhi S, Jarvinen A, Peltola H (2008) Increase in serum osmolality is possible mechanism for the beneficial effects of glycerol in childhood bacterial meningitis. Pediatr Infect Dis J 27:892–896. https://doi.org/10.1097/INF.0b013e318175d177

47. Mfinanga S, Chanda D, Kivuyo SL et al (2015) Cryptococcal meningitis screening and community-based early adherence support in people with advanced HIV infection starting antiretroviral therapy in Tanzania and Zambia: an open-label, randomised controlled trial. Lancet 385:2173–2182. https://doi.org/10.1016/S0140-6736(15)60164-7

48. Lasagna L (1979) Problems in publication of clinical trial methodology. Clin Pharmacol Ther 25:751–753

49. Robiner WN (2005) Enhancing adherence in clinical research. Contemp Clin Trials 26:59–77. https://doi.org/10.1016/j.cct.2004.11.015

50. Thoma A, Farrokhyar F, Mcknight L, Bhandari M (2010) How to optimize patient recruitment. Can J Surg 53:205–210

51. Molyneux E, Walsh A, Forsyth H et al (2002) Dexamethasone treatment in childhood bacterial meningitis in Malawi: a randomised controlled trial. Lancet 360:211–218. https://doi.org/10.1016/S0140-6736(02)09458-8

52. Haukoos JS, Newgard CD (2007) Advanced statistics: missing data in clinical research-part 1: an introduction and conceptual framework. Acad Emerg Med 14:662–668. https://doi.org/10.1197/j.aem.2006.11.037

53. Yusuf S, Wittes J, Probstfield J, Tyroler HA (1991) Analysis and interpretation of treatment effects in subgroups of patients in randomized clinical trials. JAMA 266:93–98

54. Pocock SJ, Assmann SE, Enos LE, Kasten LE (2002) Subgroup analysis, covariate adjustment and baseline comparisons in clinical trial reporting: current practice and problems. Stat Med 21:2917–2930. https://doi.org/10.1002/sim.1296

Chapter 8

Multiple Sclerosis

Tarek Nafee, Rodrigo Watanabe, and Felipe Fregni

Abstract

Multiple sclerosis is proposed to be a neurological syndrome that may have a significant impact on patients' functionality and quality of life. Due to the complexity of clinical trials in multiple sclerosis, conducting research studies in this field poses critical challenges to investigators. A review of the literature was conducted in the online database "Web of Science" for identifying the 100 most cited trials in multiple sclerosis. Articles published 2010 and 2015 were included. This chapter discusses the main insights that emerged from this review and presents fruitful suggestions for advancing the research scope in multiple sclerosis.

Key words Multiple sclerosis, MS, Disseminated sclerosis, Review, Clinical trial

1 Introduction

Multiple Sclerosis (MS) is proposed to be an immune-mediated demyelinating disease that damages central nervous system neurons. The most common sites of demyelination include the periventricular white matter, brain stem, spinal cord, and optic nerve [1–3]. Worldwide, multiple sclerosis is the most common permanently disabling disease that affects young adults. A German study reported that the annual incidence is 0.3/100,000 in children up to the age of 16 [4]. In the general population, the prevalence of MS has been reported to be from 47.2 to 295 per 100,000 depending on ethnicity and geographic region [5, 6]. Multiple Sclerosis can progress to have a tremendous effect on the functionality and quality of life of patients who are afflicted with this disease. This includes temporary or permanent visual disturbance or blindness, deafness, sensory deficit or paresthesia, motor deficit or paralysis, urinary incontinence, cognitive impairment, or peripheral pain [7]. MS initially presents as clinically isolated syndromes affecting specific organ systems. The subsequent course of the disease differentiates among the subtypes. Relapse remitting MS (RRMS) typically

Felipe Fregni (ed.), *Clinical Trials in Neurology*, Neuromethods, vol. 138,
https://doi.org/10.1007/978-1-4939-7880-9_8, © Springer Science+Business Media, LLC, part of Springer Nature 2018

presents with acute attacks known as relapses which remit with or without treatment. Progressive relapsing MS (PRMS) similarly presents with acute relapses; however, symptoms do not fully remit. Patients with PRMS retain some degree of symptoms which culminate to a more permanent disability. Regardless of the subtype, MS has been shown to progress to affect the patient's ability to work and perform daily functional tasks [8].

The exact pathophysiology of multiple sclerosis remains unknown and there is still no cure for this disease. Therefore, most of the current clinical trials focus on diagnostic modalities, treatments to reduce the frequency and severity of acute exacerbations as well as reductions in disease progression and improvements in quality of life and functional abilities of the patients. The appropriate understanding of clinical trial methodology in multiple sclerosis is essential to effectively study and explore potential treatments' effects.

Multiple Sclerosis clinical trials present some unique challenges [9]. Firstly, due to the multiplicity of trials with different treatment protocols, trial designs, choice of comparators and outcomes, it is very difficult to compare results between studies. Secondly, positive results do not always translate to real impact in clinical practice due to studies of short duration. Thirdly, there are numerous MS clinical trials performed globally that enroll a limited pool of patients due to resource limitations. Subsequently, researchers recruit for patients across continents; this drastically increases the complexity and costs of the trials. Heterogeneity in multinational trials is largely due to different methodological standards (e.g., MRI protocols) and ethical standards in regard to compensation to patients. These factors can bias the patient selection in each country; making the results less generalizable and, in turn, negatively impact external validity.

To address some of these unique design challenges and in an attempt to guide the decision making processes of multiple sclerosis investigators, this chapter summarizes and discusses the relevant characteristics of the 100 most highly cited clinical trials in MS since 2010 including, but not limited to, trial designs, study setting, trial phases, interventions, sample selection, choice of outcomes, and statistical analyses aiming to give an overview and discussion of current methodology applied in MS clinical trials.

2   Methods

In this chapter, the aim is to describe and discuss methodological aspects of the most highly cited interventional clinical trials in multiple sclerosis from 2010 to 2015. To identify these trials, a search was performed in all Web of Science databases' titles for

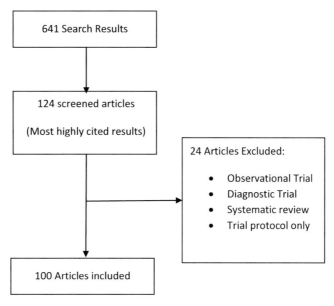

Fig. 1 Schematic illustration of search results and exclusions

"Multiple Sclerosis." The search was limited from 2010 to 2015 and document type refined to only include clinical trials. This search yielded 641 results. The results were sorted from highest to lowest cited.

Titles and abstracts were screened to include interventional trials only. To achieve 100 publications that met inclusion criteria, 124 were screened. The selection process and the 24 exclusions are schematically represented in Fig. 1.

3   Initial Glance (Review Methodology)

Data was collected about the features of the top 100 cited clinical trials from 2010 to 2015 according to CONSORT criteria for clinical trials. Table 1 summarizes the general characteristics of these articles. Seventy-one percent of the included articles were published in journals that endorse CONSORT criteria, making the data collection more straight forward and our results more robust. The remaining 29% did not explicitly describe all the items on the CONSORT checklist and data was either deduced by the authors or otherwise reported as 'not specified'.

About 25% of the articles were published in the highest impact journals, with impact factors of 20 and above. According to Fig. 2, over 50% of the cumulative numbers of citations were generated from the top six highest impact factor journals. Another two spikes

Table 1
General characteristic of cited articles

			# (Total $N = 100$)
Year		2010	35
		2011	29
		2012	23
		2013	8
		2014	5
		2015	0
Phase		Pilot	1
		Phase I	4
		Phase I/II	2
		Phase II	29
		Phase II/III	2
		Phase III	38
		Phase IV	1
		Exploratory	19
		Pivotal	5
Trial design	RCT	2 parallel arms	54
		3 parallel arms	23
		4 or more parallel arms	3
		Factorial	0
		Cross over	2
	Non RCT		18
Sample size		0–20	13
		21–50	18
		51–100	13
		101–500	33
		501–1000	6
		1001+ Above	17
		Not applicable	1
Type of intervention		Drug	76
		Device	1
		Procedure	23
Placebo controlled		Yes	59
		No	41
# of citations		20–30	45
		31–40	27
		41–50	7
		51–60	2
		61–70	3
		71–100	4
		101–200	8
		201–745	4

(continued)

Table 1
(continued)

		# (Total *N* = 100)
Journal impact factor	1–5	46
	5–10	30
	10–20	0
	more than 20	24
Consort criteria	Yes	71
	No	29
Positivity for primary outcome	Yes	72
	No	26
	Not Applicable	2

are seen in mid-high impact factor range with Journal of Neurology Neurosurgery & Psychiatry, and Multiple Sclerosis Journal (MSJ). MSJ papers, in particular, represent 19% of the search result and contribute 16.7% of the total cumulative citations. This is due to the relevance of the journal to this field of MS (Table 2).

The highest number of publications tended to be in the earlier years. The longer the period since publication, the more likely the paper has been cited, thus making it appear higher in our search results (Table 3).

Another interesting finding to note is that 72% of the publications reported results that were positive for their primary outcome. Since we are looking specifically at the 100 most cited papers in this field of research, this figure is expected to be higher than the mean which already has an inherent publication bias toward positive result publications (Table 4).

Perhaps a more relevant relationship to discuss is shown in Fig. 3. Although 72% of publications have shown positive results; 83% of the total citations are derived from them. This may indicate that achieving positive results increases the likelihood of being cited—a characteristic we deem to be a measure of relevance and impact in the field. This is particularly noteworthy in MS research since there is no cure to the disease, and its impact on patient functionality is enormous. This creates a high demand for efficacious treatments to reduce the relapse rate and slow down the progression of this disease. Immunomodulatory drugs and monoclonal antibodies that have shown positive results in reducing annualized relapse rates have received the highest citations per article, and are usually published in the highest impact journals. Alternatively, several trials exploring the efficacy of vitamin D supplementation and

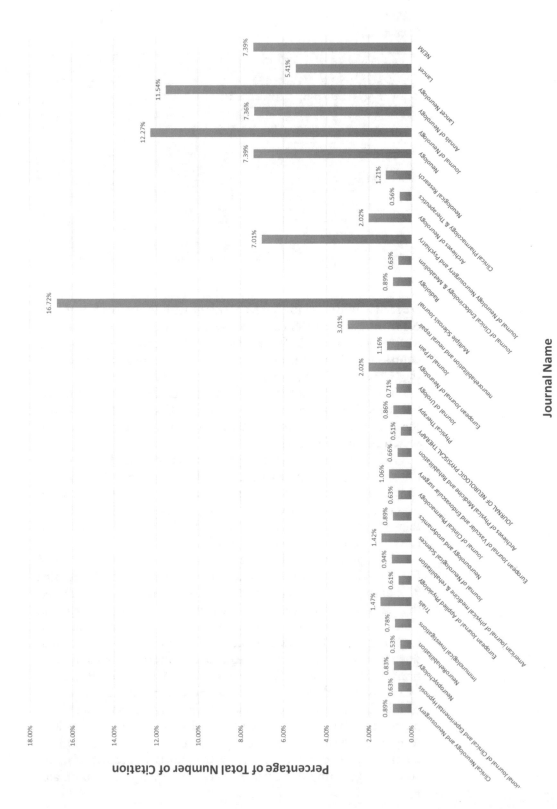

Fig. 2 Total citation per journal ordered by impact factor

Table 2
Relationship of trial design to the phase of the trial

Trial design		Study setting	Phases							
			Pilot	I	I/II	II	II/III	III	IV	N = 95[a]
RCT	2 parallel arms	Single center	2	1	2	15	0	0	0	52
		Multi center	0	1	0	9	1	20	1	
	3 parallel arms	Single center	0	1	0	3	0	0	0	23
		Multi center	0	0	0	4	0	15	0	
	4 or more parallel arms	Single center	0	0	0	0	0	0	0	3
		Multi center	0	0	0	1	0	2	0	
	Factorial	Single center	0	0	0	0	0	0	0	0
		Multi center	0	0	0	0	0	0	0	
	Cross over	Single center	1	0	0	1	0	0	0	2
		Multi center	0	0	0	0	0	0	0	
Non RCT (Quasi-experimental)		Single center	2	1	0	4	1	0	0	15
		Multi center	1	1	0	3	0	2	0	

[a]Five trials not specified

Table 3
Distribution of patient indication in MS trials

RRMS	PRMS	Pediatric MS	MS
53	6	1	40

RRPM - Relapse Remitting Multiple Sclerosis; *PRMS* - Progressive Relapsing Multiple Sclerosis; *MS* - Multiple Sclerosis

Table 4
Eligibility criteria employed in multiple sclerosis trials by study phase

	I	I/II	II	II/III	III	IV	Pivotal	Exploratory
Annualized Relapse Rate	0	0	2	0	14	1	0	0
Confirmed Disability Progression on Expanded Disability Status Scale	0	0	0	0	4	0	0	0
Immune biochemical markers	0	0	3	0	0	0	0	0
long term safety	0	0	0	0	0	0	0	0
MRI	1	0	14	0	4	0	0	0
muscle strength	0	0	0	0	0	0	1	1
Neuropsychological Assessment	0	0	0	0	0	0	0	2
pain NRS	0	0	0	0	1	0	0	1
Quality of Life	0	0	0	1	0	0	1	2
SAD and MRI	1	0	0	0	0	0	0	0
safety	1	0	3	0	1	0	0	0
spasticity NRS	0	0	0	0	2	0	0	0
walking speed	0	0	0	0	1	0	2	2

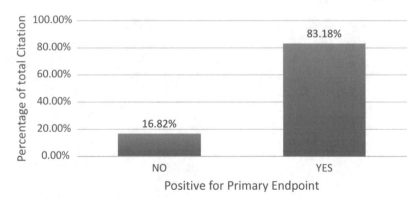

Fig. 3 Percentage of citation vs. positivity of results

Table 5
Duration of follow-up in MS clinical trials by study phase

	I	I/II	II	II/III	III	IV	Pivotal	Exploratory
≤1 month	1	0	2	0	0	0	0	0
>1 and ≤8	0	0	10	1	7	0	4	16
>8 and ≤15	2	2	8	0	9	1	1	1
>15 and ≤25	1	0	7	1	14	0	0	0
>25 and <36	0	0	1	0	4	0	0	0
>36	0	0	1	0	3	0	0	0

rehabilitation techniques have negative or weakly positive results (statistically significant, but not clinically significant), and have not been as highly cited among the top 100. One consideration to make is that a highly efficacious treatment such as fingolimod, natalizumab, teriflunomide, alemtuzumab, or BG-12; tends to have multiple positive clinical trials in different phases published and highly cited in the top 100; possibly magnifying the representation of the positive results (Table 5) [10–29].

4   Phase and Study Setting

An important aspect of a clinical trial that shapes its planning and implementation is the trial phase. The goals of a trial determine the phase. Since the goal is most directly represented by the underlying hypothesis, one can expect that the phase of the trial and the types of hypotheses are strongly correlated. The trial

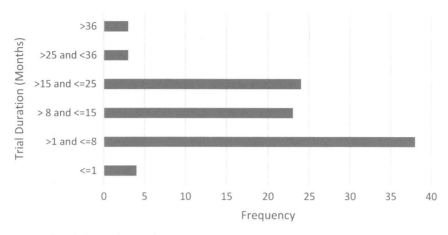

Fig. 4 Frequency of trials by study duration

Table 6
Relationship between trial design and positivity for primary outcome in MS trials

Positive for primary outcome	≤1 month	>1 and ≤8	> 8 and ≤15	>15 and ≤25	>25 and <36	>36
Yes	4	25	19	15	3	0
No	0	13	4	6	0	3

design and study setting are key aspects in testing the researcher's hypothesis, and are naturally dictated by the phase (Fig. 4).

Among the 100 selected trials, there is an evident trend that shows that as phase progresses, more trials are multicentered than single centered. When pilot, phase I, phase I/II, and exploratory trials are grouped together, it is observed that there are three multicenter studies and ten single center studies. Phase II trials show a more balanced distribution with 23 single center studies and 17 multicenter studies. Finally, when phase II/III, phase III, phase IV, and pivotal trials are grouped, we find one single center study and 41 multicenter studies. As you can see, as the trial phase progresses, the proportion of multicentered studies increases. This finding is expected because in later phases, the objective of the trials is to generalize the results and thus the designs demand a different study setting to achieve these goals and to test the respective hypotheses (Table 6).

As the phases progress, the number of patients that is required for the study increases in order to achieve adequate power to detect a difference in efficacy and effectiveness outcomes required in later phases. To test effectiveness in phase III trials, the burden of proof falls on the generalizability of the results, which naturally require multicenter trials to demonstrate this across different populations. This differs from early phase safety and proof-of-concept trials. In these trials, a small number of patients in a single center is more ethically sound and allows for an easy and focused monitoring of adverse events. This is especially significant in MS drug trials, as many of the treatments involve immunosuppressive and immunomodulatory drugs that carry a wide array of mild to severe adverse effects. In addition, smaller trials are often both a sufficient and cost-effective way to test a preliminary safety hypothesis. The MS researcher must be clear about their trial's objectives, the resources, and the sample size and study setting required to achieve success in their trial.

That being said, we did find three exceptions to this trend that are interesting to note. Soilu-Hannien et al., Ghezzi et al., and Zamboni et al. conducted multicenter early phase trials [18, 30, 31]. Zamboni et al. designed a six-center pilot study testing a surgical procedure for the management of chronic venous insufficiency as a possible underlying cause of MS. Soilu-Hannien et al. designed a nine-center phase I drug trial testing vitamin D3 supplementation to standard therapy. Ghezzi et al. designed a two-center phase I trial evaluating the safety and efficacy of Natalizumab in pediatric onset MS. We should note that although they were multicenter trials, they only involved six, nine, and two centers respectively in 1–2 countries, as opposed to the later phase trials that often involved 100+ centers in over 10–20 countries. Reasons for using a multicenter design in these early phase trials in special circumstances may be necessary. In the pediatric MS trial, there is a particularly low prevalence of this presentation of the disease, hence requiring multiple centers to facilitate recruitment. In the surgical trial, again, due to the difficulty in recruiting patients that would consent to an experimental surgical treatment for a disease that typically has a lower risk medical management as a standard therapy. The MS researcher must evaluate their resources and the feasibility of accessing the target population. While generalizability is not the primary objective in early phase trials, a multicenter approach may prove useful in easing the burden of recruitment at any one site. One must carefully balance the increased cost of such an endeavor with the utility in recruitment efforts.

4.1 Relationship Between Phase and Trial Design

The relationship between trial design and phase is also quite interesting. Among nonrandomized clinical trials; pilot, exploratory, phase I and phase I/II collectively correspond to 33.33% (5 of 15), and phase II correspond to 46.67% (7 of 15). Finally, only three investigators designed a nonrandomized clinical trial in the grouping of pivotal, phase II/III, phase III, and phase IV studies. The latter observation is nontraditional and warrants further discussion of these three trials.

Gold et al. designed an open label phase 3b/4 trial to assess the cardiac safety of fingolimod treatment in a real-world RRMS population in 285 sites worldwide [19, 32]. Since this was an after-market safety trial, a single arm nonrandomized open label design may be deemed acceptable to evaluate the effect of the drug on hemostatic stability. Endpoints of this trial included vital sign assessment and various cardiac biomarkers and adverse event reporting. Additionally, patients enrolled to this trial were those who completed a previous pivotal trial with fingolimod and likely demonstrated efficacy with the drug and thus continued to the subsequent longitudinal open label trial. This is a unique case where the preceding trial demonstrated outstanding efficacy and generally acceptable safety. Further exploration of safety concerns in this trial design may be acceptable under specific circumstances.

Khan et al. also designed an open label, single arm, phase II/III, nonrandomized clinical trial to assess the long term effect of detrusor injections of botulinum neurotoxin A on quality of life in patients with MS-induced urinary incontinence [32]. In this case, the rationale for the open label design lay in the invasive nature of the intervention. Repeated detrusor injections were administered by cystoscopy in an outpatient or day care setting. While the procedure is considered minimally invasive and generally safe, it does predispose patients to urinary tract infections and it was deemed more ethical to only subject patients to the procedure if they were receiving an active treatment regimen. This represents another scenario where a nonrandomized trial design may be considered in a later phase trial.

Putzki et al. evaluated the efficacy of natalizumab as a second line therapy for RRMS in an open-label, late phase trial [33]. Prior to this trial, no other investigators evaluated the monotherapy of natalizumab in a treatment resistant population. This trial was conducted in five sites in Germany and Switzerland. It relied on registry data and relied on retrospective enrolment of patients who were on natalizumab for at least 12 months after treatment failure with traditional therapy such as Interferon beta and prospectively followed the patients for an additional year from the time of enrollment. The trial design was atypical and in this particular scenario, nonrandomized data was robust enough for a "real life" clinical scenario in a drug that was already approved for a first-line indication.

It is thus important to critically assess the hypothesis and the purposes of the trial and weigh the level of robustness of the data required to test the predetermined hypothesis. In Western and Northern European countries, the use of clinical trial registries has become common practice and is a cost effective way to collect robust data on therapies in randomized and nonrandomized designs.

When we look at RCTs only 10% of studies are composed of early phase Pilot, Exploratory, Phase I, and Phase I/II (8 of 80), but 41.5% (33 of 80) and 48.75% (39 of 80) fall under Phase II and a grouping of Pivotal, Phase II/III, III, and IV, respectively.

This pattern of RCT having proportionally more late-phase trials than non-RCT can be explained by recognizing that later phases require more controlled settings and a randomized allocation to test and confirm their hypotheses. On the other hand, non-RCTs provide relatively less robust data and are not typically powered to answer pivotal effectiveness research questions. Thus, NRCT's primary objectives are more in line with those of pilot, exploratory, and early phase safety and preliminary efficacy questions. When developing a novel hypothesis with little or no precedent in the literature and, more importantly, limited resources a researcher may be well guided to consider a non-RCT design at the expense of highly robust data. In later phase trials, it is highly advised to consider RCTs as the method of choice to produce high quality data that is reliable in such a pivotal stage in drug or medical device development.

While these observations are consistent with expert clinical trialists' opinions, there are limitations to consider in this analysis since only 17 NRCT studies were included in this analysis as compared with 83 RCT studies. This, naturally, makes the description of the phase distributions in NRCTs less robust than those of RCTs.

5 Trial Design

In the analyzed trials, 82% were randomized clinical trials (RCT), and 18% were nonrandomized clinical trials (NRCT). Additionally, 86% of trials were phase II, III, and pivotal trials, while 14% include all other phases.

Within RCTs, one important aspect for the investigator to decide is the number of study arms and the choice of control. In MS drug trials, two parallel arm RCTs tend to most often use placebo as a comparator. In the case of rehabilitation therapies, two-arm RCTs are more often observed. They often test some form of exercise or technique such as yoga, sports climbing, or Kinesio Taping of an affected area compared to no treatment at all. Although a reliable sham control for rehabilitation procedures is challenging, the investigator needs to make all attempts to find a suitable sham group as to control for placebo effects.

As investigators add a third arm, we begin to see a large proportion of drug trials having two different doses of the experimental drug being compared to a placebo. About 22% of three-arm RCTs also included an active control arm. Finally, the only four-arm RCT in our results compared two doses of the active drug to both placebo, and an active control. When designing a clinical trial, the MS researcher must consider the choice of comparators when deciding the number of treatment arms to include. Again, increase in costs and feasibility by adding another arm needs to be carefully considered before adding more arms to test further drug dosages. Or alternatively smaller trials with surrogate outcomes can be considered in there is a real need to test different dosages. The larger the number of arms in a trial, the larger the required sample size to test the hypothesis and researchers must take this into consideration when planning their trials.

An active comparator should be employed with ethical considerations for the consequences of not receiving treatment, if the treatment is meant to treat an acute MS attack. Alternatively, a placebo comparator may be acceptable if there is no efficacious standard therapy for the indication of the experimental drug.

5.1 Changes to Trial Design

Three articles noted that they changed the design of their trial after study initiation. While this is a small proportion of the sample, it is important to note since this represents an important concept in clinical trial design. Two trials changed their design due to ethical reasons, while one was purely for logistical reasons.

Firstly, Kappos et al. had the highest cited article in our search and had a change to their phase III trial design [22]. The trial began as a three-parallel arm RCT comparing two different doses of fingolimod to placebo. Interim data analysis reviewed by the DSMB recommended eliminating the higher dose 1.25 mg arm of the trial due to negligible improvement in outcome, with a significant increase in adverse effects. This information was sent to the FDA for guidance, and they recommended eliminating the removal of this arm for ethical reasons.

Secondly, Calabresi et al. conducted a phase III trial testing the effectiveness of two different dosing frequencies of PEGylated Interferon beta against placebo in RRMS patients [34]. In this case, at interim analysis, the drug was deemed to be extremely efficacious in reducing the annualized relapse rate and therefore it would be unethical to randomize patients into a placebo arm. The placebo arm was eliminated and the trial continued as a two parallel arm trial.

Finally, in a 2012 phase III trial by Coles et al., two different doses of alemtuzumab were being compared to active control [13]. The higher dose arm of alemtuzumab was eliminated after study initiation due to recruitment challenges. The higher the number of arms in a trial, the higher the required sample size to detect a difference between the arms. Therefore, for logistical reasons, the elimination of this arm of the trial eased the resources burden of the trial.

These three trials represent classical examples of reasons why a researcher might assess carefully the study design a priori and also when new information is available the need to eliminate an arm of the trial due to ethical or logistical reasons. However, it is important to underscore that this needs to be done in a systematic way and a priori determined. These examples thus highlight the need of a good plan for a data safety monitoring board as to analyze interim analysis especially for safety as to assist these recommendations.

In MS clinical trials, the consequences of administering a treatment that is not efficacious or administering placebo in light of a highly efficacious experimental or standard therapy have quite a high impact on disease progression, relapse rate and quality of life of the patients. Therefore, ethical considerations must be made at interim analysis, especially in trials of large sample sizes. Additionally, as with clinical trials in all areas, when resources are limited and recruitment becomes a logistical barrier to completion of the study, a researcher must have the foresight to weigh and balance the benefits of completing the trial with design modifications against continuing with the current design and risking futility of the trial.

6   Population

Since the natural history of MS often begins with a nonspecific clinical presentation of clinically isolated syndromes, there is rarely a specific date of diagnosis pinpointing the clinically definitive onset of the disease. This makes it difficult to define diagnostic criteria and to choose selection criteria that are specific enough for the desired population and general enough not to impede recruitment.

In the top 100 cited MS trials, over 50% required a specific diagnosis of RRMS, while only 6% and 1% required a diagnosis of progressive remitting MS, and pediatric onset MS respectively. The remaining 40% simply required a diagnosis of MS, with no specifically defined presentation. This latter finding makes it easier to recruit patients into those trials; however, may not be feasible if the intervention is aimed to reduce relapse rate (e.g., in RRMS patients), or employs a primary outcome that is otherwise specific to a targeted subset of the MS population. It is imperative to ensure that the intervention is being tested in the appropriate and specifically defined population to which the treatment may be indicated. Eligibility criteria that are too general may not be applicable to a specific subset of MS patients. Conversely, eligibility criteria that are too specific may not show clinically significant results in this population and in reality may have utility in a different subset of the MS population.

The most common selection criteria include a specific Expanded Disability Scale Score (EDSS), McDonald's criteria as a diagnostic tool, a confirmed number of relapses in a given time period before the trial, and MRI for diagnosis. While some of these inclusions are part of McDonald's diagnostic criteria, they were explicitly mentioned in addition to a diagnosis by McDonald's criteria in many protocols. Other inclusion criteria of note were measures of motor functionality which were of particular interest in motor rehabilitation trials for MS symptoms. Additionally, previous symptomatic standard treatment futility or dissatisfaction is commonly an inclusion criterion for trials exploring the treatment of clinically isolated syndromes of MS such as optic neuritis, OAB and spasticity.

We observed that several trials testing drugs as adjuvant to standard disease modifying drug (DMD) therapy. Naturally, these trials had an inclusion criterion that specified the previous use of DMD, and requirement to be on a stable regimen prior to administration of study drugs. Some researchers relied on medical records to satisfy this criterion, while others administered the standard therapy for the protocol required period of time prior to receiving the experimental drug. The lengths to which researchers went to ensure participants received the therapy that met the inclusion criteria of the study were not disclosed. MS researchers must consider the reliability of the medical records in their population and the time or resources available at their expenditure. Standardizing the protocol specified criteria by providing standard therapy prior to initiating experimental treatment provides a more controlled setting and more reliable efficacy results that are in line with a phase II objective. Alternatively, relying on patient accounts or medical records may be more effective in a setting of limited resources or in a large phase III trial when such practice would be unfeasible. Additionally, relying on patient accounts or medical records may more closely mimic a real life clinical scenario and provide the generalizability required for a phase III trial. It is important to underscore that this decision (how detailed the diagnostic criteria will be) is also based on study phase. In an earlier phase, the investigator needs to consider decreasing sources of variability such as wrong diagnosis; however in later phases, the goal is to increase generalizability to a more real life scenario.

Key exclusion criteria include the use of specific disease modifying drugs that were deemed to affect the primary outcome or to have similar mechanisms of action as the experimental drug. Another commonly used exclusion criteria included specification of a time period in which no relapses were acceptable prior to the trial for the patient to be deemed in stable condition. Determining the eligibility criteria in a trial requires insight and a deep understanding of the patient population. Referencing previous trials with similar drug families and study designs is a good starting point for establishing precedent and is beneficial for the comparability of results.

It is interesting to note that there did not seem to be a change in the population across different trial phases, including phase I trials. This is contrary to typical phase I studies that include healthy subjects. This practice is commonly seen in oncology trials in order to facilitate the drug development process and utility of potentially effective experimental treatments in a critically ill population. In the setting of MS trials, this finding may be an indication of potential safety issues to be aware of when investigating MS drugs in healthy subjects.

7 Trial Duration

Deciding the length of a trial requires many considerations. The amount of time where an effect is expected to occur depends on the type of intervention and the purpose of the trial. In our data, trials of all phases mostly lasted 15 months or less except for phase III, which tended to skew toward the higher end (i.e., >15 weeks). This is expected as late phase trials typically aim to evaluate the long term effectiveness of a treatment. In MS trials, the long term efficacy is significant in evaluating the prevention of an acute relapse. Trials of shorter duration are appropriate for evaluation of safety and efficacy in remitting an acute attack or reduction of active symptomology. The MS researcher must be able to estimate the expected treatment duration of the chosen intervention. The investigator must also clearly indicate the primary objective of the trial and set a follow-up duration consistent with the trial goals. Any extension in the duration of the trial may further increase cost and resource expenditure and yield marginal improvements in meaningful data. This is particularly important in trials with a time-dependent primary outcome in survival analysis trials evaluating time to confirmed relapse.

Interestingly, trials that were positive for their primary outcomes tended to have longer durations on average than negative trials. This is consistent with the previously mentioned concept considering time to relapse being an important clinical outcome in MS patients. For this reason, the median trial duration of the top 20 most cited is significantly higher than the rest of the sample;

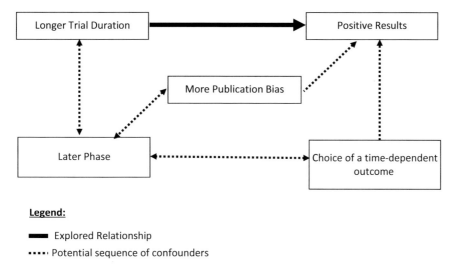

Fig. 5 Relationship between longer trial duration and positivity for primary outcome

these trials utilized time to event analysis and are thus deemed more meaningful in a relapse remitting chronic disease such as multiple sclerosis. Trials with longer trial duration have a higher chance of demonstrating an effect, and are regarded as more clinically relevant results.

One should also consider that study phase may be a confounder in this analysis. Since the top 20 trials are predominantly phase III trials which tend to be of longer durations, it is worthy to consider the possibility of a higher association of positive results in this subgroup of trials. Therefore, since the type of primary outcome a researcher employs is strongly associated with the phase, what we may be seeing here is an underlying relationship between a late phase time-dependent outcome and the trial duration which will be discussed in Sect. 9 of this chapter. This relationship is schematically displayed in the figure below (Fig. 5).

Since MS is a chronic disease, the long-term follow-up of remissions and MRI lesion changes is of particular importance to clinicians and researchers. In our sample of trials, we found about 15% offered patients an optional extension to the trial, often for 24 or 48 additional months as a long-term assessment of efficacy. In three and four-arm RCT, if a placebo arm exists, it is usually eliminated and patients from that arm are randomized into the different active treatment arms. This option is naturally particularly attractive to patients who had efficacious results and is often considered exploratory, though some authors have employed specific clinical outcomes for the extension period as well.

Three trials of note had durations of 8.5, 15, and 16 years [14, 35, 36]. They were extensions of previous trials that continued to a long-term extension and follow-up of the patients within a modified study design. Authors often reported interim analysis results at multiple time points. In a few cases, more than one follow-up publication from the same trial was included in our search. This may be an attractive option for researchers who have already demonstrated safety and efficacy in their intervention and wish to demonstrate long term effects in their population. The costs associated with such an extension may be minimized by widely spaced follow-up intervals or even telephone follow-up. Here therefore the investigator needs to assess costs of adding these extensions vs. benefits of what these data (from extensions) would produce. Given that MS is a chronic disease associated with long-term outcomes (remission and relapse), this may be a condition where extensions need to be carefully considered.

8   Interventions

In this section, we discuss the distribution of interventions employed by MS researchers in the reviewed trials. While this is primarily a descriptive analysis, a new MS researcher may use this information to learn about the type of intervention that is highly regarded in this field and specific examples of successful therapies within every type of intervention. The investigator may also use this information to find examples of clinical trial designs of drug and nondrug trials that are highly cited.

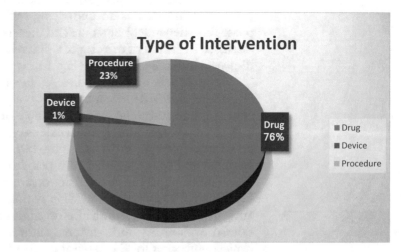

Fig. 6 Type of intervention explored in multiple sclerosis trials

As shown in Fig. 6, the overwhelming majority of the top 100 cited MS clinical trials are drug trials (76%), while procedure trials comprise 23%, and 1% are device trials. These figures represent the distribution of modalities of interest in the treatment of Multiple Sclerosis. Since drugs are often classified as disease modifying treatments, they work via various mechanisms to reduce the onset and relapse rate of the disease. Procedures and devices, on the other hand, often aim to reduce the severity of the symptoms associated with an acute exacerbation. For this reason, it is understandable why more resources are allocated to investigating disease modifying treatments.

Furthermore, among the 76 drug articles, 40 are within the top 50 most cited. This disproportionate amount of drug trials, within the top 50 further emphasizes the notion that MS drug trials are deemed more advantageous than other therapies. Among these, eight trials are testing various forms and dosing regimens of the approved drug Interferon Beta. As far as novel drug research; Fingolimod, BG-12, Teriflunomide, and humanized monoclonal drugs such as Natalizumab and Alemtuzumab were among the top and most commonly cited. Other drug trials of note were four trials testing the safety and efficacy of different doses of vitamin D supplementation in delaying the onset and reducing the severity of MS symptoms [30, 37–41]. Several trials used cannabinoid drugs such as Sativex to test its effect on reducing symptoms of clinically isolated syndromes of MS, namely neuropathic pain, spasticity, and overactive bladder syndrome [42–48]. One important consideration here is that given the high costs of MS trials (as discussed above: large sample sizes and long-term outcomes), research is concentrated with pharmaceutical companies that have larger budgets for drug-related clinical trials. This is an area of development: nonpharmacological trials that are disease modifying treatments.

Looking at trials studying the effects of procedures, many utilize exercise techniques and resistance training to test for improvement in motor function. Sports climbing, yoga and resistance exercise were studied for the improvement of gait deficits [49–55]. Several trials had interesting uses of technology to note. Researchers tested Nintendo Wii for improvement in balance and motor function. Patients also underwent intensive computer-assisted cognitive training or physical training [56–59].

Finally, device trials comprised the lowest proportion of interventions in our results with only one trial in the top 100 cited. While some interventions that fell under the "procedure" category (such as Nintendo Wii training, or Kinesio Taping of the lower extremities for improved balance) may be considered a device intervention by some; we only deemed trials as device trials if the interventional device was developed for the purpose of and is directly, actively involved in treating the condition and would require health authority approval before clinical use for the indication. In the single cited device trial included, transcranial direct current stimulation was studied in the management of chronic and central neuropathic pain associated with MS, as well as improvement in quality-of-life scales [60].

9   Outcomes

The choice of an appropriate primary outcome in a clinical trial is of utmost importance. A primary outcome should be clinically relevant, validated, and responsive enough to the intervention to detect a statistically significant difference. The primary outcome dictates the sample size calculation, and the entire trial is powered to detect a particular change or effect size in this outcome. Secondary outcomes are often exploratory, though some trials are powered to appropriately assess secondary outcomes as well.

In MS trials, 83% of researchers chose clinical and functional outcomes, while 12% used surrogate outcomes. Four percent used a composite clinical and surrogate endpoint (co-primary outcomes). The most commonly used surrogate outcome was an immune biochemical marker. The only three trials in the top 100 cited that used surrogate markers were all early phase II trials. Mehling et al. were evaluating the adaptive immune response against an influenza vaccine in immunosuppressed MS patients [25]. Jurynczyk et al. tested a transdermal immune modulator's effect on immune responses in the skin, lymph nodes, and peripheral blood [61]. Finally, Kimball et al. measured the effect of cholecalciferol and calcium on peripheral blood mononuclear response associated with MS [38]. These are classic examples of biomarkers. Though not commonly employed in MS research, they may of particular interest to researchers in this field. The challenges with use of surrogate biomarkers in MS studies is the lack of validation, and typically employ small sample sizes in early phases to establish a proof of concept.

The most commonly used clinical outcomes were annualized relapse rate (ARR), and specifically defined MRI changes. Combined, they were used as a primary outcome by 38% of the researchers. MRI changes included number of new lesions, changes in size of lesions, cumulative number and size of lesions, and involvement of new regions. ARR was most commonly calculated based on neurologist confirmed relapse symptoms accompanied by a change in Expanded Disability Status Score (EDSS). Trials using these outcomes were typically late phase drug trials. They were also associated with articles with significantly higher mean citations per paper. This gives the MS researcher an indication of the clinical relevance of these outcomes in MS research. Their use in late phase trials along with their high citations is indicative of the relevance of these outcomes to direct patient improvement. They should be employed as outcomes in the evaluation of efficacy and effectiveness after safety has been established in early phase trials. This is particularly relevant in trials evaluating the effect of disease modifying drugs.

Procedure and device trials employed a different distribution of outcomes given they usually have different goals (i.e., to improve neurological outcomes). Most commonly used were walking assessments, neuropsychological assessments, muscle strength, spasticity, quality of life, and pain scales. Just by browsing the outcomes, one can begin to get an idea of the objectives of these types of trials. It is noteworthy that these trials also generated high mean citations per paper. Just as important as modifying the course of the disease is modifying the quality-of-life outcome of patients. Procedure and device trials tend to focus on improvement of the symptomology and quality of life, the choice of outcome should be validated in order to achieve comparability with previous trials.

Most researchers employed numerous additional secondary outcomes in their trials. Some outcomes were well validated such as ARR, MRI outcomes and Confirmed Disability Progression by EDSS scale. Others were exploratory in nature such as time to relapse, proportion of relapse-free patients, proportion of relapses in patients, and various functional outcomes. The most frequently used secondary outcomes were time-to-event outcomes.

The importance of time-dependent outcomes in a clinical setting cannot be undermined in a disease with a relapse and remitting course like MS. While the progression of the disease needs to be closely monitored, the acute relapses and the interval to relapse result in an enormous burden on the patient's ability to lead a normal life, particularly in the early stages of the disease. This relationship is most directly evaluated by ARR and time-to-event outcomes.

10   Sample Size

Calculating the sample size of a trial can be a tiresome task. Earlier phase and exploratory trials tend to have a small sample size, while later phase pivotal trials have larger sample sizes. Many factors determine the required sample size of a trial such as outcome, the intervention, previous similar trials, power, significance level, and expected attrition. Highly powered studies with larger sample sizes require more resources and present more logistical challenges. On the other hand, the results can be more accurate and be more

Table 7
Sample size distribution in MS trials by Study Phase

	I	I/II	II	II/III	III	IV	Pivotal	Exploratory
0–20	2	0	4	0	0	0	0	7
21–50	0	2	7	0	0	0	1	7
51–100	2	0	4	0	1	0	3	3
101–500	0	0	12	2	17	1	1	1
501–1000	0	0	1	0	4	0	0	0
1001+ Above	0	0	0	0	17	0	0	0
Not applicable	0	0	1	0	0	0	0	0

highly cited. To help guide MS researchers decide on the sample size calculation method, Table 7 summarizes the findings in current MS research.

More importantly, is the method of calculating sample size. After determining the research question and the statistical tests to be utilized, there are several parameters that a researcher must consider including power, significance level, the expected attrition rate and the available literature upon which to estimate effect size. All of these factors play an important role in determining the sample size of a trial.

In our analysis, 53% of trials failed to report the method used to determine the sample size. These trials were distributed among all phases. Two characteristics are important to underscore and may help the investigator when designing a MS trial: (1) the attrition rate between 10% and 20%; which may be conservative given the longer follow-ups; though median dropout rate is about 11% according to our analysis (see discussion on adherence below); and (2) the effect sizes were mostly calculated based on previous studies and in fact none of them reported using the minimal clinically important difference (MCID), showing that that most of these trials based on actual data to calculate sample size which may have improved their predictions. A summary of the parameters used in the remaining 47 trials are detailed in Table 8.

11   Analysis and Stoppage Criteria

Discontinuation and stoppage criteria are sometimes necessary to be well defined prior to the study. Only 30% of MS trials reported this in detail. More than half of those specifically mentioned Data

Table 8
Parameters used in samples size calculations in MS clinical trials

Alpha	0.025 (one sided)	1
	0.025 (two sided)	2
	0.05 (two sided)	39
Power	80%	17
	81–90%	23
	91–95%	6
Expected Effect	Pilot	8
	Previous trials	34
	not reported	5
Expected Attrition	10%	9
	12–15%	5
	20–25%	10
	50%	1
	not reported	22

Safety Monitoring Committee oversight to guide safety stoppage criteria. Two trials described specific clinical discontinuation criteria such as an attack influencing the pyramidal system or an episode of acute relapse [62]. Four trials described specific guidelines to titrate up or titrate down the dose in case of low tolerability, and make note of the possibility of discontinuation if the recommended titration schedule fails to alleviate the adverse events [43, 45, 63]. Three trials provide the option of switching treatment arms, or reverting to standard therapy and remaining in the trial if the patient has had at least one relapse and does not find the treatment is efficacious at a predetermined time point.

The choice of emphasizing discontinuation criteria in an article must be made with knowledge of specific known or expected side effects of concern regarding the intervention. Titration schedules are particularly interesting to other researchers and allow for easy reproducibility of the work as well as shed light on the use of the intervention in a clinical setting. In phase II trials where preliminary efficacy is being tested, the eligibility criteria are often kept tight and homogeneity of a specific target population is desired. In this case, discontinuation criteria based on clinical condition to target specific severity subgroups of MS may be appropriate. In any case, safety monitoring should take place with any experimental treatment in a clinical trial, the details of which are rarely specified in the top cited MS trials.

12   Randomization

The random method by which patients get allocated to different treatment arms is called randomization. Randomization, if properly performed, eliminates selection and allocation bias and its methods help maintain the blind. More importantly, randomization is an extremely effective way for controlling for confounders by evenly distributing them across all groups. A poorly planned randomization strategy can result in type I and type II error and drastically affect the validity of the results.

Randomization strategies have three main aspects—type of randomization, sequence generation, and allocation concealment. Table 9 summarizes our findings in MS trials in these three categories. We see the use of technology simplifying the process of randomization with the use of computer generated sequence generation and allocation by IVRS or webpage. Use of envelope, coin flip, and hat draw methods are rarely in use, though they are cost effective and were used in pilot and early phase studies which typically have lower budgets.

The type of randomization depends on the complexity of the study and the variable being studied, as well as the sample size. Almost all multicentered trials stratified by study center. Other factors stratified for were baseline scores on various scales (EDSS, Severity scales), age, disease duration, and dosing frequency of IFN-B for experimental adjuvant therapies. It is pertinent for the MS researcher to identify the variables that may interfere with the primary outcome in the trial and align the randomization strategy to stratify for these variables. This will help create a balance between treatment groups which will in turn result in a more robust data set which can employ a wider range of simple or complex statistical tests. It is also important to consider the sample size of the trial when weighing the need of stratification. In small sample size studies, the use of stratification may lead to significant imbalances.

Table 9
Randomization strategies employed un MS trials

Type of randomization	Simple	37
	Block	32
	Minimization	1
	Stratified	41
Allocation concealment	Webpage	1
	IVRS	16
	Envelope	3
Randomization sequence generation	Computer generated	44
	Coin flip	1
	Hat draw	1
	Not reported	46

Performing the randomization in balanced blocks in multiples of the number of arms in the trial has several advantages. It ensures that the arms are well balanced within the blocks and throughout the trial. This way, if interim data analysis is required or if recruitment is hindered, the data set of those enrolled is balanced among all arms and the data will thus have internal comparability between arms and has more easily interpretable value. With the use of computer generated randomization sequences, blocking is more easily utilized now and helps add a layer of concealment of the treatment allocation.

Simple randomization is often sufficient in trials with large sample sizes due to the law of large numbers. Despite this, there were a few trials with low sample sizes that used simple randomization with small sample sizes <50 and there were imbalances in numbers and baseline characteristics between the groups [58, 63–66]. This practice is not recommended and a well guided researcher should account for appropriate randomization strategies according to the sample size. This will mitigate most imbalances among treatment groups and allow for more straight forward statistical analysis of the results.

13   Blinding

Data collection in clinical research may be highly influenced by the perception of the assessor of the outcome and by the perception of the patient on the response. This is called ascertainment bias and often leads to exaggerated effects of the treatment. Blinding is an effective way to reduce ascertainment bias. In addition, performance bias may occur if patients know they are taking an active drug as they may change their behavior. Keeping the treatment assignments concealed from assessors and patients reduces type I error and increases the accuracy of the results. Clinical trials can be single blinded, double blinded, or open label. In MS trials, 70% of trials explicitly described their blinding using these terms.

Most MS drug trials were double blinded, with only five single-blinded drug trials [55, 64, 65, 67, 68] and three open label studies [63, 69, 70]. One important point is that these trials use some objective outcomes such as lesions in the MRI that can be assessed in a more objective manner; regardless if subjects know they are receiving active drug they may change their lifestyle and therefore may induce bias.

Device and procedure trials on the other hand, were often single blinded (assessor blinded or patient blinded) given challenges to have a similar sham group. Furthermore, device and procedure trials generally did not define the blinding as clearly as drug trials.

The ability to manufacture matched placebo drugs at a low cost allows for straightforward blinding in drug trials, particularly since MS experimental drugs are administered by conventional oral, intravenous, and intramuscular routes of administration. In earlier phases, when safety is yet to be established, open-label designs were more prevalent.

There are some interventions in MS trials that incorporate technology to maintain the blind. A trial by Mori et al. used transcranial direct current stimulation in the treatment of chronic neuropathic pain in MS patients [60]. This device can administer a sham protocol that mimics the active treatment sensation thus blinding the patient from their treatment. This is an example of innovative use of technology that was developed to blind device trials. Most nonpharmacologic MS trials tested exercise procedures, and as thus, they used a single blind design where the assessor or rater of the progress of the primary outcomes was blinded.

The logistical difficulty in blinding your trial depends on the characteristics of the intervention and the choice of outcome. Outcomes that are more subjective in nature such as rating scales and clinical judgment can be influenced greatly by unblinded assessors and blinding is advisable if possible. This is not to say that more objective outcomes such as serology lab results or MRI outcomes are free of bias. Studies have shown that interpretation of MRI and lab results can be unconsciously influenced by knowledge of treatment allocation or patients may change behavior if they know they are receiving active treatment that may change results of lab exams.

14 Adherence

Maintaining a low attrition rate is very important in a clinical trial. Failing to do so can have serious implications on the validity of the results. Even the most impeccable trial design can fail to execute due to a high number of drop outs. Main reasons for low adherence rates include complexity of the trial or number of study visits, adverse events, or lack of efficacy. One study correlated low adherence to disease-modifying treatment with high rates of disease relapse and progression. In long-term follow-up studies (a popular design in MS studies) this is of particular importance because disease severity and relapse are highly correlated with patients lost to follow-up.

The anticipated adherence or attrition rate should be based on estimates from previous trials in the same population, if possible, and conservative estimates should be made. This helps the researchers estimate the number of patients to enroll in order to achieve the desired power for the trial. This also translates to accurate budget estimates and feasibility assessments.

In MS trials, 81% of researchers reported their adherence rates. The data demonstrated adherence from 21% to 100%, with 89% median adherence. While we expected to see a correlation between longer trial duration and higher attrition rate, the data failed to show any correlation. The same observation was found with number of study visits. We did find, however, that in trials longer than 2 years, the adherence rate significantly drops. Furthermore, adherence rates were similar in trials with positive and negative results.

In this regard, our results showed that overall adherence rates in MS trials are quite high. The only aspect of study design that significantly impacted adherence rates in MS trials was a trial duration longer than 2 years. These correlated with long follow-up trials that were often offered as extensions of the primary study. Most other MS trials were <2 years of duration and as thus had satisfactory adherence rates. The MS researcher should make note of this exceptional adherence yet tread with caution. Keeping a close eye on this, researchers are advised to maintain regular contact with study participants and engage in efforts to improve this at all costs. Reaching enrollment targets ahead of schedule and having a complete data set can save a considerable amount of resources and avoid decreased validity with complex missing data computations.

15 Recruitment Yield

Recruitment is often considered the bottleneck of a clinical trial timeline. It is frequently the most time-consuming activity and this time comes at a high cost. Monitoring the effectiveness of recruitment efforts is necessary as a feedback indication to researchers if they are accurately targeting their desired population. Recruitment yield is defined as the number of patients randomized divided by the number of patients screened. This is another variable contributing to accurate budget estimates. Theoretically, the earlier phase trials have more extensive eligibility criteria and therefore carry a lower enrollment rate. This is not true for MS trials.

Since eligibility criteria did not seem to change across different phases, this is not a factor in enrollment rates across different phases. In fact, earlier phase studies tended to have slightly higher enrolment rates than later phase trials. It is; however, noteworthy that procedure trials had an enrollment rate 15–20% lower than drug trials.

Table 10
Recruitment yield in clinical trials by study phase

Phases	Recruitment yield
I	0.88
I/II	0.92
II	0.70
II/III	0.35
III	0.78
IV	0.84
Pivotal	0.66
Exploratory	0.58

Overall, median enrollment yield was 80%, with 85% for drug trials, and 62% for device and procedure trials. This recruitment rate is actually relatively high but this may be a result of how authors calculated the denominator (or in other words, how they calculated the number of screened patients). Table 10 can be used as an estimate to guide future researchers in planning recruitment logistics and resource allocation.

16 Statistical Analysis

The choice of statistical tests utilized in a clinical trial is defined according to the type of data (continuous or categorical), the distribution of the data, and the number of groups to be compared. These factors are all part of the primary research question of the study.

Among the statistical tests used in MS trials, t-test, ANCOVA, and descriptive statistics were the most commonly utilized. T-test compares the means of two groups with normally distributed continuous data. ANCOVA offers the advantage of adjusting the results of a group comparison to an important covariate. Finally, descriptive statistics shows the distribution of the data as a range, mean with standard deviation and median with interquartile ranges.

The MS studies which utilized T-tests often compare effects of the drugs tested in decreasing the number of MRI lesions, which is a very important feature in MS diagnosis, and also to compare annualized relapse rate (ARR), to compare the efficacy of different doses of a given treatment or a treatment against standard therapy or placebo. Regression models were also frequently used in studies with ARR as a primary outcome in order to control for covariates and determine the associations of some factors on the effects of the treatments.

Given that outcome in disease-modifying studies were often time to event (for instance time to relapse), a good number of studies used survival outcomes such as time to relapse or time to disability progression, Kaplan–Meier curves with log rank tests and Cox proportional hazards models.

Regression models, Cox proportional hazards models, and ANCOVA tests are used in different circumstances but all serve the purpose of controlling for covariates. This occurs frequently in MS trials due to the large variability in the presentation and baseline features, severity and symptoms of the disease as well as different levels of progression. Many researchers were only able to control for one or two of these factors by stratifying the randomization among the treatment arms, and therefore, utilized these statistical techniques to their advantage to control for these potential confounders on the outcomes.

17   Limitations

Regardless of the knowledge and experience of the researcher, rarely a clinical trial has no limitations. In MS trials, limitations of trials were discussed in 43% of papers. Table 11 summarizes the findings. Analyzing a summary of the limitations reported by the most cited papers in multiple sclerosis can be a helpful resource for a researcher currently planning a trial in multiple sclerosis.

The most commonly reported limitations were sample size, outcomes, trial duration and design, and population characteristics. Nineteen percent of trials reported not being powered enough for appropriate data analysis. This was usually due to low sample size. Several trials went one step further to describe a higher than expected attrition rate and a low recruitment yield. The outcomes in many MS trials are based on quality-of-life scales. This is quite important for this chronic debilitating disease. Thus, self-reported quality-of-life outcomes are often used, and this has its own limitation about the reliability of the results.

Trial duration was often reported to be too short, and longer term studies were recommended by the researchers. The highest cited trials usually went on to perform long-term follow-up studies using a modified study design. This is important in chronic diseases such as MS. Issues with trial design and blinding included lack of

Table 11
Limitations discussed in MS clinical trials

Limitations discussed in the articles	Trial duration	Adherence	Power (sample size)	Trial design	Outcomes	Choice of comparators	Unblinded	Population characteristics	Not Specified
Frequency	9	4	19	7	12	3	5	7	57

healthy controls, open label studies, or single blinded studies as potential sources of bias.

Finally, population characteristics posed a challenge for several researchers due to variability in disease status, and baseline characteristics. Sometimes the populations were considered too homogeneous in baseline disease features and not representative of the spectrum of presentations of the target population. Due to the high variance in the natural history of multiple sclerosis and its various forms of presentations, severities, and progression outcomes, it is understandably challenging for any sample to represent the entire MS population.

References

1. Newcombe J et al (1991) Histopathology of multiple sclerosis lesions detected by magnetic resonance imaging in unfixed postmortem central nervous system tissue. Brain 114(Pt 2):1013–1023

2. Filippi M, Rocca MA (2011) MR imaging of multiple sclerosis. Radiology 259(3):659–681

3. Rocca MA, Messina R, Filippi M (2013) Multiple sclerosis imaging: recent advances. J Neurol 260(3):929–935

4. Pohl D et al (2007) Paediatric multiple sclerosis and acute disseminated encephalomyelitis in Germany: results of a nationwide survey. Eur J Pediatr 166(5):405–412

5. Mackenzie IS et al (2014) Incidence and prevalence of multiple sclerosis in the UK 1990-2010: a descriptive study in the General Practice Research Database. J Neurol Neurosurg Psychiatry 85(1):76–84

6. Noonan CW et al (2010) The prevalence of multiple sclerosis in 3 US communities. Prev Chronic Dis 7(1):A12

7. Kinsinger SW, Lattie E, Mohr DC (2010) Relationship between depression, fatigue, subjective cognitive impairment, and objective neuropsychological functioning in patients with multiple sclerosis. Neuropsychology 24(5):573–580

8. Noseworthy JH et al (2000) Multiple sclerosis. N Engl J Med 343(13):938–952

9. Montalban X (2011) Review of methodological issues of clinical trials in multiple sclerosis. J Neurol Sci 311(Suppl 1):S35–S42

10. Cohen JA et al (2010) Oral fingolimod or intramuscular interferon for relapsing multiple sclerosis. N Engl J Med 362(5):402–415

11. Calabresi PA et al (2014) Safety and efficacy of fingolimod in patients with relapsing-remitting multiple sclerosis (FREEDOMS II): a double-blind, randomised, placebo-controlled, phase 3 trial. Lancet Neurol 13(6):545–556

12. Cohen JA et al (2012) Alemtuzumab versus interferon beta 1a as first-line treatment for patients with relapsing-remitting multiple sclerosis: a randomised controlled phase 3 trial. Lancet 380(9856):1819–1828

13. Coles AJ et al (2012) Alemtuzumab for patients with relapsing multiple sclerosis after disease-modifying therapy: a randomised controlled phase 3 trial. Lancet 380(9856):1829–1839

14. Confavreux C et al (2012) Long-term follow-up of a phase 2 study of oral teriflunomide in relapsing multiple sclerosis: safety and efficacy results up to 8.5 years. Mult Scler 18(9):1278–1289

15. Confavreux C et al (2014) Oral teriflunomide for patients with relapsing multiple sclerosis (TOWER): a randomised, double-blind, placebo-controlled, phase 3 trial. Lancet Neurol 13(3):247–256

16. Fox RJ et al (2012) Placebo-controlled phase 3 study of oral BG-12 or glatiramer in multiple sclerosis. N Engl J Med 367(12):1087–1097

17. Freedman MS et al (2012) Teriflunomide added to interferon-beta in relapsing multiple sclerosis: a randomized phase II trial. Neurology 78(23):1877–1885

18. Ghezzi A et al (2010) Safety and efficacy of natalizumab in children with multiple sclerosis. Neurology 75(10):912–917

19. Gold R et al (2014) Assessment of cardiac safety during fingolimod treatment initiation in a real-world relapsing multiple sclerosis population: a phase 3b, open-label study. J Neurol 261(2):267–276

20. Gold R et al (2012) Placebo-controlled phase 3 study of oral BG-12 for relapsing multiple sclerosis. N Engl J Med 367(12):1098–1107

21. Kappos L et al (2012) Effect of BG-12 on contrast-enhanced lesions in patients with relapsing--remitting multiple sclerosis: subgroup analyses from the phase 2b study. Mult Scler 18(3):314–321

22. Kappos L et al (2010) A placebo-controlled trial of oral fingolimod in relapsing multiple sclerosis. N Engl J Med 362(5):387–401

23. Khatri B et al (2011) Comparison of fingolimod with interferon beta-1a in relapsing-remitting multiple sclerosis: a randomised extension of the TRANSFORMS study. Lancet Neurol 10(6):520–529

24. MacManus DG et al (2011) BG-12 reduces evolution of new enhancing lesions to T1-hypointense lesions in patients with multiple sclerosis. J Neurol 258(3):449–456

25. Mehling M et al (2011) Antigen-specific adaptive immune responses in fingolimod-treated multiple sclerosis patients. Ann Neurol 69(2):408–413

26. Mellergard J et al (2010) Natalizumab treatment in multiple sclerosis: marked decline of chemokines and cytokines in cerebrospinal fluid. Mult Scler 16(2):208–217

27. Phillips JT et al (2011) Sustained improvement in Expanded Disability Status Scale as a new efficacy measure of neurological change in multiple sclerosis: treatment effects with natalizumab in patients with relapsing multiple sclerosis. Mult Scler 17(8):970–979

28. O'Connor P et al (2011) Randomized trial of oral teriflunomide for relapsing multiple sclerosis. N Engl J Med 365(14):1293–1303

29. Vermersch P et al (2014) Teriflunomide versus subcutaneous interferon beta-1a in patients with relapsing multiple sclerosis: a randomised, controlled phase 3 trial. Mult Scler 20(6):705–716

30. Soilu-Hanninen M et al (2012) A randomised, double blind, placebo controlled trial with vitamin D3 as an add on treatment to interferon beta-1b in patients with multiple sclerosis. J Neurol Neurosurg Psychiatry 83(5):565–571

31. Zamboni P et al (2012) Venous angioplasty in patients with multiple sclerosis: results of a pilot study. Eur J Vasc Endovasc Surg 43(1):116–122

32. Khan S et al (2011) Long-term effect on quality of life of repeat detrusor injections of botulinum neurotoxin-A for detrusor overactivity in patients with multiple sclerosis. J Urol 185(4):1344–1349

33. Putzki N et al (2010) Efficacy of natalizumab in second line therapy of relapsing-remitting multiple sclerosis: results from a multi-center study in German speaking countries. Eur J Neurol 17(1):31–37

34. Calabresi PA et al (2014) Pegylated interferon beta-1a for relapsing-remitting multiple sclerosis (ADVANCE): a randomised, phase 3, double-blind study. Lancet Neurol 13(7):657–665

35. Bermel RA et al (2010) Intramuscular interferon beta-1a therapy in patients with relapsing-remitting multiple sclerosis: a 15-year follow-up study. Mult Scler 16(5):588–596

36. Goodin DS et al (2012) Relationship between early clinical characteristics and long term disability outcomes: 16 year cohort study (follow-up) of the pivotal interferon beta-1b trial in multiple sclerosis. J Neurol Neurosurg Psychiatry 83(3):282–287

37. Dorr J et al (2012) Efficacy of vitamin D supplementation in multiple sclerosis (EVIDIMS Trial): study protocol for a randomized controlled trial. Trials 13:15

38. Kimball S et al (2011) Cholecalciferol plus calcium suppresses abnormal PBMC reactivity in patients with multiple sclerosis. J Clin Endocrinol Metab 96(9):2826–2834

39. Mosayebi G et al (2011) Therapeutic effect of vitamin D3 in multiple sclerosis patients. Immunol Invest 40(6):627–639

40. Smolders J et al (2011) Efficacy of vitamin D3 as add-on therapy in patients with relapsing-remitting multiple sclerosis receiving subcutaneous interferon beta-1a: a Phase II, multicenter, double-blind, randomized, placebo-controlled trial. J Neurol Sci 311(1-2):44–49

41. Stein MS et al (2011) A randomized trial of high-dose vitamin D2 in relapsing-remitting multiple sclerosis. Neurology 77(17):1611–1618

42. Serpell MG, Notcutt W, Collin C (2013) Sativex long-term use: an open-label trial in patients with spasticity due to multiple sclerosis. J Neurol 260(1):285–295

43. Kavia RB et al (2010) Randomized controlled trial of Sativex to treat detrusor overactivity in multiple sclerosis. Mult Scler 16(11):1349–1359

44. Collin C et al (2010) A double-blind, randomized, placebo-controlled, parallel-group study of Sativex, in subjects with symptoms of spasticity due to multiple sclerosis. Neurol Res 32(5):451–459

45. Zajicek JP et al (2012) Multiple sclerosis and extract of cannabis: results of the MUSEC trial. J Neurol Neurosurg Psychiatry 83(11):1125–1132

46. Novotna A et al (2011) A randomized, double-blind, placebo-controlled, parallel-group, enriched-design study of nabiximols* (Sativex((R))), as add-on therapy, in subjects with refractory spasticity caused by multiple sclerosis. Eur J Neurol 18(9):1122–1131

47. Notcutt W et al (2012) A placebo-controlled, parallel-group, randomized withdrawal study of subjects with symptoms of spasticity due to multiple sclerosis who are receiving long-term Sativex(R) (nabiximols). Mult Scler 18(2):219–228

48. Langford RM et al (2013) A double-blind, randomized, placebo-controlled, parallel-group study of THC/CBD oromucosal spray in combination with the existing treatment regimen, in the relief of central neuropathic pain in patients with multiple sclerosis. J Neurol 260(4):984–997

49. Broekmans T et al (2011) Effects of long-term resistance training and simultaneous electro-stimulation on muscle strength and functional mobility in multiple sclerosis. Mult Scler 17(4):468–477

50. Cakt BD et al (2010) Cycling progressive resistance training for people with multiple sclerosis: a randomized controlled study. Am J Phys Med Rehabil 89(6):446–457

51. Collett J et al (2011) Exercise for multiple sclerosis: a single-blind randomized trial comparing three exercise intensities. Mult Scler 17(5):594–603

52. Conklyn D et al (2010) A home-based walking program using rhythmic auditory stimulation improves gait performance in patients with multiple sclerosis: a pilot study. Neurorehabil Neural Repair 24(9):835–842

53. Dodd KJ et al (2011) Progressive resistance training did not improve walking but can improve muscle performance, quality of life and fatigue in adults with multiple sclerosis: a randomized controlled trial. Mult Scler 17(11):1362–1374

54. Fimland MS et al (2010) Enhanced neural drive after maximal strength training in multiple sclerosis patients. Eur J Appl Physiol 110(2):435–443

55. Hebert JR et al (2011) Effects of vestibular rehabilitation on multiple sclerosis-related fatigue and upright postural control: a randomized controlled trial. Phys Ther 91(8):1166–1183

56. Prosperini L et al (2013) Home-based balance training using the Wii balance board: a randomized, crossover pilot study in multiple sclerosis. Neurorehabil Neural Repair 27(6):516–525

57. Prosperini L et al (2010) Visuo-proprioceptive training reduces risk of falls in patients with multiple sclerosis. Mult Scler 16(4):491–499

58. Motl RW et al (2011) Internet intervention for increasing physical activity in persons with multiple sclerosis. Mult Scler 17(1):116–128

59. Nilsagard YE, Forsberg AS, von Koch L (2013) Balance exercise for persons with multiple sclerosis using Wii games: a randomised, controlled multi-centre study. Mult Scler 19(2):209–216

60. Mori F et al (2010) Effects of anodal transcranial direct current stimulation on chronic neuropathic pain in patients with multiple sclerosis. J Pain 11(5):436–442

61. Jurynczyk M et al (2010) Immune regulation of multiple sclerosis by transdermally applied myelin peptides. Ann Neurol 68(5):593–601

62. Dalgas U et al (2010) Muscle fiber size increases following resistance training in multiple sclerosis. Mult Scler 16(11):1367–1376

63. Burton JM et al (2010) A phase I/II dose-escalation trial of vitamin D3 and calcium in multiple sclerosis. Neurology 74(23):1852–1859

64. Filippi M et al (2012) Multiple sclerosis: effects of cognitive rehabilitation on structural and functional MR imaging measures--an explorative study. Radiology 262(3):932–940

65. Starck M et al (2010) Acquired pendular nystagmus in multiple sclerosis: an examiner-blind cross-over treatment study of memantine and gabapentin. J Neurol 257(3):322–327

66. Rice CM et al (2010) Safety and feasibility of autologous bone marrow cellular therapy in relapsing-progressive multiple sclerosis. Clin Pharmacol Ther 87(6):679–685

67. Velikonja O et al (2010) Influence of sports climbing and yoga on spasticity, cognitive function, mood and fatigue in patients with multiple sclerosis. Clin Neurol Neurosurg 112(7):597–601

68. Mattioli F et al (2010) Efficacy and specificity of intensive cognitive rehabilitation of attention and executive functions in multiple sclerosis. J Neurol Sci 288(1-2):101–105

69. Naismith RT et al (2010) Rituximab add-on therapy for breakthrough relapsing multiple sclerosis: a 52-week phase II trial. Neurology 74(23):1860–1867

70. Jensen MP et al (2011) Effects of self-hypnosis training and cognitive restructuring on daily pain intensity and catastrophizing in individuals with multiple sclerosis and chronic pain. Int J Clin Exp Hypn 59(1):45–63

Chapter 9

Alzheimer's Disease

Aura M. Hurtado-Puerto, Cristina Russo, and Felipe Fregni

Abstract

This chapter reviews the detailed characteristics of 100 interventional clinical trials in Alzheimer's disease, published between 2010 and 2015, with the aim of providing conceptual tools for the optimization of design and condition of future clinical trials on this population. For this purpose, the literature review of the 100 most highly cited interventional clinical trials between 2010 and 2015, was performed using the search engine Web of Science. Based on the guidelines by the CONSORT group and the criteria contained in them, relevant practices in these clinical trials are discussed and suggestions are made for the improvement on the design and conduction of further clinical studies in the field.

Key words Alzheimer, Dementia, Study design, Review, Clinical trial

1 Introduction

Alzheimer's disease (AD) is a chronic neurodegenerative disorder that mainly affects older adults. At the age of 60 years, the risk of developing AD is estimated to be around 1%, doubling every 5 years thereon, reaching 30–50% by the age of 85 years [1]. The chronic degeneration process of the illness progressively affects speech, orientation and mood. Nonetheless, the most prominent feature in early phases of AD is an impairment in the selective memory domain which progressively leads to dementia. Although AD is considered the most common cause of dementia worldwide, there is yet no clear cause of this chronic disorder and the underpinning pathophysiological mechanisms are still under investigation, which also adds additional difficulty to development of novel treatments as well as the conduction of clinical trials.

The prevalence and the devastating consequences of this illness represent a major problem for health and society [2]. In fact, every 3 s there is one new case of dementia around the world. Additionally, the total number of people with dementia is estimated to be around 47 million in 2015 and is projected to increase to 135 million by

Felipe Fregni (ed.), *Clinical Trials in Neurology*, Neuromethods, vol. 138,
https://doi.org/10.1007/978-1-4939-7880-9_9, © Springer Science+Business Media, LLC, part of Springer Nature 2018

2050. Moreover, AD is associated with an economic and emotional burden that affects society in general: as the number of individuals with Alzheimer's disease increases, the number of autonomous individuals decreases. This means that AD affects, not only the individual with the diagnosis, but also the caregivers and family members, since with the progression of the illness patients need assistance in performing daily-life activities. On the other hand, it is estimated that the cost of Alzheimer's disease is equal to the compound burden of cancer, heart disease, and stroke [2]. Specifically, it is estimated that the its current economic burden around 818 billion US$ and that it will rise up to US$ 2 trillion by 2050. Given the characteristics of the disease and the current absence of a cure, researchers are putting their efforts in studying the underlying pathological mechanisms, attempting to slow down its progression or to halt its course. In addition, prevention efforts are at very early stages, with limited results so far.

2   Challenges to Design Clinical Trials in Alzheimer's Disease

Researchers who design clinical trials in AD face several challenges due to the nature of the disease and the current means of diagnosing clinical and preclinical stages.

1. Lack of consistent biomarkers: decline in memory is part of the normal aging continuum; it becomes a cognitive sign of AD when is clinically relevant and impairs normal functioning and quality of life. Consequently, in order to identify populations at risk, it is fundamental to find markers as precursors of the disease. The advent of new biomarkers for this disease could bring more clarity to the field and to the selection of specific study populations. Beta amyloid, tau protein, electroencephalography, and magnetic resonance imaging volumetric and functional connectivity analyses could prove useful for this purpose in the future; however, they are still under investigation.

2. Difficulty in characterizing pathophysiology and stages of the disease: AD is a chronic degenerative disorder, characterized by a progressive, nonreversible deterioration, in which identification of biomarkers of progression could be of immense value to accurately assess, and biologically treat patients according to their stage of the disease.

3. AD progresses at different rates across patients [3]: it has been shown that age is a determining factor of the rate of progression of AD, with older patients progressing more slowly. This introduces a source of variability in the study samples that could be addressed by including even more baseline characteristics in covariate analyses, or simply restricting populations.

4. AD is very frequently accompanied by comorbidities: AD is a multidimensional disorder, with neurophysiological, psychological, and social aspects that need to be taken into consideration when conceiving the appropriate interventions.

5. Patients with AD are very often dependent on caretakers: this poses a major challenge in recruitment, adherence, and consistency of treatments in general as there needs to be awareness, education and availability of time from the caretaker, as well from the participant.

In the past decades, the rising awareness of the need for effective treatments has led to an increasing number of diagnostic and interventional studies. Indeed, even with this increased effort, there has been little advance for the treatment of AD in the past decade. We carried out a nonexhaustive search in Web of Science with search criteria for clinical trials in Alzheimer's disease, showing a trend of thousand percent increase in the number of works published from 1970 to 1990 versus those published from 1990 to 2015. In the present chapter, we focus on the 100 most highly cited interventional trials published over the last 5 years (2010–2015) with the aim of giving a methodological overview of the prevention and treatment trials in Alzheimer's disease. We present and discuss the main features of these publications with the purpose of offering the most updated literature in the field and aiding researchers in the design and planning of feasible studies in AD.

3   Methods

We conducted a search of the 100 most highly cited articles describing interventional clinical trials on the prevention and treatment of Alzheimer's disease, published between 2010 and 2015. For the analysis of the core features of each study, we followed the checklist elaborated by CONSORT (Consolidate Standards of Reporting Trials) in 2010: a set of recommendations for reporting clinical trials in a detailed and standardized consensual way, elaborated and revised by a scientific committee and endorsed by many well-known peer reviewed journals [4]. This statement enforces transparency for the critical appraisal of clinical trials. Furthermore, we focused on the specific challenges in conducting clinical trials in the AD population, giving a critical overview of how they have been addressed so fare. Since there are particular differences between interventional and diagnostic trials, we focused only on the former, given the aims of this book. Thus, studies with diagnostic purposes were out of the scope of research and are not discussed in this chapter.

To identify eligible articles, a literature search was performed using Web of Science, with "Alzheimer*" as key search term in the title. Then, the findings were restricted only to document file "Clinical trial" from 2010 to 2015. From all databases, the search result was of 488 articles through which we then screened (using a descending citation order) until we found 100 articles describing interventional trials. We screened, therefore, the first 148. Of these, 48 were excluded for the following reasons: 20 were clinical trials with the aim of finding a diagnostic marker; 19 of them were observational trials; 4 of them were excluded because, in spite of attempting contact with the authors, we were not able to retrieve the original article; 2 of them were not with AD patients, 2 were protocol articles and 1 was a review. The resulting 100 most cited clinical trials represent the corpus of data that are discussed in this chapter. This process and the exclusions are schematically represented in Fig. 1 [5–104].

4   Overview

Table 1 summarizes the main features of our search result. As stated above, our inclusion criteria were interventional studies published in the last 5 years (from 2010 and 2015) with Alzheimer's disease patients or at risk population. Given the fact that we sorted them by number of citations, the majority of these trials were conducted between 2010 and 2013 and five were published in 2014. Most of the selected works (65%) were published in journals with an impact factor between 1 and 5, while 12 of them were published in journals with impact factor of more than 20.

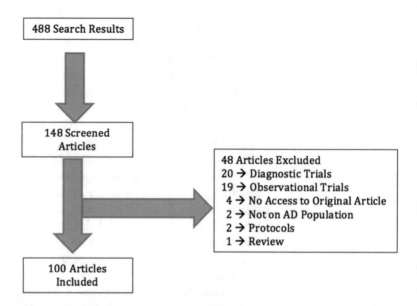

Fig. 1 Schematic illustration of the search results, inclusions, and exclusions

Table 1
General characteristics of selected articles

		Total
		N = 100
Year	2010	32
	2011	28
	2012	24
	2013	11
	2014	5
	2015	0
Number of citations	0–20	25
	21–30	31
	31–40	8
	41–50	4
	51–60	6
	61–70	5
	71–100	9
	101–200	10
	201–745	2
Journal impact factor	1–5	65
	5–10	15
	10–20	8
	>20	12
Consort criteria	Yes	42
	No	58

Following the CONSORT checklist, we systematically analyzed the characteristics of the selected studies. Furthermore, for each article we checked whether the journal in which it was published endorsed CONSORT. We found that 42 articles were published in such journals and, of the remaining 58, 10 detailed their methods so completely, that they fitted each item of the checklist. This means that, even when the journal they were published in did not formally endorse the CONSORT guidelines, authors did follow these guidelines. On the other hand, other articles published in journals endorsing CONSORT lacked details in their description of the design, in the specification of data analysis, in the sample size calculation and in the critical discussion to the limitations. By strongly agreeing with the need of clear and transparent reporting, other researchers have the possibility to understand, replicate, and appraise the applicability and limitations of a study. While transparency in the reporting of clinical trials is paramount to the advance of research, lack of transparency is associated with bias [105]. Generally, publication bias (i.e. the tendency to not publish negative results) is a relevant problem, especially when reporting the design, conduction and results of clinical trials. Lack of evidence regarding the effects of a treatment can have a direct negative impact on populations. It is, therefore, highly recommended that investigators (especially beginners) follow these guidelines, not only for reporting purposes (CONSORT), but also to design a trial, such as the SPIRIT guidelines (http://www.spirit-statement.org/).

Another relevant point is the results of these trials. Many of the articles here reviewed were not positive for their primary outcome (see Table 2). This means that trials with both nonconclusive and negative results have been published and still have a high number of citations. Thus, these trials help to elucidate which interventions and strategies lead to meaningful conclusions, advancing research in the field. These encouraging findings can reassure researchers that the best understanding of mechanisms of disease and treatments is under way.

5   Studies

The selection of the studies included in this analysis is described above. Table 2 summarizes their main features. The total sample of studies is 107, because seven articles described 2 studies each.

Table 2
General characteristics of analyzed studies, contained in selected articles

		Total
		N = 107
Sample size	0–20	14
	21–50	27
	51–100	15
	101–500	34
	501–1000	4
	1001 + Above	13
Type of intervention	Drug	85
	Device	5
	Procedure	17
Positive for primary outcome	Yes	69
	No	33
	Partially	5

There is an additional remark for the trial carried out by Harrington and coworkers on the effects of Rosiglitazone on cognition and global functioning [42]. This work describes three studies; we analyzed only two because there were not enough elements about the procedures used for one of them (REFLECT 4) in order to collect all the chosen information.

6   Trial Design

In this section, we refer to "trial design" as the experimental plan for the study (parallel, crossover, factorial, etc.). There are other features regarding these types of designs, such as the number of groups in the sample, the size of the groups (e.g., uneven allocation ratios), and whether there are several dosages of the same medication being tested within each group or between groups (e.g., dose escalation and dose titration). Table 3 shows the distribution of the trials by phase and design: drug trials are sorted from phase I through IV, while procedure and device trials are sorted by feasibility and pivotal phases. In our sample, we found various types of trial designs including randomized controlled trials and open label studies with no control group. In this section, 'Open Label trials' are those that did not use blinding.

The most frequent study design was parallel, with no additional remarks. Adaptive parallel designs were those that used some sort of algorithm to define different interventions within participants as the trial was conducted, and accumulating information was collected [106]. Among randomized controlled trials, we found

Table 3
Trials sorted by design and phase

| | Open label | Randomized clinical trials—parallel designs | | | | | Total |
| | | Parallel (no further specification) | Additional designs | | | | |
			Adaptive	Factorial	Dose escalation	Dose titration	
Phase I	5	10	1	1	8	2	27
Phase II	0	17	0	0	15	3	35
Phase III	0	10	0	2	9	1	22
Phase IV	0	0	0	0	1	0	1
Feasibility	4	9	0	0	0	0	11
Pivotal	0	8	0	0	1	0	11
Total	9	54	1	3	34	6	107

simple parallel designs and parallel designs with some variations: adaptive, factorial, dose escalation and dose titration. Briefly, in dose escalation trials, various fixed dosages of the same medication are tested against each other [107]. Most frequently, dose escalation trials compared groups with different but fixed dosages. In contrast, dose titration designs allowed for a flexible dose range within groups [108]. This kind of design is frequently used when the most appropriate dose—often the highest tolerable dose—is sought in early phase studies, in order to build a safety and efficacy profile. Customarily, only drug trials are included in this category, as dosages can be varied. Nonetheless, other kinds of interventions, such as exercise programs or brain stimulation techniques can use dose escalation designs. For example, the intensity of exercise routines can be adjusted and the number of pulses or the stimulation frequency of a brain stimulation technique can be altered [53].

As stated above, drug interventions are the most suitable for dose escalation designs, and the most frequent in this sample. We found a considerable amount of dose escalation trials in this description, typically in phase I and phase II trials. However, it is remarkable that this design has also been used in phase III and even in phase IV. Significantly, the only Phase IV trial implemented a dose escalation strategy. Figure 2 shows the 85 drug trials distributed by phase and design.

In addition, seven dose escalation trials modified dosages within the groups. This is notable because traditionally, trials have been designed to test fixed dosages in order to control for this variable, potentially leading to the consequence of the reduction of adherence due to poor tolerability or poor response. Consequently,

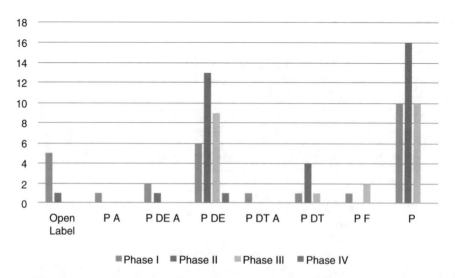

Fig. 2 Trials sorted by design and Phase. *P A* parallel adaptive, *P DE A* parallel dose escalation adaptive, *P DE* parallel dose escalation, *P DT A* parallel dose titration adaptive, *P DT* parallel dose titration, *P F* parallel factorial, *P* parallel

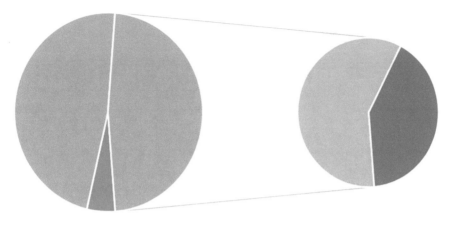

■ open label ■ rct ■ same dose ■ dose variation

Fig. 3 Drug trials sorted by design: open label trials (shown in pink) compared to randomized controlled trials (RCT, in blue). Of the latter, the distribution of those maintaining the same dose throughout the trial (in green) in contrast to the ones that used dose variation (in violet)

the flexibility in dosage can allow for more tolerability, adherence, and the analysis of more data, increasing the generalizability. In turn, this would more faithfully resemble a real-life scenario.

Most of the drug trials described in this sample are placebo-controlled, meaning the experimental arm is compared with placebo as a control condition. This characteristic increases internal validity. On the other hand, protocols often require participants to stop all medications for their condition before enrollment which, in turn, may reduce the feasibility and the applicability of the results to the target population. Additionally, in most cases, this requires that there is no standard care medication available in order for the principle of equipoise to exist (see Chap. 2 for a discussion on this topic). As shown in Fig. 3, a great amount of the selected trials used a drug as the control. The latter situation better reflects the condition of Alzheimer's disease patients who are usually polymedicated; this makes the conduction of the trial more feasible and increases its generalizability.

Furthermore, there are three studies that were designed with no intervention as a control—meaning the control was not placebo, but the absence of an intervention (Table 4). This approach is uncommon for drug controlled trials and the reasons for choosing this design must be well justified, as it is difficult to rule out placebo effects or natural history of diseases in studies with no intervention as a control. In the case of the trials we analyzed, those circumstances were not clearly explained.

The gold-standard for randomized controlled trials is the design with a placebo control under blinded conditions in order to control for placebo effect. This can be observed in a variety of

Table 4
Drug trials sorted by phase and type of control

	Controlled designs—type of control			Noncontrolled designs	Total
	Placebo	Drug	No intervention		
Phase I	19	4	2	2	27
Phase II	28	6	1	0	35
Phase III	18	4	0	0	22
Phase IV	1	0	0	0	1
Total	66	14	3	2	85

Noncontrolled designs used no comparator. Controlled designs with no intervention, were labeled "No Intervention" as their comparator

interventions and varies greatly from intervention to intervention interfering with the measurement of the effect of the intervention and with the interpretation of results [109–113]. Therefore, a simple and robust design such as the parallel design with a placebo control is still the gold standard; however, ethical and feasibility issues need to be considered.

7   Changes to Trial Design

Unforeseen circumstances may prompt the change the design of a trial amidst its conduction. Some of these reasons are: new evidence on the pathophysiology of the disease; that the intervention becomes widely available, making the conduction of the study futile; that the monitoring board and/or the external ethical committee considers the intervention to be unsafe before the end of recruitment. A general recommendation by the CONSORT group is that protocols of clinical trials should be very detailed, with clarity and transparency in the description of any changes. Therefore, having a well-defined Data and Safety Monitoring Board (DSMB) is very important, especially in trials with a large number of subjects in which what was learned at the beginning of the trial can be used to avoid unnecessary risk for remaining subjects.

Regarding this analyzed sample, a small number of studies ($n = 8$) changed their design after starting recruitment.

Rinne and collaborators aimed to determine the effects of bapineuzumab, a humanised anti-amyloid-β monoclonal antibody, in patients with mild to moderate Alzheimer's disease [5]. Patients in the experimental groups were assigned to receive ascending intravenous doses (0.5, 1.0, 2.0 mg/kg) with placebo as a control.

While the study was ongoing, new information became available from another trial [114] about the negative consequences of the 2.0 mg/kg dose of bapineuzumab. Consequently, there was a stoppage of enrolment and allocation of participants in that group.

Similarly, Salloway and collaborators designed two phase-III clinical trials evaluating the effects of bapineuzumab in three groups receiving different doses (0.5, 1.0 and 2.0 mg/kg) with placebo as a control group. Soon after the starting of the study, an external monitoring committee recommended to discontinue the highest dose based on the finding of effusion or edema related to amyloid as evaluated by imaging [10].

One study was halted due to the results of an interim analysis leading to the conclusion that the intervention was futile [16], at this stage [14]. Another trial held off the administration of the fourth dose of the study drug due to external evidence pointing towards unsafety [20]. When it was decided that the study could be resumed, the window for the planned dose at the beginning of the trial had already passed for patients in this cohort.

Moreover, during a phase III trial [29] there was a discontinuation of the two high-dose groups in a four arm placebo controlled trial so, the primary analysis was based on 166 patients in two arms of the study. There was also a case in which the analysis plan needed to be changed due to new evidence that could increase attrition rates. Consequently, only timepoints with less than 30% of attrition were considered for analysis in order to minimize the possibility of error [74].

Changes in protocol can also occur when other drugs are approved for prescription in that particular population during the study. The ongoing study of Geldamcher and coworker was modified to allow patients to begin memantine if prescribed, thus not compromising equipoise. The use of the drug was then considered as a covariate to run the analysis [31].

To sum up, in our sample most changes in the designs of trials occurred when the intervention was a drug, after the occurrence of adverse events in the study or as suggested by the emergence of new evidence in the literature. These cases are more frequent in large phase III trials. Importantly, these recommendations are endorsed or suggested by a DSMB. Also, there were cases in which the investigator pursued an interim analysis for futility, i.e.,, to check whether preliminary data indicated that the trial would not be successful after completion. It is fundamental for clinical practice and research activity, to report this information and the causes for the adjustments. The need to modify the design of the experiment according to continuously changing circumstances represents an important challenge for clinical trials. Furthermore, caution in the design of experiments is advised to researchers, assessing potential pitfalls beforehand and keeping up to date with the latest available information.

8   Study Setting

When it comes to the design of a clinical trial, the setting from which participants are recruited has many implications: recruitment, consistency of the measures for the outcomes, adherence and external generalizability are important for a successful trial. Specifically, characteristics such as economic, social and demographic features of the setting have the potential to affect external validity. In a similar way, cultural differences and even organizational culture, can lead to variability in the collection of data, in the interaction with participants, etc. [115, 116]. Altogether, study setting can influence these aspects, as they are direct consequences of the particularities of the institutions or communities from where the study population is recruited.

In our sample, as grossly depicted in Fig. 4, about half of the trials reported either inpatient or outpatient settings. Most of these, detailed that their populations were recruited from outpatient centers, being consistent with research in Alzheimer's disease population, mostly found in outpatient situation. Moreover, recruitment of institutionalized Alzheimer's disease patients becomes challenging as they frequently have severe comorbidities. This adds variability to the study and complexity to the design. In spite of these difficulties, research for this population should neither be discouraged nor neglected. On the other hand, clinical trials in AD could be carried out in smaller samples with less variability. As stated by CONSORT, the information about the setting is crucial because it provides further background to the findings and can help the reader in judging the feasibility and generalizability of the results for further works. The great majority of studies missed to clearly report the origin of their

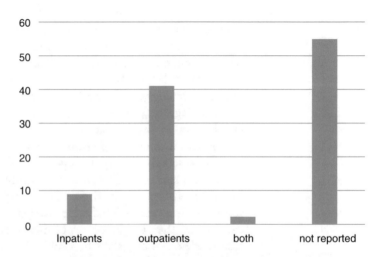

Fig. 4 Setting of clinical trials sorted by inpatient and outpatients

population as inpatients or outpatients, and one of the reasons is possibly the use of mixed population, thus not adequately outlined. Once more, this information is important to infer generalizability. Scenarios such as inpatient and outpatient have implications on caregiver involvement, treatment culture, socioeconomic background, adherence to treatment and other interventions, engagement on the research being conducted, access to specialized medical care, etc. Furthermore, these factors, potentially impact the results of the trials and their applicability [116, 117].

As outlaid in Table 5, more than half of the trials in our sample were multicenter, namely they were conducted simultaneously in more than one center. One interesting aspect in AD research is the need for multicenter studies even in phases I and II (see Table 5). In fact, given the small effect sizes in AD trials, there is a need of a large sample, thus, usually requiring a multicenter approach. In fact, the mean sample size of single center trials was 62, and that of multi-center trials was 442. Furthermore, 13 of them were run in different countries. In particular, multicenter trials conducted in a single country had a mean sample size of 370 participants, while in multicenter trials with more countries involved, the mean was 740.

In spite of regulatory challenges particular to each country and of organizational culture and policy differences, many research groups chose to recruit in this manner, with several implications. First, it can help researchers combine efforts towards the same goal; second, it contributes to collaborative work and sharing of expertise; third, it improves the generalizability of the results. Conversely, 37 studies in our sample were carried out in a single center. Additionally, three studies did not specify the setting of their trial. Table 5 summarizes these findings, along with the distribution of trials by phase and settings as single center, multicenter, multicenter in different countries, or not specified. Drug trials are sorted from phase I

Table 5
Study settings—multicenter, single-center, and multicenter in different countries, sorted by phase

	Single-center	Multicenter	Multicenter in different countries	N/S	Total
Phase I	14	11	2	0	27
Phase II	9	20	4	2	35
Phase III	0	15	7	0	22
Phase IV	0	1	0	0	1
Feasibility	11	2	0	0	13
Pivotal	3	5	0	1	9
Total	37	54	13	3	107

through IV while procedure and device trials are sorted by feasibility and pivotal as described in Sect. 5.

9   Participants

One of the important question readers can ask themselves when reading the results of a study is: are these applicable to other populations with the same disease? In other words, what is the relevance of these results? This responds to external generalizability. Rothwell [116] details the importance of all aspects surrounding external validity and how the choice of inclusion and exclusion criteria has strong repercussions on it. Another detail important to this evaluation is whether participants have been recruited with targeted or non-targeted strategies (e.g., newspaper ad, invitation letter) with direct influence on the potential characteristics of the population—educational level, adherence, access to medical attention, availability of economic resources, motivation, etc. However, this information is often not reported in the description of participants in the selected publications.

It is worth noticing that, often during the screening process, when inclusion criteria leave room for interpretation, it may add undesired variability. This could entail selection bias and the observed effect may be unrepresentative of the target population and thus misleading.

As described above, AD is a multidimensional syndrome whose complexity and lack of a clear diagnosis with clinical and biomarker correlation makes it difficult to differentiate from other population with similar clinical features (e.g., vascular dementia, Lewy body dementia, chronic traumatic encephalopathy, depression, etc.). Consequently, and due to the progressive nature of the disease, there is a huge variety of inclusion criteria for cognitive stages in AD trials. Additionally, the most frequently used scales do not evaluate the multidimensional aspects of the disease, nor do they have the specificity to exclude similar diagnoses. Consequently, participants of the studies in our sample have been recruited using different criteria. A good part of these studies ($n = 30$) used Minimental State Examination (MMSE) scores as inclusion criteria to identify dementia and select a specific spectrum of cognitive impairment. Moreover, 11 trials used National Institute of Neurological and Communicable Disorders and Stroke and Alzheimer's Disease and Related Disorders Association (NINCDS-ARDA) scores for probable and/or possible AD. Fifty-one studies used both criteria. Four studies used different measures and four other did not specify which was the instrument used to assess AD.

Remarkably, studies using the MMSE stratification as inclusion criteria employed different score ranges to target mild, moderate, and severe AD, or a combination of these stages. Additionally,

MMSE provides gross evaluation of mental function with no assessment of functional status and is susceptible to educational, cultural, and language conditions [118]. Other studies coupled cognitive indicators with genetic biomarkers to stratify participants with imaging in order to correlate clinical with surrogate measures. The studies that used a stratification strategy are discussed in the randomization and statistical analysis sections, as pertinent (Sects. 14 and 18, respectively). These studies found the stratification strategy useful because there was a difference in disease progression in the strata with positive or negative expression of Apolopoprotein A, allelic isoform E4 (APOE e4). For example, the two studies reported by Salloway and coworkers, were designed in the same way except for the fact that one included APOE e4 positive AD patients and the other one included APOE e4 negative AD patients [10].

Other studies used imaging techniques—head computed tomography (CT) or magnetic resonance imaging (MRI)—to more accurately include participants according to the stages of disease clinically and morphologically. Yet, there is no single or compound surrogate measurement to diagnose Alzheimer's disease and this is another challenge of conducting trials with such a population. In our sample, 23 studies used some kind of imaging technique to confirm diagnosis of AD for their inclusion criteria. This becomes particularly useful when the interventions are aiming at a pathophysiological target in the disease that has been studied in animal models or humans with AD. On the other hand, it has less impact when studying populations where the interventions aim at relieving symptoms or quality of life because, if effective, these interventions can have a potential in nonrestricted populations as well. Furthermore, seven studies used these instruments as tools for the exclusion of other brain abnormalities such as vascular involvement [67, 119], other potential causes of dementia [32], brain abnormalities [29, 92, 93], or being inconsistent with AD diagnosis [18]. This becomes more relevant when the intervention is in early study phases and it might be of added risk to populations with the same clinical diagnosis but a different etiology. Two studies [18, 29] used these tools for both inclusion and exclusion purposes.

If you are planning your study in AD, what to do then? It depends on the phase of development. For phase I and II studies, it is also important to understand the mechanisms of the intervention so as to refine it. In this context, it is important to be as specific as possible. For phase III studies, wider and broader inclusion criteria are more accepted, based on a clinical diagnosis.

Future studies should include more specific inclusion criteria, as there is currently a challenge in the specific selection of patients with AD. In our sample about half of the trials included patients with AD with mild to moderate cognitive impairment and were mostly designed to slow the rate of cognitive decline. Given the fact that AD is still diagnosed clinically, patients are symptomatic at the time of diagnosis rendering most trials palliative in character. Three studies in our sample included patients in preliminary stages of disease, namely participants with Mild Cognitive Impairment (MCI). However, it is important to remark that the specificity of the diagnosis is uncertain, as not all MCI will evolve to AD. Only one study in our sample considered only patients with MCI as participants. On the other hand, only three studies were done with healthy subjects aiming to prevent their evolution to AD. The huge impact of the disease on the society provides a niche to invest efforts towards the prevention on asymptomatic and individuals diagnosed with MCI.

Unfortunately, some articles (13%) did not specify the characteristics of their population with clear eligibility criteria. As mentioned before, the lack of this information undermines the applicability of results and the reproducibility for future research. This is particularly relevant in cases of advanced phase research trials (III and IV), whose results have direct implications on clinical decisions. It is also critical in drug trials where disease staging constitute radically different populations. All sorts of pathophysiological processes take place in different stages of the disease and, consequently, different drug effects are expected.

10   Duration and Follow-Up

The intended duration of the participation of a subject in a clinical trial is an important part of the design of a clinical trial because it plays a central role in the risk of attrition and in the impact of the results. Phase III or pivotal trials usually intend to evaluate the intervention during long terms to assess long-term exposure or the after effects of a limited exposure (as to learn more about clinical effects). Conversely, early stage trials may aim to evaluate immediate effects of an intervention as measured by an extensive array of assessments, and they may be designed to do so in very few days or even in a single visit. On the other hand, attrition is highly dependent on the amount of time a participant spends in a trial as longer durations make it more challenging to keep a participant motivated, thus, risking the data collection as their participation approaches the end. This is an important point to keep in mind when designing trials in which the power provided by the sample size can cause problems. Consequently, all these aspects ought to be considered when designing a study in which various interventions are assessed at several stages: exercise routines in a feasibility trial, existing medications in a phase III trial or IV infusions of immunotherapy in phase I studies are examples.

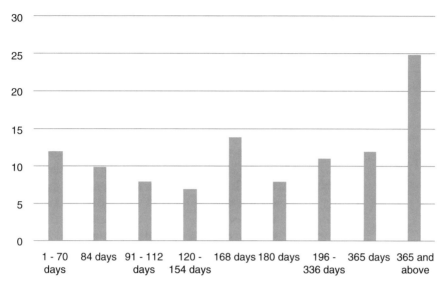

Fig. 5 Duration of participation in studies

The mean duration of the trials we included is 304 days while the minimum duration was 1 single day, and the maximum duration was 1825 days (Fig. 5). The aim of the trial with the shortest duration was to assess the role of a single session of music for 30 min on encoding and recognition of verbal information while comparing the difference between preintervention and postintervention assessments [44]. On the other hand, the trial with the longest duration for participants was designed to evaluate the long-term effects of a medication compared to a placebo, in the prevention of occurrence of AD in neurologically healthy subjects. The effects of the intervention were followed up for 5 years [25]. Additionally, 24 studies were designed for subjects to participate for more than 1 year, with a mean of 696 days. This is an interesting observation since it depicts that clinical trials in AD focused on the evolution of the disorder and its neurodegenerative features, requiring a considerable commitment from subjects involved.

Furthermore, few of all trials ($n = 37$) performed follow-ups after the intervention had concluded. The majority of these ($n = 20$) followed up participants between 30 and 180 days with 7 of them involving more than one follow-up. Conversely, most of the studies we analyzed did not require any follow-up after the intervention concluded ($n = 70$). Once more, the need for planning follow-ups as a part of the study is a result of the aim of the study and can be carried out in several ways including visits, phone calls, and registration of diaries. The possibility to reassess subjects after the intervention is relevant when long-term effects on multiple aspects of the population are of interest. Finally, it is important to keep in mind that the validity and generalizability of results may depend both on continuing evaluation and on the results of the evaluation itself [120].

11   Intervention: Drug/Device/Procedure

When designing a trial, the first step is to understand the intervention. How is it hypothesized to be useful—which are the mechanisms behind it and to which population is it applicable? For how long should it be implemented in order to observe a significant effect? Is it possible to use a control that allows for double blinding? What are the outcomes that can measure the effects of this intervention? All of these questions depend on the intervention, whether it is a new drug, a drug already approved for other uses, a device or a procedure. In this section we differentiate device from procedure as per their definition by the FDA [121].

It should be pointed out that all details of the intervention need to be planned in advance, and later described in the article along with the results of the trial. The level of detail should include parameters, routes of administration of the active intervention and control, titration and dose escalation procedures. Frequently in AD trials interventions involve usual care. In these cases, it should be clarified what this treatment entails.

In the sample we examined, the majority ($n = 85$) of studies consisted of drug clinical trials. Five of them evaluated devices and 17 evaluated procedures. Among the device trials, one used encapsulated cell biodelivery (ECB) that delivered nerve growth factor (NGF) [26], one used Deep Brain Stimulation (DBS) [8], three evaluated the effects of repetitive Transcranial Magnetic Stimulation (TMS), associated with linguistic tasks [39], memory performance [101] and cognitive function and cortical excitability [53]. There is also a place for invasive and noninvasive stimulation studies in this population, which could induce changes in both the cognitive and clinical domains. Of the 17 procedure trials, the following interventions were adopted: cognitive rehabilitation [26], aerobic exercises [43], music [44], light and guided sleep education [51], walking [59], at-home exercises [68], cognitive intervention [70], lifestyle intervention and medication [71], movement based training [73], exercise routines [77], psychosocial intervention [79], specific care plan [83], technology-aided verbal instructions [91], cognitive rehabilitation [96], activity-specific exercise program [98], and cognitive therapy [100]. These interventions are usually implemented for a short period of time because attrition can become problematic. In contrast, drug interventions, in which the additional effort for the participation in the study is not as costly as that of procedure and device trials, there is a greater investment of time. In the case of surgical procedures, it may add an additional risk to increase duration.

Most of the trials we analyzed used drugs as interventions. Likewise, pharmacologic sponsors supported most of these drug trials. Some of the pharmacologic agents used in these studies were: agents targeting amyloid plaques or amyloid precursors including antibodies, intravenous immunoglobulin, γ-secretase inhibitors, Tau kinase inhibitors and antibody production stimulators; neurotrophic agents; Neurotransmitter modulators including antipsychotic agents, antiepileptics, stimulants, antidepressants, cholinesterase inhibitors and N-methyl D-aspartate (NMDA) receptor antagonists; antidiabetics including oral antidiabetics and insulin; Lipid lowering agents; Cardiovascular agents such as: antiplatelet agents, vasodilators, nonsteroidal anti-inflammatory drugs, calcium channel blockers and beta blockers; Vitamins, lipids, and supplements; Hormones; Chemical species acting as ionophores.

The level of planning needed for the execution of such a trial include the following questions: which is the control to be used and how is it to be implemented? What are the procedures surrounding its implementation? How similar are the controls and other interventions and how will this be overcome in order to maintain blinding? In case blinding is not feasible or ethical, what are the reasons? And is it possible to change circumstances in order to make it feasible and ethical? All these factors surrounding the intervention are crucial for the achievement of a valid trial that can impact the field, regardless of the results. For longer duration trials, it is important that the time point of measurement of the main outcome (main endpoint) is well defined. There are two strategies: to measure the main outcome early and make subsequent measurements as follow-up assessments, or vice versa: to measure the main outcome late and make earlier assessments, as secondary outcomes. Here, a deep analysis needs to be made considering when most of the effects will be seen, and also to what extent the attrition rate would affect this. For the analyzed trials, we noticed that overall, studies using clinical outcomes as main endpoints, measured primary and secondary outcomes concomitantly and regularly (e.g., every 3 months, for a year). Generally, the longer the duration of the studies, the later the measurements of primary outcomes were performed [19, 25]. For phase I trials, when the main outcome was safety and the study duration was overall shorter, measurements of the primary endpoint were made earlier, while secondary measures (clinical effects) were made later on [20].

12   Primary Outcomes

The choice for the primary outcome should be the product of careful thought based on the aim of the study and the expected changes in the selected population. How to detect and measure these changes? How to answer the study question? Outcomes should be selected based on the answers to these queries. As mentioned before, in AD trials arena, outcomes are mostly aimed to measure clinical changes. This is reflected by the 72 trials selecting clinical criteria as their primary outcomes in an exclusive way. Of the remaining studies, three chose functional outcomes and eight used surrogate and clinical measures jointly as their primary outcomes. To note, many trials used, not only one primary outcome, but two or more "co-primary" outcomes (Fig. 6). It is interesting to highlight that AD research has very few validated surrogate outcomes, and the ones being currently tested are not reliable and/or not easy to be measured.

Functional outcomes in our sample included measures of sleep quantity and quality, namely, total wake time based on wrist actigraphy and caregiver ratings of participant sleep quality on the Sleep Disorders Inventory. These outcomes were selected in a trial aimed at evaluating the effects of walking and light exposure in AD patients [51]. In addition, one group measured the ability to perform bed transfers and mobility using acute care index of function [98]. Some of the trials evaluating safety were classified as implementing both clinical and surrogate outcomes because safety outcomes consisted of both the occurrence of adverse events and the combination of physiological and biochemical measures. Figure 7 shows the distribution of the frequency of choice for primary outcomes of all studies included in the analysis. For this figure, the term "biochemical measures" refers to parameters measured in

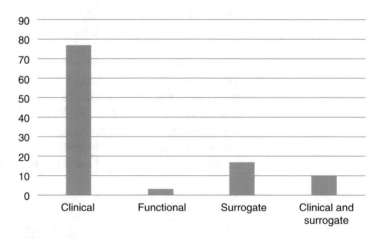

Fig. 6 Frequency of types of primary outcome in the studies

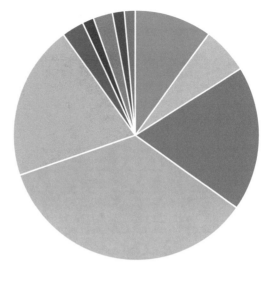

Fig. 7 Categories of primary outcomes. *Cog–Cl* cognitive and clinical, *Saf–Cl* safety and clinical, *Bio–cog* biochemical and cognitive, *Cog–Saf* cognitive and safety, *Cog–Im* cognitive and imaging, *Bio–Cl* biochemical and clinical

tissues or fluids (blood, cerebrospinal fluid, urine, saliva, etc.) through chemical means. They include, but are not limited to, pharmacokinetics, pharmacodynamics, and bioavailability [122].

The fact that there are hardly any biomarkers that orient the evolution or diagnosis of AD makes it even more challenging for phase I and phase II trials to convey significance as measured by surrogate biomarkers, which is the standard for research in other conditions which do have surrogate biomarkers. In this way, clinical outcomes become the only tool these trial designs have in order to prove efficacy and safety. Consequently, clinical outcomes appropriate for the conduct of phase III and phase IV studies are also prominent in early phase trials, as grossly depicted in Table 6.

Surrogate outcomes reflecting clinical function are rare in Alzheimer's disease. Some trials correlate neurological function with neuroimaging. Also, other surrogate outcomes used are intended to measure safety outcomes. There is currently no biomarker that reflects severity or that can be used as a diagnostic tool for AD. This is one of the main challenges to be overcome while making the distinction of trial phases sometimes unclear. Here, it is also important to differentiate the surrogate markers (those that can provide information related to clinical changes) from

Table 6
Primary outcomes, co-primary outcomes and phases

	Phase I	Phase II	Phase III	Phase IV	Feasibility	Pivotal	Total
Biochemical	5		0	0		0	7
Cognitive	4	12	18	0	5	1	40
Imaging						0	4
Safety	8	3	0	0	1	0	12
Cognitive-other	2	5	2	0	2	2	13
Safety-other	2						
Biochemical-cognitive	0	1	0	0	0	0	1
Cognitive-safety	0						
Cognitive-imaging	0	0	0	0	1	0	1
Biochemical-other	0						
Total	27	35	22	1	13	9	107

predictive markers (those that can provide information about the likelihood of response) and prognostic (or diagnostic) markers (those that can provide information regarding current or future presence of disease). It is important to notice that these markers also depend on the type of treatment being tested.

13 Sample Size Calculation

All clinical trials are strongly encouraged to report how they planned their sample sizes and report all the information following CONSORT recommendations. However, approximately half of the studies in the sample did not report how they calculated it ($n = 49$). Two trials corresponded to post hoc analyses and were deemed nonapplicable, since this calculation was ought to be made a priori. Conversely, 56 studies did mention in varying degrees of detail how the sample size was calculated. Twenty articles did not detail this method. Out of those describing some sort of method of calculation, minimal clinically important difference and literature review were used equally (15 studies each). Importantly, literature reviews for sample size calculation purposes should point towards finding meaningful values, which do not mean statistical significance alone. In fact, the concept of statistical significance does not necessarily translate into a meaningful clinical difference. When there is no available information on the field, or insufficient evidence on a specific outcome, a pilot study can be the most compelling approach to start with. Some of the studies included were pilot studies themselves, while six trials used pilot study information in order to calculate the size of their sample.

Our results reflect those of a systematic review [123] that points out common mistakes and omissions in the description of sample size calculations in RCTs. On the other hand, a recent systematic review in trials on anesthesia underlines that a good number of articles describe how researchers calculated the sample sizes of their studies, while failing to report important information about essential parameters, namely, standard deviation, Type I error, minimal clinically important difference and power [124]. In a similar manner, we evaluated whether the trials we analyzed reported these parameters. We found that, in 56 trials reporting a description of their sample size calculation, only 10 of these reported all elements properly. Considering the remaining studies, the least reported parameter was the standard deviation—described in only 18 studies, followed by Type I error—described in 31 studies, the minimal clinically important difference—reported in 38 studies—and the power—reported in 44 studies.

14   Randomization: Generation, Process, and Implementation

Randomization is the core of controlled clinical trials, as it aims to divide groups with similar characteristics and, thus, it can be assumed that any effects of the intervention being tested are due to the intervention and not to differences in characteristics across groups. In fact, individual characteristics may vary in the sample and can be influenced by the intervention. Indeed, this is the main reason for randomization, as the characteristics of the population—covariates—can be balanced across groups. Several types of randomization can be considered: simple, blocked, stratified, adaptive, etc. These are described in more detail in Chap. 1. Who generates this sequence and how, is core for maintaining its randomness. Roughly 44 trials mentioned a method of randomization sequence generation the most common being through computer generation. Nine trials had an open label one arm design, so no blinding or randomization was needed. The remaining studies did not mention any methods.

The other related and important issue is allocation concealment. This refers to the process of keeping the allocation of participants a secret from investigators responsible for recruitment, so the sequence cannot be predicted and they are unaware of the treatment assignment of the next participant. On the other hand, the perils of failing to conceal allocation can lead to an overestimation of the treatment effect [125]. Among the selected trials, allocation concealment was mostly done by an automatic voice response system (mentioned in 33 trials), followed in frequency by sealed envelopes. Once more, the majority of the studies mentioned no allocation concealment method at all.

The implementation of randomization refers to the methods used to carry out and keep the recruitment, randomization, and allocation processes independent. In other words, separate personnel should take part in each one of these procedures to prevent bias. Importantly, reporting this item is necessary for the readers' assessment of the possibility of bias. Once more, these details were all described only in 20 studies. This leads to believe that this item is considered unimportant when reporting the procedures carried out in these trials, or were ignored during the trial, showing potential sources of bias in these studies. Nowadays, with current technological methods, it is fairly easy to implement allocation concealment methods by using a computer, or even an app in a mobile device.

14.1 Allocation Ratio   Participants can be allocated to groups with a different probability, in unequal allocation ratios—for example, in a 2:1 ratio of allocation of an active intervention versus a placebo intervention. This method has become frequent in clinical trials, as reflected in our findings. Some of the reasons claimed by trialists to choose unequal allocation ratios are: to minimize the loss of statistical power if a dropout happens, to decrease costs in the study and to maximize information about the intervention while conducting the trial, to mitigate patient withdrawal and to hasten recruitment as it might seem more appealing to participants [126, 127] On the other hand, unequal allocation ratio could lead to overburdening of study population. Additionally, therapeutic misestimation can occur, while the reason for conducting the trial in the first place is that there is no evidence of the clinical effectiveness of the treatment. Another important reason to reconsider when choosing this strategy is that it needs a greater sample size to adjust for statistical power. Still, the change in power is minimal if a 2:1, or even 3:1 ration randomization strategy is adopted (see the following references for further discussion [128, 129]). Nonetheless, in early study phases (I and II), unequal allocation strategies might be a good alternative, as opposed to late study phases (II and IV), given the fact that early study phases have the objective of learning about the treatment dimensions [127].

15   Blinding, Assessment of Blinding and Similarity of Interventions

Blinding is one of the cornerstones of controlled trials. This process aims to prevent bias both from assessors and participants. In the first case, participants can be influenced by knowledge of their allocation by having different expectations and behaviors during the trial, further impacting, for instance, the adherence. In the second situation, unblinded assessors can potentially evaluate or treat participants differently depending on the group they are allocated to. In AD trials, open label designs have been carried out by many researchers including Laxton and coworkers [8] who performed a study aiming to prove feasibility of deep brain stimulation with no sham control due to the stage of the trial. Nonsurgical studies also implement open label strategies in early phase designs. These may include drug trials or interventions in which participants are inevitably aware of their allocation (e.g., exercise routines and lifestyle changes). An example of a drug trial with an open label design is the work done by Padala and coworkers [72] evaluating the effect of statins on cognitive function while using a crossover design. Worth noting, these studies mention the lack of blinding as a limitation of their results in order to caution readers on their interpretation. In fact, although these studies are often not considered as definitive evidence to change clinical practice, they can still generate important data for further confirmation.

Description and planning of blinding procedures, along with their implementation, is important in order to gauge and depict the extent of the bias that could be introduced in the trial due to the awareness of treatment either by subjects or by the assessors. Of all the studies included the sample, 80 reported some type of blinding. Of these, 63 reported being double-blinded and 17 were single blinded. Of the latter, all were assessor blinded. Only 14 trials had an open label design. Remarkably, 14 of the studies did not mention blinding or treatment awareness of assessors or participants. Apart from the need to report blinding procedures on a article, it is necessary to detail and prepare them in advance while planning the study.

Another important consideration when designing the protocol is to address unexpected situations, such as the occurrence of serious adverse events, the occurrence of medical emergencies, or the determination of the Data Safety Monitoring Boarding (DSMB) to unblind. Particularly in the studies we included, the DSMB or ethics committee required to unblind in special circumstances, described in Sect. 7.

It is necessary to assess, and later convey, whether the active and control interventions are similar enough to allow blinding of participants. In the case of drug trials, characteristics such as the package, as well as the smell, the taste, and the presentation, aid in this assessment. Some of the studies ($n = 34$) mentioned that the interventions were similar in one way or another showing some sort of similarity between them. Designing a double-blind study becomes even more challenging when there is more than one intervention being tested and when the interventions differ in "presentation" in a radical way. As an example, in drug trials with a double-blind design, when the interventions consist of two drugs

Table 7
Similarity of interventions sorted by type of intervention

	Similar	Not Similar	Not Specified	Not applicable	Total
Drug	33	0	46	6	85
Device	2	0	1	2	5
Procedure	0	5	10	2	17
Total	34	7	57	9	107

that have two different administration routes there is a need to design a placebo intervention for both presentations. This is known as double-dummy design and is clearly depicted in studies such as the one carried out by Alvarez and coworkers [86] in which the intervention consisted of the administration of Cerebrolysin given intravenously, and Donepezil given orally. Of all studies in the sample, five reported to implement double or triple dummy designs.

When a study is designed to have some type of blinding, it improves its internal validity. Table 7 depicts the number of trials that described the degree of similarity of their interventions sorted by the type of intervention. The column "Similar" sorts those blinded studies that mention their interventions resembling in one way or another; the column "Not Similar" sorts blinded studies that mentioned their interventions being different in some way—all were procedures, which required a blinded assessor; the column "Not Specified" sorts studies that mentioned no degree of similarity, nor was it possible to assess this with the information provided by the articles; the column "Not Applicable" sorts trials with no blinding—open label designs.

Only a small part of the studies ($n = 4$) assessed the effectiveness of blinding. In particular, in one trial physicians were asked to guess the treatment assignment, while in three other studies caregivers, study coordinators, psychometrics assessors and partners were asked through questionnaires. In general, this item is not described in articles: its value has been questioned, as the possibility of subjects guessing active treatment may be related to clinical improvement [130].

16   Recruitment Yield and Adherence

The recruitment yield refers to the proportion of participants that were effectively randomized, out of all the screened subjects. On the other hand, adherence refers to the proportion of randomized participants who actually completed their participation, as indicated, in their assigned group.

Among the studies that reported the number of screened subjects and the number of randomized subjects, recruitment yield was calculated. A notable distinction should be made: prescreened individuals correspond to those who reached out to, or were reached by the investigators for preliminary assessment of eligibility; while screened subjects are those who, after this preliminary assessment, underwent a thorough evaluation, according to the inclusion criteria. Many of the articles did not describe the number of subjects screened, let alone those who were prescreened. Articles describing the flow of participants usually account for those who were randomized, out of the ones who underwent full screening, and, in this sense, the numbers of recruitment yield discussed in the following lines correspond to this description (i.e., do not include prescreened individuals). We advise that this be interpreted carefully, as it may be overestimating actual recruitment yield. Information about the flow of participants aids understanding how the design of the trial favors or hinders feasibility of recruitment and conduction of a study.

Only half of the studies we examined reported the number of screened subjects ($n = 55$). Four articles reported post hoc analyses of trials described in separate articles [11, 76, 89, 102] and three more were sub-studies analyzing data of already recruited populations [17, 48, 71]. Since these were based on subjects who were already enrolled, they were not taken into account for this part of the analysis. Twenty-one trials reported its recruitment yield to be between 70% and 97%, 24 trials reported it to be between 50% and 70% and 5 trials described it to be between 30% and 45%. Remarkably, four trials reported their recruitment rate to be between 8% and 25% which may be closer of the actual numbers. It is difficult to compare numbers given that denominator may be calculated differently across these studies. Taking a close look at their description of recruitment procedures, it becomes clear that this probably depicts more accurately the flow of participants since initial contact in most scenarios. Three of those articles detail the flow of participants starting from the available random sample, going through the number of subjects who had contact with the research team and had some sort of prescreening, and finally, those who passed full screening and were enrolled [51, 72, 77].

Some interesting conclusions arise. First, most of the subjects who were not included did not meet eligibility criteria and only a small percentage declined to participate after screening. Other reasons involved the caregivers, whose consent is often needed, as well as that of participants. In some cases, it was possible for participants to pass a first screening, while still needing to pass a second stage in order to be enrolled (such is the case of studies with run-in phases, discussed below); thus, failing the second screening was a possibility. This information is valuable in order to increase the awareness of the potential factors that prevent participants from being enrolled in trials.

In particular, recruitment yield can be considered an indirect indicator of how representative the final sample of the population is, in other words: are the included participants representative of the whole group of patients? This useful information comes directly from the enrolment stage and represents an important cue to assess external validity; it also helps future researchers to plan their recruitment strategies.

Once subjects are enrolled, they can desist from their participation for a number of reasons. For example, they can withdraw consent because they need to receive other interventions that are not compatible with eligibility criteria. They may also lose motivation as the trial advances, especially when allocated in control groups where it is often possible no significant symptom differences are detected by patients or their family members. In many cases, the consent of the caregiver is needed for enrolment as well. In these cases, caregivers themselves might be affected by the aforementioned circumstances. Finally, in exceptional cases, the study physician may remove participants from the trial due to safety concerns (rapid cognitive decline, adverse events, etc.). In some cases, the possibility of high attrition is a concern to the researchers, such as when the intervention is uncomfortable (medication side effects or physical exercise), or when it entails a significant routine or change of habits. In these scenarios, researchers may choose to preselect highly adherent subjects with a run-in phase, as some researchers did [14, 60, 94], including Maher-Edwards [15, 71, 81] and coworkers [56, 75]. Their flow of participants describes their attrition rate as 19%—once the run-in phase was over, adherent subjects were selected to continue to randomization. If the whole screened population had been included the attrition rate could have gone up to 29%.

There were several cases of unequal proportions in adherence between the control and the experimental group. Salloway and coworkers [10] enrolled 1121 participants in a 78-week duration trial. 75.4% in the placebo arm and 69.5% in the experimental arm successfully completed the study. Notably, this study reported adverse events in the highest dose group, which was eventually suspended. Similarly, in the second study reported in the same document, adherence of 1311 participants was negatively associated with the dose—71.2% in the placebo arm versus 70.6%, 68.7%, and 67.4% in the following dosage groups of Bapinezumab 0.5 mg/kg, 1 mg/kg and 2 mg/kg. Along the same lines, the 24 week study run by Coric and colleagues with 209 participants, showed a better adherence of the placebo group (80.9%), as compared to the experimental group of Avagacestat, starting from the lowest dose of 25 mg/day (78.5% of adherence) and rising to the highest dose of 125 mg/day (50% of adherence) [34].

An opposite trend is observed in the 52-week factorial design study run by Howard and those of the initiation of Memantine. Out of a total of 295 patients, the highest attrition rate was seen in the control arm (60% attrition observed), the lowest attrition rate was observed in the full intervention group (30%) while attrition of partial intervention groups was 31–45% [13]. It is important to know this information a priori to understand how this would affect data; it should be used for planning clinical trials to address study power, as well as potential bias when attrition is unequal across treatment groups.

Another interesting example is the study by Kaisu and coworkers on the effects of two different types of exercises [77]. Participants were home dwelling patients with their spousal caregiver, and were randomly assigned to receive home based exercises by a physiotherapist, 1 h, twice a week, or a group based exercise with ten other participants and a physiotherapist, during 1 h, twice a week. The effects were compared to that of a third group of subjects receiving community care. While the rationale of this trial is to prove that intense, long-term exercise has a promising role on function and mobility, adherence was higher in the home-based arm (97%), compared to the group exercise arm (87%). In this case, participants were home dwelling and dependent upon caregivers, so the necessity of going to the research center for physiotherapy could be negatively impacting adherence to the experimental design.

Handling adherence to the protocol in drug trials represents a big challenge since it is not possible to ensure compliance of participants. In these studies, compliance was monitored via self-report, by asking subjects to fill out drug diaries and by counting the remaining pills at the end of each assessment visit. Fortunately, there are several methods that can improve adherence such as simplifying the characteristics of the regimen, educating patients and caregivers, addressing their beliefs and fears and improving communication with subjects [131]. Only 14 studies described the implementation of methods to improve adherence. Among them, five used phone calls and visits to subjects and five included open label, run-in phases, with further selection of adherent subjects for enrolment in the randomized trial. Other strategies included supervision of self-commitment tasks [26], organized psychosocial classes for patients and their families [55] and external reinforcements, such as cookies, at the end of the sessions [59].

17 Interim Analysis and Stoppage Criteria

All clinical trials should have criteria to stop participation of subjects for safety purposes and experimental consistency, regardless of their size. For instance, even in a trial with ten subjects, if the second subject has a severe adverse effect such as death, then the investigator and the ethical committee need to analyze whether the study should proceed. Moreover, these criteria need to be implemented, as a part of a plan, by a designated monitoring party in larger studies (e.g., data safety monitoring board, etc.). This section focuses on these strategies and how they were described and employed by the analyzed trials. Only 32 studies described, in varying degrees of detail, how they included stoppage criteria as part of their plan. Furthermore, 21 trials mentioned the advice of a Data Safety Monitory Boarding (DSMB). Additionally, three studies reported interim analyses and seven reported the occurrence of adverse events as stoppage criteria; therefore, it was inferred that such trials had planned a revision committee of some sort. Ideally, this information should be stated clearly, as it accounts for safety, proper conduction and integrity of decisions.

There is a particular example of the detailed description of these procedures in a study monitored by an Ethical Committee. It aimed at assessing the feasibility of Deep Brain Stimulation (DBS) as a procedure and was monitored by the Research Ethics Board (REB) to obtain information about the two enrolled participants 6 months after the surgical implant, before enrolling a second cohort of four patients [8]. Thirty-one of the articles of this sample conveyed, one way or another, that an ethics committee or data safety monitoring board was supervising those trials. In ten of these articles, this was inferred from the authors mentioning some type of interim analysis. In particular, in a dose escalation trial, the DSMB assessed the safety of the current treatment before the succeeding dose group was approved to start [5]. Furthermore, in this case, the article well specified the number of people involved in the decision and that the committee was independent from the research team.

Additionally, the preestablished frequency of the monitoring organ meetings can be determined by number of subjects recruited or by periods of time. For example, the DSMB reviewed the safety data semiannually in a 4-year study aimed at assessing intranasal insulin therapy in AD and MCI patients [6]. Similarly, the DSMB met quarterly to evaluate safety of Solenazumab in mild to moderate AD patients [12], as occurred in a trial aimed at assessing the safety and tolerability of Avagacestat [34].

The implementation of interim analyses needs appropriate statistical methods in order to adjust for the probability of false positive results. For example, in an 18-month study evaluating the effects of intravenous infusion of Solenezumab or placebo once every 4 weeks, an interim analysis was planned after 50% of the participants had completed 12 months of treatment [12]. In a study aimed as assessing the efficacy of a medical food on mild AD, the team programmed a prespecified blinded interim analysis after 84 patients had completed the core 12-week study. The results were reviewed by an independent DSMB to assess sample size and safety concerns [21]. In other cases, the interim analysis was performed after the completion of a considerable number of participants. For example, an independent DSMB determined that the study could proceed as planned after an interim tolerability analysis "was conducted without statistical testing after 400 subjects" [36]. Only one study specified the method used to plan the interim analysis for treatment efficacy using the method of Haybittle, designed to be flexible and allow the DMC to evaluate on request [41]. In fact, in AD research, given the uncertainty of treatment options and large sample size, a DSMB is often needed.

18 Statistical Methods

The statistical analysis is the way data are processed in order to be interpreted. Therefore, it should entail a careful evaluation of appropriate methods. The same dataset can tell completely different stories, depending on the methods that have been used to analyze it. Consequently, these strategies should be reported faithfully in order to facilitate reproducibility and transparency. The International Committee of Medical Journal Editors is, as they describe themselves, a group of medical journal editors who work for the improvement of quality and reporting of medical science. Their recommendation on the reporting of statistical methods is to describe it with "enough detail to enable a knowledgeable reader with access to the original data to judge its appropriateness for the study and to verify the reported results" [132]. The great majority of the trials in our sample missed to report their statistical analysis in such detail so as to allow for reproducibility. Many reported they performed a series of statistical tests considered "as appropriate"; however, there was no description of the procedure used to process the data. This is not suitable for reproducibility or transparency. Thus, the description of these methods should be done to the extent to understand how the analysis that is reported was obtained. To be noted, descriptions such as "statistical analysis was done using Mann-Whittney U-tests, T-tests, Chi-squared tests and ANOVA, when suitable" seem insufficient to understand how data was treated: What was the sequence used? Which were the explanatory variables, with which covariates were included? and What were the assumptions made? are always questions worth answering in this section.

The purpose of conducting a randomized trial is to estimate the effect of an intervention with the least amount of bias in order to state with a good degree of confidence that the intervention is effective. The purpose of estimating the treatment effects is to contrast the outcomes of the different groups. This can be expressed as confidence interval, of which 95% is the customary value that expresses the range of values expected to contain the true mean for that population in 95 out of 100 studies with similar populations. In our sample of trials in AD the confidence interval was not frequently reported. Statistical significance can also be detailed; it is customarily the most addressed statistical measure in clinical research and represented as the p-value. In this case, there is an arbitrary value of 0.05, which determines whether the difference that was found is likely to be due to chance. This is the reason the recommendation by CONSORT is to report the actual p-value as opposed to a range (e.g. $p < 0.05$). Careful attention should be paid to this value though; it is only supported by a hypothesis-driven study as it can be of no value otherwise [133].

There are customary methods for analyzing data according to the type of variables (continuous, dichotomous, categorical, etc.) and the amount of measurements taken from each individual of the sample. Nonconventional or novel complex statistical methods such as multilevel modeling and methods to analyze imaging data are out of the scope of this chapter. Trialists aiming to conduct such trials may sink deeper into these topics with apposite references [134]. However, it should be highlighted that well designed parallel RCTs do not need complex modeling to analyze the data; even simple statistical tests may suffice.

All studies included some type of description of statistical methods used to process the data they collected. As expected, the degree of detail of this section differs greatly from study to study and this, of course, depends on many factors including: the complexity of the outcomes, the number of groups and the type of intervention performed. Regardless of the complexity of the design, this description should include the necessary amount of detail in order to favor reproducibility. Examples of such descriptions are found in the article by Valen Sendstat et al.. [84] and in the article by Craft et al. [6]. Worth mentioning, the latter includes and analysis comprising several steps, variables and covariates. Indeed, the most complex analyses were described in studies with samples in which there was some kind of dose escalation. Especially in the ones in which dose escalation or dose titration was done within subjects.

As mentioned before, small samples were most frequently obtained in phase I trials. Data collected in these were mostly treated with descriptive analyses [18, 20, 45]—for safety outcomes—or with nonparametric tests [17]. It was also frequent to find pooled data analyses [19] and logarithmic transformations [18] in order to address the main characteristic of small samples: nonnormal distributions.

Mixed effects models are becoming more frequent when analyzing data collected over a period of time, in which changes in the response or the relationship of the response to subjects' characteristics are suspected to affect outcomes. In fact, in AD research, given the duration of trials and sometimes blinding status, some multivariate modeling is often necessary as to address confounding and effects of covariates. Cheng et al. [134] suggest a strategy for the construction of more accurate covariance models, key to the construction of mixed effects models, while aiming to minimize type I error (an error related to obtaining false positives). They suggest using two approaches: carrying out exploratory and confirmatory analyses in a separate way—with different data sets—or by completely specifying the limited tests and the testing sequence while carrying out multiple comparisons to control type I error. Some examples of studies that used mixed effects models in the sample of our studies.

19 Limitations

Almost one third of the analyzed studies ($N = 32$) did not report any limitations of the trial. The readers of the articles are left to judge by themselves. Although this is a common trend [135], it is recommended that authors advance in efforts to recognize the weakness of their studies, thus guiding other research works on the shortcomings of applying results liberally while contributing to overcome common problems. Beyond limitations inherent to each study, there are other general pitfalls reported frequently. They are often related to the sample, the intervention, and the outcome.

The sample size was too small for one-third of the studies in our sample ($N = 31$). The well-known problem of a small sample size is often associated with the problem of the power. Studies with small sample size are often thought to lack power, but this is not always the case [136]. In fact, one of the trials had a small sample size, but was well powered [67]. On the other hand, five studies were not powered enough Regardless of power, small sample sizes limits generalizability; hence, power should be considered carefully. In some cases, subjects are chosen out of a convenience sample, meaning in turn, that it represented a group of selected motivated volunteers. Authors of one trial [6] cautioned for sampling bias because, in this study, some individuals volunteered to undergo invasive procedures, thus, it was possible this subgroup was not representative of the entire sample. Furthermore, some studies recognized unbalanced groups in their baseline performances ($n = 4$). Attrition was reported as a limitation by several studies of our sample; in particular, two RCTs remarked it was higher than expected [51, 55].

Another very important limitation in AD research is the assumption that all individuals experiencing cognitive decline have AD [65]. This was partially addressed in Sect. 9. In particular, the diagnosis of MCI remains controversial, as there is no consensus on the classification system and there are still doubts on the reliability and the validity of the construct [70]. On the other hand, there is great variability in the cognitive performance during the advanced stages of AD, many times, unrelated to treatment [90].

Regarding cognitive tests, ADCS-ADL may not be applicable in studies and populations across the world due to differences in daily living in practices of caregivers [36]. On the other hand, ADAS COG was developed to test the efficacy of Acetyl Cholinesterase Inhibitors (AChEIs) in subjects with AD, and has been used for different interventions [42]. This test has been used widely in the selected trials, including those that used interventions different from AChEIs. Furthermore, this scale is not as sensitive detecting changes in performance in mild AD subjects [14, 56, 96]. These authors recognize the need for a more comprehensive assessment [46]. Similar limitations can be observed in the MMSE, which shows floor effects [36]. The rate of the decline in neuropsychological function is variable [35] and could be a plateau of stabilization of cognitive impairment [93, 103].

Additionally, the duration of the intervention is a critical variable, especially when considering the safety, tolerability and effects of an intervention. The studies in our sample recognize as a limitation the short duration of the intervention.

There are limitations that are not apparent at first, but are the product of keen observation from researchers. Two studies recognized as a limitation the lack of information about decliners, patients who failed the screening and patients who dropped out early [46, 63]. Finally, another well-known limitation is the generalization of the results from experimental to clinical settings [17]. Similarly, results from medical centers, might not be generalizable to other setting, such as primary care. As mentioned in the beginning of this section, limitations may be general to the AD arena or particular to each study. What is important is to recognize and disclose these so as to not overestimate the results of a given study.

20 Conclusion

Researchers face several challenges when planning and conducting clinical trials in Alzheimer's disease, which are directly associated with the nature of the disease and the stage of our understanding of it. AD is chronic degenerative disease with many components, for which current diagnostic criteria are still insufficient in terms of specificity and sensitivity [137]. It should be kept in mind that, even when a study's goal is to assess a particular symptom of the disease, AD affects many aspects of the health, including not only memory and cognitive performance but also quality of life, mood and self-sufficiency. All these aspects deserve attention by researchers considering their populations while conducting specific interventions, and designing them. Furthermore, when defining the inclusion criteria for the population, too strict criteria yield to a sample that is not representative of the real population. On the other hand, too broad criteria can lead to a nonspecific sample, including diseases with similar symptomatology. This selection needs to be balanced with the intrinsic aims of the study phase and its intended impact. Dubois and colleagues [137] suggested a series of biomarkers that can be used in conjunction with the clinical NINCDS–ARDRA workgroup criteria. These can be implemented in research, although feasibility may still be limited in some settings. The difficulties in diagnosis require a standard in the assessment: not conventional shared method AD leads to difficulties in comparing the studies but also in difficulties in selecting homogeneous cohorts for the same study.

Regarding recruitment, in all stages of the disease, with special attention for advanced stages, it is important to consider the patients and their support system, the caregivers. Thus, recruitment strategies and the methods to improve the adherence should be planned accordingly, for example providing attention, support, and educational interventions to the family. Another well recognized challenge, especially in advanced phase trials, was the need to follow-up patients for a long period of time in order to evaluate the long-term effects of the intervention, taking into consideration the progressive character of the disease. This results in an increased risk of attrition, which can be high for longer studies. However, our data show that there are potentially successful strategies to enhance adherence. Once more, these strategies should involve not only the participants, but also the families/caregivers. About attrition, it is important to note that it should be expected and the sample size calculated accordingly, especially in studies with some kind of sequential design (e.g., dose escalation or adaptive designs) where there is a greater impact on statistical analysis as the groups are at greater risk of being unbalanced.

Furthermore, the conduction of multisetting—multicenter studies is suggested whenever possible, in order to convey efforts and share knowledge when possible. Also, multicenter studies in multiple countries may also improve the generalizability of results. Some studies mentioned limitations related to the generalizability of their samples, explaining that their resulting populations were very specific groups. This can potentially be improved by the implementation of multicenter and multi-country studies. Thus, resulting in samples that potentially include demographically diverse individuals. In fact, AD research usually requires a large sample given the small effect sizes being sought; thus, increasing costs and decreasing feasibility of further research. This is a problem that could be partially addressed with the development of surrogate outcomes.

Finally, the elements in the CONSORT checklist are not addressed properly in the articles. This can be either due to incomplete planning during the design of the study, or to incomplete reporting. The authors of this chapter strongly encourage the appropriate description of these elements in order to improve conduction of clinical trials in the present and to aid in the design of future studies. Particular attention should be paid to the description of the trial limitations, since no one better than the research team knows these particulars in detail.

Strong clinical research in AD is much needed in order to improve therapeutics for this condition, we hope this chapter was helpful for investigators to understand the number of challenges that can also be at the same time opportunities to develop novel clinical trials that will change the landscape of AD treatment in the future.

References

1. Geldmacher DS, Whitehouse PJ (1997) Differential diagnosis of Alzheimer's disease. Neurology 48:S2–S9

2. Prince M, Wimo A, Guerchet M et al (2015) The global impact of dementia. World Alzheimer Report. Alzheimer's Disease International (ADI), London, pp 3–5

3. Bernick C, Cummings J, Raman R et al (2012) Age and rate of cognitive decline in Alzheimer disease: implications for clinical trials. Arch Neurol 69:901–905. https://doi.org/10.1001/archneurol.2011.3758

4. The Lancet (2010) Consort 2010. Lancet 375:1136. https://doi.org/10.1016/S0140-6736(10)60456-4

5. Rinne JO, Brooks DJ, Rossor MN et al (2010) 11C-PiB PET assessment of change in fibrillar amyloid-β load in patients with Alzheimer's disease treated with bapineuzumab: a phase 2, double-blind, placebo-controlled, ascending-dose study. Lancet Neurol 9:363–372. https://doi.org/10.1016/S1474-4422(10)70043-0

6. Craft S (2012) Intranasal insulin therapy for Alzheimer disease and amnestic mild cognitive impairment. Arch Neurol 69:29. https://doi.org/10.1001/archneurol.2011.233

7. Quinn JF, Raman R, Thomas RG et al (2010) Docosahexaenoic acid supplementation and cognitive decline in Alzheimer disease. JAMA 304:1903–1911. https://doi.org/10.1001/jama.2010.1510

8. Laxton AW, Tang-Wai DF, McAndrews MP et al (2010) A phase I trial of deep brain stimulation of memory circuits in Alzheimer's disease. Ann Neurol 68:521–534. https://doi.org/10.1002/ana.22089

9. Feldman HH, Doody RS, Kivipelto M et al (2010) Randomized controlled trial of atorvastatin in mild to moderate Alzheimer disease. Neurology 74:956–964. https://doi.org/10.1212/WNL.0b013e3181d6476a

10. Salloway S, Sperling R, Fox NC et al (2014) Two phase 3 trials of bapineuzumab in mild-to-moderate Alzheimer's disease. N Engl J Med 370:322–333. https://doi.org/10.1056/NEJMoa1304839

11. Faux NG, Ritchie CW, Gunn A et al (2010) PBT2 rapidly improves cognition in alzheimer's disease: additional phase II analyses. J Alzheimers Dis 20:509–516. https://doi.org/10.3233/JAD-2010-1390

12. Doody RS, Thomas RG, Farlow M et al (2014) Phase 3 trials of solanezumab for mild-to-moderate Alzheimer's disease. N Engl J Med 370:311–321. https://doi.org/10.1056/NEJMoa1312889

13. Howard R, McShane R, Lindesay J et al (2012) Donepezil and Memantine for moderate-to-severe Alzheimer's disease. N Engl J Med 366:893–903. https://doi.org/10.1056/NEJMoa1106668

14. Gold M, Alderton C, Zvartau-Hind M et al (2010) Rosiglitazone monotherapy in mild-to-moderate Alzheimer's disease: results from a randomized, double-blind, placebo-controlled phase III study. Dement Geriatr Cogn Disord 30:131–146. https://doi.org/10.1159/000318845

15. Sato T, Hanyu H, Hirao K et al (2011) Efficacy of PPAR-γ agonist pioglitazone in mild Alzheimer disease. Neurobiol Aging 32:1626–1633. https://doi.org/10.1016/j.neurobiolaging.2009.10.009

16. Doody RS, Raman R, Farlow M et al (2013) A phase 3 trial of Semagacestat for treatment of Alzheimer's disease. N Engl J Med 369:341–350. https://doi.org/10.1056/NEJMoa1210951

17. Ostrowitzki S (2012) Mechanism of amyloid removal in patients with Alzheimer disease treated with Gantenerumab. Arch Neurol 69:198. https://doi.org/10.1001/archneurol.2011.1538

18. Siemers ER, Friedrich S, Dean RA et al (2010) Safety and changes in plasma and cerebrospinal fluid amyloid beta after a single administration of an amyloid beta monoclonal antibody in subjects with Alzheimer disease. Clin Neuropharmacol 33:67–73. https://doi.org/10.1097/WNF.0b013e3181cb577a

19. Blennow K, Zetterberg H, Rinne JO et al (2012) Effect of immunotherapy with bapineuzumab on cerebrospinal fluid biomarker levels in patients with mild to moderate Alzheimer disease. Arch Neurol 69:1002–1010. https://doi.org/10.1001/archneurol.2012.90

20. Winblad B, Andreasen N, Minthon L et al (2012) Safety, tolerability, and antibody response of active Aβ immunotherapy with CAD106 in patients with Alzheimer's disease: randomised, double-blind, placebo-controlled, first-in-human study. Lancet Neurol 11:597–604. https://doi.org/10.1016/S1474-4422(12)70140-0

21. Scheltens P, Kamphuis PJGH, Verhey FRJ et al (2010) Efficacy of a medical food in mild Alzheimer's disease: a randomized, controlled trial. Alzheimer's Dement 6:1–10.e1. https://doi.org/10.1016/j.jalz.2009.10.003

22. Sano MM, Bell KLKL, Galasko DD et al (2011) A randomized, double-blind, placebo-controlled trial of simvastatin to treat Alzheimer disease. Neurology 77:556–563

23. Rosenberg PB, Drye LT, Martin BK et al (2010) Sertraline for the treatment of depression in Alzheimer disease. Am J Geriatr Psychiatry 18:136–145. https://doi.org/10.1097/JGP.0b013e3181c796eb

24. Farlow M, Arnold SE, Van Dyck CH et al (2012) Safety and biomarker effects of solanezumab in patients with Alzheimer's disease. Alzheimers Dement 8:261–271. https://doi.org/10.1016/j.jalz.2011.09.224

25. Vellas B, Coley N, Ousset PJ et al (2012) Long-term use of standardised Ginkgo biloba extract for the prevention of Alzheimer's disease (GuidAge): a randomised placebo-controlled trial. Lancet Neurol 11:851–859. https://doi.org/10.1016/S1474-4422(12)70206-5

26. Clare L, Linden DEJ, Woods RT et al (2010) Goal-oriented cognitive rehabilitation for people with early-stage Alzheimer disease: a single-blind randomized controlled trial of clinical efficacy. Am J Geriatr Psychiatry 18:928–939. https://doi.org/10.1097/JGP.0b013e3181d5792a

27. Dodel R, Rominger A, Bartenstein P et al (2013) Intravenous immunoglobulin for treatment of mild-to-moderate Alzheimer's disease: a phase 2, randomised, double-blind, placebo-controlled, dose-finding trial. Lancet Neurol 12:233–243. https://doi.org/10.1016/S1474-4422(13)70014-0

28. Galasko DR, Peskind E, Clark CM et al (2012) Antioxidants for Alzheimer disease. Arch Neurol 69:836. https://doi.org/10.1001/archneurol.2012.85

29. Salloway S, Sperling R, Keren R et al (2011) A phase 2 randomized trial of ELND005, scyllo-inositol, in mild to moderate Alzheimer disease. Neurology 77:1253–1262. https://doi.org/10.1212/WNL.0b013e3182309fa5

30. Vigen CLP, Mack WJ, Keefe RSE et al (2011) Cognitive effects of atypical antipsychotic medications in patients with Alzheimer's disease: outcomes from CATIE-AD. Am J Psychiatry 168:831–839. https://doi.org/10.1176/appi.ajp.2011.08121844

31. Geldmacher DS, Fritsch T, McClendon MJ, Landreth G (2011) A randomized pilot clinical trial of the safety of pioglitazone in treatment of patients with Alzheimer disease. Arch Neurol 68:45–50. https://doi.org/10.1001/archneurol.2010.229

32. Scheltens P, Twisk JWR, Blesa R et al (2012) Efficacy of souvenaid in mild Alzheimer's disease: results from a randomized, controlled trial. J Alzheimers Dis 31:225–236. https://doi.org/10.3233/JAD-2012-121189

33. Black RS, Sperling RA, Safirstein B et al (2010) A single ascending dose study of bapineuzumab in patients with Alzheimer disease. Alzheimer Dis Assoc Disord 24:198–203. https://doi.org/10.1097/WAD.0b013e3181c53b00

34. Coric V, van Dyck CH, Salloway S et al (2012) Safety and tolerability of the γ-secretase inhibitor Avagacestat in a phase 2 study of mild to moderate Alzheimer disease. Arch Neurol 69:1430. https://doi.org/10.1001/archneurol.2012.2194

35. Rafii M, Walsh S, Little J et al (2011) A phase II trial of huperzine A in mild to moderate Alzheimer disease. Neurology 76:1389–1394

36. Farlow MR, Salloway S, Tariot PN et al (2010) Effectiveness and tolerability of high-dose (23 mg/d) versus standard-dose (10 mg/d) donepezil in moderate to severe Alzheimer's disease: a 24-week, randomized, double-blind study. Clin Ther 32:1234–1251. https://doi.org/10.1016/j.clinthera.2010.06.019

37. Weintraub D, Rosenberg PB, Martin BK et al (2010) Sertraline for the treatment of depression in Alzheimer disease: week-24 outcomes.

Am J Geriatr Psychiatry 18:332–340. https://doi.org/10.1097/JGP.0b013e3181cc0333

38. Douaud G, Refsum H, de Jager CA et al (2013) Preventing Alzheimer's disease-related gray matter atrophy by B-vitamin treatment. Proc Natl Acad Sci U S A 110:9523–9528. https://doi.org/10.1073/pnas.1301816110

39. Cotelli M, Calabria M, Manenti R et al (2011) Improved language performance in Alzheimer disease following brain stimulation. J Neurol Neurosurg Psychiatry 82:794–797. https://doi.org/10.1136/jnnp.2009.197848

40. Tariot PN, Schneider LS, Cummings J et al (2011) Chronic divalproex sodium to attenuate agitation and clinical progression of Alzheimer disease. Arch Gen Psychiatry 68:853–861. https://doi.org/10.1001/archgenpsychiatry.2011.72

41. Dysken MW, Sano M, Asthana S et al (2014) Effect of vitamin E and Memantine on functional decline in Alzheimer disease. JAMA 311:353. https://doi.org/10.1001/jama.2013.282834

42. Harrington C, Sawchak S, Chiang C, Davies J, Donovan C, Saunders AM, Irizarry M, Jeter B, Zvartau-Hind M, van Dyck CH, Gold M (2011) Rosiglitazone does not improve cognition or global function when used as adjunctive therapy to AChE inhibitors in mild-to-moderate Alzheimer's disease: two phase 3 studies. Curr Alzheimer Res 8:592–606. BSP/CAR/0168 [pii]

43. Baker LD, Frank LL, Foster-Schubert K et al (2010) Aerobic exercise improves cognition for older adults with glucose intolerance, a risk factor for Alzheimer's disease. J Alzheimers Dis 22:569–579. https://doi.org/10.3233/JAD-2010-100768

44. Simmons-Stern NR, Budson AE, Ally BA (2010) Music as a memory enhancer in patients with Alzheimer's disease. Neuropsychologia 48:3164–3167. https://doi.org/10.1016/j.neuropsychologia.2010.04.033

45. Sabbagh MN, Agro A, Bell J et al (2011) PF-04494700, an oral inhibitor of receptor for advanced glycation end products (RAGE), in Alzheimer disease. Alzheimer Dis Assoc Disord 25:206–212. https://doi.org/10.1097/WAD.0b013e318204b550

46. Porsteinsson AP, Drye LT, Pollock BG et al (2014) Effect of citalopram on agitation in Alzheimer disease. JAMA 311:682. https://doi.org/10.1001/jama.2014.93

47. Del Ser T, Steinwachs KC, Gertz HJ et al (2013) Treatment of Alzheimer's disease with the GSK-3 inhibitor Tideglusib: a pilot study. J Alzheimers Dis 33:205–215. https://doi.org/10.3233/JAD-2012-120805

48. Cornelli U (2010) Treatment of Alzheimer's disease with a cholinesterase inhibitor combined with antioxidants. Neurodegener Dis 7:193–202. https://doi.org/10.1159/000295663

49. Modrego PJ, Fayed N, Errea JM et al (2010) Memantine versus donepezil in mild to moderate Alzheimer's disease: a randomized trial with magnetic resonance spectroscopy. Eur J Neurol 17:405–412. https://doi.org/10.1111/j.1468-1331.2009.02816.x

50. Akhondzadeh S, Shafiee Sabet M, Harirchian MH et al (2010) A 22-week, multicenter, randomized, double-blind controlled trial of Crocus sativus in the treatment of mild-to-moderate Alzheimer's disease. Psychopharmacology 207:637–643. https://doi.org/10.1007/s00213-009-1706-1

51. McCurry SM, Pike KC, Vitiello MV et al (2011) Increasing walking and bright light exposure to improve sleep in community-dwelling persons with Alzheimer's disease: results of a randomized, controlled trial. J Am Geriatr Soc 59:1393–1402. https://doi.org/10.1111/j.1532-5415.2011.03519.x

52. Alvarez XA, Cacabelos R, Sampedro C et al (2011) Efficacy and safety of Cerebrolysin in moderate to moderately severe Alzheimer's disease: results of a randomized, double-blind, controlled trial investigating three dosages of Cerebrolysin. Eur J Neurol 18:59–68. https://doi.org/10.1111/j.1468-1331.2010.03092.x

53. Ahmed MA, Darwish ES, Khedr EM et al (2012) Effects of low versus high frequencies of repetitive transcranial magnetic stimulation on cognitive function and cortical excitability in Alzheimer's dementia. J Neurol 259:83–92. https://doi.org/10.1007/s00415-011-6128-4

54. Fox C, Crugel M, Maidment I et al (2012) Efficacy of Memantine for agitation in Alzheimer's dementia: a randomised double-blind placebo controlled trial. PLoS One 7:e35185. https://doi.org/10.1371/journal.pone.0035185

55. Fleisher AS, Truran D, Mai JT et al (2011) Chronic divalproex sodium use and brain atrophy in Alzheimer disease. Neurology 77:1263–1271. https://doi.org/10.1212/WNL.0b013e318230a16c

56. Maher-Edwards G, Dixon R, Hunter J et al (2011) SB-742457 and donepezil in Alzheimer disease: a randomized, placebo-controlled study. Int J Geriatr Psychiatry 26:536–544. https://doi.org/10.1002/gps.2562

57. Cummings J, Froelich L, Black SE et al (2012) Randomized, double-blind, parallel-group, 48-week study for efficacy and safety of a higher-dose rivastigmine patch (15 vs. 10 cm)

in Alzheimer's disease. Dement Geriatr Cogn Disord 33:341–353. https://doi.org/10.1159/000340056

58. Akhondzadeh S, Sabet MS, Harirchian MH et al (2010) ORIGINAL ARTICLE: Saffron in the treatment of patients with mild to moderate Alzheimer's disease: a 16-week, randomized and placebo-controlled trial. J Clin Pharm Ther 35:581–588. https://doi.org/10.1111/j.1365-2710.2009.01133.x

59. Venturelli M, Scarsini R, Schena F (2011) Six-month walking program changes cognitive and ADL performance in patients with Alzheimer. Am J Alzheimers Dis Other Demen 26:381–388. https://doi.org/10.1177/1533317511418956

60. Stein MS, Scherer SC, Ladd KS, Harrison LC (2011) A randomized controlled trial of high-dose vitamin D2 followed by intranasal insulin in Alzheimer's disease. J Alzheimers Dis 26:477–484. https://doi.org/10.3233/JAD-2011-110149

61. Nunes MA, Viel TA, Buck HS et al (2013) Microdose lithium treatment stabilized cognitive impairment in patients with Alzheimer's disease. Curr Alzheimer Res 10:104–107

62. Padala PR, Burke WJ, Shostrom VK et al (2010) Methylphenidate for apathy and functional status in dementia of the Alzheimer type. Am J Geriatr Psychiatry 18:371–374. https://doi.org/10.1097/JGP.0b013e3181cabcf6

63. Devanand DP, Mintzer J, Schultz SK et al (2012) Relapse risk after discontinuation of risperidone in Alzheimer's disease. N Engl J Med 367:1497–1507. https://doi.org/10.1056/NEJMoa1114058

64. Wahlberg LU, Lind G, Almqvist PM et al (2012) Targeted delivery of nerve growth factor via encapsulated cell biodelivery in Alzheimer disease: a technology platform for restorative neurosurgery. J Neurosurg 117:340–347. https://doi.org/10.3171/2012.2.JNS11714

65. Leoutsakos J-MS, Muthen BO, Breitner JCS, Lyketsos CG (2011) Effects of non-steroidal anti-inflammatory drug treatments on cognitive decline vary by phase of pre-clinical Alzheimer disease: findings from the randomized controlled Alzheimer's disease anti-inflammatory prevention trial. Int J Geriatr Psychiatry. https://doi.org/10.1002/gps.2723

66. Okahara K, Ishida Y, Hayashi Y et al (2010) Effects of Yokukansan on behavioral and psychological symptoms of dementia in regular treatment for Alzheimer's disease. Prog Neuropsychopharmacol Biol Psychiatry 34:532–536. https://doi.org/10.1016/j.pnpbp.2010.02.013

67. Penner J, Rupsingh R, Smith M et al (2010) Increased glutamate in the hippocampus after galantamine treatment for Alzheimer disease. Prog Neuropsychopharmacol Biol Psychiatry 34:104–110. https://doi.org/10.1016/j.pnpbp.2009.10.007

68. Vreugdenhil A, Cannell J, Davies A, Razay G (2012) A community-based exercise programme to improve functional ability in people with Alzheimer's disease: a randomized controlled trial. Scand J Caring Sci 26:12–19. https://doi.org/10.1111/j.1471-6712.2011.00895.x

69. Burns A, Perry E, Holmes C et al (2011) A double-blind placebo-controlled randomized trial of *Melissa officinalis* oil and donepezil for the treatment of agitation in Alzheimer's disease. Dement Geriatr Cogn Disord 31:158–164. https://doi.org/10.1159/000324438

70. Buschert VC, Friese U, Teipel SJ et al (2011) Effects of a newly developed cognitive intervention in amnestic mild cognitive impairment and mild alzheimer's disease: a pilot study. J Alzheimers Dis 25:679–694. https://doi.org/10.3233/JAD-2011-100999

71. Richard E, Gouw AA, Scheltens P, van Gool WA (2010) Vascular care in patients with Alzheimer disease with cerebrovascular lesions slows progression of white matter lesions on MRI: the evaluation of vascular care in Alzheimer's disease (EVA) study. Stroke 41:554–556. https://doi.org/10.1161/STROKEAHA.109.571281

72. Padala KP, Padala PR, McNeilly DP et al (2012) The effect of HMG-CoA reductase inhibitors on cognition in patients with Alzheimer's dementia: a prospective withdrawal and rechallenge pilot study. Am J Geriatr Pharmacother 10:296–302. https://doi.org/10.1016/j.amjopharm.2012.08.002

73. Yáguez L, Shaw KN, Morris R, Matthews D (2011) The effects on cognitive functions of a movement-based intervention in patients with Alzheimer's type dementia: a pilot study. Int J Geriatr Psychiatry 26:173–181. https://doi.org/10.1002/gps.2510

74. Wharton W, Baker LD, Gleason CE et al (2011) Short-term hormone therapy with transdermal estradiol improves cognition for postmenopausal women with Alzheimer's disease: results of a randomized controlled trial. J Alzheimers Dis 26:495–505. https://doi.org/10.3233/JAD-2011-110341

75. Cumbo E, Ligori LD (2010) Levetiracetam, lamotrigine, and phenobarbital in patients with epileptic seizures and Alzheimer's disease. Epilepsy Behav 17:461–466. https://doi.org/10.1016/j.yebeh.2010.01.015

76. Farlow MR, Alva G, Meng X, Olin JT (2010) A 25-week, open-label trial investigating rivastigmine transdermal patches with concomitant memantine in mild-to-moderate Alzheimer's disease: a post hoc analysis. Curr Med Res Opin 26:263–269. https://doi.org/10.1185/03007990903434914

77. Pitkälä KH, Pöysti MM, Laakkonen M-L et al (2013) Effects of the Finnish Alzheimer disease exercise trial (FINALEX). JAMA Intern Med 173:894. https://doi.org/10.1001/jamainternmed.2013.359

78. Hayashi Y, Ishida Y, Inoue T et al (2010) Treatment of behavioral and psychological symptoms of Alzheimer-type dementia with Yokukansan in clinical practice. Prog Neuropsychopharmacol Biol Psychiatry 34:541–545. https://doi.org/10.1016/j.pnpbp.2010.02.016

79. Waldorff FB, Buss DV, Eckermann A et al (2012) Efficacy of psychosocial intervention in patients with mild Alzheimer's disease: the multicentre, rater blinded, randomised Danish Alzheimer intervention study (DAISY). BMJ 345:e4693–e4693. https://doi.org/10.1136/bmj.e4693

80. Nathan PJ, Boardley RL, Scott N et al (2013) The safety, tolerability, pharmacokinetics and cognitive effects of GSK239512, a selective histamine H3 receptor antagonist in patients with mild to moderate Alzheimer's disease: a preliminary investigation. Curr Alzheimer Res 10:240–251. https://doi.org/10.2174/1567205011310030003

81. Kennelly SP, Abdullah L, Paris D et al (2011) Demonstration of safety in Alzheimer's patients for intervention with an anti-hypertensive drug Nilvadipine: results from a 6-week open label study. Int J Geriatr Psychiatry 26:1038–1045. https://doi.org/10.1002/gps.2638

82. Serrano-Pozo A, Vega GL, Lutjohann D et al (2010) Effects of simvastatin on cholesterol metabolism and Alzheimer disease biomarkers. Alzheimer Dis Assoc Disord 24:220–226. https://doi.org/10.1097/WAD.0b013e3181d61fea

83. Nourhashemi F, Andrieu S, Gillette-Guyonnet S et al (2010) Effectiveness of a specific care plan in patients with Alzheimer's disease: cluster randomised trial (PLASA study). BMJ 340:c2466–c2466. https://doi.org/10.1136/bmj.c2466

84. Valen-Sendstad A, Engedal K, Stray-Pedersen B et al (2010) Effects of hormone therapy on depressive symptoms and cognitive functions in women with Alzheimer disease: a 12 month randomized, double-blind, placebo-controlled study of low-dose estradiol and norethisterone. Am J Geriatr Psychiatry 18:11–20. https://doi.org/10.1097/JGP.0b013e3181beaaf4

85. Ihl R, Tribanek M, Bachinskaya N (2012) Efficacy and tolerability of a once daily formulation of Ginkgo biloba extract EGb 761(R) in Alzheimer's disease and vascular dementia: results from a randomised controlled trial. Pharmacopsychiatry 45:41–46. https://doi.org/10.1055/s-0031-1291217

86. Alvarez XA, Cacabelos R, Sampedro C et al (2011) Combination treatment in Alzheimer's disease: results of a randomized, controlled trial with cerebrolysin and donepezil. Curr Alzheimer Res 8:583–591

87. Shinto L, Quinn J, Montine T et al (2014) A randomized placebo-controlled pilot trial of omega-3 fatty acids and alpha lipoic acid in Alzheimer's disease. J Alzheimers Dis 38:111–120. https://doi.org/10.3233/JAD-130722

88. Devanand DP, Pelton GH, Cunqueiro K et al (2011) A 6-month, randomized, double-blind, placebo-controlled pilot discontinuation trial following response to haloperidol treatment of psychosis and agitation in Alzheimer's disease. Int J Geriatr Psychiatry 26:937–943. https://doi.org/10.1002/gps.2630

89. Kamphuis PJGH, Verhey FRJ, Olde Rikkert MGM et al (2011) Efficacy of a medical food on cognition in Alzheimer's disease: results from secondary analyses of a randomized, controlled trial. J Nutr Health Aging 15:720–724. https://doi.org/10.1007/s12603-011-0105-6

90. Lorenzi M, Beltramello A, Mercuri NB et al (2011) Effect of memantine on resting state default mode network activity in Alzheimer's disease. Drugs Aging 28:205–217. https://doi.org/10.2165/11586440-000000000-00000

91. Lancioni GE, Singh NN, O'Reilly MF et al (2010) Technology-aided verbal instructions to help persons with mild or moderate Alzheimer's disease perform daily activities. Res Dev Disabil 31:1240–1250. https://doi.org/10.1016/j.ridd.2010.07.021

92. Landen JW, Zhao Q, Cohen S et al (2013) Safety and pharmacology of a single intravenous dose of Ponezumab in subjects with mild-to-moderate Alzheimer disease. Clin Neuropharmacol 36:14–23. https://doi.org/10.1097/WNF.0b013e31827db49b

93. Kume K, Hanyu H, Sakurai H et al (2012) Effects of telmisartan on cognition and regional cerebral blood flow in hypertensive patients with Alzheimer's disease. Geriatr Gerontol Int 12:207–214. https://doi.org/10.1111/j.1447-0594.2011.00746.x

94. Frölich L, Ashwood T, Nilsson J, Eckerwall G (2011) Effects of AZD3480 on cognition in patients with mild-to-moderate Alzheimer's disease: a phase IIb dose-finding study. J Alzheimers Dis 24:363–374. https://doi.org/10.3233/JAD-2011-101554

95. Winblad B, Giacobini E, Frölich L et al (2010) Phenserine efficacy in Alzheimer's disease. J Alzheimers Dis 22:1201–1208. https://doi.org/10.3233/JAD-2010-101311

96. Kurz A, Thöne-Otto A, Cramer B et al (2012) CORDIAL: cognitive rehabilitation and cognitive-behavioral treatment for early dementia in Alzheimer disease: a multicenter, randomized, controlled trial. Alzheimer Dis Assoc Disord 26:246–253. https://doi.org/10.1097/WAD.0b013e318231e46e

97. Uenaka K, Nakano M, Willis BA et al (2012) Comparison of pharmacokinetics, pharmacodynamics, safety, and tolerability of the amyloid β monoclonal antibody solanezumab in Japanese and white patients with mild to moderate Alzheimer disease. Clin Neuropharmacol 35:25–29. https://doi.org/10.1097/WNF.0b013e31823a13d3

98. Roach KKE, Tappen RMR, Kirk-Sanchez N et al (2011) A randomized controlled trial of an activity specific exercise program for individuals with Alzheimer disease in long-term care settings. J Geriatr Phys Ther 34:50–56. https://doi.org/10.1519/JPT.0b013e31820aab9c.A

99. Farlow M, Veloso F, Moline M et al (2011) Safety and tolerability of donepezil 23 mg in moderate to severe Alzheimer's disease. BMC Neurol 11:57. https://doi.org/10.1186/1471-2377-11-57

100. Niu Y-X, Tan J-P, Guan J-Q et al (2010) Cognitive stimulation therapy in the treatment of neuropsychiatric symptoms in Alzheimer's disease: a randomized controlled trial. Clin Rehabil 24:1102–1111

101. Rabey JM, Dobronevsky E, Aichenbaum S et al (2013) Repetitive transcranial magnetic stimulation combined with cognitive training is a safe and effective modality for the treatment of Alzheimer's disease: a randomized, double-blind study. J Neural Transm 120:813–819. https://doi.org/10.1007/s00702-012-0902-z

102. Claxton A, Baker LD et al (2013) Sex and ApoE genotype differences in treatment response to two doses of intranasal insulin in adults with mild cognitive impairment or Alzheimer's disease. J Alzheimers Dis 35:789–797. https://doi.org/10.3233/JAD-122308.Sex

103. Sakurai H, Hanyu H, Sato T et al (2013) Effects of cilostazol on cognition and regional cerebral blood flow in patients with Alzheimer's disease and cerebrovascular disease: a pilot study. Geriatr Gerontol Int 13:90–97. https://doi.org/10.1111/j.1447-0594.2012.00866.x

104. Munro CA, Longmire CF, Drye LT et al (2012) Cognitive outcomes after sertaline treatment in patients with depression of Alzheimer disease. Am J Geriatr Psychiatry 20:1036–1044. https://doi.org/10.1097/JGP.0b013e31826ce4c5

105. Rennie D (2001) CONSORT revised—improving the reporting of randomized trials. JAMA 285:2006–2007

106. Cummings J, Gould H, Zhong K (2012) Advances in designs for Alzheimer's disease clinical trials. Am J Neurodegener Dis 1:205–216

107. Le Tourneau C, Lee JJ, Siu LL (2009) Dose escalation methods in phase I cancer clinical trials. J Natl Cancer Inst 101:708–720. https://doi.org/10.1093/jnci/djp079

108. FDA (1994) European Medicines Agency, ICH Topic E4, Dose response information to support drug registration, Reference No. CPMP/ICH/378/95, Published May 1994. Guidel Ind 2–4

109. Benedetti F, Mayberg HS, Wager TD et al (2005) Neurobiological mechanisms of the placebo effect. J Neurosci 25:10390–10402. https://doi.org/10.1523/JNEUROSCI.3458-05.2005

110. Hróbjartsson A, Gøtzsche PC (2004) Is the placebo powerless? Update of a systematic review with 52 new randomized trials comparing placebo with no treatment. J Intern Med 256:91–100. https://doi.org/10.1111/j.1365-2796.2004.01355.x

111. Hoffman GA, Harrington A, Fields HL (2005) Pain and the placebo: what we have learned. Perspect Biol Med 48:248–265. https://doi.org/10.1353/pbm.2005.0054

112. Kamper SJ, Machado LAC, Herbert RD et al (2008) Trial methodology and patient characteristics did not influence the size of placebo effects on pain. J Clin Epidemiol 61:256–260. https://doi.org/10.1016/j.jclinepi.2007.03.017

113. Turner JA, Deyo RA, Loeser JD et al (1994) The importance of placebo effects in pain treatment and research. JAMA 271:1609–1614. https://doi.org/10.1001/jama.1994.03510440069036

114. Salloway S, Sperling R, Gilman S et al (2009) A phase 2 multiple ascending dose trial of bapineuzumab in mild to moderate Alzheimer disease. Neurology 73:2061–2070. https://doi.org/10.1212/WNL.0b013e3181c67808

115. Cole SR, Stuart EA (2010) Generalizing evidence from randomized clinical trials to target populations: the ACTG 320 trial. Am

J Epidemiol 172:107–115. https://doi.org/10.1093/aje/kwq084

116. Rothwell PM (2005) External validity of randomised controlled trials: "to whom do the results of this trial apply?". Lancet 365:82–93. https://doi.org/10.1016/S0140-6736(04)17670-8

117. Shapiro SH, Weijer C, Freedman B (2000) Reporting the study populations of clinical trials clear transmission or static on the line? J Clin Epidemiol 53:973–979. https://doi.org/10.1016/S0895-4356(00)00227-4

118. Ng T-P, Niti M, Chiam P-C, Kua E-H (2007) Ethnic and educational differences in cognitive test performance on mini-mental state examination in Asians. Am J Geriatr Psychiatry 15:130–139. https://doi.org/10.1097/01.JGP.0000235710.17450.9a

119. Eriksdotter-Jönhagen M, Linderoth B, Lind G et al (2012) Encapsulated cell biodelivery of nerve growth factor to the basal forebrain in patients with Alzheimer's disease. Dement Geriatr Cogn Disord 33:18–28. https://doi.org/10.1159/000336051

120. Shuster JJ (1991) Median follow-up in clinical trials. J Clin Oncol 9:191–192

121. FDA, (Office of Combination Products in the Office of the Commissioner (OCP), the Center for Biologics Evaluation and Research (CBER), the Center for Drug Evaluation and Research (CDER) and the C for D and RH (CDRH)) (2011) Classification of products as drugs and devices & additional product classification issues. Guidance for Industry. FDA. p. 3

122. Aronson JK (2005) Biomarkers and surrogate endpoints. Br J Clin Pharmacol 59:491–494. https://doi.org/10.1111/j.1365-2125.2005.02435.x

123. Charles P, Giraudeau B, Dechartres A et al (2009) Reporting of sample size calculation in randomised controlled trials: review. BMJ 338:b1732. https://doi.org/10.1136/bmj.b1732

124. Abdulatif M, Mukhtar A, Obayah G, Hardman JG (2015) Pitfalls in reporting sample size calculation in randomized controlled trials published in leading anaesthesia journals: a systematic review. Br J Anaesth 115:699–707. https://doi.org/10.1093/bja/aev166

125. Chalmers TC, Celano P, Sacks HS, Smith H (1983) Bias in treatment assignment in controlled clinical trials. N Engl J Med 309:1358–1361. https://doi.org/10.1056/NEJM198312013092204

126. Dumville JC, Hahn S, Miles JN, Torgerson DJ (2006) The use of unequal randomisation ratios in clinical trials: a review. Contemp Clin Trials 27:1–12. https://doi.org/10.1016/j.cct.2005.08.003

127. Hey SP, Kimmelman J (2014) The questionable use of unequal allocation in confirmatory trials. Neurology 82:77–79. https://doi.org/10.1212/01.wnl.0000438226.10353.1c

128. Ard MC, Edland SD (2011) Power calculations for clinical trials in Alzheimer's disease. J Alzheimers Dis 26:369–377. https://doi.org/10.3233/JAD-2011-0062

129. Vozdolska R, Sano M, Aisen P, Edland SD (2009) The net effect of alternative allocation ratios on recruitment time and trial cost. Clin Trials 6:126–132. https://doi.org/10.1177/1740774509103485

130. Moscucci M, Byrne L, Weintraub M, Cox C (1987) Blinding, unblinding, and the placebo effect: an analysis of patients' guesses of treatment assignment in a double-blind clinical trial. Clin Pharmacol Ther 41:259–265

131. Atreja A, Bellam N, Levy SR (2005) Strategies to enhance patient adherence: making it simple. MedGenMed 7(1):4. https://doi.org/10.2147/RMHP.S19801

132. Rosenberg J, Bauchner H, Backus J et al (2014) The new ICMJE recommendations. Indian J Med Microbiol 32:219–220. https://doi.org/10.4103/0255-0857.136545

133. Hoffmann TC, Thomas ST, Shin PNH, Glasziou PP (2014) Cross-sectional analysis of the reporting of continuous outcome measures and clinical significance of results in randomized trials of non-pharmacological interventions. Trials 15:362. https://doi.org/10.1186/1745-6215-15-362

134. Cheng J, Edwards LJ, Maldonado-Molina MM et al (2010) Real longitudinal data analysis for real people: building a good enough mixed model. Stat Med 29:504–520. https://doi.org/10.1002/sim.3775

135. Ioannidis JPA, Evans SJW, Gøtzsche PC et al (2004) Better reporting of harms in randomized trials: an extension of the CONSORT statement. Ann Intern Med 141:781–788. https://doi.org/10.7326/0003-4819-141-10-200411160-00009

136. Gandhi M, Ameli N, Bacchetti P et al (2005) Eligibility criteria for HIV clinical trials and generalizability of results: the gap between published reports and study protocols. AIDS 19:1885–1896. https://doi.org/10.1097/01.aids.0000189866.67182.f7

137. Dubois B, Feldman HH, Jacova C et al (2007) Research criteria for the diagnosis of Alzheimer's disease: revising the NINCDS-ADRDA criteria. Lancet Neurol 6:734–746. https://doi.org/10.1016/S1474-4422(07)70178-3

Chapter 10

Tumors of the Central Nervous System

Aline Patrícia Briet, Beatriz Teixeira Costa, Isadora Santos Ferreira, Rivadávio Fernandes Batista de Amorin, and Felipe Fregni

Abstract

Tumors of the central nervous system have a significant impact on the daily living of affected individuals and research studies on this field are essential for understanding their diagnostic techniques and therapeutic effects. This book chapter discusses the findings of a review of the literature conducted in 2015 by using the database engine "Web of Science," which searched for clinical trials in brain tumors published between 2010 and 2015. The 90 most cited articles in this field were included in the review. Important methodological aspects of these trials are deeply explored in this chapter with the goal of improving the conduction of future research studies.

Key words Brain tumor, Brain neoplasm, Glioblastoma, Meningioma, Clinical trial, Review

1 Introduction

Tumors originated in the brain have a meaningful effect in the population due to their significant impact on mortality and morbidity after diagnosis. Numerous types of brain tumors are often associated with a high rate of recurrence followed by an excessive disability rate. Moreover, the capacity of brain tumors (BTs) to directly influence physical, psychological, and cognitive status of patients highlights the importance of conducting clinical trials in this field. Indeed, research studies may allow a greater understanding of the pathological aspects of brain tumors, also their diagnostics and prognostics features.

Although BTs and other nervous system cancers are not the most prevalent in the world's population, they account for at least 1.4% of all new cancer cases in the USA, according to the National Cancer Institute. Considering all the primary Central Nervous System (CNS) neoplasms, the prevalence of primary BTs represents 85–90% among these cancers [1]. Additionally, there is a large variation in the prevalence and incidence among the various types of brain tumors and it directly depends on particular fea-

Felipe Fregni (ed.), *Clinical Trials in Neurology*, Neuromethods, vol. 138,
https://doi.org/10.1007/978-1-4939-7880-9_10, © Springer Science+Business Media, LLC, part of Springer Nature 2018

tures, such as age of onset, tumor histology, and patient's gender, race, and behavior. The survival rate of patients with brain cancer also varies according to the type of tumor, but it generally decreases with older age at diagnosis [2].

In 2016, the World Health Organization (WHO) presented the practical Classification of Tumors of the Central Nervous System, as an advance over the last version of 2007. The WHO has chosen to use molecular mechanisms in addition to histology to define the various brain tumor entities [3]. Hence, taking into account anatomical, histological, and molecular characteristics of brain cancers, it becomes possible to identify a wide variety of tumor entities. For instance, with this classification, meningiomas can be differentiated from gliomas and pituitary tumors. Also, glioblastomas may be differentiated from other subtypes of gliomas.

Among the primary brain tumors, meningioma (MG) is the one with larger incidence in adults (~2 in 100,000) [4], accounting for 36% of these cancers [5]. Meningiomas are originated in the layers of tissue that cover the brain and spinal cord, called as meninges. Their prevalence in adults is difficult to estimate, as their lesions are usually asymptomatic and incidentally diagnosed [5]. Most meningiomas have a benign and slow-growing nature, usually not affecting nearby healthy brain tissue. Only 1–3% of these tumors have an aggressive nature and invade the brain, eventually spreading to other organs [1]. Additionally, there is approximately a 2:1 female to male predominance of meningiomas [5], possibly due to the fact that most of these tumors are progesterone- and estrogen-sensitive [6].

According to the WHO, meningiomas are mainly classified into three categories [7]: Grade I—slow growth and benign features; Grade II—Atypical characteristics, which correspond to a mixed pattern of noncancerous and cancerous elements; and Grade III—Fast growing and malignant nature [8]. From these categories, meningiomas are also differentiated according to specific histologic subtypes. Certainly, each of these types displays a particular behavior, clinical presentation and prognosis. Therefore, understanding how different types of meningiomas proliferate in the brain and respond to pharmacological or surgical therapies, has been a crucial challenge for clinical researchers.

Alike meningiomas, glioblastomas (GBs) are also widely investigated in research studies, due to their high incidence in adult populations. To date, these tumors correspond to the most common primary malignant BTs, causing at least 3% of all deaths cancer-related [8]. GBs account for at least 15% of all primary brain tumors in adults [5], but represent approximately 75% of all malignant brain tumors together with other malignant gliomas. In addition, GBs have a poor overall survival rate, with only 0.05–4.7% surviving 5 years after diagnosis [9]. Due to these epidemiologic aspects of glioblastomas, there are several reports in the literature assessing the behavior of these tumors, the most effective therapies and their effects on patients' quality of life.

Research studies involving patients with primary brain tumors are usually focused on improving diagnostic, therapeutic, and prognosis estimation methods. As the affected population is significantly heterogeneous, information on how to conduct clinical trials involving these patients is mandatory for improving daily clinical practice in oncology. Therefore, this chapter aims to review the most cited research studies of brain tumors in the literature, in order to guide investigators and gather useful information regarding every step of oncology clinical trials. As meningiomas and glioblastomas are the most common adult primary brain tumors (36% and 15%, respectively) [5], this chapter exclusively focuses on these two types of cancer.

2   Methodology

A review of the literature was carried out through a comprehensive search of all Web of Science Database using the keywords: "Meningioma" or "Glioblastoma." Only Randomized Clinical Trials (RCT) available between 2008 and 2015 were included in this review. Articles available before 2008 were not included due to the publication of the WHO Classification of Brain Tumors in 2007. Also, research studies were excluded if they presented a different design rather than clinical trial and if they were not focused on assessing meningiomas or glioblastomas in the population.

Initially, a total of 46 clinical trials of meningioma and 311 of glioblastoma were found in the literature search (Fig. 1), of which 26 and 231 were excluded, respectively. The goal was to include in this review the most cited clinical trials, which corresponded to a total of 90 articles. Hence, as there was a different proportion between meningioma and glioblastoma studies, only 20 out of 90 were of meningioma and the remaining 70 were of glioblastoma.

3   Overview

3.1   Standardizing Trial Methodology

The journals in which the articles of this review were published had different publishing policies. For instance, the majority of journals did not adopt CONSORT guidelines as a requirement to publish research studies. As mentioned in previous chapters of this book, the CONSORT (Consolidated Standard of Reporting Trials) is a crucial tool in clinical research, as it raises the essential criteria for conducting studies with high quality methodology and it guides investigators on reporting RCTs. Of those journals, only 12 (41%) used CONSORT as a prerequisite to the publication of articles. This information highlights that several clinical trials involving individuals

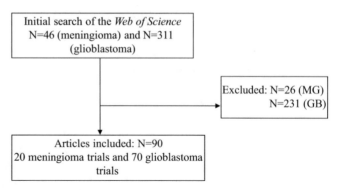

Fig. 1 Flowchart of article search methodology

with meningioma or glioblastoma have not been following report guidelines and thus may not present reliable results. Therefore, researchers conducting oncologic trials must be encouraged to adhere to CONSORT in order to show the best level of evidence.

3.2 Journals Impact Factor and Number of Citations

Most of the articles included in this review were published in journals with low impact factor (IF) and there was no distinct pattern when comparing journals of meningioma and glioblastoma trials. Generally, the journals IF varied from 0.7 to 54.4 and the majority of them (53.3%) were under 5.0. In regard to meningioma trials, 13 studies out of 20 were published in journals with an IF under 5.0, while only 7 relied between 5.0 and 20. None of the journals for meningioma studies had an IF over 20, which may be directly related to their lack of adherence to CONSORT guidelines. Comparatively, most of the journals in which glioblastoma articles were published had an IF under 5.0, accounting for 50% of them. Additionally, an IF between 5 and 20 corresponded to a total of 28, while only 7 studies were reported in journals with an IF over 20.

The number of citations considerably varied between both types of brain tumor studies, which may be related to the satisfactory methodology of a few trials, as well as to their relevance and adherence to CONSORT guidelines. It was notable that studies of glioblastoma presented a larger number of citations in comparison to meningioma studies. For instance, 28 out 70 glioblastoma trials (40%) had more than 100 citations, while all the meningioma studies had less than 50 citations. Besides the fact that numerous glioblastoma trials were cited more than 100 times, most of them (67.1%) had less than 50 citations.

Table 1 presents an overview of the 90 most cited clinical trials related to brain tumors, according to year of publication, study phase, sample size, type of intervention, number of citations and journal impact factor. Generally, looking at the articles, it is clear that most of the clinical trials of meningioma and glioblastoma are phase II studies, which means that they aim to evaluate the efficacy and safety aspects of a specific intervention. While assessing safety, these trials mainly estimate the extent of adverse events over the study duration. Another important information extracted from Table 1 is that most studies evaluated the effects of a drug for treating meningioma or glioblastoma, instead of assessing a certain device or procedure. The importance and utility of pharmaceutical therapies for treating cancer, either alone or in combination, may justify the large amount of studies assessing drugs. The primary goal of these studies is to evaluate the chemical composition of drugs, their effects on cancer cells, usefulness in treating tumors and common side effects.

4   Trial Design

The most common design performed by the studies was a parallel design, varying only in regard to the number of study arms. A total of 23 articles corresponded to parallel two-arm studies, in which an intervention was compared to a placebo group or a standard therapy. Only 6 out of the 90 trials used a three arms design, comparing two different interventions or a combination of them. The remaining trials referred to one-arm design or failed to clearly report their design. There were no clinical trials with a crossover of factorial design.

4.1   Phases of Meningioma and Glioblastoma Trials

In regard to the study phases, a greater number of articles corresponded to phase II trials, accounting for 55.6% of all studies. The predominance of this type of design was evident in both meningioma and glioblastoma studies (Table 2). This may be due to the constant need of introducing new therapeutic methods for brain neoplasms and consequently improving patient's survival and quality of life. In second place, phase I trials accounted for 13.3%, while phase III corresponded to 11.1% of all studies. Accordingly, only a few research studies of brain tumors were conducted with a view to give a definitive judgment about how effective an intervention is. The lack of phase III studies of glioblastoma may be justified by the low overall survival of patients with this tumor, as these trials have a comparatively long duration and usually demand a larger number of participants.

Eight out of the 90 studies corresponded to phase I/II, in which investigators published the results of both phases I and II in the same study. In these phase I/II trials, the investigators mostly assessed safety parameters, dose escalation, and efficacy of an experimental drug. Furthermore, several trials did not specify or clearly stated the study design (14.4%), thus compromising their interpretation of the results.

An important aspect to highlight is the lack of phase IV studies in both types of brain tumors. This type of study design aims to assess long-term adverse events of an intervention after its approval.

Table 1
General characteristics of the clinical trials included in this review

		N
Year		
	2008–2010	60
	2011–2013	24
	2014–2015	6
Phase		
	I	12
	II	53
	I and II	8
	III	9
	IV	0
	NS	8
Trial design		
	RCT Parallel 2 arms	23
	RCT Parallel 3 arms	6
	RCT Parallel 4 arms	0
	Single-arm nonrandomized	48
	NS	13
Site		
	Single-center	53
	Multicenter	37
Intervention		
	Device	0
	Diagnostic methods	4
	Drug	59
	Procedure	20
	Drug and procedure	7
Primary outcome		
	Clinical	63
	Surrogate	16
	Clinical and surrogate	11
Sample size		
	<50	55
	51–100	15
	101–500	14
	>500	6
Number of citations		
	<50	47
	50–100	15
	>100	28
Impact factor		
	<5	48
	5–20	35
	>20	7

Table 2
Phases of trials for MG and GB

Meningioma					Glioblastoma				
Phase I	Phase II	Phase I/II	Phase III	NS	Phase I	Phase II	Phase I/II	Phase III	NS
N = 2 (10%)	N = 9 (45%)	N = 1 (5%)	N = 0 (0%)	N = 8 (40%)	N = 10 (14.2%)	N = 44 (62.8%)	N = 7 (10%)	N = 9 (12.8%)	N = 0 (0%)

Not specified

Hence, phase IV trials are imperative to determine whether a given experimental drug may induce unexpected side effects in patients, even after successful research studies have proven its benefits. Accordingly, the lack of phase IV trials represents a remarkable issue, especially in neurology, as this reflects the limited advance on the approval of new therapies.

4.2 Single Center Versus Multicenter Trials

Meningioma and glioblastoma trials also varied in regard to the number of study sites. While single center trials were mostly performed with a small sample size, multiple centers were applied in larger scale studies. Indeed, the decision of conducting a multicenter trial is often related to the need of recruiting numerous participants or to the aim of achieving a more heterogeneous study sample. In this review, most clinical trials were single-center (58.9%), which complies with the larger number of small-scale studies. The remaining trials were multicenter and they recruited a sample size bigger than most single-center studies. For instance, a multicenter study of Tsien et al. [10] recruited a total of 209 participants in order to assess early and late toxicities of treatment with *carmustine*, a chemotherapeutic drug, for glioblastoma. On the other hand, a single-center trial of Lee et al. [11] recruited 26 patients and assessed tumor progression after treatment with *telozomide* in glioblastoma patients. Furthermore, a shorter number of meningioma trials were multicenter ($n = 4$), in comparison to glioblastoma. For this type of brain tumor, 33 out of the 90 trials were multicenter.

5 Population

Choosing the population of a study is a critical step to conduct successful research. In order to define a target population, numerous aspects must be considered, especially epidemiologic data about the disease on investigation. For instance, the usual age of the disease onset determines the average age of the target population. When conducting oncologic studies, as the population affected by brain tumors is not homogenous in most circumstances, precise eligibility criteria must be established in order to recruit patients that truly represent the target population.

The articles of this review used different inclusion and exclusion criteria for selecting participants. Also, there were only a few distinct eligibility criteria between studies of meningioma and glioblastoma. The articles of meningioma mostly included adult participants over 18 years old (18%). About 30% of the meningioma articles (32.4%) included patients with a histological diagnosis while the other 67.6% used different methods to confirm such as image exams (CT and/or MRI) or included patients based on Karnofsky performance scores. Additionally, in a few articles (14.4%), patients with adequate renal, hematological, and hepatic function were also included. For instance, a study by Reardon et al. [12], required participants to have a histologically proven recurrent meningioma and a satisfactory renal function assessed by serum creatinine levels, among other criteria. Furthermore, a few studies applied the Karnofsky Performance Status Scale (KPS) as criteria to include/exclude potential participants. KPS is a commonly used method for assessing the amount of functional impairment presented by patients. It is a subjective scale applied over the past 60 years, in order to evaluate the impact of a disease on patient's daily life. Among researchers and clinicians, it is believed that KPS works as a meaningful scale to predict the prognosis and response to therapy. Patients with brain tumor that present a KPS score between 60 and 100 were generally considered eligible to participate in the studies [13].

Alike the criteria used in studies of meningioma, glioblastoma articles mostly included adults over 18 years old (67.5%), with a histological diagnosis (60.75%) and adequate renal, hematological, or hepatic function (51.75%). Additionally, a few trials of glioblastoma (16.7%) defined an acceptable life expectancy to allow patients' enrollment, which varied from 8 weeks to 4 months. A study of Wick et al. [14], for instance, included patients with histologically confirmed WHO grade IV glioblastoma and a life expectancy of at least 8 weeks. Another example includes the study of Brandes et al. [15], which enrolled patients who had a progressive glioblastoma proven by imaging exam and a life expectancy of at least 3 months. The decision of admitting only patients over 18 years old is potentially due to the fact that meningiomas and glioblastomas have a higher incidence in adult populations. These studies also considered KPS (52.88%) as eligibility criteria.

Another important challenge when conducting oncology research is to recruit patients who have a well-established diagnosis of cancer. To date, there are different histopathological, cytological, and molecular methods that are essential in the diagnosis and estimation of cancers prognosis. Also, imaging exams, such as MRI and CT, play an important role in this process, although they are usually not able to establish a definitive diagnosis. Indeed, the confirmatory diagnosis of brain tumors is usually obtained after a histological examination of the tumor through directed biopsy or mass removal in surgical procedure.

6 Interventions

The treatment of brain tumors directly depends on their histological type, size, and location, as these aspects define tumor's severity and propensity to respond to a given therapy. There are numerous therapeutic methods available to patients, each of them being specific to a certain type of brain tumor. Hence, the particularity of each treatment demonstrates the usefulness of clinical trials investigating their proper effects and indications. In this review, the studies investigated different interventions, which included pharmaceutical drugs, diagnostic methods, therapeutic procedures, or a combination of these. The majority of the trials (65.5%) chose to evaluate pharmacological treatments, testing different drug dosages and administration schemes. In second place, therapeutic procedures, such as radiation therapy, were also explored by several studies, accounting for 22.2% of the total amount. Diagnostic methods were used as an intervention in only four studies, which basically assessed imaging exams and predictive biomarkers.

The treatment of meningioma is commonly based on surgical removal of the tumor followed by radiotherapy, combined or not with chemotherapy in case of recurrence or incomplete resection [16]. Accordingly, as observed in Table 3, most meningioma clinical trials assessed the efficacy of therapeutic procedures (50%), such as fractionated stereotactic radiotherapy (FSRT). For instance, a study of Combs et al. [17] used FSRT and ion carbon radiation therapy as interventions and assessed patients overall survival after these treatments. Although 40% of meningioma trials explored the effects of pharmaceutical drugs, only two studies assessed imaging diagnostic methods.

Table 3
Interventions applied in GB and MG trials

	Drug	Diagnostic methods	Therapeutic procedures	Drug + therapeutic procedure
Meningioma	$N = 8/40\%$	$N = 2/10\%$	$N = 10/50\%$	$N = 0$
Glioblastoma	$N = 51/72.9\%$	$N = 2/2.9\%$	$N = 10/14.2\%$	$N = 7/10\%$

Table 4
Primary outcomes used for GB and MG trials

BT Primary outcomes	Glioblastoma (%)	Meningioma (%)
PFS	30	30
Overall survival	24.28	10
Tumor progression/response rate	12.85	25
Maximum tolerated dose	4.28	5

Table 5
Secondary outcomes GB and MG trials

BT Secondary outcomes	Glioblastoma (%)	Meningioma (%)
PFS	21.42	5
Overall survival	28.57	25
Tumor progression/response rate	21.42	20
Toxicity	25.71	20
Safety	20	25
Quality of life	12.85	–
Side effects	4.28	–

Similar to meningioma studies, the majority of glioblastoma articles (72.9%) investigated safety and efficacy of pharmacotherapy. According to Hottinger et al. [18], the combination of radiotherapy and chemotherapy has become one of the first-line treatments for glioblastoma, which agrees with the large number of studies being conducted on chemotherapeutic drugs. Notably, patients with glioblastoma are mostly treated with radiation and pharmacotherapy, instead of surgical removal, possibly due to the usual malignant nature of this tumor. A few trials chose to evaluate a combination of treatments, such as pharmaceutical drug plus a therapeutic procedure. A study of Stupp et al. [19], for example, compared standard radiation therapy (SRT) alone with a combination of *Temozolomide* plus SRT. Also, a study of Malmstrom et al. [20] investigated a combination of interventions, comparing *Temozolomide* vs. SRT vs. hypofractionated radiotherapy. An overview of the interventions used in the clinical trials of this review is presented in Tables 4 and 5.

7   Outcomes

There is a wide variety of primary endpoints commonly used in clinical trials, especially in the oncology field, as each type of cancer has its particular features. Besides the large number of alternatives, carefully defining the primary outcome of a study is essential for acquiring meaningful and significant results. In this review, most articles chose tumor shrinkage in response to treatment ($n = 19$), progression-free survival ($n = 27$) and overall survivor ($n = 19$) as their primary outcomes. Additionally, a few studies also analyzed two aspects, such as progression-free survival and overall survivor ($n = 11$) combined. For instance, Minniti et al. [21] defined tumor control according to MRI results and overall survival as the main outcome, while Stummer et al [22] described their primary outcome as radiographic complete resections (CRs) and 6-month progression-free survival.

Regarding brain tumors studies particularly, the selection of a precise and correct primary outcome is directly related to the type of cancer being investigated. As glioblastomas are associated with a higher mortality rate when compared to meningiomas, their studies often investigate survival: either progression-free (PFS), overall survivor or both. Within the 90 studies reviewed, 48 were related to glioblastoma and referred to survival as the primary outcome. On the contrary, only nine meningioma studies mentioned PSF, overall survivor or both as their main outcome. However, it is important to notice that the percentage of PSF in both glioblastoma and meningioma corresponded to 30% (Table 4).

As cancer progression represents a major threat for patient's quality of life and survival, researchers believe that these outcomes provide valid and useful information to the oncology practice. Following the same reasoning, the use of PFS and overall survival are also believed to be important outcomes in oncologic studies, as improving life expectancy of patients with specific treatments is one of the primary concerns of oncologists. Indeed, as Venok et al. [23] stated in their study, the longer the progression of cancer, the longer may be a patient's life. Furthermore, as most brain tumor studies correspond to phase I and II trials and are especially focused on safety, the most common outcomes are related to this topic. For instance, a few studies assessed the toxicity of the intervention under investigation ($n = 22$), although these were mostly used as one of the secondary endpoints. Also, a minority of studies used the maximum tolerated dose as the primary outcome, which shows, once again, a main focus on assessing the safety aspects of an intervention (Table 6).

Regarding the primary outcome, an impressive 70% of the studies presented clinical endpoints, while 17.7% used biomarkers and 12.2% chose both clinical and surrogate outcomes. However, when analyzed individually, only one meningioma study used a clinical outcome, which was conducted by Sluzewski et al. [24] and only one glioblastoma study [25] used a surrogate outcome

Table 6
Stoppage criteria

Stoppage criteria	Glioblastoma (%)	Meningioma
Toxicity	27.14	35%
Disease progression	4.28	5%
Side effects	1.42	5%
Death	21.42	NA
Safety	5.71	20%
Not specified	40	40%

NA not available

Table 7
Statistical analysis

Primary outcome	Glioblastoma	Meningioma
Survival analysis	82.86%	50%
T-test	4.28%	20%
Wilcoxon	NA	5%
Chi-square	1.42%	10%
Fisher	1.42%	5%
Pearson correlation	1.42%	NA
Spearman correlation	NA	5%
Linear regression	2.85%	NA
Logistic regression	2.85%	NA
Kruskal–Wallis	1.42%	NA
Not specified	11.14%	15%

as the main endpoint. Therefore, it is possible to conclude that, in this review, most primary outcomes in meningioma and glioblastoma studies were surrogate and clinical, respectively.

Finally, the most commonly used secondary outcomes were overall survival ($n = 25$), toxicity ($n = 22$), safety ($n = 19$), and PFS ($n = 16$). In articles of glioblastoma, the most frequent used secondary outcome was overall survival (28.57%), followed by toxicity (25.71%). Some examples are Jonhson et al. [26] who described PSF and overall survivor as their secondary outcome, Izumoto et al. [27] who analyzed hematological and nonhematological toxicity, and Hasselbalch et al. [28] who investigated skin toxicity. On the other hand, studies of meningioma mainly used overall survival (25%), safety (25%), tumor progression/response rate (20%), and toxicity (20%) as confirmed by Table 7. For instance, Kreissl et al.

and Trippa et al. [29, 30] both described safety and side effects as their secondary outcomes. In addition, no meningioma studies mentioned quality of life or safety as secondary outcomes.

8 Sample Size

The sample size of clinical trials directly depends on study design and particular parameters, such as statistical power, acceptable level of significance, and expected effect size. These parameters must be clearly defined during the study planning as to ensure reliable sample size estimation. Also, researchers must clearly report in articles the methods and estimated measures applied to calculate sample size, in order to guarantee that their calculation is reproducible. In other words, any other investigator must be able to recalculate a study sample size, thus verifying researcher's estimation.

In this review, many studies did not report which parameters were applied to calculate the sample size. The lack of this information makes it difficult to reproduce the calculations and decreases the trial's validity. A total of 40 clinical trials failed to report sample size calculation parameters, accounting for 44.5% of the studies. In regard to the remaining articles, some of them stated which level of significance were used (37.8%) and a few reported the statistical power (28.9%). In general, there was a variety of alpha values used in the studies. While 24 trials applied an alpha-level of 0.05, 7 used $\alpha = 0.1$ and the remaining three use alpha equals to 0.025 or 0.55. Taal et al. [31], for instance, achieved a total of 44 patients per group based on an α and β of 0.1. The trials that applied alpha as 0.05 varied concerning interventions and outcomes, however, most of them assessed pharmaceutical drugs and used overall survival, PFS, and radiographic response rate as outcomes.

In addition, most trials considered 90% a reasonable study power, while 80% was the second most applied. Other power values also used in these trials were 82%, 85%, 88%, and 92%. There was no considerable difference between glioblastoma and meningioma articles concerning the use of different power values.

As most of the studies corresponded to phase I and II trials, the studies' sample sizes were mostly small to moderate and varied from 7 to 833. Indeed, there were only 20 studies with a sample size over 100 patients, which complies with the small number of phase III trials of both meningioma and glioblastoma. Thus, most studies achieved a sample size under 100 patients ($n = 70$), and, from these, 55 trials had a sample size <50 (61.11%).

By analyzing meningioma and glioblastoma trials separately, it is worth mentioning that the most observed sample size was also under 50 (meningioma: 80%; glioblastoma: 55.71%). Although six glioblastoma trials presented a sample size over 500 (8.57%), no meningioma studies had a similar or an equivalent sample size. This could also be related to the fact that most studies included in this review corresponded to phase I and phase II trials. Therefore, there is still a lot of information to be collected on both conditions as to allow researchers to develop following phases of clinical trials.

9   Interim Analysis and Stoppage Criteria

Although an adequate statistical analysis may provide evidence that the trial is headed towards the right direction, the monitoring process goes beyond statistical warnings for stopping a study. In fact, the superiority/inferiority of a trial has an important influence and many different factors should be accounted for in order to prevent possible harm of the subjects involved in research studies.

Generally, some of the circumstances that could lead to stopping a trial are poor data quality, poor adherence, lack of resources, severe adverse effects, fraud and the discovery of new information that might question the relevance or ethics of the study. As the entire process is not easy and objective most of the times, the decision usually lies with an independent data monitoring committee which uses a prespecified statistical stopping method as to reach a decision. The most common stoppage criteria overall in both glioblastoma and meningioma trials were toxicity (28.88%) and death (16.67%). However, when analyzed individually, there were no meningioma studies, which reported death, as listed in Table 8. In addition, the most common stoppage criterion in both was toxicity and the percentage between disease progression and side effects in meningioma studies was the same (5%).

Carpentier et al. [32] mentioned a few not commonly used stoppage criteria, such as inefficacy (if the estimated success rate was <15%) and futility, if the expected gain from further inclusions in the reliability of the estimated success rate was 0.05 or less. Butowski et al. [33] also stated that "If there was a greater than 20% rate for discontinuation of poly-ICLC during RT, the study was to stop and the design reassessed." On the other hand, Malmström et al. [20] described more than one aspect in terms of stoppage criteria, such as radiological progression, clinical progression, both radiological and clinical progression, unacceptable adverse effects, or until a physician or patient chose to discontinue treatment.

One alternative in terms of monitoring is doing an interim analysis, which consists in conducting a data analysis before data collection has been completed. Malmström et al. [20], for an example, performed two interim analyses to enable exclusion of a treatment group if results were inferior to the two others at a 1% level. If an investigator plan on doing this type of analysis, the statistical stopping approach should be prespecified in the protocol and, preferably, it should be conducted by an independent

Table 8
Statistical analysis

Secondary outcome	Glioblastoma	Meningioma
Survival analysis	52.87%	30%
T-test	2.85%	10%
Wilcoxon	5.71%	5%
Chi-square	2.85%	5%
Fisher	2.85%	NA
Pearson correlation	1.42%	NA
Spearman correlation	NA	5%
Linear regression	NA	NA
Logistic regression	NA	NA
Kruskal–Wallis	1.42%	NA
Not specified	38.57%	50%

NA not available

trial statistician. For instance, besides also planning two interim analyses, Chinot et al. [34] stated that the real-time monitoring of safety events was overseen by an independent data and safety monitoring board. However, a transparent reporting rarely happens and, when it does, a lack of information is commonly noticed. In this review, 40% of the studies did not report the stoppage criteria, 7 studies of meningioma and 28 of glioblastoma. As a result, investigators and even readers remain unaware of these statistical aspects.

Therefore, both investigators and readers should be attentive for reporting the interim analysis and for unreported interim analysis, respectively. Poor reporting might camouflage the interim looks that the investigators did and influence sample size calculation, p value and other different aspects of the study.

10   Statistical Analysis

The statistical analysis of a clinical trial corresponds to one of the most important steps towards finding whether a certain intervention has shown positive results or not. In fact, it allows a researcher to either reject or confirm the null hypothesis, by analyzing patterns and trends of a specific sample and providing interpretable data. Although this element has a high importance and most studies do report which statistical tests were used in their analysis, several researchers are still hesitant and indecisive about choosing the most appropriate statistical test to match the characteristics of the sample, the variables being analyzed, the study design and the condition under investigation.

Overall, a great number of studies properly reported their statistical analysis and which test was used in their hypothesis testing. Among all the tests used, some were more common than others such as survival analysis, t-test, Wilcoxon, and chi-square. Regarding the primary outcome of meningioma and glioblastoma trials, 75.56% performed a survival analysis and, of these, 73.34% used a Kaplan–Meier curve. On the other hand, although several trials reported all the necessary information about the statistical analysis of primary outcomes, 12.22% of them did not provide sufficient or any information.

As to secondary outcomes, in general, the most used statistical test was survival analysis (47.77%) as well, followed by Wilcoxon test (5.56%), T-test (4.45%), chi-square (3.33%), and Fisher exact test (2.22%). Some other tests were also mentioned, such as ANOVA, Spearman correlation, Pearson correlation, and Kruskal–Wallis. In addition, 41.11% of them failed to mention the necessary information about this topic.

10.1 Primary Outcome

Most MG trials reported the statistical test for both primary and secondary outcomes. In terms of the primary outcome, from 20 MG trials, 50% performed a survival analysis, thus corresponding to the most used statistical test. Of these, ten trials used the Kaplan–Meier curve to assess and represent the data collected, two the log-rank test, and two the Cox proportional hazard model. In addition, four studies used a t-test, and two the chi-square test, and a few studies mentioned the binomial test, Spearman test, Wilcoxon test, and Pearson correlation (Table 7). Some studies, however, failed to provide specific details about this section, which corresponded to 15% of the MG trials.

Considering the 70 GB trials, an impressive 82.86% corresponded to survival analysis, from which 80% also mentioned a Kaplan–Meier curve, 31.42% the log-rank test, and 18.57% the Cox proportional hazard model. Gilbert et al. [35], for instance, estimated overall survivor and progression-free survivor by using the Kaplan–Meier method, and the differences between treatment groups were tested by using the log-rank test. Also, the Cox proportional hazards model was used to estimate the treatment hazard ratios (HRs) associated with each endpoint while adjusting for stratification factors. A few other studies also performed the Wilcoxon test (4.28%), a linear regression (2.85%), and a logistic regression (2.85%). Different tests such as Pearson correlation, Kruskal–Wallis, Fisher, and chi-Square were also used, however, in a smaller frequency (1.42%).

Finally, eight GB studies did not properly discuss the statistical analysis process, corresponding to 11.14% of the trials. Due to the importance of this topic, especially in regard to the primary outcome, this percentage corresponds to a concerning number.

10.2 Secondary Outcome

The most used statistical test for the analysis of the secondary outcome also corresponded to survival analysis. From the six trials that chose this method, 100% used the Kaplan–Meier curve, two trials used the log-rank test and two used the Cox proportional hazard model. Moreover, two studies used the T-test and a few mentioned the Spearman test, Wilcoxon test, and chi-square (Table 8). For an example, Sluzewski et al. [24] used a chi-square test for comparing proportions and an unpaired T-test for comparison of means. Differently from the primary outcomes, 50% of the MG studies failed to mention how the analysis of the secondary outcomes was performed and which test was used.

According to Table 8, the secondary outcomes of GB trials were mostly analyzed through survival analysis, which corresponded to 52.87% of the trials. Additionally, 42.85% of these trials also used a Kaplan–Meier curve, 22.82% performed a log-rank test, and 7.14% adopted a Cox proportional hazard model. For instance, Gorlia et al. [36] performed a log-rank test for binary variables, and log-rank trend test for ordered categories. Some other statistical tests were also mentioned, such as Wilcoxon (5.71%), T-test (2.85%), chi-Square (2.85%), Fisher (2.85%), Kruskal–Wallis (1.42%), and Pearson correlation (1.42%). On the contrary, 38.57% of the trials did not specify how the statistical analysis was conducted.

11 Randomization

According to the basic concepts of RCTs, the randomization process is one of the primary steps towards the achievement of excellence and high methodological standards in clinical research. It is considered a respected and trustworthy bias-reducing technique as it has the unique capability of assigning participants by chance, rather than by choice. Thus, it allows the use of probability theory to demonstrate whether a cause–effect relation truly exists.

Indeed, there are several numbers of benefits when it comes to randomization and allocation concealment. However, these aspects still remain one of the least-understood elements of a trial. Depending on the circumstances and ethical aspects involved, randomizing and concealing the allocation of participants may not be an easy task. For instance, as brain tumors are a life-threatening condition and the best specific treatment is not always clear, finding new interventions which are able to control its growth and thus conducting new trials with that aim not always allow an adequate randomization process.

Regarding GB trials, 26 studies ($n = 25/37.14\%$) were not clear about the randomization process or did not specify it and 31

(n = 30/44.28%) were not randomized. On the other hand, there were 13 randomized studies (n = 13/18.57%) in which participants were assigned to two or more groups, either receiving different types of drugs, different combinations or different drug dosages. In one of the trials, for instance, patients were randomized in a 1:1:1 ratio to receive bevacizumab as a single agent, lomustine alone, or bevacizumab and lomustine combined [31]. Chinot et al., however, conducted a phase III trial in which patients were randomly assigned, in a 1:1 ratio, to bevacizumab or placebo [34]. Some other trials also mentioned block randomization [20], stratification [34], and stratification plus randomly permuted blocks method [35].

Within MG trials, 14 (n = 14/70%) were not clear about the randomization process or did not specify it. In five trials (n = 5/25%), randomization was not applicable (three open labels, one convenience sampling, and one consecutive sampling) and, finally, there was only one randomized trial, which opted for one of the most commonly used randomization method: Simple randomization 1:1. In this trial by Tang et al. [37], 232 random numbers were sorted from least to greatest. The first 116 patients were enrolled in the active group and the remaining ones to the Control Group, thus every patient obtained a random number upon admission.

In terms of the chosen methods to generate the randomization sequence, most trials did not properly mention how the generation was conducted or failed to mention completely. No studies of MG and only 4.28% of GB provided this information. In one of the trials by Malmström et al., patients were assigned according to a computer-generated randomization schedule, stratified by center [20]. In addition, Chinot et al. [34] randomized patients with the use of an interactive voice-response system, with stratification according to study region (Western Europe, Eastern Europe, Asia, USA, or other) and recursive partitioning analysis class (III, IV, or V).

12   Blinding

According to CONSORT guidelines, clinical researchers should clearly report the blinding type applied in the study (single-, double- or triple-blinded) and state who in the trial was blinded and how [1]. Although this is an evident requirement, numerous studies fail to report the blinding process and to define whether patients, study assessors and/or data analysts were blinded. To that end, the study validity is commonly affected if blinding information is not well described in articles, which compromises the reliability of results.

In this review, a large number of clinical trials used an open-label design (57.8%), in which neither patients nor study assessors are blinded. The predominance of open-label studies comprises with the large number of phase I and II trials observed in this review. Although the report of blinding aspects is mandatory, a significant amount of the trials (30%) did not provide any information about the blinding type or main strategies applied to ensure blinding. The remaining trials were double-blinded, accounting for 10% of the articles, and only two studies were single-blinded. The predominance of double-blinded trials over single-blinded is also often observed in research studies of other medical fields, as double blinding is more efficient than single blinding for preventing potential biases. There were no triple-blinded clinical trials of meningioma or glioblastoma.

Particularly referring to meningioma studies, the majority of them failed to discuss practical aspects of the blinding process (85%), such as blinding type and assessment strategy. In addition, among these trials, only one article reported a double-blind design, in which the investigators estimated progression-free survival in patient receiving interferon alpha over a 6-month period [38]. Furthermore, out of the 20 meningioma articles, only 2 studies used an open-label design, including a study of Johnson et al. [26] that assessed radiographic response rate in patients who received subcutaneous *octreotide* for 6 months or more. Due to the lack of blinding report in these trials, no further conclusion in this regard could be made.

Differently from meningioma studies, most of the glioblastoma clinical trials reported the blinding type, which corresponded to open-label, accounting for 72.9%. In second place, double-blind studies corresponded to 11.4% of the 70 glioblastoma trials, followed by single-blind, which accounted for 2.85%. There were no triple-blind studies in this review. Moreover, 12.8% of glioblastoma trials did not further describe blinding information, such as treatment characteristics (drug color, odor, taste and administration) and allocation concealment. Also, blinded trials failed reporting whether or not blinding was successful.

13 Study Duration

There was a significant variation regarding the duration of each trial and no specific pattern was identified in this review neither to meningioma nor glioblastoma studies. This disparity may be due to diverse reasons, such as the constant development of new protocols; researchers' concern about safety aspects of the intervention; and different designs in each study. In this review, as there was a predominance of phase II trials, the duration of the studies mostly varied from 1 to 12 months. Also, few studies had a longer length, during more than a year, although the study phases and type of outcome did not significantly vary.

According to Fig. 2, the majority of the trials had a total duration of less than 6 months, accounting for 52.2% of the 90 articles.

Duration of Study

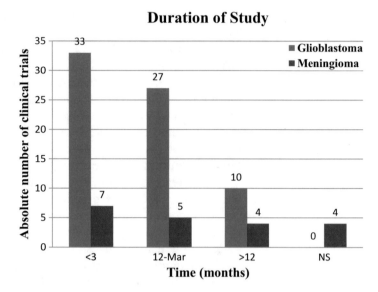

Fig. 2 Overview of brain tumor trial duration

Besides that, multiple studies were conducted over a period of 6–12 months, corresponding to 27.8% of the total. A shorter number of studies was conducted during a period of longer than 12 months (15.6%), which comprises with the predominance of phase I and II trials. For instance, a phase II study of Maier-Hauff et al. [39] assessed the effects of intratumoral therapy in patients with glioblastoma over a period of 13 months.

In articles of glioblastoma, the follow-up duration was related to the time required for performing the intervention (procedure or device) or assessing the drug efficacy/safety (commonly up to 36 months). A study of Hasselbach et al. [28] is an example, as patients received either *cetuximab* plus *bevacizumab* or *cetuximab* plus *irinotecan* and the followed period varied from 7 to 25 months. Moreover, while investigating drug effects, researchers mostly focused on the number of treatment cycles, which varied from 1 to 6 cycles, in order to define study duration. The average acceptable cycles duration used in the clinical trials was 4 weeks (28 days). A study conducted in 2014 [40], for instance, evaluated the effects of 1–8 cycles of *octreotide*, each with a total duration of 28 days.

The studies of meningioma presented similar characteristics to glioblastoma studies. However, there was a significant difference regarding the most applied follow-up duration, which was up to 72 months, instead of 36 months, as observed in glioblastoma trials. In general, there was not standard criteria in regard to number of visits and time points, as each study followed its own protocol. Nonetheless, in studies evaluating pharmacotherapy, risk of toxicity and serious adverse events considerably influenced the total number of study visits.

14   Limitations

In spite of having a strong design or excellent statistical results, any researcher is susceptible to face unanticipated factors, which might affect the study somehow. As to address these factors, such as a lack of adherence, a high dropout rate, or being too population-specific, it is fundamental that the researcher acknowledges them, which is not always the most common. Consequently, a detailed report followed by an explanation of what was done or what could have been done, in order to prevent the limitations from happening, are also of most importance. In fact, besides all that information, the investigator must describe to the readers how the limitations impacted or influenced the interpretation of findings.

In our review, 50% of both glioblastoma and meningioma trials did not mention any limitations nor challenges faced during the entire process. Therefore, it becomes hard for readers to know whether or not there were any limitations or if researchers just failed to mention them. In case the second option was true indeed, the interpretations of the results might alter as well as the conclusions drawn from a specific study.

Regarding the articles that did mention their limitations, most of them were related to the lack of previous studies to which the results could be compared; a short follow-up period; heterogeneity among groups, which may have led to confounders and possible biases; limited external validity; small sample size; and, in case of MG trials, histological grade of meningiomas (grade I, II, and III). Johnson et al. [26], in their phase II meningioma trial, reported their population consisted primarily of patients with aggressive tumors and, therefore, may not reflect the general population of patients. Additionally, they reported a small sample size, which might have affected the statistical power of the study. Simó et al. [40] also reported that the population's characteristics were a limitation given its heterogeneity. Finally, Minniti et al. [21] mentioned the relatively short follow-up as a clear limitation of the study.

One limitation commonly mentioned in GB trials was the heterogeneity of recurrent glioblastomas and malignant brain tumors in general, and how this may affect the patient's response to the intervention. However, according to 17.14% of the glioblastoma studies, the most cited limitation was a small sample size, which may have resulted in a smaller statistical power. Other studies, such as the one conducted by Iwamoto et al. [41], also referred to the difficulties in translating laboratory findings to clinical studies due to the limitation of commonly used glioma preclinical models. As some of the meningioma studies, the phase II glioblastoma trial by Desjardins et al. [42] also referred to the heterogeneity of the population as their limitation.

Overall, the most common limitation of both meningioma and glioblastoma trials was the heterogeneity of the population and the restricted sample size. In addition, it was noticeable an intriguing lack of reporting and discussion towards the studies limitations and of how they could have affected the results. As most guidelines encourage authors to describe the main pitfalls of the study, investigators should take an extra effort to discuss not only the results but also the difficulties they faced so as the problems can be addressed and possibly solved in future studies.

15 Conclusion

In spite of how complex brain tumors are, different types of treatment are available for patients with GB and MG. Some treatments are considered standard, such as chemotherapy, and some are still being tested in clinical trials. Consequently, many of those trials intend to help deepen the investigation of current treatments or to obtain accurate scientific information on new treatments in order to improve patient's quality of life. Their importance is undeniable; however, when it comes to brain tumors, conducting research is still a challenging task. Whether it is the small number of patients, the unique response to a certain intervention or the lack of information available in the literature, investigators often face several difficulties.

Although there was a distinctive variation among most study designs and interventions given the variation of types, administration and combination between therapies, the outcomes overall were very similar. Indeed, progression-free survival, tumor response rate and toxicity were commonly observed in both meningioma and glioblastoma trials. Most studies corresponded to phase II trials, and thus there is still a lot of information to be gathered and data to be collected in order to obtain approval for new therapies.

In addition, when researchers do develop trials, they commonly do not report all the required information to a proper interpretation of the results, or even fail to mention it completely. As a result, the planning of future studies is compromised, as researchers have restricted information to rely on. Hence, it is fundamental that investigators are able to overcome limitations, promote critical thinking, and provide a detailed report of vital information as to improve health care.

References

1. Ostrom QT, Gittleman H, Farah P, Ondracek A, Chen Y, Wolinsky Y et al (2013) CBTRUS statistical report: primary brain and central nervous system tumors diagnosed in the United States in 2006–2010. Neuro Oncol 15(suppl 2):ii1–ii56. https://academic.oup.com/neuro-oncology/article-lookup/doi/10.1093/neuonc/not151

2. Ostrom QT, Gittleman H, Liao P, Rouse C, Chen Y, Dowling J et al (2014) CBTRUS statistical report: primary brain and central nervous system tumors diagnosed in the United States in 2007–2011. Neuro Oncol 16:iv1–iv63

3. Louis DN, Perry A, Reifenberger G, von Deimling A, Figarella-Branger D, Cavenee WK et al (2016) The 2016 World Health Organization classification of tumors of the central nervous system: a summary. Acta Neuropathol 131(6):803–820

4. Rogers L, Barani I, Chamberlain M, Kaley TJ, McDermott M, Raizer J et al (2015) Meningiomas: knowledge base, treatment outcomes, and uncertainties. A RANO review. J Neurosurg 122(1):4–23. http://thejns.org/doi/10.3171/2014.7.JNS131644

5. McNeill KA (2016) Epidemiology of brain tumors. Neurol Clin 34(4):981–998. https://doi.org/10.1016/j.ncl.2016.06.014

6. Blitshteyn S, Crook JE, Jaeckle KA (2008) Is there an association between meningioma and hormone replacement therapy? J Clin Oncol 26(2):279–282

7. Walcott BP, Nahed BV, Brastianos PK, Loeffler JS (2013) Radiation treatment for WHO grade II and III meningiomas. Front Oncol 3. http://journal.frontiersin.org/article/10.3389/fonc.2013.00227/abstract

8. Louis DN, Ohgaki H, Wiestler OD, Cavenee WK, Burger PC, Jouvet A et al (2007) The 2007 WHO classification of tumours of the central nervous system. Acta Neuropathol 114(2):97–109. http://link.springer.com/10.1007/s00401-007-0243-4

9. Ostrom QT, Bauchet L, Davis FG, Deltour I, Fisher JL, Langer CE et al (2014) The epidemiology of glioma in adults: a state of the science review. Neuro Oncol 16(7):896–913

10. Tsien C, Moughan J, Michalski JM, Gilbert MR, Purdy J, Simpson J et al (2009) Phase I three-dimensional conformal radiation dose escalation study in newly diagnosed glioblastoma: radiation therapy oncology group trial 98–03. Int J Radiat Oncol Biol Phys 73(3):699–708

11. Lee IH, Piert M, Gomez-Hassan D, Junck L, Rogers L, Hayman J et al (2009) Association of 11C-methionine PET uptake with site of failure after concurrent temozolomide and radiation for primary glioblastoma multiforme. Int J Radiat Oncol Biol Phys 73(2):479–485

12. Reardon DA, Norden AD, Desjardins A, Vredenburgh JJ, Herndon JE II, Coan A et al (2012) Phase II study of Gleevec(R) plus hydroxyurea (HU) in adults with progressive or recurrent meningioma. J Neurooncol 106(2):409–415

13. Terret C, Albrand G, Moncenix G, Droz JP (2011) Karnofsky Performance Scale (KPS) or Physical Performance Test (PPT)? That is the question. Crit Rev Oncol Hematol 77(2):142–147. http://linkinghub.elsevier.com/retrieve/pii/S1040842810000302

14. Wick W, Puduvalli VK, Chamberlain MC, Van Den Bent MJ, Carpentier AF, Cher LM et al (2010) Phase III study of enzastaurin compared with lomustine in the treatment of recurrent intracranial glioblastoma. J Clin Oncol 28(7):1168–1174

15. Brandes AA, Tosoni A, Franceschi E, Blatt V, Santoro A, Faedi M et al (2009) Fotemustine as second-line treatment for recurrent or progressive glioblastoma after concomitant and/or adjuvant temozolomide: a phase II trial of Gruppo Italiano Cooperativo di Neuro-Oncologia (GICNO). Cancer Chemother Pharmacol 64(4):769–775

16. Saraf S, McCarthy BJ, Villano JL (2011) Update on meningiomas. Oncologist 16(11):1604–1613. http://www.ncbi.nlm.nih.gov/pubmed/22028341%5Cnhttp://www.pubmedcentral.nih.gov/articlerender.fcgi?artid=PMC3233296

17. Combs SE, Hartmann C, Nikoghosyan A, Jäkel O, Karger CP, Haberer T et al (2010) Carbon ion radiation therapy for high-risk meningiomas. Radiother Oncol 95(1):54–59

18. Hottinger AF, Stupp R, Homicsko K (2014) Standards of care and novel approaches in the management of glioblastoma multiforme. Chin J Cancer 33(1):32–39. http://www.cjcsysu.cn/abstract.asp?fr=doi&idno=20706

19. Stupp R, Hegi ME, Mason WP, van den Bent MJ, Taphoorn MJ, Janzer RC et al (2009) Effects of radiotherapy with concomitant and adjuvant temozolomide versus radiotherapy alone on survival in glioblastoma in a randomised phase III study: 5-year analysis of the EORTC-NCIC trial. Lancet Oncol 10(5):459–466

20. Malmström A, Grønberg BH, Marosi C, Stupp R, Frappaz D, Schultz H et al (2012) Temozolomide versus standard 6-week radio-

therapy versus hypofractionated radiotherapy in patients older than 60 years with glioblastoma: the Nordic randomised, phase 3 trial. Lancet Oncol 13(9):916–926

21. Minniti G, Clarke E, Cavallo L, Osti M, Esposito V, Cantore G et al (2011) Fractionated stereotactic conformal radiotherapy for large benign skull base meningiomas. Radiat Oncol 6(1):36. http://ro-journal.biomedcentral.com/articles/10.1186/1748-717X-6-36

22. Stummer W, Reulen HJ, Meinel T, Pichlmeier U, Schumacher W, Tonn JC et al (2008) Extent of resection and survival in glioblastoma multiforme: identification of and adjustment for bias. Neurosurgery 62(3):564–574

23. Venook AP, Tabernero J (2015) Progression-free survival: helpful biomarker or clinically meaningless end point? J Clin Oncol 33(1):4–6. http://ascopubs.org/doi/10.1200/JCO.2014.57.9557

24. Sluzewski M, Van Rooij WJ, Lohle PN, Beute GN, Peluso JP (2013) Embolization of meningiomas: comparison of safety between calibrated microspheres and polyvinyl-alcohol particles as embolic agents. Am J Neuroradiol 34(4):727–729

25. Laprie A, Catalaa I, Cassol E, McKnight TR, Berchery D, Marre D et al (2008) Proton magnetic resonance spectroscopic imaging in newly diagnosed glioblastoma: predictive value for the site of postradiotherapy relapse in a prospective longitudinal study. Int J Radiat Oncol Biol Phys 70(3):773–781

26. Johnson DR, Kimmel DW, Burch PA, Cascino TL, Giannini C, Wu W et al (2011) Phase II study of subcutaneous octreotide in adults with recurrent or progressive meningioma and meningeal hemangiopericytoma. Neuro Oncol. 13(5):530–535

27. Izumoto S, Tsuboi A, Oka Y, Suzuki T, Hashiba T, Kagawa N et al (2008) Phase II clinical trial of Wilms tumor 1 peptide vaccination for patients with recurrent glioblastoma multiforme. J Neurosurg 108(5):963–971. http://thejns.org/doi/10.3171/JNS/2008/108/5/0963

28. Hasselbalch B, Lassen U, Hansen S, Holmberg M, Sørensen M, Kosteljanetz M et al (2010) Cetuximab, bevacizumab, and irinotecan for patients with primary glioblastoma and progression after radiation therapy and temozolomide: a phase II trial. Neuro Oncol 12(5):508–516

29. Kreissl MC, Hänscheid H, Löhr M, Verburg FA, Schiller M, Lassmann M et al (2012) Combination of peptide receptor radionuclide therapy with fractionated external beam radiotherapy for treatment of advanced symptomatic meningioma. Radiat Oncol 7:99. http://www.

pubmedcentral.nih.gov/articlerender.fcgi?artid=3439242&tool=pmcentrez&rendertype=abstract

30. Trippa F, Maranzano E, Costantini S, Giorni C (2009) Hypofractionated stereotactic radiotherapy for intracranial meningiomas: preliminary results of a feasible trial. J Neurosurg Sci 53(1):7–11

31. Taal W, Oosterkamp HM, Walenkamp AME, Dubbink HJ, Beerepoot LV, Hanse MCJ et al (2014) Single-agent bevacizumab or lomustine versus a combination of bevacizumab plus lomustine in patients with recurrent glioblastoma (BELOB trial): a randomised controlled phase 2 trial. Lancet Oncol 15(9):943–953

32. Carpentier A, Metellus P, Ursu R, Zohar S, Lafitte F, Barrié M et al (2010) Intracerebral administration of CpG oligonucleotide for patients with recurrent glioblastoma: a phase II study. Neuro Oncol 12(4):401–408

33. Butowski N, Chang SM, Junck L, DeAngelis LM, Abrey L, Fink K et al (2009) A phase II clinical trial of poly-ICLC with radiation for adult patients with newly diagnosed supratentorial glioblastoma: a North American Brain Tumor Consortium (NABTC01-05). J Neurooncol 91(2):175–182

34. Chinot OL, Wick W, Mason W, Henriksson R, Saran F, Nishikawa R et al (2014) Bevacizumab plus radiotherapy–temozolomide for newly diagnosed glioblastoma. N Engl J Med 370(8):709–722. http://www.nejm.org/doi/10.1056/NEJMoa1308345

35. Gilbert MR, Wang M, Aldape KD, Stupp R, Hegi ME, Jaeckle KA et al (2013) Dose-dense temozolomide for newly diagnosed glioblastoma: a randomized phase III clinical trial. J Clin Oncol 31(32):4085–4091

36. Gorlia T, van den Bent MJ, Hegi ME, Mirimanoff RO, Weller M, Cairncross JG et al (2008) Nomograms for predicting survival of patients with newly diagnosed glioblastoma: prognostic factor analysis of EORTC and NCIC trial 26981-22981/CE.3. Lancet Oncol 9(1):29–38

37. Tang X, Yin X, Zhang T, Peng H (2011) The effect of hyperbaric oxygen on clinical outcome of patients after resection of meningiomas with conspicuous peritumoral brain edema. Undersea Hyperb Med 38(2):109–115

38. Chamberlain MC, Glantz MJ (2008) Interferon-alpha for recurrent World Health Organization grade 1 intracranial meningiomas. Cancer 113(8):2146–2151. http://www.ncbi.nlm.nih.gov/pubmed/18756531

39. Maier-Hauff K, Ulrich F, Nestler D, Niehoff H, Wust P, Thiesen B et al (2010) Efficacy and

safety of intratumoral thermotherapy using magnetic iron-oxide nanoparticles combined with external beam radiotherapy on patients with recurrent glioblastoma multiforme. J Neurooncol. http://www.ncbi.nlm.nih.gov/pubmed/20845061

40. Simó M, Argyriou AA, Macià M, Plans G, Majós C, Vidal N et al (2014) Recurrent high-grade meningioma: a phase II trial with somatostatin analogue therapy. Cancer Chemother Pharmacol 73(5):919–923

41. Iwamoto FM, Lamborn KR, Robins HI, Mehta MP, Chang SM, Butowski NA et al (2010) Phase II trial of pazopanib (GW786034), an oral multi-targeted angiogenesis inhibitor, for adults with recurrent glioblastoma (North American Brain Tumor Consortium Study 06-02). Neuro Oncol 12(8):855–861

42. Desjardins A, Reardon DA, Coan A, Marcello J, Herndon JE, Bailey L et al (2012) Bevacizumab and daily temozolomide for recurrent glioblastoma. Cancer 118(5):1302–1312

Chapter 11

Spinal Cord Injury

Faddi Ghassan Saleh Velez, Camila Bonin Pinto, and Felipe Fregni

Abstract

This book chapter discusses detailed aspects that were analyzed regarding the design and methodology of 110 most highly cited clinical trials in Spinal Cord Injury. The aim of this book chapter is to provide the reader a summarized analysis of detailed aspects of clinical trial design in this group of population in order to optimize the development of future trials. A literature search was performer utilizing the search engine "web of science" in order to collect the data of the 110 most cited articles in SCI for the last 10 years. A detailed discussion based on the criteria of the consort guidelines is provided with the most common findings as well as suggestions for the improvement on the design of future SCI clinical trials.

Key words Spinal cord injury, Methodology, Clinical trials in neurology, Scientific research, CONSORT guidelines

1 Introduction

In spinal cord injury (SCI), a lesion in the spinal cord pathways due to a traumatic (fracture or a dislocation) or degenerative (i.e., infection, cancer, metastatic, etc.) process [1–3] may cause a variety of symptoms ranging from motor to sensorial to autonomic deficits. The level and extension by which spinal tracts are affected determine the severity of these deficits [4–7]. When there is a total absence of sensory and motor function in the lowest sacral nerves (S4 and S5) it is known as a complete lesion while in an incomplete injury there is partial function preserved in the same segments of the sacral spinal cord [8, 9].

SCI is a disease with a major worldwide impact. According to the NIH, the prevalence of SCI in the USA is approximately 250,000, with an incidence of approximately 12,000 cases per year. Approximately, 90% of the causes of SCI are due to a traumatic injury [5, 10]. Vehicle accidents represent about 36.5% of the cases, followed by around 25% due to falls and the remaining of them due to violence [1]. Among the affected population 80% are men. The US health care system spends more than three billion US

Felipe Fregni (ed.), *Clinical Trials in Neurology*, Neuromethods, vol. 138,
https://doi.org/10.1007/978-1-4939-7880-9_11, © Springer Science+Business Media, LLC, part of Springer Nature 2018

dollars a year in the management and rehabilitation of patients with SCI [1] (WHO). Moreover, the lack of reporting in developing countries signalize for an even higher impact of SCI worldwide.

As the world evolves and the rates of life expectancy increase among all the population, the prevalence of patients with spinal cord injury increase proportionally and the number of years that the injured patient will live with all the sequels raise rapidly; therefore, giving critical importance to the developing of adequate clinical trials for the treatment and specially rehabilitation post SCI.

2 Challenges Faced in Clinical Spinal Cord Injury Trials

Due to the wide variety mechanisms of trauma or degeneration of a spinal cord injury there are specific challenges that a researcher may face to develop a valid clinical trial. We described below some challenges regarding the design and execution of clinical trials in SCI patients:

1. Sample—After a traumatic injury, the vast majority of SCI cases happen due to an uncontrolled accident resulting in multiple kinds of lesions such as cavitations, bone fragmentation, root avulsion, spine dislocation, vascular damage, or spine fracture. This heterogeneity can be a challenge since it leads to important difficulties to generalize and analyze this group of participants[11]. It also in some cases combines with lesions in other levels of the neuroaxis such as brain structures.

    (a) Outcome—The lack of surrogate functional endpoints is another challenge to design a trial in SCI. Although magnetic resonance imaging (MRI) is one of the best tools to diagnose and assess structural changes in patients with SCI, this technique is still considered a poor predictor of outcome [11]. Functional neurophysiological markers such as fMRI and EEG have not been developed for spinal cord.

    (b) Blinding—blinding is an important issue, especially due to the type of intervention required (usually surgery procedure) what makes more difficult for the investigator to be blinded [11].

These aspects can make the design and implementation of a study in SCI highly variable. Moreover, in the past years a few preclinical trials were successfully translated into efficacious clinical trials in SCI subjects. Among those, just one was able to show that methylprednisolone given between the first 8 h post injury had statistically significant efficacy to treat SCI; however, it is not currently accepted as standard of care worldwide [11].

It is critical therefore prior to starting a trial to have a clear and valid hypothesis, primary endpoint, and adequate inclusion and exclusion criteria to select the most representative SCI population that can give a generalized idea of the global population affected [12].

3   Methodology

A systematic review of the literature was conducted in order to find the 100 most highly cited clinical trials. An online search on the web, "web of science" was conducted with the MeSH term "spinal cord injury" in the title between the years 2010 and 2015, and we collected the initial 100 most highly cited articles; however, the mean of citations for those 100 initial studies was considered low (11.6). Due to the lack of highly cited clinical trials, a second search was necessary in order to achieve a greater number of citations; therefore, increasing the impact of the information collected. The second search included the ten most highly cited articles over the last 10 years (Fig. 1).

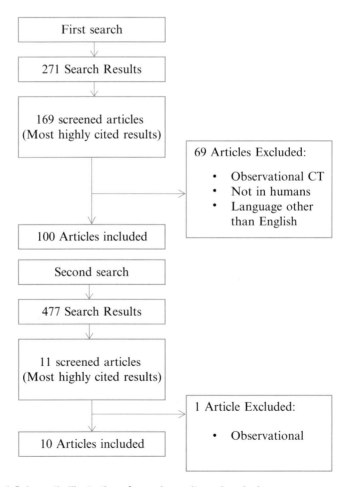

Fig. 1 Schematic illustration of search results and exclusions

4   Initial Results

4.1   Citations and Consort Criteria

The data of the 110 included studies was collected and analyzed according to the framework denoted by the CONSORT criteria headings for clinical trials [13]. We found that among the highest cited articles for spinal cord injury just 48% were published following the CONSORT criteria headings.

Initially we analyzed the data for the 100 most highly cited articles over the last 5 years in which the mean number of citations was 11.6. However for the second search that included the ten most highly cited articles over the last 10 years a mean of 152.9 citations was found. A summary of the general characteristics observed among the total of studies analyzed is available in Table 1.

4.2   Number of Citations and Positivity for Primary Outcome

In this review, we analyzed the positive result for the primary outcome. We found that out of 110 reviewed articles, 78 of them showed positive results and at the same time, the number of citations was higher in this group in comparison with the 28 studies that showed negative results for the primary outcome (Graph 1). This finding is not surprising, however is relevant to point out that positive findings are as good as negative findings if the trials performed were well designed and well carried for the research group.

5   Trial Design, Study Setting and Phase of a Clinical Trial

In this section, we discuss the general aspects of trial design and the relevance between different phases of clinical trials. During our search two major groups of studies were analyzed, trials with drugs as intervention were classified by phases ranging from pilot through phase I, II, III and IV. Furthermore, clinical trials with devices or procedures as the intervention were classified utilizing the NIH classification, and thus divided in Exploratory—Pilot; Pivotal; and Postmarketing.

5.1   Phases of Clinical Trials Using Drug as Intervention

In our search, we found 39 clinical studies in which a drug was utilized as the intervention. In the following paragraphs, we discuss those research studies subdividing them by the phase (Tables 2 and 3).

Pilot studies are small scale preliminary studies developed to evaluate feasibility, adverse events, possible effect size, probable costs, and preliminary analysis prior to developing a phase II clinical trial. It is often used for a drug that has already been approved for a different indication. Usually pilot studies should be avoided unless there is an important reason to run such trials. A trial with a defined safety and efficacy goal has a much greater value. In our SCI review, we found four pilot studies. These four studies evaluated

Table 1
General characteristics of cited articles

		# (Total *N* = 110)
Year	2005–2009	10
	2010–2012	66
	2013–2015	34
Phase	Pilot	4
	Phase I	14
	Phase I/II	5
	Phase II	15
	Phase III	1
	Phase IV	0
	Exploratory pilot	60
	Pivotal	11
	Post market	0
Trial design RCT	2 parallel arms	42
	3 parallel arms	5
	4 or more parallel arms	4
	Factorial	1
	Cross over	20
	Open label	38
Sample size	0–20	50
	21–50	36
	51–100	13
	101–500	11
Type of intervention	Drug	39
	Device	34
	Procedure	37
# of citations	0–10	66
	11–20	19
	21–50	12
	51–70	3
	71–100	2
	100–200	6
	200–300	2
Journal impact factor	1–5	97
	5–10	13
	10–20	0
	More than 20	0
Consort criteria	Yes	48
	No	62

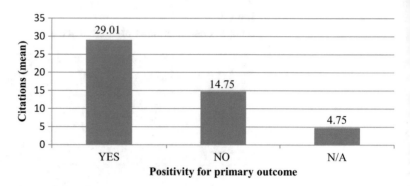

Graph 1 Citations vs. Positivity for primary outcome

Table 2
Trial design, study setting, phase

Trial design		$N = 39$	Pilot	I	I/II	II	II/III	III	IV	Study setting
							Phases			
RCT	2 parallel arms	13	0	2	0	5	0	0	0	Single center
			0	0	1	4	0	1	0	Multicenter
	3 parallel arms	1	0	0	0	1	0	0	0	Single center
			0	0	0	0	0	0	0	Multicenter
	4 or more parallel arms	1	0	0	0	0	0	0	0	Single center
			0	0	0	1	0	0	0	Multicenter
	Factorial	0	0	0	0	0	0	0	0	Single center
			0	0	0	0	0	0	0	Multicenter
	Cross over	7	1	3	0	3	0	0	0	Single center
			0	0	0	0	0	0	0	Multicenter
Non-RCT (quasi-experimental)		17	3	8	2	0	0	0	0	Single center
			0	1	2	1	0	0	0	Multicenter

drugs such as Dronabinol for the management of pain, botulinum toxin injection for the relief of neurogenic bladder symptomatology; as well as feasibility of injection of harvested bone marrow autologous cells from the iliac crest or safety and feasibility of implantation of olfactory mucosa autografts into the participants injured spinal cord [14–17]. The trials gave the following reasons to perform a pilot study: (1) little is known as regards the efficacy and safety of cannabinoids for the treatment of chronic nonmalignant pain in patients with SCI; (2) There is evidence of the efficacy of botulinum toxin A for the management of detrusor overactivity; however, there is no evidence on its effects to prevent UTIs; (3) There is no evidence on the growth kinetics of BM MSC in patient with SCI as well as it safety; (4) There is lack of evidence on the

Table 3
Trial design, study setting, phase

Trial Design		N = 71	Device and procedures trials phases			Study setting
			Exploratory pilot	Pivotal	Postmarketing	
RCT	2 parallel arms	29	20	2	0	Single center
			2	5	0	Multicenter
	3 parallel arms	4	3	1	0	Single center
			0	0	0	Multicenter
	4 or more parallel arms	3	1	2	0	Single center
			0	0	0	Multicenter
	Factorial	1	1	0	0	Single center
			0	0	0	Multicenter
	Cross over	13	11	0	0	Single center
			2	0	0	Multicenter
Non-RCT (quasi-experimental)		21	17	0	0	Single center
			3	1	0	Multicenter

safety and feasibility of the use of olfactory ensheanting cells for neural repair in SCI.

Phase I clinical trials are designed to evaluate safety, adequate dosage range and identification of side effects of a novel intervention in small populations. In this SCI review, we found 14 phase I clinical trials. Some of the studies evaluated the safety of relatively novel therapies that applied either bone marrow cells, olfactory cells, fetus cells, or autologous monocytes in the participant spinal cord through a surgical procedure or via injection [18–20]. Additionally, eight studies tested safety of medications such as Nabilone, Riluzole, Duloxetine, Lithium, or Botulinum toxin [21–24]. Given the main goal of phase I is to establish initial safety, it is important to define well the safety outcomes. These outcomes depend on the drug being tested, but we list the most used safety outcomes in these protocols: Physical examination at baseline (i.e., heart rate, blood pressure, temperature, motor and sensorial examination), adverse events scales, neurologic examination, CSF laboratory test, hematology tests (i.e., RBC counts, WBC counts, platelets, BUN, creatinine, etc.) as well as ASIA score as it can measure neurologic deterioration [25].

Phase II clinical trials require a larger group of participants and are designed to evaluate effectiveness and safety of the intervention. In this review, we found 15 phase II clinical trials. All of these were randomized clinical trials. The studies were designed to evaluate the effectiveness of drugs such as Pregabalin, Carbamazepine, and Nicotine patches for the management of neuropathic pain; as well as 4-AP, 5-HT antagonists in recovery of motor function [26–29].

Phase III clinical trials aim to confirm effectiveness, monitor side effects, and usually compare an intervention with standard of care; in addition, phase III trials require a larger number of participants. We found just one phase III clinical study. The trial conducted by Cardenas et al., evaluated the effects of the drug Fampridine in motor recovery utilizing the Ashworth scale as outcome [30].

Phase IV clinical trials are developed after approval and marketing of the drug or treatment aiming to collect information about side effects with long term use. In our search we didn't find phase IV clinical trials.

5.2 Phases of Clinical Trials Using Devices or Procedures as Intervention

In our search, we found 71 articles testing a device (34 trials) or a procedure (37 trials) as the intervention. For clinical trials utilizing devices or procedures as the intervention the classification of phases differed when compared with the one applied for drug trials. In this sub-section we discuss the studies that we found in our SCI search and how we classified them as exploratory: pilot; pivotal; or postmarketing [31, 32].

Exploratory pilot studies evaluate safety and improvement of device's design/parameters in a small group of subjects and define tolerability limits. We found a good number of studies in this category: 60 exploratory pilot studies. These studies tested on average 22.8 subjects (min of 3 subjects and max of 68 subjects) and assessed motor recovery, pain modulation as the main topics; additional aspects that are of relevance to evaluate in the field of SCI are autonomic nervous system modifications, urinary tract alterations, bone density changes, and psychological problems, among others. We found that studies that evaluated the effects of diverse stimulation techniques such as transcranial direct current stimulation (tDCS), transcranial magnetic stimulation (TMS); transcutaneal electrical nerve stimulation (TENS); or cranial electric stimulation (CES) [33–37] for the management of pain or motor recovery, as well as training programs utilizing robotic body weight supports devices, brain computer interface programs or traditional therapy combined with robotic training to improve motor function as well as quality of life of individuals with SCI [34, 38–42].

Interestingly we found that the only two clinical trials that included pediatric population as participants evaluated a similar intervention (functional electric stimulation (FES) combined with a cycling exercise training program) for motor recovery which show that some intervention protocols could be more easy to apply in pediatric participants such as cycling instead of conventional therapy [43, 44]. Additionally we found a large amount of trials utilizing a wide variety of training programs (procedure) as intervention in order to improve motor recovery [45–48]; this in fact is an important point in terms of developing a novel trial as the field of rehabilitation allow us the development of a vast group of different rehabilitation approaches specific for each typo of affected area.

Pivotal trials test safety and effectiveness of either the device or procedure in a larger population. The fact of selecting a larger sample is required in order to prove efficacy. We found 11 pivotal studies. Among those we would like to point out the study by Nussbaum et al., evaluating management of pressure ulcers with ultraviolet C-irradiation (UVC) [49], specially due to the fact of being a novel therapy as well as the application of it after the surgical debridement of the injured region. A pivotal trial is usually performed after successful results in exploratory studies. In the case of this therapy, there was a preliminary trial performed in 1994 showing that ultrasound and UVC treatment have a major effect on wound healing either combined or alone [50]; therefore, allowing the development of further studies evaluating this effect in specific situations such after surgical debridement of a pressure ulcer in SCI patients.

Post marketing studies are designed as surveillance trials aiming to identify long-term effects and monitor adverse effects in a large and heterogeneous population. In our search we also didn't find any clinical trial in this phase, which in fact point out the necessity of evolving and continue developing and evaluating novel therapies for rehabilitation, pain management and other sequels caused due to a SCI.

5.3 Study Settings

SCI has the challenge of being mostly caused by a sudden traumatic event, due to this, and the impressive capacity of the spinal cord to recover, the study setting plays an important role while planning and developing clinical trials for this population. As an example, multicenter trials are able to enroll a greater amount of participants with acute traumatic injury when compared with single center trials. However, in this review, 86 out of 110 studies were single center studies, being most of them exploratory pilot clinical trials (61.6%). The only phase III trial found in our search was multicenter, as well as the majority of pivotal trials (54.5%). The importance of developing multicenter studies rely basically on improving recruitment, which in fact is a frequent factor that may potentially hinder the proper development of a clinical study with this population.

6   Participants

In this section we focus in describing the inclusion and exclusion criteria, since this aspect is important in order to achieve an adequate and representative group of participants for a SCI clinical trial. Therefore, having a right balance between internal and external validity is essential to allow the generalization of the results in a wider population of patients that suffered spinal injury.

Among the great majority of analyzed articles, most of them selected SCI subjects on basis of the scores of ASIA classification scale. In fact among the 110 clinical trials, 60 used the ASIA classification as part of the inclusion criteria to categorize the specific population required for the study. Table 4 shows the ASIA classification utilized in the great majority of studies to describe the amount of neurological loss in subjects with SCI [51–53].

There is a good number of studies that included subjects classified as ASIA B, C or D (Table 4). This was found especially in rehabilitation studies that involved the application of robotic training devices due to the necessity of having at least little amount of remaining mobility to participate in the active training groups. On the other hand, subjects with ASIA A are mostly included in studies that require the implantation of novel therapies aiming to regenerate spinal cord tissue and generate functional improvement of symptoms; such a trials testing the transplantation of autologous bone marrow cells or olfactory ensheathing cells implantation [16, 18].

Another relevant aspect is the time length between the spinal cord injury event and the enrollment in the clinical trial. There were only few clinical trials including acute spinal cord injuries subjects. In our search we found only two studies. The first one recruited patients with less than 48 h of onset of SCI to evaluate the Neuroprotective effects of granulocyte colony-stimulating Factor [54]. The second one recruited subject within 12 h post injury, this study was a randomized placebo controlled clinical trial that evaluated the effects of Mynocicline on SCI recovery [55]. However, a total of 11 clinical trials utilized subject with SCI up to

Table 4
ASIA classification

Grade	
ASIA A	Complete lack of motor and sensory function below the level of injury (including the anal area)
ASIA B	Some sensation below the level of the injury (including anal sensation)
ASIA C	Some muscle movement is spared below the level of injury, but 50% of the muscles below the level of injury cannot move against gravity
ASIA D	Most (more than 50%) of the muscles that are spared below the level of injury are strong enough to move against gravity
ASIA E	All neurologic function has returned

1 month from the lesion date (acute SCI). Furthermore, 35 studies were designed including individuals with SCI for a period greater than 6 months. This pattern was associated to the intervention, which aimed to evaluate effectiveness on the management of neuropathic pain; as well as training programs for recovery of motor and sensory function. Therefore, there is still a lack of high impact studies assessing therapies in the acute stages of SCI.

As described before, the localization of injury is correlated with the degree of symptomatology and more specifically with the affected region of the body. Fifty seven of the 110 (51.8%) reviewed clinical trials defined the specific location of injury as a requirement to be enrolled in the study. Among those, 33 (57.89%) of the clinical trials required cervical or thoracic injuries; followed by 10 (17.54%) cervical-thoracic-lumbar; 6 (10.52%) cervical alone and 4 (7.01%) thoracic alone. The distribution was mostly correlated to the type of intervention in which subjects with lower limb deficits were required for trials utilizing specific robotic training devices while subjects with a higher level of injury were included in trials evaluating neuropathic pain management or upper limb specific training protocols.

Even though the great majority of SCI are caused due to traumatic events, out of the 110 studies just 14 (12.7%) specified traumatic etiology as an inclusion criteria. For the vast majority of studies (87.3%) the etiology was not clear or did not specified as a relevant point in the selection of participants. Table 5 summarizes the most frequent characteristic utilized to select the desired population evaluated in the analyzed 110 studies.

6.1 Exclusion Criteria

For the exclusion criteria, most of the exclusion criteria were based on ensuring that subjects had the specific diagnosis being tested. For instance neuropathic pain studies excluded subjects with pain from other etiologies. Additionally is important to point out that for some clinical trials that aimed to test interventions that required certain amount of residual movement; therefore, excluding patients with tetraplegia or patients that were unable to complete a baseline task for the trial.

7 Trial Duration and Follow-Up Duration

In this section, we discuss the average duration of the interventions that were given to the subjects with SCI in this literature search. Among all the studies we found a large percentage of interventions in which the subject was exposed to a procedure during a single session (Table 6); most of the interventions were novel therapies that included injection of autologous bone marrow cells as well as olfactory mucosa ensheathing cells [16, 18, 19]. However, the great majority of trials were characterized by interventions that lasted between 1 month and 3 months (Table 6).

Table 5
Eligibility criteria

Eligibility criteria	N	Phases drug studies				Phases device—procedure		
		Pilot	I	I/II	II	III	Expl: pilot	Pivotal
ASIA	60	3	7	3	7	1	33	5
Specific location of injury								
Cervical	6	0	1	0	0	0	5	0
Thoracic	4	0	0	0	1	0	3	0
Cervical–Thoracic	33	2	4	0	2	1	23	1
Cervical–thoracic–lumbar	10	0	1	0	3	0	3	3
Thoracic–lumbar	3	0	0	0	1	0	2	0
Lumbar	1	0	0	0	0	0	0	1
Age specific								
Adults	108	4	11	4	14	1	63	11
Children	2	0	0	0	0	0	2	0
Pain scales								
VAS for pain	13	1	1	0	2	0	8	1
Etiology								
Traumatic	14	0	0	1	1	0	12	0
Traumatic and nontraumatic	1	0	0	0	0	0	1	0
Nonspecific	95	4	11	3	13	1	52	11
Time since injury								
Acute	5	0	1	2	2	0	0	0
Intermediate	9	0	0	0	2	0	4	1	2
Chronic	42	2	6	1	5	0	14	12	2

Table 6
Duration of intervention

Duration of intervention	Number of trials	%
Just surgical procedure	25	22.72
1 week	9	8.19
>1 weeks to 1 month	19	17.28
>1 month to 3 months	35	31.81
>3 months to 6 months	15	13.63
>6 months	7	6.37

Due the wide variety of interventions reviewed in this search, the follow-up periods are very heterogeneous. However, we also encountered a large percentage of clinical trials in which the participants were not followed up and the data was collected immediately after the end of the intervention.

The follow-up periods varies depending on the intervention, for instance in studies in which the subject was exposed to a single session of therapy, a subsequent short-term follow-up period was used to assess each participant [39]. On the other hand, studies in which the subject had multiple weeks or months of therapy to improve recovery and functionality the participants were followed in long term (6 months to a year). Multiple follow-up visits are helpful to determine long-lasting effects of an intervention.

We found 22 studies in which there was no follow-up visits and the outcomes were measured right after the end of the intervention. These studies are characterized by the evaluation of specific procedures or single sessions of stimulation (i.e., TMS, tDCS, FES, electroejaculation) and the outcome measurements were assessed pre and post intervention or in other cases that required longer interventions (i.e., specific rehabilitation therapy protocols) the outcomes measurements were also performed at baseline, during the visits and at the end of the intervention while no extra follow-ups were applied. This type of evaluation gives us some insight on the changes generated before and after; however, it does not provide information in regard to the long-term duration of those effects.

Seventeen of the studies followed the subject for less than a month while the great majority followed the subject for more than a month to as much as a year after the end of the intervention (49 studies). The large number of studies with long follow-up periods may be because most of the interventions tested were associated with long term effects of physical therapies; therefore long-lasting effects were possible (Table 7).

One point that the SCI researcher should consider when designing a trial is the natural recovery especially for subjects with incomplete lesions. In the majority of patients, during the first 3 months post injury some extent of natural recovery is expected; however in some cases spontaneous recovery can be observed even up to the first 18 months [12]. Therefore clinical trials in which the primary outcome is the evaluation of motor or sensory recovery, the research group should consider performing multiple follow-ups in order to avoid confounding factors that can affect the real result of an intervention. However, if multiple follow-ups are not feasible the research group should consider extending the time between the end of intervention and the last follow-up for more than 3 months.

Finally, the number of follow-ups is important to be considered since the great majority of studies just followed up their participants in one single visit (71.8%) while four studies (3.64%) used

Table 7
Follow-up duration

Follow-up duration	Number of studies	%
End of intervention	22	20
<1 week	10	9.09
>1 weeks and <2 weeks	4	3.64
>2 weeks and <1 month	3	2.72
>1 month and <6 months	22	20
>6 months and 1 year	27	24.55
>1 year to 2 years	7	6.36
>2 years	4	3.64
No follow-up	11	10

Table 8
Number of follow-up visits

Number of follow-up visits	Number of studies	%
1 visit	79	71.81
2 visits	4	3.64
>2 visits	5	4.55
No follow-up	22	20

two follow-up visits and five studies followed their subjects in more than two visits (2.55%) (Table 8). The studies that followed participants in multiple time point aimed to asses long term effects of the evaluated intervention.

8 Intervention: Drug/Device/Procedure

In this section, we discuss types of intervention observed in our search. As discussed before, SCI patients experience a wide variety of symptoms such as neuropathic pain, hyperalgesia or anesthesia as well as a multiple sequels such as paresis or hemiplegia. Due to this, a broad group of approaches can be studied and different types of interventions (i.e., drugs, devices, or procedures) can be applied (Graph 2). In one hand, the research group can focus on working with subjects that had a recent injury to understand and evaluate the response to novel therapies aiming at neuroprotection

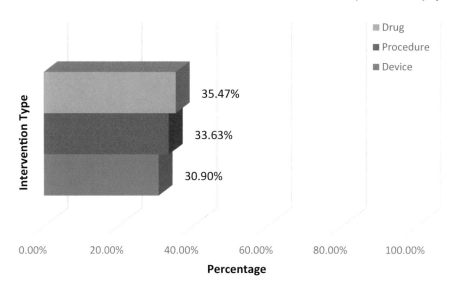

Graph 2 Types of intervention

that had being already tested in animal studies. In the other hand, the efforts can be directed towards a subjects with chronic spinal cord injury in which training programs, robotic devices can be studied for enhancing plasticity and potentially motor recovery.

Among all the studies, 15 (13.6%) evaluated the effects of the intervention for the treatment of pain (chronic, neuropathic). These trials assessed the effects of either drugs (i.e., pregabalin; carbamazepine; nicotine patches; escitalopram) or devices such as tDCS, TMS, TENS, and CES. The remaining amount of studies using drug as interventions evaluated its effects for the management and improvement of spasticity (Lithium, Riluzole) as well as prevention of bone demineralization due to long term immobility (Zolendronic acid). Additionally, some trials evaluated safety and efficacy of novel rehabilitation therapies with autologous bone marrow cells, isolated monocytes injection, olfactory mucosa cells grafts. Moreover, one particular trial evaluated the safety and tolerability of escalating doses of the medication BA—201 in order to define the optimal dose for future management of motor recovery [56].

Furthermore, it is an important to focus on rehabilitation after a SCI; therefore, the development of devices and training programs (procedures) aiming to facilitate motor and sensory recovery is still a focus of interest and great amount of the available resources should be applied.

For intervention with devices, as discussed above, noninvasive brain stimulation techniques (NIBS) have been gaining great attention recently despite the fact of being available for decades. The use of NIBS techniques for the management of pain is well established [57, 58]; therefore, finding multiples trials in SCI population with this kind of intervention was not surprising. In addition, multiple studies evaluated the effects of devices such as

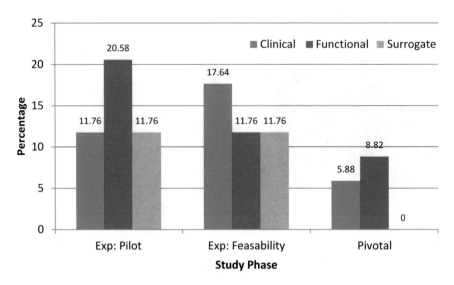

Graph 3 Device and types of outcomes (%)

treadmills, robotic body weight supported treadmill-training programs versus occupational therapy training programs indexing changes in motor recovery and spasticity. Even though we found three clinical trials that used tilt reclining chairs as intervention it is significant to point out its importance given that simple aspects such as the degree of positioning of a subject with SCI could potentially prevent the development of pressure ulcers as well as increase blood flow in certain skin areas.

The SCI researcher needs to consider the differences when designing a trial with medical devices versus a trial with drugs or a trial with a behavioral intervention. The following issues need to be considered: (1) Blinding could be easier in a drug trial when compared with devices or procedures; however, most of devices have sham mechanisms that help during this task; (2) Adherence is also higher in studies that require a one-time intervention (i.e., surgical procedure or one single tDCS session) when compared with trials requiring multiple visits; (3) In the same way, adherence to long term usage of a medication can potentially decrease adherence in a study; (4) the inclusion and exclusion criteria selection is also dependent on the type of medication, procedure or device tested.

9   Outcomes

In this section we discuss a number of endpoints that were most frequently used by the researches groups in the 110 reviewed articles.

Outcomes applied in a clinical study are usually classified in three separate groups:

- *Clinical Outcomes*: this type of outcome focus in providing direct measurements on how the participant feel on respect to diverse assessed measurements (i.e., vas pain). They are valuable measures that gives a clear perspective on how the patients perceives the effects of an intervention and in the same way allows to evaluate the efficacy of it.

- *Functional Outcomes*: This type of outcome plays an important role in SCI clinical trials because it provides direct measurements of functional changes that the participant experience in activities of the daily living (i.e., walking speed).

- *Surrogate Outcomes*: This type of outcome evaluates the clinical status of the participant throughout indirect measurements (i.e., MRI of the spinal cord) which provide valid estimates that could be faster and easier to perceive.

Overall, for device studies, there was a balanced choice of outcomes between these three classes for the preliminary studies. However, as expected, pivotal trials chose only clinical and functional outcomes (Graph 3). The most commonly used clinical outcomes were the visual analogue scale for pain (VAS), side effects questionnaires and adverse event reports; for the case of functional outcomes the most common were the ASIA scale, the Ashworth scale for spasticity and walking performance measures (i.e., walking speed, walking distance).

When choosing the outcome, the SCI investigator needs to consider the duration of effects of the intervention being tested. For instance, ASIA scale should only be used when it is expected that a large effect is observed, then likely over a long period of time. Even though the reliability of the ASIA scale is questionable, in this review we found nine studies (8.18%) that used changes in the ASIA scales as primary outcomes independent on the time of onset of the spinal cord injury.

Interesting, a good number of studies used simple measurements such as walking speed and walking parameters such as 10 m walking time or transfer time (functional outcomes). It is important that the investigator considers choosing an outcome that has been used in other studies as to compare the results.

In addition to that, 15 studies aimed to evaluate the effects of determined intervention to relief neuropathic pain. Among those, most of them (80%) used the visual analogue scale for pain (VAS), while two (13.3%) of these studies use the numerical rating scale for pain (NRS) and the remaining utilized chronic grade pain scale.

Drug trials used on the other hand more clinical outcomes as compared to device trials. There are few reasons to explain this difference, but maybe the most important is the more diffuse effects of drugs as compared to devices that may have a more focal effect and therefore allowing the use of surrogate outcomes more often.

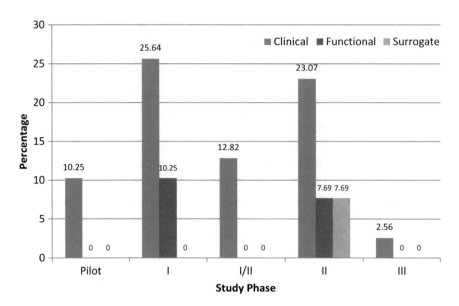

Graph 4 Drugs and types of outcome (%)

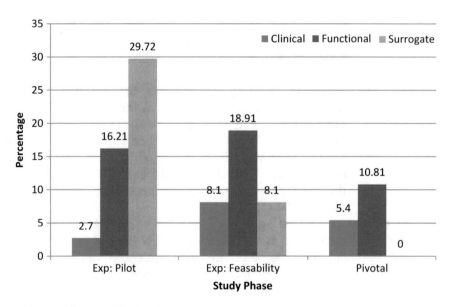

Graph 5 Procedure and types of outcome (%)

This is an interesting area of development: better surrogate outcomes in SCI clinical trials (Graph 4).

Similarly to device trials, trials testing procedures (such as olfactory ensheathing cell transplantation, Olfactory Mucosa Autografts application or bone marrow mesenchymal cell transplantation among others) also used in a good number of times, surrogate outcomes such as magnetic resonance imaging changes, skin perfusion Doppler and bone mass density [17, 18, 59, 60] (Graph 5).

10   Sample Size

In this section, we focus on analyzing and discussing the most relevant aspects related with the adequate selection of sample size. For any clinical research, determining the adequate sample size is one of the most relevant steps.

One of the parameters that it is used to calculate the sample size is the expected effect size, which in SCI trial design could be challenging to interpreted and in the same way to apply. As discussed before, SCI patients present some degree of spontaneous recovery; especially during the first 3 months after injury [12]. This is critical when it comes to estimate the expected effect size of a treatment, since most studies utilize ASIA as a primary outcome. In this context, the International Campaign for Cure of Spinal Cord Injury Paralysis (ICCP) established guidelines for statistical power and sample size calculations on clinical trials in SCI [12]. This document suggests that the greater the time after injury the greater the power of an interventional study. Consequently, studies that include patients with long-term injury require smaller sample sizes. In contrast, trials designed to test early interventions require relatively more subjects to achieve positive results [12].

In 21 out of 110 (19%) articles, the method utilized to perform the sample size calculation was described. Among those, ten clinical trials gave complete information, detailing aspects such as alpha, power, control group expected effect, active group expected effect and attrition. The data of the 21 articles is summarized in Table 9.

Overall, we found that the great majority (88.2%) utilized 0.5 alpha level of significance; however we found two trials in which a smaller level of alpha was applied (0.025 and 0.1) [33, 61]. In fact this can be expected for trials that aim to be more strict as well as to provide a better estimate that the results are not due to chance [33, 61]. Furthermore, we also found that for the 17 trials reporting the sample size calculation, 15 of them utilized a power of 80%, which in fact is commonly used in other trials.

In addition, to calculate the expected effect size for a study, the research group is required to analyze the difference between control group response and the expected effect for the intervention that will be tested. For this task, the principal source utilized to collect the data for expected effect of a control group came from preliminary data; in the other hand, the active group expected effect values were based on previous literature and pilot studies.

11   Analysis and Stoppage Criteria

In this section, we focused in understanding the mechanism applied to perform interim analysis and the stoppage criteria found in the review of SCI clinical trials.

Table 9
Sample size calculations

	Total
Alpha	*17*
0.05	15
Two-sided	3
One-sided	2
Not specified	10
0.25 one-sided	1
0.10	1
Power	*17*
80%	12
90%	3
Other	2
Control group expected effect	*21*
Observational data	1
Preliminary data/pilot study	13
Previous clinical trials	2
Not reported	5
Active group expected effect	*21*
Minimally clinical significance	1
Observational data	1
Pilot studies	4
Previous literature	8
Previous clinical trials	2
Not reported	5
Attrition	*21*
10–20%	4
>20–30%	3
>30–60%	3
Not specified	11

Data monitoring during the development of a trial can enhance the efficiency and affect the results. Therefore it is ethically suggested that if there is substantial evidence that one of the interventions given has great superiority among the others or if the intervention is harmful to the subjects participating in the study an early termination or change in trial design is expected [62, 63].

Given that, interim analysis are performed usually when about 50% of the participants have been randomized and completed at least one follow-up. This task should be performed by and independent statistician that must be blinded for the treatment allocation and afterwards a data and safety monitoring committee will analyze and decide the pertinent next steps for the trial [62, 64].

Among all the 110 trials analyzed none of them provide a clear specification about their analysis and stoppage criteria, even though this finding may look surprising, it is somehow expected because usually a more complete explanation about interim analysis and stoppage criteria is provided in the study protocol and not as frequent in the published clinical trial. Nevertheless, the consort guidelines strongly advise to add a detailed paragraph providing this information [13].

Particularly for this chapter above all the 110 reviewed clinical trials none of the studies reported plans to stop the study due to safety reasons or describe an interim analysis even though some of the experimental treatments used could potentially be harmful. Overall, there were no severe adverse events for most of the therapeutical interventions. Among all the studies, one clinical trial stopped early due to lack of funding having as a consequence a smaller sample size than the one expected [35].

12   Randomization

In this section we discuss the most common randomization methods implemented in the 110 reviewed SCI clinical trials, in particular, we discuss the type of randomization, randomization sequence generation method and we analyze some aspects of the allocation concealment.

Randomization is the allocation process of participants in the particular arms of an intervention, more important is that this process must be completed by chance. There are different methods to randomize and each one will be useful to specific trial designs and determined sample sizes. Simple randomization is a method comparable to the one produced when you flip a coin in which you could have heads or tails; in the same way, a subject could be allocated to either the treatment or placebo group. It is important to point out that in this type of randomization the sample size required must be large to assure adequate balance between groups. The second possible procedure it is known as block randomization. In this type of randomization instead of randomizing each participant individually, the research group randomizes groups of participants assuring a certain fixed amount of participants allocated in each group, therefore a balanced population in each arm of the study. Even though this method ensures that a same amount of participants are in each group, characteristics between group can vary widely, having said that, the next available step would be stratify each group by characteristics that will help to balance the groups, this process is known as stratified randomization [65–69].

In our search we found that over the 110 articles, 72 of them were randomized (65.4%). There is a general lack of reporting of randomization methods. From the few that have reported the

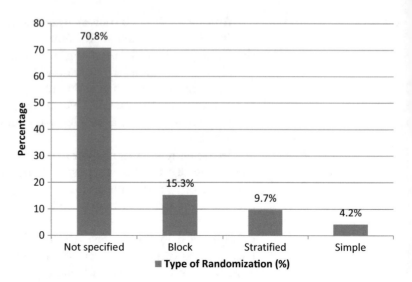

Graph 6 Types of randomization

randomization methods (20 out of 72), most used blocked randomization. This randomization is especially important for small sample size trials, which has been a characteristic of SCI trials. There was a good relative number of stratified randomization. In this case, an important issue is which stratification factor should be chosen. Most of the trials chose one factor (i.e., enrollment center, ulcer location, Asia score) and one additional study had two factors (gender and severity of injury) (Graph 6). Only few studies used the simple randomization method as this method is not valid for small sample sizes due to the risk of imbalance.

There were also a good number of studies that did not have randomization (38 out of 110 trials). These studies included surgical procedures with limitations to perform an adequate randomization or ethical considerations to randomize to other treatments.

12.1 Allocation Concealment

Despite the adequate generation of a randomization list, a common pitfall of clinical studies is an inadequate method to prevent that both the researcher that provides or assesses the intervention and the participant remain blinded to the specific intervention applied. An adequate allocation concealment ensures that participants remain blinded to the intervention, therefore results are less biased [67, 70, 71]. Among the 110 analyzed studies only 16 reported the allocation concealment method. The most commonly used (and also easy method) is the use of sealed opaque envelopes.

12.2 Randomization Sequence Generation

Another important issue is the generation of the randomization list. Most of the SCI trials used software programs such as clorandm.exe, PTClinResNet, randperm.m function in Matlab, Epidat 3.1 statistical software [71].

13   Blinding

In this section, we focus on discussing the relevance of blinding and blinding methods utilized most frequently in our review of SCI clinical trials. This aspect of the design of a trial is critical in order to avoid observer and also performance bias. A research group should ensure that both the participant and if possible the assessor are unaware of the intervention applied for each subject [72, 73].

It is remarkable but not surprising the large amount of non-blinded studies in spinal cord research. In this search we found 43 open label clinical trials; of those, 36 were nonrandomized (83.7%), while 5 were randomized two parallel arms clinical trial in which a proper blinding was challenging due to the special characteristic of the intervention. As an example Norrbrink et al. 2011 evaluated the effects of acupuncture vs. massage therapy using VAS pain as the primary outcome measure, the major differences between the intervention made the blinding not feasible[74]. Due to these difficulties encountered while blinding some of interventions it is common to find a great number of nonblinded studies. Moreover, we also found one unblinded RCT with three parallel arms and one RCT with a factorial design in which the research group studied the differences in skin perfusion over the ischial tuberosity after exposure to a wide variety of degrees of inclination in a tilt wheelchair [75]. Besides that, our search showed that of the blinded studies, 24 where double blind and 32 single blinded (Graph 7).

These finding were expected, especially because a greater amount of clinical trials found in this search had training therapies or rehabilitation techniques that are difficult to mask for either the subject or the co investigator applying the intervention. However, adding a

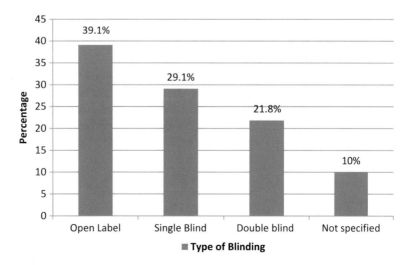

Graph 7 Types of blinding

blinded researcher to analyze and collect the data can help to avoid bias. In fact for the case of surgical procedures it is encouraged to utilize sham surgeries; however, the risk-benefit percentage should be addressed in a case by case manner. For the case of the research group, despite the necessity of having the surgeon unblinded, other integrants of the group such as surgical assistant, nurse and postsurgical team could be blinded without compromising the health of the patient [76]. Behavioral therapies remain as one of the hardest intervention to blind and specific protocols should be developed in order to provide an effective blinding method [77].

13.1 Assessment of Blinding

Although it is controversial, assessment of blinding has been performed in few trials. Out of the 110 of the reviewed studies just only 4 studies specified if the integrity of the blinding methods was assessed. A study by Boswell-Ruys et al. asked the participants to avoid discussing their training or group allocation with the assessor of the study and the authors reported that the success of the blinding was recorded after the last assessment for each subject but there was no clear specification of the method implemented assess it [46]. In addition another study by Richardson et al. utilizing nicotine patches vs. placebo developed a small questionnaire for the patient to specify in which group they believed to be part of [28].

Most of studies assessing blinding did so through the use of questionnaires. These questionnaires varies but most of them include simples questions like (1) do you think you received sham or active or do not know? Some questionnaires also asked about the confidence of guessing utilizing numeric scales in order to rate the blinding efficacy. An important aspect is to avoid introducing bias when using these questionnaires as simply asking the subjects may prime subjects to be more attentive and try to guess especially in a cross over trials.

14 Adherence: Recruitment Yield

Adherence in SCI is a topic that deserves special attention when designing a trial with SCI. These individuals have difficulties with locomotion; therefore transportation to a research center is challenging. In addition other factors such as comorbidities and the priority to solve them was a relevant aspect that decreased adherence rates in this population.

Overall for this review, we found three clinical trials that specified the methods that were implemented to improve the adherence. Tables 10 and 11 summarize the findings regarding adherence and the phase of clinical trials.

The most important factor to increase adherence was the economic compensation offered to each participant once enrolled in the trial. In one of the clinical trials the subjects were compensated

Table 10
Adherence and phase of drug trials

Adherence	Pilot	I	I/II	II	III	IV
0–50%	0	0	0	0	0	0
50–69%	0	1	0	2	1	0
70–89%	2	0	1	6	0	0
90–100%	2	7	3	2	0	0
N/S	0	6	1	5	0	0

Table 11
Adherence and phases of device and procedure trials

Adherence	Expl: pilot	Pivotal	Postmarketing
0–50%	1	1	0
50–69%	3	4	0
70–89%	15	4	0
90–100%	25	1	0
N/S	16	1	0

with 20 dollars vouchers to reimburse the travel costs from each visit, besides that subjects were contacted weekly by telephone for one of the co investigators in order to increase compliance [14]. Even though this method requires a great amount of time and dedication, it is a cost effective way to strengthen the relationship between co investigator and subjects, as well as increase adherence during the entire intervention. Additionally a randomized study by Otomanelli et al. with a follow-up visit a year after the beginning of the study showed an adherence rate of 87.6%, a value considered adequate especially for a study with a long-term follow-up [78]. During this study the intervention applied was an Evidence-based practice supported employment method in which SCI patient were supported in their job throughout an entire year, therefore receiving a constant income during the entire time of the study. Another study used a similar payment method than the one described before in which each subject received two 50 dollars paychecks covering the entire study participation; the first was given during baseline visit, right after the enrollment and the second one during the assessment/follow-up visit, 4 weeks post intervention. This type of compensation method may increase the subject's probability to complete the intervention and follow-ups required [79]. This evidence shows that some types of monetary compensation are important to enhance adherence.

Another method to improve adherence is allowing subjects to continue with their treatment plan (which may also be needed for ethical reasons). A trial by Heutink et al. evaluating pain indexed by Chronic Pain Grade Questionnaire (CPG) in SCI subjects with chronic neuropathic pain, allowed the subjects to continue with their current pain treatment but also requested them to avoid modifying those treatments regimens during the intervention of the study [80].

Acknowledging the previously mentioned cases it is important to recognize how similar methods to improve adherence can have different responses. Indeed it is even more important to understand how essential is to plan in advance strategies to maintain a good adherence during the entire course of a clinical trial in subjects with spinal injury.

15   Statistical Analysis

The international committee of medical journal editors in 1988 published guidelines for an adequate presentation of statistical analysis in any clinical article [81]. In fact what they were aiming was to have future clinical trials that could report their statistical methods with enough detail to enable a knowledgeable reader to verify and understand the veracity of the results reported [81, 82]. In the same way it is important for a research group if possible to quantify the corresponding findings and show them with appropriate indicators for measurement error or uncertainty. Moreover it is also relevant to discuss the eligibility of the experimental subjects to clarify and exclude possible bias. In summary the study group should add a detailed description of the statistical methods and data reported in the results section of the final article [81]. Graph 8 summarizes all the statistical analysis used by these authors.

The great majority of trials applied an ANOVA in order to perform analysis of the primary outcomes of the study comparing baseline measures with post intervention as well as with follow-up findings. The second most common statistical analysis test was the t-test, these clinical trials usually evaluated functional outcomes after interventions such as aerobic training or lokomat training.

For nonparametrical sample distribution the preferred statistical analysis method was Wilcoxon sign rank or sum test (or Mann–Whitney), most of these trials analyzed clinical outcomes such as VAS for pain in subjects with neuropathic pain after chronic SCI comparing baseline findings with postintervention results.

For categorical outcome, most of the clinical trials utilized chi square in order to compare respose in either functional or clinical type of outcome before and after the intervention. It's important to point out that some trials categorized scales that are normally considered as continue such as VAS for pain or Ashworth scale for spasticity in order to decrease the sample variability.

In fact, most of the SCI trials utilized simple statistics to analyze the data. In addition, only one trial analyzed survival function utilizing Kaplan–Meier estimator considering the survival probabilities to a new event to occur. During that trial, the research group evaluated the effects of an intervention with Intermittent Catheterization With a Hydrophilic-Coated Catheter measuring recurrent urinary tract infections as the event [83].

16 Limitations

In this section we discuss the major limitations encountered and reported in the reviewed articles. For every research group the limitation section aim to detail the probable pros and cons that they encountered during the development of the study, in the same way discuss the aspects that can limit the validity of the results obtained. It is in fact extremely important to discuss weakness of the study design that could interfere with reliable measurement of outcomes and therefore with the results [13].

Overall, a small sample size was the most common limitation reported in the reviewed articles, up to 27 clinical trials discussed that in order to provide more reliable results that consequently could be generalized to the global SCI population, the sample size required should be greater; therefore, encouraging future research groups to develop further studies with a bigger number of participants.

In addition to that, four studies considered that the length of the intervention was not enough to provide the expected effect size. Of those four studies, two highlighted this important limitation. The first one evaluated the effects on endurance indexed by

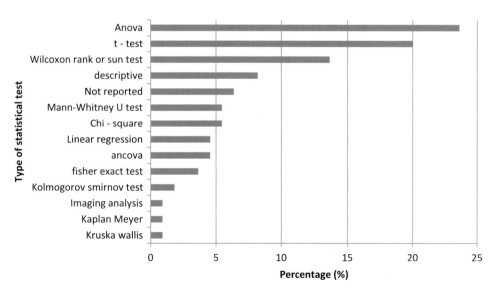

Graph 8 Types of statistical test

Plasmatic levels of total antioxidant as well as erythrocyte glutathione peroxidase activity after 12 weeks of arm cracking exercise therapy, in this clinical trial the research group considered that longer trials with longer intervention period will provide stronger results [84]. The second study developed by the same research group applied the same intervention (arm cracking exercise), however the main outcome evaluated was the measurement of specific biochemical markers [85].

Furthermore, among the other frequent limitations reported, we found three clinical trials which were nonblinded, three studies in which the large number of dropouts was the main weakness, two trials without control group. Additionally, three studies were nonrandomized, this in fact decreased the study results credibility as well as two extra studies in which the outcomes were self-reported providing subjective results. Moreover, we analyzed a trial that recruited as participants, a specific population of male veterans; therefore, the observed results were not applicable and generalizable to the SCI population [78]. This is an important factor to take into consideration while developing a SCI trial due to the great diversity of affected subjects especially due to the wide degree of severity and locations f the injuries.

Table 12
Study limitations

Limitation	N
Small sample size	27
Short duration of intervention	4
Lack of blinding	3
Large amount of drop outs	3
Lack of randomization	3
Lack of control group	3
Lack of long-term follow-up	2
Outcomes were self-reported	2
Not generalizable population	2
The device used to measure the outcome varied from baseline to end of intervention	1
Lack of follow-up	1
No control of lifestyle factor	1
Differences between control group and intervention group were to high	1
Intervention was in one institution and control group in another institution	1

The complete list of limitations is summarized in Table 12. To finalize, is important for a research group to consider a broad amount of possible limitations and encourage them to think ahead to prevent that major weakness can affect the execution of a properly designed SCI clinical trial.

17   Conclusions

Reviewing the most cited clinical trials in SCI, and thus those with greater impact, it is clear that that are significant limitations with these trials and the investigator should carefully consider the special challenges when designing a SCI trial. We highlight some of these challenges:

- Adherence: this is a special challenge that needs to be carefully considered. Still from the trials analyzed, there was no particular and novel method that seem to address this issue effectively.

- Small sample size trials: when designing a small sample size trial, the investigators needs to be aware of the power of their sample size to respond the question they are asking. If it is not possible to access a larger sample, the investigators need to consider the use of other outcomes such as surrogate outcomes.

- Blinding: given the large number of trials testing behavioral interventions or surgical procedures, blinding may not be possible. In this case, the investigator needs to consider other methods such as use of surrogate or objective clinical markers, which indeed were remarkably not used frequently in the trials reviewed here.

As in other neurological areas, there is an unmet clinical need for the development of novel interventions in SCI field. We hope that this chapter will help the investigator to identify the challenging areas of this field as to design the most optimal clinical trial and thus advance this area of research.

References

1. Cheriyan T, Ryan DJ, Weinreb JH et al (2014) Spinal cord injury models: a review. Spinal Cord 52:588–595. https://doi.org/10.1038/sc.2014.91

2. McDonald JW, Sadowsky C (2002) Spinal-cord injury. Lancet 359:417–425. https://doi.org/10.1016/S0140-6736(02)07603-1

3. Sadowsky C, Volshteyn O, Schultz L, McDonald JW (2002) Spinal cord injury. Disabil Rehabil 24:680–687. https://doi.org/10.1080/09638280110110640

4. Sköld C, Levi R, Seiger A (1999) Spasticity after traumatic spinal cord injury: nature, severity, and location. Arch Phys Med Rehabil 80:1548–1557. https://doi.org/10.1016/S0003-9993(99)90329-5

5. Devivo MJ (2012) Epidemiology of traumatic spinal cord injury: trends and future implications. Spinal Cord 50:365–372. https://doi.org/10.1038/sc.2011.178

6. White JP, Thumbikat P (2012) Acute spinal cord injury. Surgery 30:326–332. https://doi.org/10.1016/j.mpsur.2012.05.005

7. Proctor MR (2002) Spinal cord injury. Crit Care Med 30:S489–S499. https://doi.org/10.1097/01.CCM.0000034132.14832.6D

8. Dietz V, Fouad K (2014) Restoration of sensorimotor functions after spinal cord injury. Brain 137:654–667. https://doi.org/10.1093/brain/awt262

9. van den Brand R, Heutschi J, Barraud Q et al (2012) Restoring voluntary control of locomotion after paralyzing spinal cord injury. Science 336:1182–1185. https://doi.org/10.1126/science.1217416

10. Van Den Berg MEL, Castellote JM, Mahillo-Fernandez I, De Pedro-Cuesta J (2010) Incidence of spinal cord injury worldwide: a systematic review. Neuroepidemiology 34:184–192. https://doi.org/10.1159/000279335

11. Reier PJ, Lane MA, Hall ED et al (2012) Translational spinal cord injury research: preclinical guidelines and challenges. Spinal Cord Inj 109:411–433. https://doi.org/10.1016/B978-0-444-52137-8.00026-7

12. Fawcett JW, Curt A, Steeves JD et al (2007) Guidelines for the conduct of clinical trials for spinal cord injury as developed by the ICCP panel: spontaneous recovery after spinal cord injury and statistical power needed for therapeutic clinical trials. Spinal Cord 45:190–205. https://doi.org/10.1038/sj.sc.3102007

13. Schulz KF, Altman DG, Moher D, Group C (2010) CONSORT 2010 Statement: updated guidelines for reporting parallel group randomised trials. BMC Med 8:18. https://doi.org/10.1186/1741-7015-8-18

14. Tan G, Holmes SA (2010) Effect of Dronabinol on Central neuropathic pain after spinal cord injury: a pilot study. Am J Phys Med Rehabil 89:840–848. https://doi.org/10.1097/PHM.0b013e3181f1c4ec

15. Jia C, Liao L-M, Chen G, Sui Y (2013) Detrusor botulinum toxin A injection significantly decreased urinary tract infection in patients with traumatic spinal cord injury. Spinal Cord 51:487–490. https://doi.org/10.1038/sc.2012.180

16. Pal R, Venkataramana NK, Jaan M et al (2009) Ex vivo-expanded autologous bone marrow-derived mesenchymal stromal cells in human spinal cord injury/paraplegia: a pilot clinical study. Cytotherapy 11:897–911. https://doi.org/10.3109/14653240903253857

17. Lima C, Pratas-Vital J, Escada P et al (2006) Olfactory mucosa autografts in human spinal cord injury: a pilot clinical study. J Spinal Cord Med 29(3):191–203

18. Féron F, Perry C, Cochrane J, Licina P et al (2005) Autologous olfactory ensheathing cell transplantation in human spinal cord injury. Brain 128:2951–2960. https://doi.org/10.1093/brain/awh657

19. Knoller N, Auerbach G, Fulga V, Zelig G, Attias J, Bakimer R, Marder JB, Yoles E, Belkin M, Schwartz M, Hadani M (2005) Clinical experience using incubated autologous macrophages as a treatment for complete spinal cord injury: phase I study results. J Neurosurg Spine 3:173–181

20. Karamouzian S, Nematollahi-mahani SN, Nakhaee N (2012) Clinical safety and primary efficacy of bone marrow mesenchymal cell transplantation in subacute spinal cord injured patients. Clin Neurol Neurosurg 114:935–939. https://doi.org/10.1016/j.clineuro.2012.02.003

21. Theiss RD, Hornby TG, Rymer WZ, Schmit BD (2011) Riluzole decreases flexion withdrawal reflex but not voluntary ankle torque in human chronic spinal cord injury. J Neurophysiol 105:2781–2790. https://doi.org/10.1152/jn.00570.2010

22. Wong YW, Tam S, So KF et al (2011) ORIGINAL ARTICLE: A three-month, open-label, single-arm trial evaluating the safety and pharmacokinetics of oral lithium in patients with chronic spinal cord injury. Spinal Cord 49:94–98. https://doi.org/10.1038/sc.2010.69

23. Yang ML, Li JJ, So KF et al (2012) Efficacy and safety of lithium carbonate treatment of chronic spinal cord injuries: a double-blind, randomized, placebo-controlled clinical trial. Spinal Cord 50:141–146. https://doi.org/10.1038/sc.2011.126

24. Pooyania S, Ethans K, Szturm T et al (2010) A randomized, double-blinded, crossover pilot study assessing the effect of nabilone on spasticity in persons with spinal cord injury. Arch Phys Med Rehabil 91:703–707. https://doi.org/10.1016/j.apmr.2009.12.025

25. Steeves JD, Lammertse D, Curt A et al (2007) Guidelines for the conduct of clinical trials for spinal cord injury (SCI) as developed by the ICCP panel: clinical trial outcome measures. Spinal Cord 45:206–221. https://doi.org/10.1038/sj.sc.3102008

26. Siddall PJ, Cousins MJ, Otte A et al (2006) Pregabalin in central neuropathic pain associated with spinal cord injury: a placebo-controlled trial. Neurology 67(10):1792–1800

27. Cardenas DD, Nieshoff EC, Whalen E, Scavone JM (2013) A randomized trial of pregabalin in patients with neuropathic pain due to spinal cord injury. Neurology 80(6):533–539

28. Richardson EJ, Ness TJ, Redden DT et al (2012) Effects of nicotine on spinal cord injury pain vary among subtypes of pain and smoking status: results from a randomized, controlled experiment. J Pain 13:1206–1214. https://doi.org/10.1016/j.jpain.2012.09.005

29. Grijalva I, Guizar-sahag G, Casta G (2010) High doses of 4-aminopyridine improve functionality in chronic complete spinal cord injury patients with MRI evidence of cord continuity. Arch Med Res. https://doi.org/10.1016/j.arcmed.2010.10.001

30. Cardenas DD, Ditunno JF, Graziani V et al (2013) Two phase 3, multicenter, randomized, placebo-controlled clinical trials of fampridine-SR for treatment of spasticity in chronic spinal cord injury. Spinal Cord 52:70–76. https://doi.org/10.1038/sc.2013.137

31. Medina LA., Wysk RA, Okudan Kremer GE (2011) A review of design for X methods for medical devices: the introduction of a design for FDA approach. In: Vol. 9 23rd Int. Conf. Des. Theory Methodol. 16th Des. Manuf. Life Cycle Conf. pp 849–861

32. Burlington DB (1996) FDA regulation of medical devices. FDA perspective. Ann Thorac Surg 61:482–484; discussion 484–498

33. Fregni F, Boggio PS, Lima MC et al (2006) A sham-controlled, phase II trial of transcranial direct current stimulation for the treatment of central pain in traumatic spinal cord injury. Pain 122:197–209. https://doi.org/10.1016/j.pain.2006.02.023

34. Soler MD, Kumru H, Pelayo R et al (2010) Effectiveness of transcranial direct current stimulation and visual illusion on neuropathic pain in spinal cord injury. Brain 133:2565–2577. https://doi.org/10.1093/brain/awq184

35. Tan G, Rintala DH, Jensen MP et al (2011) Efficacy of cranial electrotherapy stimulation for neuropathic pain following spinal cord injury: a multi-site randomized controlled trial with a secondary 6-month open-label phase. J Spinal Cord Med 34:285–296

36. Wrigley PJ, Gustin SM, Mcindoe LN et al (2013) Longstanding neuropathic pain after spinal cord injury is refractory to transcranial direct current stimulation: a randomized controlled trial. Pain 154:2178–2184. https://doi.org/10.1016/j.pain.2013.06.045

37. Celik EC, Erhan B, Gunduz B, Lakse E (2013) The effect of low-frequency TENS in the treatment of neuropathic pain in patients with spinal cord injury. Spinal Cord 51:334–337. https://doi.org/10.1038/sc.2012.159

38. Müller-Putz GR, Daly I, Kaiser V (2014) Motor imagery-induced EEG patterns in individuals with spinal cord injury and their impact on brain-computer interface accuracy. J Neural Eng 11:35011. https://doi.org/10.1088/1741-2560/11/3/035011

39. Jetté F, Côté I, Meziane HB, Mercier C (2013) Effect of single-session repetitive transcranial magnetic stimulation applied over the hand versus leg motor area on pain after spinal cord injury. Neurorehabil Neural Repair 27:636–643. https://doi.org/10.1177/1545968313484810

40. Thomas SL, Gorassini MA, Sarah L (2005) Increases in corticospinal tract function by treadmill training after incomplete spinal cord injury. J Neurophysiol 94:2844–2855. https://doi.org/10.1152/jn.00532.2005.

41. Winchester P, Mccoll R, Querry R et al (2015) Changes in supraspinal activation patterns following robotic locomotor therapy in motor-incomplete spinal cord injury. Neurorehabil Neural Repair 19:313–324. https://doi.org/10.1177/1545968305281515

42. Zariffa J, Kapadia N, Kramer JLK et al (2011) Feasibility and efficacy of upper limb robotic rehabilitation in a subacute cervical spinal cord injury population. Spinal Cord 50:220–226. https://doi.org/10.1038/sc.2011.104

43. Johnston TE, Modlesky CM, Betz RR, Lauer RT (2011) Muscle changes following cycling and/or electrical stimulation in pediatric spinal cord injury. Arch Phys Med Rehabil 92:1937–1943. https://doi.org/10.1016/j.apmr.2011.06.031

44. Lauer RT, Smith BT, Mulcahey MJ et al (2011) Effects of cycling and/or electrical stimulation on bone mineral density in children with spinal cord injury. Spinal Cord 49:917–923. https://doi.org/10.1038/sc.2011.19

45. Roth EJ, Stenson KW, Powley S et al (2010) Expiratory muscle training in spinal cord injury: a randomized controlled trial. Arch Phys Med Rehabil 91:857–861. https://doi.org/10.1016/j.apmr.2010.02.012

46. Boswell-Ruys CL, Harvey LA, Barker JJ et al (2009) Training unsupported sitting in people with chronic spinal cord injuries: a randomized controlled trial. Spinal Cord 48:138–143. https://doi.org/10.1038/sc.2009.88

47. Mcbain RA, Boswell-Ruys CL, Lee BB et al (2013) Abdominal muscle training can enhance cough after spinal cord injury. Neurorehabil Neural Repair 27:834–843. https://doi.org/10.1177/1545968313496324

48. Bakkum AJT, de Groot S, van der Woude LHV, Janssen TWJ (2012) The effects of hybrid cycle training in inactive people with long-term spinal cord injury: design of a multicenter randomized controlled trial. Disabil

Rehabil 35:1–6. https://doi.org/10.3109/09
638288.2012.715719

49. Nussbaum EL, Flett H et al (2013) Ultraviolet-C irradiation in the management of pressure ulcers in people with spinal cord injury: a randomized, placebo controlled trial. Arch Phys Med Rehabil 94:650–659. https://doi.org/10.1016/j.apmr.2012.12.003

50. Nussbaum EL, Biemann I, Mustard B (1994) Comparison of ultrasound/ultraviolet-C and laser for treatment of pressure ulcers in patients with spinal cord injury. Phys Ther 74:812–823; discussion 823–825

51. Kirshblum SC, Waring W, Biering-Sorensen F et al (2011) Reference for the 2011 revision of the international standards for neurological classification of spinal cord injury. J Spinal Cord Med 34:547–554. https://doi.org/10.1179/107902611X13186000420242

52. Ho CH, Wuermser LA, Priebe MM et al (2007) Spinal cord injury medicine. 1. Epidemiology and classification. Arch Phys Med Rehabil. https://doi.org/10.1016/j.apmr.2006.12.001

53. El Masry WS, Tsubo M, Katoh S et al (1996) Validation of the American Spinal Injury Association (ASIA) motor score and the National Acute Spinal Cord Injury Study (NASCIS) motor score. Spine (Phila Pa 1976) 21:614–619. https://doi.org/10.1097/00007632-199603010-00015

54. Takahashi H, Yamazaki M, Okawa A, Sakuma T (2012) Neuroprotective therapy using granulocyte colony-stimulating factor for acute spinal cord injury: a phase I/IIa clinical trial. Eur Spine J 21:2580–2587. https://doi.org/10.1007/s00586-012-2213-3

55. Casha S, Zygun D, Mcgowan MD et al (2012) Results of a phase II placebo-controlled randomized trial of minocycline in acute spinal cord injury. Brain 135:1224–1236. https://doi.org/10.1093/brain/aws072

56. Fehlings MG, Theodore N, Harrop J et al (2011) A phase I/IIa clinical trial of a recombinant Rho protein antagonist in acute spinal cord injury. J Neurotrauma 28:787–796. https://doi.org/10.1089/neu.2011.1765

57. Fregni F, Freedman S, Pascual-Leone A (2007) Recent advances in the treatment of chronic pain with non-invasive brain stimulation techniques. Lancet Neurol 6:188–191. https://doi.org/10.1016/S1474-4422(07)70032-7

58. Fregni F, Gimenes R, Valle AC et al (2006) A randomized, sham-controlled, proof of principle study of transcranial direct current stimulation for treatment of pain in fibromyalgia. Arthritis Rheum 54:3988–3998. https://doi.org/10.1002/art.22195

59. Bubbear JS, Gall A, Middleton FRI et al (2011) Early treatment with zoledronic acid prevents bone loss at the hip following acute spinal cord injury. Osteoporos Int 22:271–279. https://doi.org/10.1007/s00198-010-1221-6

60. Groah SL, Lichy AM, Libin AV, Ljungberg I (2010) Intensive electrical stimulation attenuates femoral bone loss in acute spinal cord injury. PM R 2:1080–1087. https://doi.org/10.1016/j.pmrj.2010.08.003

61. Domurath B, Kutzenberger J, Kurze I, Knoth HS (2011) Clinical evaluation of a newly developed catheter (SpeediCath Compact Male) in men with spinal cord injury: residual urine and user evaluation. Spinal Cord 49:817–821. https://doi.org/10.1038/sc.2011.14

62. Chan AW, Tetzlaff JM, Altman DG et al (2013) SPIRIT 2013 statement: defining standard protocol items for clinical trials. Ann Intern Med 158:200–207. https://doi.org/10.7507/1672-2531.20130256

63. Sankoh AJ (1999) Interim analyses: an update of an FDA reviewer's experience and perspective. Drug Inf J 33:165–176. https://doi.org/10.1177/009286159903300120

64. Geller NL, Pocock SJ (1987) Interim analyses in randomized clinical trials: ramifications and guidelines for practitioners. Biometrics 43:213–223. https://doi.org/10.2307/2531962

65. Vickers AJ (2008) How to randomize. J Soc Integr Oncol 4:194–198. https://doi.org/10.1016/j.bbi.2008.05.010

66. Schulz KF, Grimes DA (2002) 1-s2.0-S0140673602080297-main. 359:966–970

67. Schulz KF, Grimes DA (2002) Allocation concealment in randomised trials: defending against deciphering. Lancet 359:614–618. https://doi.org/10.1016/S0140-6736(02)07750-4

68. Schulz KF (1995) Subverting randomization in controlled trials. JAMA 274:1456–1458. https://doi.org/10.1001/jama.274.18.1456

69. Schulz KF, Chalmers I, Hayes RJ, Altman DG (1995) Empirical evidence of bias with estimates of treatment effects in controlled trials. Unequal group sizes in randomised trials: guarding against guessing. The Lancet JAMA 273:408–412

70. Pildal J (2005) Comparison of descriptions of allocation concealment in trial protocols and the published reports: cohort study. BMJ 330:1049–1050. https://doi.org/10.1136/bmj.38414.422650.8F

71. Schulz KF, Grimes DA (2002) Generation of allocation sequences in randomised trials: chance, not choice. Lancet 359:515–519. https://doi.org/10.1016/S0140-6736(02)07683-3

72. Schulz KF, Grimes DA (2002) Blinding in randomised trials: hiding who got what. Lancet 359:696–700. https://doi.org/10.1016/S0140-6736(02)07816-9

73. Landscape T (2015) Medical writings

74. Norrbrink C, Lundeberg T (2011) Acupuncture and massage therapy for neuropathic pain following spinal cord injury: an exploratory study. Acupunct Med 29:108–115. https://doi.org/10.1136/aim.2010.003269

75. Jan Y, Liao F, Jones MA et al (2013) Effect of durations of wheelchair tilt-in-space and recline on skin perfusion over the ischial tuberosity in people with spinal cord injury. Arch Phys Med Rehabil 94:667–672. https://doi.org/10.1016/j.apmr.2012.11.019

76. Schulz KF, Chalmers I, Altman DG (2002) The landscape and lexicon of blinding in randomized trials. Ann Intern Med 136:254–9

77. Viera AJ, Bangdiwala SI (2007) Eliminating bias in randomized controlled trials: importance of allocation concealment and masking. Fam Med 39:132–137

78. Ottomanelli L, Goetz LL, Suris A et al (2012) Effectiveness of supported employment for veterans with spinal cord injuries: results from a randomized multisite study. Arch Phys Med Rehabil 93:740–747. https://doi.org/10.1016/j.apmr.2012.01.002

79. Mulroy SJ, Thompson L, Kemp B et al (2011) Strengthening and optimal movements for painful shoulders (STOMPS) in chronic spinal cord injury: a randomized controlled trial. Phys Ther 91:305–324. https://doi.org/10.2522/ptj.20100182

80. Heutink M, Post MWM, Bongers-Janssen HMH et al (2012) The CONECSI trial: results of a randomized controlled trial of a multidisciplinary cognitive behavioral program for coping with chronic neuropathic pain after spinal cord injury. Pain 153:120–128. https://doi.org/10.1016/j.pain.2011.09.029

81. Bailar JC, Mosteller F (1988) Guidelines for reporting in articles for medical journals. Ann Intern Med 108:266–273

82. Yusuf S, Wittes J, Probstfield J, Tyroler HA (1991) Analysis and interpretation of treatment effects in subgroups of patients in randomized clinical trials. JAMA 266:93–98

83. Cardenas DD, Moore KN, Dannels-McClure A et al (2011) Intermittent catheterization with a hydrophilic-coated catheter delays urinary tract infections in acute spinal cord injury: a prospective, randomized, multicenter trial. Pm R 3:408–417. https://doi.org/10.1016/j.pmrj.2011.01.001

84. Ordonez FJ, Rosety MA, Camacho A et al (2013) Arm-cranking exercise reduced oxidative damage in adults with chronic spinal cord injury. Arch Phys Med Rehabil 94:2336–2341. https://doi.org/10.1016/j.apmr.2013.05.029

85. Rosety-Rodriguez M, Camacho A, Rosety I et al (2014) Low-grade systemic inflammation and leptin levels were improved by arm cranking exercise in adults with chronic spinal cord injury. Arch Phys Med Rehabil 95:297–302. https://doi.org/10.1016/j.apmr.2013.08.246

Chapter 12

Peripheral Neuropathy

Rafaelly Stavale, Felipe Jones, Alejandra Malavera, and Felipe Fregni

Abstract

Peripheral neuropathy is a condition that involves different symptoms and can be caused by different pathologies, especially diabetes mellitus. Due to its high incidence and the great number of challenges faced by researchers in this field, this chapter provides an extensive discussion of the 100 most cited clinical trials between 2005 and 2015. A systematic review was performed, based on the CONSORT guidelines, by analyzing Web of Science database as to improve the planning of future trials and addressing the main aspects involved with Peripheral Neuropathy clinical trials.

Key words Peripheral neuropathy, Research, Review, CONSORT guidelines, Methodology

1 Introduction

Peripheral neuropathy (PN) is a disease that affects motor, sensory, and/or autonomic fibers of the peripheral nervous system, which may induce different symptomology depending of the affected fibers. This condition is caused by a variety of etiologies, ranging from diabetes mellitus (DM), tumors, vitamin deficiencies, autoimmune diseases, to medications and alcoholism, among others [1]. The symptoms include weakness, pain, paresthesia, sensory loss, and orthostatic hypotension [1]. In this context, PN can severely compromise daily activities, leading to a reduced quality of life and long-lasting disability [2, 3].

One of the most common causes of peripheral neuropathy is DM type I and II, with 66% and 59% incidence on these populations, respectively [4]. Diabetic peripheral neuropathy (DPN) affects sensory and motor fibers, resulting in atrophy of limb muscles and ulceration. In fact, the progress of PN is the primary risk factor for diabetes foot ulcer, a condition responsible for 85% of lower-extremity amputation in people with PN [1]. Additionally, the increased awareness about the pathophysiology of DPN has led to new investigations on early prevention, symptomatology management, and therapies to slow down the disease progression.

Felipe Fregni (ed.), *Clinical Trials in Neurology*, Neuromethods, vol. 138,
https://doi.org/10.1007/978-1-4939-7880-9_12, © Springer Science+Business Media, LLC, part of Springer Nature 2018

Another frequent cause of PN is associated with chemotherapy and antiretroviral (ARVs) neurotoxicity. The incidence of chemotherapy-induced peripheral neuropathy (CIPN) is 68.1% [4]. Thus, PN has been studied as a main outcome on clinical trials involving neurotoxic agents. The major consequence of drug-induced PN are dose reduction or treatment cessation by exposing patients to their primary disease morbidity and mortality [5]. Although the survival rates of oncological and HIV-infected patients have increased with the development of more effective therapies, there is still a concern about severe disabling adverse effects due to treatment, such as PN. For instance, a meta-analysis reported an incidence of CIPN in 30% of patients after 6 months of chemotherapy [4]. In this context, research studies for protective therapies and treatment for PN have been conducted for this type of population.

Finally, regardless of the etiology, PN therapeutics relies on three mainstays: prevention, symptom management, and changes in the disease status. For each of them, the causality of PN will guide the most appropriate treatment option for the clinical practice. Moreover, drug-induced neuropathy trials often target protective therapies or the reversibility of the disease—focus on protection and changes in the disease status. Interestingly, most of the investigations for DM, HIV, renal and kidney diseases emphasize on effective treatment, pathophysiology, and genetic studies. Nonetheless, research for effective pain control and rehabilitation are common interests for all PN etiologies.

This chapter aims to guide clinical researchers on PN by considering unique features of clinical trials in this field. For this purpose, a systematic review of the 100 most cited papers of PN since 2005 was conducted. The topics covered herein are based on the CONSORT guidelines and include, trial design and settings, phases of study, sample selection, intervention, choice of outcomes, sample size, randomization, blinding, adherence, statistical analysis, assessment of biases, and confounding.

2   Methods

A systematic review of the 100 highly cited clinical trials on peripheral neuropathy was conducted at all Web of Science database. Our search strategy focused on a 5-year interval of publications with the keyword "Peripheral Neuropathy" present on the publication tittle, organized from the most cited. In addition, we refined our search by document type, selecting clinical trials. From 2010 to 2015, the results were not enough to retrieve 100 trials. Thus, our research was extended to the period of 2005–2015 and furthermore from 2000 to 2015 when we finally achieved the targeted number of articles. Published papers in English of interventional clinical trials were included. We screened 222 papers and excluded 122. Among reasons for exclusion were: articles published in another language rather than English (20); reviews (8); study protocol (1); repeated articles (5); noninterventional and observational studies (49). Figure 1 illustrates our inclusion and exclusion criteria results.

3 Initial Glance

3.1 Etiologies of Peripheral Neuropathy

Diabetic peripheral neuropathy (DPN) and chemotherapy-induced peripheral neuropathy (CIPN) represent 83% of the retrieved studies. Hence, this chapter gives special attention to these two etiologies. Though, other causes are also evaluated throughout all the sections. Table 1 summarizes the amount of studies by etiology in our search.

3.2 CONSORT Endorsement by Number of Citation

The CONSORT statement was created to minimize issues due to inappropriate report of randomized clinical trials. It is a 25-item checklist to guide authors to achieve a transparent and complete report of their research. In addition, the endorsement of CONSORT by a scientific journal implies quality on RCT's report.

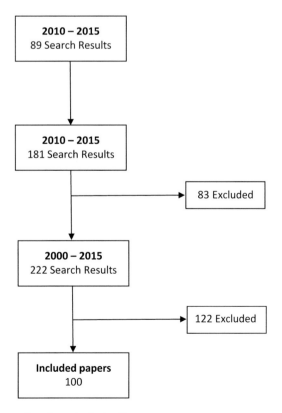

Fig. 1 Diagram flow for search results

Table 1

Etiologies of peripheral neuropathy

	N
Diabetic peripheral neuropathy—DPN	48
Chemotherapy-induced peripheral neuropathy—CIPN	35
HIV peripheral neuropathy—HIVPN	2
Multiple etiologies	3
Undefined	3
Uremic peripheral neuropathy	2
Hemodialysis peripheral neuropathy	2
Posttraumatic	1
Mixed cryoglobulinemia peripheral neuropathy	1
Isotretinoin induced	1
Diabetic kidney disease	1
Leprosy	1
Total	100

In our systematic review, we noticed a balanced trade-off on number of citations among articles published at journals endorsed (49%) and not (51%) by CONSORT. Because the majority of citations are among articles published at journals not endorsed by CONSORT we can presume two things: bad quality of report among the articles selected for this review or use of CONSORT regardless of a journal endorsing it.

3.3 Results for Primary Endpoint and Number of Citation

When analyzing the most cited articles we found positive and negative results among the 100 most cited papers regarding peripheral neuropathy. In our review we expected to have a minority of negative results published due to having a relatively small number of phases II or I studies exploring complete new therapies or interventions. The difference between negative and positive results among the analyzed studies can be observed in Fig. 2. Additionally, due to the lack of a standard therapy, often trials showed positive results at early phases and failed to have a statistical significant result at late phases. An example of it is the series of studies testing vitamin E for Chemotherapy Induced Peripheral Neuropathy. In 2006, Argyriou and colleagues performed a phase II, parallel two-arm RCT, to evaluate safety and efficacy of vitamin E to prevent cisplatin CIPN [6]. Their study showed a statistically significant result. However, in 2011, Kottchad and colleagues [7] conducted a phase III parallel two-arm RCT to prevent CIPN of platinum

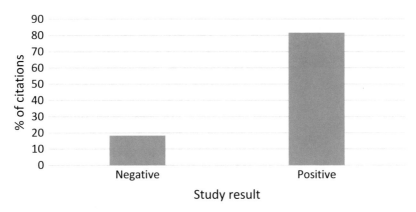

Fig. 2 Proportion of citations of peripheral neuropathy trials with positive and negative results

and, or taxanes compounds with a nonsignificant result for their primary hypothesis. Both studies add equally important scientific contributions and must be taken into account when planning a trial in this field, regardless of its results.

Table 2 illustrates the main characteristics of the 100 papers of our review by its frequency such as publication year range; study phase; trial design; type of intervention; primary outcome results; publication at a journal endorsed by CONSORT; sample size; number of citations; journal impact factor. Those topics will be better discussed through this chapter.

4   Trial Design, Study Phase, and Setting

The study design is carefully chosen among all the potential options given the optimal balance between the research question, ethical concerns, and feasibility. The current phase of development of a given drug or device is also important when choosing the study design.

4.1   Trial Design

An important part of a clinical trial is the study design to achieve systematically reliable results. Flaws on the study design imply waste of resources and unnecessary risk exposure of subjects. In addition, the decision of which trial design suits better the selected research question must be ethically and scientifically supported by previous findings. A more appropriate design will depend on the study aim, if the trial aims to test a treatment for symptom management, for reversibility, for prevention or if it is a dose adjustment study. Therefore, the three mainstays previously mentioned must be taken into consideration.

Our review resulted in 71% of randomized clinical trials (RCT) and 29% of nonrandomized clinical trials (NRCT) with a considerable spectrum of designs. The following topics will address each type of study design we encounter in this review.

Table 2
Summary of the included clinical trials

		N
Year		
	2000–2005	20
	2006–2010	37
	2011–2015	43
Phase		
	I	3
	I–II	1
	II	56
	III	16
	IV	1
	Exploratory	23
	Pivotal	0
	Postmarketing	0
Trial design		
	RCT parallel 2 arms	52
	RCT parallel 3 arms	8
	RCT parallel 4 arms	1
	RCT parallel 5 arms	1
	RCT crossover, 2 groups	7
	Nonrandomized single arm	21
	Nonrandomized 2 arms	8
Drug/device/procedure		
	Device	13
	Drug	73
	Procedure	14
Primary outcome positive		
	Yes	27
	No	73
Journal follows consort guideline		
	Yes	44
	Not in the endorsement list	56
Sample size		
	0–30	27
	30–60	27
	60–200	20
	>200	26
Number of citations		
	0–10	36
	10–20	17
	20–60	26
	>60	21
Impact factor		
	1–5	62
	5–10	20
	10–20	1
	>20	17

4.1.1   RCT Parallel Group Design

This study design represented 62% of the reviewed articles. It refers to trials wherein participants are signed randomly to one or more treatment arm [8]. In fact, this is the preferred design in several situations as this is a simple and robust design. Here, it is possible to assume that the comparison groups—study arms—are balanced. Randomization advantages will be further discussed in this chapter. As for this study design concern, RCT's parallel group usually compares active treatment and placebo, having two arms. However, it can have different number of arms that are compared directly, for instance different dosages with or without additional placebo arm.

In fact, RCT parallel studies varied from two treatment arms up to five. Seventy percent of the parallel three, four, and five arm study cluster were designed to test different dosages of the same intervention compared to a matching placebo. Except for three studies [9–11], one to be mentioned is the phase III trial conducted by Razazian and colleagues, 2014, to compare safety and efficacy of pregabalin, venlafaxine and carbamazepine directly without placebo group.

4.1.2   RCT Crossover Design

This study design is suitable if the disease or subject condition won't change appreciably over time [12]. For it, participants are randomly signed to a treatment arm and after a washout period they switch arms. In that way, subjects will be their own control group.

In order to avoid a treatment effect being carryover to the other arm, when switching treatment protocols, participants remain a period with no active intervention. This is known as the washout period; moreover it allows a common baseline for each treatment [12].

As for studies with PN, this design was used for CIPN, post-traumatic PN, hemodialysis, and multiple conditions of PN. They all had two groups comparison and a washout period. The trials for CIPN considered participants eligible if they had PN symptom duration for over 3 months; studies with hemodialysis used the same parameter; studies with traumatic PN selected participants with over 3 months PN after resolution of trauma; and studies with PN multiple etiologies selected those with over 5 years symptoms. This is to illustrate how researches selected participants with a stable condition for a crossover design [13].

Furthermore, of the seven trials using this design, only two were phase III, the remaining were phase II trials. It was commonly reported a 2-week washout period regardless of the PN etiology, however this decision must be accordingly to the treatment tested.

4.1.3   Studies with Run-In Phase

Between RCTs, we found 12% using run-in phases. This type of design is composed by a period in which subjects participate in the trial as to determine further eligibility. It assists on the determination of an effect size between active and placebo treatment; increases power especially for single arm trials; and decrease required sample size.

As an example, Vinik and colleagues, 2014, conducted a trial to evaluate efficacy and tolerability of Tapentadol extended release among patients with chronic DPN [14]. Their study consisted of a 3-week open-label titration to optimal dose in terms of pain reduction and tolerability. Participants with improvement equal or bigger than 1 in pain intensity (11-point numerical scale)—from the pre titration period to the last 3 days of the open-label phase—were randomly signed to receive active treatment or placebo. Participants who did not meet this criterion were withdrawal of the study. The next phase of their trial was a 12-week double-blinded, parallel two-arm. Participants with optimal dose kept their dose till the end of the study if they were signed to active treatment; those signed to placebo had their dose tapered down in a blinded way.

Moreover, Raskin and colleagues, 2014, conducted a trial to evaluate pregabalin for the treatment of DPN [15]. It was a phase III, multicenter study with an initial baseline phase, followed by a single-blinded and a double-blinded one. The 1-week baseline was to access eligible participants to entry the trial. The single blinded phase last for 6-weeks, wherein all received pregabalin beginning at 150 mg/day titrated up to 300 mg/day. Only participants who achieved a response equal or greater than 30% at the 11-pont numerical scale were randomized either to receive pregabalin or matched placebo—they entered the double blinded study phase. Participants who didn't meet this criterion were withdrawal of the study.

This is an interesting strategy to avoid participants dropping out due to adverse effects, it may improve adherence since all participants receive treatment initially, and assist assignment to adequate doses of medication. Additionally, only dose with a minimum initial response remain in the study.

4.1.4 Quasi Experimental: Non-RCT

It refers to studies without randomization, often with nonequivalent groups differing not only by the treatment received. For peripheral neuropathy, the majority of researches, 72%, used single arm quasi-experimental designs. The studies were meant to test only one intervention. Single arm quasi-experimental studies were conducted for exploratory, phase I and II.

Quasi-experimental studies with parallel two arm compared mono therapy and adjuvant therapy. For instance, Vahdat and colleagues [16] compared paclitaxel versus paclitaxel + glutamine in order to evaluate if adjuvant therapy would reduce CIPN. They conducted a nonrandomized single center phase II. In 2005, Stubblefield and colleagues, conducted a similar trial to assess neuro protective effect of glutamine by comparing its use alone and with paclitaxel [17].

4.2 Phases and Trial Design for Peripheral Neuropathy

Translational research integrates basic biological knowledge to clinical trials auxiliary on the comprehension of drug targets and disease biomarkers [18]. The famous "bench to bedside" paves the way for unifying genomic studies with pharmacology in order to pursuit an efficacious treatment for cancer. One translational research of a preclinical and phase II trial was among the 100 most cited papers of our systematic review. The study aimed to test an alternative treatment for CIPN, in a single center setting. In order to establish an ideal phase II trial, Coriat and colleagues [19] used a model based on results of their preclinical research. This strategy increases the chances of positive results and decrease patient exposure to risk; however, for late study phases it might be challenging to develop it.

Phase I clinical trials are designed to access safety of a new therapy. These studies have a small sample size of subjects often healthy [20]. However, for peripheral neuropathy trials, it is unethical to expose healthy participants to a neurotoxic agent. Thus, most of the Phase I research on this field targets diseased subjects. Furthermore, the objective of this phase is to determine the ideal dose-response and identify side effects of the new therapy. Through our review, three articles were Phase I trials. Two were studies regarding pharmacokinetics and dose adjustment of chemotherapy agents conducted by Mielke and colleagues [21]; Verstappen and colleagues [22], respectively. Both were randomized parallel two arms studies.

The third publication was a dose escalation nonrandomized single arm research for DPN by Thisted and colleagues [23]. During this study participants had a 2 weeks medication washout period prior to receiving active treatment. Afterwards, intervention consisted of one capsule of DM3OIQ30 up to four capsules as tolerate, with an interval of 1 week before dose escalation.

Phase II studies represented a significant amount of publications in our search. Usually they are conducted to evaluate the feasibility of exploring a new intervention, therapy or device [24]. Potential strengths and limitations of a design are often identified during this initial stage. Phase II studies represented 56% of our evidence. Initial research supports further safety and efficacy trials. Moreover, it provides more advanced information about the new drug, device or procedure. Therefore, this finding is coherent to the need of exploring new therapies for PN. For instance, Paice and colleagues [25] evaluated the use of topical Capsaicin for HIV peripheral neuropathy. Although there are several studies about its use for painful PN, no study with a robust design was performed for HIVPN until them. Because of it, their trial was able to evaluate strengths and limitations of conducting a robust study with Capsaicin. They notified the challenges on keeping blinding to the burn sensation Capsaicin provoke compared to the placebo cream. Therefore, their trial is of great value for further investigations in this field, they provide information that can be used for future studies not only regarding their results but also about study design and planning.

Phase III clinical trials aim to test the effectiveness of interventions in a broader population. During this phase, the design favors generalizability of the study findings [19]. Sixteen papers were phase III research. Indicating a careful progress to generalizability in studies for PN. This may be related to the etiology of the disease and how the particularities of each one of those may affect generalizability. Rao and collegaues, published a double blind crossover design RCT, to evaluate the efficacy of gabapentin in the management of CIPN [26]. The study consisted of a 6-week intervention, wherein participants could receive either placebo or active treatment, followed by a 2-week washout prior to an additional 6 week intervention—note participants switch treatment arm at this point.

Phase IV clinical trials or postmarketing are studies to gather additional information of adverse events of a therapy already available at the market. One study was at this stage of research. Hotta et al. conducted a large multicenter trial with DPN for 3 years, to evaluate the long-term effect of epalrestat [27]. The study consisted of a multicenter parallel two-arm design, it selected subjects with mild DPN using electrophysiological studies as primary outcome and screening strategy (this will be discussed further in detail at Sect. 8).

Study phases for medical devices are classified as exploratory, pivotal, and postmarket. Exploratory trials correspond to phase I–II studies, while pivotal resemble phase III and postmarket, phase IV. In this context, our review involved research-involving exploratory phase that corresponded to 23% of the trials. Table 3 summarizes our findings for this section. Table 3 summarizes our findings for this section.

4.3 Study Settings

The decision between a multicenter and a single center setting, have impact on generalizability of the study findings. Therefore, some important considerations must be taken into account before

Table 3
Trial design aspects

Trial design		N = 100	Study setting	I	I–II	II	III	IV	Exploratory	Translational
RCT	2 parallel arms	52	Single center	1	–	14	1	–	11	1
			Multicenter	1	1	13	9	1	–	–
	3 parallel arms	8	Single center	–	–	3	1	–	–	–
			Multicenter	–	–	4	–	–	–	–
	4 or more parallel arms	2	Single center	–	–	–	–	–	–	–
			Multicenter	–	–	–	2	–	–	–
	Withdraw	12	Single center	–	–	–	–	–	–	–
			Multicenter	–	–	–	1	–	–	–
	Crossover	7	Single center	–	–	–	1	–	–	–
			Multicenter	–	–	5	1	–	–	–
Non-RCT	Quasi experimental	29	Single center	–	–	10	–	–	14	–
			Multicenter	1	–	5	–	–	1	–

choosing a suitable setting plan for each trial. Researchers must evaluate logistic and funding feasibility to conduct multicenter experiments; evaluate available resources, such as site center; balance accessible population with the required sample size to conduct a powerful investigation; and finally, the main focus of the study phase—internal or external validity.

4.3.1   Single Center

Single center trials are conducted at a unique facility. As for our review, 60% of the studies were single center. When assuming phases I, I–II, II and exploratory as a unique block, it represents 43 out of 60 papers of this study setting.

In order to correlate study setting, and sample size we used descriptive statistics to evaluate the sample size for single center trials among our review. It resulted in an average sample size of proximally 114 with a standard deviation of proximally 139, a minimum N equal to 8 and the maximum equal to 257, with only 10 studies with N larger than 80, and mode of $N = 20$. That being said, the majority of these studies had small samples.

Studies of initial phases trials are often single given they are usually more complex designs, with limited funding and more exploratory—as discussed above—and have small sample size. An example is a non-RCT conducted by Strempska and colleagues to evaluate the effect of high-tone external muscle stimulation on symptoms and electrophysiological parameters of uremic peripheral neuropathy [28]. Their study had 28 participants; it was single center, single arm and a phase II study. However, there are exceptions. Some trials may have a large sample size, even though it is an early phase study, and external validity may be also a goal. For instance, the non-RCT conducted by Bestrad and colleagues [29] with 249 participants. It was an open-label phase II, parallel two-arm study to compare Nabilone and Gabapentin as adjuvant and mono therapy.

4.3.2   Multicenter

Multicenter trials are known to favor generalizability and external validity. Another important factor is to give attention on the purpose of the trial been designed; for instance, late phases designs usually have generalizability as a goal making a multicenter study more appropriate. However, same considerations we mentioned above for single center settings must be taken into account when deciding rather or not to conduct a multicenter study. That being addressed, in our review, when comparing study phase with setting, the majority of phase III trials are multicenter with a parallel two-arm design.

A descriptive analysis of the sample size for this study setting was also performed to assist to co-relate both aspects. It resulted in an average of proximally 126 and a standard deviation of proximally 147. The minimum N is equal to 20, the maximum N is equal to 707 and the mode of $N = 256$. Tuttle and colleagues [30] conducted the larger multicenter trial, in order to evaluate riboxistaurin effect on GFR among patients with diabetic kidney

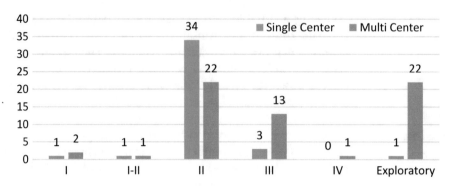

Fig. 3 Study phases and settings

disease peripheral neuropathy. It was a RCT parallel two-arm phase III study. In conclusion, multicenter studies usually have a larger sample size also to achieve desirable generalizability.

The graph (Fig. 3) is a comparison between study setting and phase. As predictable, late phases of PN researches are often multi-center trials, in order to obtain generalizability and assess a larger sample size. Initial phases, such as exploratory and phase I, represented 2.5% of the single center investigations. The majority of our evidence is of phase II trials; among this group of studies approximately 60% were conducted in a single center setting. This finding agrees with the absence of standard treatment for peripheral neuropathy, and the need of testing new therapies. In addition, it illustrates the difficulties of finding a therapy effective enough to support further phase studies.

5 Population

The inclusion and exclusion criteria for a trial have impact on the generalizability and internal validity of its findings. Generalizability is associated to how much the study population represents the target population. Internal validity refers to the extent that the trial can respond to the original question proposed truthfully. It reflects a balanced and homogeneous study population avoiding confounders.

Peripheral neuropathy is associated with multiple comorbidities, medication, genetic predispose, and behavioral factors. The spectrum of causes for PN can be controlled with an inclusion and exclusion criteria, randomization or statistically. The last two possibilities will be discussed further.

We observed that some trials emphasize on selecting the population according to the primary etiology of the disease. Thus, in this section we presented some etiologies for PN and the main criteria to include subjects in clinical research.

5.1 Diabetic Peripheral Neuropathy

Diabetic Peripheral Neuropathy (DPN) research involves pain therapy that can possibly affect response to the new intervention tested in a trial. In addition, this population consists of several comorbidities that can affect disease manifestation or response to the new tested therapy. Therefore, we aim to address the use of inclusion criteria in order to avoid confounders. For instance, in our research, trials controlled the use of pain medication by selecting patients who passed through a washout period. Also, nondiabetic neurological disorders were used as an exclusion factor such as alcoholic neuropathy, peripheral vascular disease, renal disorder, leprosy, arteriosclerosis obliterans, sequelae due to cerebral vascular accident, and exposure to chemotherapy agents. In addition, risk factors such as alcohol or drug abuse that affect PN symptoms were among exclusion criteria.

Baseline characteristics of DPN were among 30% of the inclusion criteria of the searched trials. Moreover, DM types II and I diagnose with acceptable glycemic control, HbA1C lower or equal 11%, were common baseline feature of 90% of the studies. Presence of foot ulcer or diabetic foot, even though it is a common sign for diabetes, was an exclusion factor at 40% of the trials.

Skin condition was also considered for selecting the population, for instance a multicenter double-blind study using QR-333 cream for DPN, Valensi and colleagues [31], excluded participants with skin allergies or dermatological conditions that would interfere on the protocol.

As for safety, 16% of trials DPN used a negative pregnancy test and required the use of appropriate contraceptive method as criteria to enter the study. Additionally, women could not be lactating. A common practice among the studies in our review was to submit participants to anamneses, clinical and laboratorial exams before entering the trial. An example of It implications is the noninclusion of participants wherein ECG and creatinine clearance was abnormal—30% of the papers reviewed. Wherein VDT (Vibration Detection Threshold) or F-wave was not detected participant were also excluded of the trial.

Another important consideration for DPN is the disease's impact on different peripheral structures such as arms, legs, hands, feet, and other areas. Hence, it was common to mention the area affected and evaluated through the trial as an inclusion feature.

5.2 Chemotherapy-Induced Peripheral Neuropathy

Inclusion criteria for CIPN encompass a wide range of aspects: type of chemotherapy prior to the study or during it, life expectation, age, CIPN baseline characteristics, PN symptom duration, histological and laboratorial findings, pain medication or intervention use. In this regard, there were few practices on inclusion criteria to avoid confounder. For instance, over 90% of the reviewed trials conducted to test a new therapy for CIPN reported the determined chemotherapy agent allowed for entering the study and the diagnosed type of cancer. Another common feature with potential to avoid confounder was the selection of subjects without PN medication or taking stable doses of painkillers. Except for researches evaluating protective effect of a new therapy, additional strategies were the inclusion of patients without other possible causes for PN, such as DM, HIV, and alcohol abuse.

As for safety, researchers used platelet, hemoglobin count and serum creatinine findings to evaluate if participants could enter the study. These laboratory exams also are used to assess if chemotherapy can go on or must be interrupted.

For baseline characteristics, we acknowledge trials selecting participants without PN, with mild or severe signs. The reported symptoms duration of CIPN varied from 0 up to 3 months; baseline life expectancy range was from a minimum of 3 months up to a minimum of 6 months. Furthermore, previous treatment with chemotherapy and, or for peripheral neuropathy was part of the main inclusion/exclusion criteria for this etiology.

5.3 Uremic and Renal Disease Peripheral Neuropathy

For the following sections, we included uremic diabetic peripheral neuropathy (UDPN), diabetic kidney disease peripheral neuropathy and hemodialysis peripheral neuropathy, which represent 5% of our searched papers. Those etiologies are related to nephrology and uremic functions in some extent.

Hemodialysis peripheral neuropathy trials used 40 mm pain score in Short Form of McGill Pain Questionnaire (mild pain) as an eligibility criteria. Among this etiology, participants with hepatic, cardiopulmonary and uncontrolled psychiatric disorders were excluded. Additionally, others pain syndrome causes rather than peripheral neuropathy would prevent participants to join the trial.

The others etiologies did not report the use of a standardized instrument to select a baseline PN characteristic However, presence of pain or participants undergoing hemodialysis was enough to consider eligibility. None of these studies reported an inclusion criterion to control for other causes of PN, or mentioned restrictions for the use of pain medication. In fact, one study included patients undergoing treatment for pain relief.

5.4 HIV Peripheral Neuropathy

Pathogenesis of HIV associated PN are unknown. Proposed mechanisms include cytokine deregulation, viral protein produced neurotoxicity, and mitochondrial dysfunction associated with antiretroviral drugs (ARVs) [32, 33]. Although HIV has a pandemic incidence and PN affects 30–35% of this population [34, 35], only two studies were found in our research. One study, by Paice and colleagues, was design to test topical capsaicin for HIVPN management. On the other hand, Anastasi and colleagues compared true versus sham acupuncture [36]. Both used eligibility criteria to control medication use for PN and excluded participants who were undertaken topical medication on the lower extremities.

The safety exclusion criteria used by Anastasi and colleagues, was presence of hypertension, pulmonary and kidney disease. As for Paice and collegues, pregnancy and presence of lesions on feet and legs would exclude participants.

6   Intervention

6.1   The Role of Drug, Procedure, and Device Studies

Previous studies have reported findings of different strategies for the treatment of Peripheral Neuropathy. Drugs, procedures, and devices trials have different implications during the selection of a study phase and design. Regarding our findings, 73% were studies testing drugs, 14% procedures, and 13% devices trials. For drug studies, we assumed experiments of pharmacological therapies; for procedures, we considered nonpharmacological interventions such as meditation, gait pattern, acupuncture, and for devices, we included therapies using electrical stimulation or any other device responsible for providing therapy.

The three different categories of intervention were found in our research through different etiologies of PN. In order to make correlation between intervention and PN etiology, this section will address the main types of PN of the 100 selected articles.

6.2   Diabetic Peripheral Neuropathy

In the context of DPN, treatment is usually partially effective or with intolerable side effects [37]. Hence, studies with the same intervention but with different dosage comparison are not uncommon. Pregabalin was tested in six different trials [38–43]. Five of those studies were of phase II RCT parallel three arm or two arm. In fact, at least six trials with DPN were dose adjustment research. Another finding is the application of an initial preintervention titration phase for dose adjustment for drug therapy. Therefore, patients achieve the best treatment dosage without suffering adverse effects as much as if they received a fixed amount of medication. This approach has influence on adherence by reducing experience of side effects and the study duration. Both aspects will be discussed further in detail.

Among the interventions for this etiology, research for drug therapy represented 35%, devices 7%, and procedures 14%. Table 4 contains the interventions tested throughout trials for DPN.

6.3   Chemotherapy-Induced Peripheral Neuropathy (CIPN)

Therapies for chemotherapy-induced peripheral neuropathy may have different response according to the chemotherapy agent [44]. Neurotoxicity varies among chemotherapy drugs according to treatment schedule, dose, and combination or single agents. Hence it is frequent to observe trials evaluating the same treatment for CIPN in a population receiving different chemotherapies. The reason for it is that a treatment may demonstrate effects for a chemotherapy agent, while for another agent it will not consider all the variants among treatment.

Table 4
Diabetic peripheral neuropathy interventions

Drugs	N	Device	N
Tupiramate	1	*Photon stimulation*	1
Lacosemide	1	*ATS therapy*	1
Ruboxistaurin	2	*Whole body vibration therapy*	1
Ruboxistaurin mesylate	1	*Spinal cord stimulation*	1
Pregabalin	7	*Anodyne light therapy*	1
Gabapentin enacarbil	1	*Active MIRE*	2
Gabapentin	3	**Procedure**	**N**
Vitamin B1 + Vitamin B6	1	Leg strengthening exercise + motivational phone call	1
Fidarestat	1	Leg strengthening + balance exercises	1
Epalrestat	3	Motor learning of a gate pattern	1
Tapentadol ER	2	Exercises in a standing up position	1
Alpha lipoic acid	3	Exercises in a sitting or lying position	1
QR-333	1	Aerobic + strengthening exercise	1
Venlafaxine HCl	1	Diabetic foot care education	1
Minocycline	1	Exercise regime	2
Carbamazepine	1	Therapy as usual	1
Amantadite	1	Cognitive behavioral therapy	1
L-Methylfolate calcium + methylcobalamin + pyridoxal	1	Backwards walking	1
Venlafaxine	1	Meditation	1
Dextromethorphan + quinidine	1	Acupuncture	1
Tramadol	1		

The majority of the reviewed studies had chemotherapy agent involved as an intervention, in order to evaluate dose to prevent CIPN, it incidence according to chemotherapy agent or treatment. Table 5 illustrates the distribution of it.

Another intervention to prevent PN was the use of vitamins combined with chemotherapy agents. Additionally, among the interventions to avoid neurotoxicity are the use of a safe dose, with this aim, five studies had distinct comparable dosage arms. Table 6 indicates the interventions used by the 100 most cited papers.

6.4 More Etiologies of PN

In this review, a few etiologies of PN had a low frequency. In order to highlight their intervention, we organized Table 7 with each one of them. Research on drugs, procedures, and devices are mixed

Table 5
Frequency of prevention, treatment, and incidence interventional trials for CIPN

Prevention	Treatment	Incidence	Total
13	16	3	32

Table 6
Interventions used in the included clinical trials

	N
Drugs and chemotherapy agents	
Thalidomide	1
Thalidomide + Melphalan	1
Doxorubicin + cyclophosphamide + Docetaxel	1
Doxorubicin + cyclophosphamide	1
Docetaxel	1
Paclitaxel	2
Paclitaxel + Vitamin E	1
Paclitaxel + Glutamine	1
Oxilaplatin + Vitamin E	1
Vitamin E	2
Oxilaplatin + Mangafodipir	1
Ketamine + Amitriptyline	1
Bortezomib	4
Bortezomib + doxorubicin + dexamethasone	1
Lamotrigine	1
Glutathione	1
Paclitaxel + Cremophor EL + Ethanol	1
Duloxetine	1
Bacoflen + amitriptyline + Ketamine gel	1
Recombinant human leukemia inhibitor factor	1
Omega-3 fatty acid	1
Epothilone Sagopilone + Acetil L Carnitina	1
Bortezomib + Dexamethosone	1
Dexamethasone + Cyclophosphamide + Etoposide + Cys-platinum	1
Notriptyline	1

(continued)

Table 6
(continued)

	N
Vincristine (VCR)	1
Vincristine + doxorubicin + dexamethasone	1
Lafutidine	1
Dexamethasone	1
Gabapentin	1
Device	
MC5—A Calmare device	1
Scrambler therapy	2
Procedure	
Acupuncture	3

among these etiologies. Interestingly, our evidence showed acupuncture as a common procedure tested among through different causes of PN, such as HIV-PN, DPN, CIPN, and undefined PN.

7   Study Duration

During the planning phase of a clinical trial, researchers may wonder how long a study should last in order to provide useful data and respond their main research questions. In the case of peripheral neuropathy, this decision will be based on the studied etiology, treatment duration and the costs for a long trial duration. The etiology studied is relevant due to the characteristics of the population selected when someone enters the trial; this is defined as the baseline characteristics. For instance, a study conducted for Chemotherapy-Induced Peripheral neuropathy, must take into account life expectancy related to the disease and treatment the participant is undergoing prior to or within the study. It makes no sense, to conduct a trial of 12-month duration or 5-month treatment with population who has 3 months of life expectancy. Therefore, the treatment duration, the etiology of PN and the population features must be aligned during the study planning. Considering the important role of these topics to the study duration we will explore the trial duration for each etiology of PN on the following subsections.

Table 7
Low-frequency etiologies

Low-frequency etiologies interventions	N
Uremic PN	
High-tone external muscle stimulation	2
Diabetic kidney disease PN	
Ruboxistaurin	1
Hemodialysis PN	
Gabapentin and Pregabalin	2
Undefined PN	
Acupuncture	1
Tai chi	1
Multiple etiologies of PN	
Ondansetron	1
Pulse signal therapy	1
MeCbl	1
HIV	
Acupuncture	1
Acupunture + Moxibustion	1
Capsaicin	1
Posttraumatic PN	
Lidocain	1
Isotretinoin induced	
Isotretionin	1

7.1 Chemotherapy-Induced Peripheral Neuropathy

In the context of CIPN, the study duration must take into account number of chemotherapy cycles and life expectancy of participants. Therefore, this stage of designing relies not only on choice of outcomes endpoints and intervention but on the study population as well. Previously, we reported the use a 3-month life expectancy as common eligibility criteria for CIPN research. The impact of it can be noticed in the study length, in which most of trials last at least 3 months. According to Fig. 4, among 27 studies on CIPN, 4 had 14 weeks of duration. Of these, two trials used a crossover design with 2 weeks wash out period between 6 weeks intervention, leading to 14 weeks trial duration. The other two defined the study period according to the chemotherapy treatment cycles. In

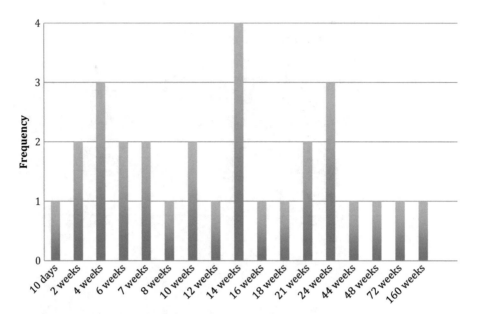

Fig. 4 Chemotherapy-induced peripheral neuropathy

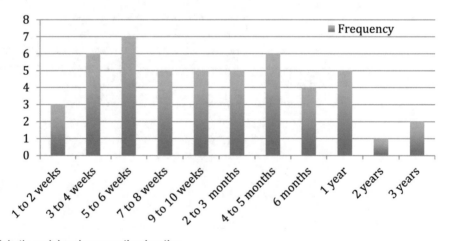

Fig. 5 Diabetic peripheral neuropathy duration

addition, depending on the research phase, the trial may have a longer or shorter length. This agrees with the long lasting 160 weeks study, being a phase III. The graph of CIP Study Duration summarizes our research findings in this manner.

7.2 Diabetic Peripheral Neuropathy

Researchers have a larger spectrum of time to conduct their trials in the context of DPN. This is a chronic condition associated with comorbidities and behavioral risk factors but not with reduced survival time expectancy as CIPN. Thus, in our review studies length varied from 11 days to 3 years duration (Fig. 5). Durable trials aimed to evaluate long-lasting effect of treatments while short-range ones target initial safety and efficacy results.

Table 8
Uremic and renal disease peripheral neuropathy length

	Hemodialysis PN	Uremic PN	Diabetic kidney disease
Study duration (days)	98	90	730
Frequency	2	2	1

The main challenge for DPN is to establish the ideal study length to capture clinical and statistical significance. Hence, the decision of the study duration must be aligned with the accuracy of outcomes to detect immediately and, or posterior effect of a treatment. In addition, the period for a therapy to show effect must be pondered. As mentioned in Sect. 6, a titration or taper phase for drug research has an influence on the study duration as well.

The type of intervention and phase will affect DPN study duration. Trials with devices tend to last a short period when the main goal is to evaluate safety and efficacy in a more immediately time frame. However, several authors made observations about the need of long lasting trials to observe results of a device or procedure through time. Therefore, when planning a study with DPN, it is important to take into account what was previously done and the actual needs for this field.

7.3 Uremic and Renal Disease Peripheral Neuropathy

Uremic PN had two trials with the same study duration. Both exploratory researches aimed to test the same device. Perhaps because one was precursor of the other, the study duration was kept the same as a standard. Similarly, it was found for hemodialysis PN. In the case of diabetic kidney disease, we found a 3-year post hoc study that evaluated the protective effect of Ruboxistaurin. It was the longest trial of this group of etiologies due to a phase III. Table 8 summaries our review findings for Uremic and Renal Disease peripheral neuropathy study length.

7.4 HIV Peripheral Neuropathy

HIV-PN studies had two different time duration, 4 and 15 weeks. The 4 weeks trial was a drug phase II, while the 15 weeks was a procedure exploratory research. In our review, there was not enough evidence to make a rational of study length for this etiology. Although further investigation is needed, general considerations for this topic may be applied as well.

8 Outcomes

The choice of outcomes is highly relevant when designing a study. It can be challenging depending on the disease, intervention, and study aim to select the optimal measurement (if there is such

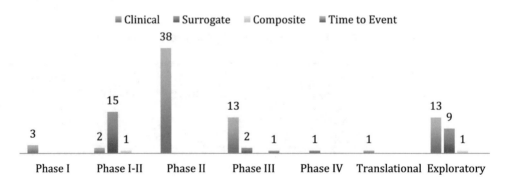

Fig. 6 Types of outcomes related to study phase

measurement) [45]. Therefore, awareness of types of outcomes and its implications for research need to be contextualized.

8.1 Types of Outcomes

For our data collection, we classified outcomes as clinical, functional, and surrogate. Clinical outcomes provide information about patient feelings, habits, and behavior. It is a subjective measurement often collected using scales and questionnaires. Functional outcomes are directly related to daily leaving activities. Surrogate outcomes are objective measurements of physiological function, for example laboratory exams [45, 46]. The graph (Fig. 6) represents the distribution of type of primary outcomes for the 100 articles accordingly to study phase.

8.2 Choosing Outcomes for Peripheral Neuropathy Research

It is important to select a validated tool to measure what your research aim to answer. Trials on PN often use subjective scales to measure pain as a main outcome. For instance, 32% of our searched trials used VAS and/or 11-point-NRS to assess pain, and established a standardized cutoff point in order to assess clinical significance. In addition, evaluation of severity of disease is among the primary goal of PN studies, which clinical and surrogate outcomes can be used.

Additionally, consider the use of surrogate outcomes. They are more sensitive to detect a minimal difference without the need of several event rates. Consequently, the sample size required for the study may be reduced and the accuracy of results is increased. Surrogate outcomes play an important role for PN research because this disorder is characterized by physiological changes in neuronal function. Therefore, electrophysiological studies are interesting to evaluate affected nerves and what is the disorder extension. Electrophysiological studies such as Minimum F-wave Latency test, Vibration Perception Threshold (VPT), Sensory Nerve Action Potential (SNAP) and, or Nerve Conduction Velocity (NCV) were used in 23% of our searched trials for screening patients according to their baseline characteristics and for the primary endpoint. In fact, 63% of the papers reviewed using surrogate outcomes had at least one of those outcomes previously mentioned.

Nerve Conduction Studies (NCS) and EMG (needle electromyography) are the main tests for electrophysiological studies used at researches in this review. Sensory and motor NCS test usually includes specialized parameters to be analyzed, such as amplitude, latency and velocity of the stimulus conducted through the nerve. Those features indicate the state of myelination and presence or not of axon loss.

Also, comorbidities of PN such as cancer, HIV, diabetes and kidney diseases have important surrogate outcomes as indicators of its status. Following comorbidities activity is important because it affects PN and may affect the treatment success as well.

When selecting clinical and functional outcomes, a decision must be done regarding how to collect data. The National Cancer Institute—Common Toxicity Criteria (NCI—CTC), Pain Disability Index (PDI) and Neuropathy Pain Scale (NPS) are examples of means to access PN either with patient self-report or with a physician based assessment. Through the 100 most cited papers, studies used patient self report in a diary as data collection, while other studies used a trained physician to assess data. Both strategies have advantages and disadvantages. Patient-assessment or physician-assessment of PN scale instruments may over or underestimate severity of disease [47, 48]. Patient-assessment reduces costs but it may affect the quality of information, since they may not report everything or under or overestimate the importance of a finding. The use of a physician to assess data must take into account the professional capability of doing it. One paper had its results questioned due to outcome assessment being made by a professional with not enough training. Thus, this must be considered for study planning and interpretation of results of a clinical research.

The three main clinical outcomes for PN encompasses the Brief Pain Inventory (BPI); Numeric Rating Scale (NRS) and Visual Analogue Scale. The BPI scale is validated and widely used to assess pain. It is able to detect sensory and reactive dimensions of pain. Wherein sensory dimension stands for intensity or severity (worst, least, average) of pain and reaction stands for daily function impairment (mood, walking, sleeping etc.).

The VAS is an instrument to evaluate if participants have no pain up to unbearable pain by indicating their pain status in a scale. This is a visual scale graded in millimeters, meaning 0 mm is equivalent to absence of pain and 100 mm meaning severe pain. It can be used a 0 to 10 scale, drawings representing a range of expressions of pain. This instrument can be used to measure other characteristics rather than pain. Finally, the NRS scale is also not specific used for pain. It varies from 0 to 10, having up to 11 points to measure pain. Both scales are often used for self-report of pain.

Table 9
Reported screening instrument

DPN	CIPN	HIV	Hemodyalisis PN
• Diagnosis and classification of diabetes mellitus and its complications by the World Health Organization • Michigan diabetic neuropathy score • Semmes-Weinstein monofilament • Glycosylated hemoglobin (HbA1c) • Electrophysiological studies	• National Cancer Institute—common toxicity criteria (NCI—CTC) • NCI common terminology criteria for adverse event • Electrophysiological studies • Patient neurotoxicity questionnaire (PNQ) • Biopsy	• Gracely pain scale (GPS) • Hospital Data Base	• Short form of McGill pain questionnaire

Although VAS and NRS was commonly used as primary outcome to assess pain, those instruments were seldom used alone. Outcomes such as instruments to evaluate quality of life (SF-36 form) or even pain (McGill Pain Questionnaire, BPI among others) were often utilized for secondary analysis by complementing information.

8.3 Screening or Diagnose Outcomes

Peripheral Neuropathy—PN diagnostic criteria vary across physicians. Acknowledgment of the strategy used for screening assist others researchers on the generalizability of study findings. In addition, these criteria support the identification of the primary or multiple causes of the disease and its severity. Thus, it is recommended to mention the standardized principles used for screening.

Peripheral neuropathy screening criteria is mostly based on scales for pain severity, neurotoxicity or surrogate electrophysiological findings. Among those measures clinical, functional and surrogate outcomes are used. The usage of reliable screening outcomes is fundamental to avoid bias and confounder. Furthermore, it assists on selecting baseline PN characteristics and a balanced study population. Hence, these features of outcomes are convenient for trials with stratified randomization.

Chemotherapy-Induced PN trials, of our search, often used surrogate outcomes to confirm cancer diagnoses and neurotoxicity instruments to evaluate PN. Diabetes Peripheral Neuropathy trials used glycemic control plus pain scales or electrophysiological studies to screen patients. Some articles reported the instruments used for screening; Table 9 illustrates the main tools for this purpose.

8.4 Endpoints

Endpoints refers to how a direct measure of body function, benefits detected by the participant, survival or a surrogate outcome are evaluated [45, 49]. The same outcome can be analyzed through different cutoff points that represent a range of significances.

8.5 Clinical Response Endpoint

Clinical response on PN trials refers to the assessment of signs and symptoms of the disease, for instance pain, numbness, vibration perception, quality of sleep, among others. A large number of trials target clinical response mainly due to the fact that PN have significant assessable clinical manifestations. Hence, around 72 papers used a variation of clinical outcomes to evaluate the effect of an intervention. Of those, only four had a clinical measurement as a co-primary for a surrogate outcome.

Through our review, trials using 11-point NRS considered a reduction of at least 30% or 50% a clinical significant change. Some considered at least a 20% reduction. As for Visual Analogue Scale, a change of 3 cm was assumed significant. Another approach was to evaluate difference from baseline until the last data collection. Moreover, few studies adopted a self-report diary to access clinical response from baseline to the end of the trial.

8.6 Surrogate Endpoint

Through our review, primary surrogate endpoints consisted basically of electrophysiological studies. However, some studies used laboratory measurements for safety. Specially, in the context of CIPN, the use of creatinine level, white blood counts, and hemoglobin counts were applied. For other etiologies such as DPN the surrogate outcomes were skin microvasclular blood flow (SkBF) and HbA1c levels while for hemodyalisis PN, eGFR measurements were considered.

8.7 Time to Event

Time to event is the outcome for survival analysis, a statistical method to infer about events that had occurred and have not until the end of the trial. Censoring information based on data set allows statistical analysis even when the studied event has not happened. Although it may add complication to interpretation of results, it is an interesting approach for research regarding reversibility or occurrence of an event. In the context of PN, it could be used for CIPN etiology, where a survival rate of patients also affects the study planning.

This type of measure is used when time have an important pattern for assessment of results. In our review one trial for CIPN used time to event as an outcome [11]. This particular study aimed to evaluate reversibility of CIPN due to bortezomib through time according to chemotherapy cycles. For it purpose, they created an algorithm with time to event based on the European Blood and Marrow Transplant Group criteria for disease progression.

8.8 Neurotoxicity

Chemotherapy-Induced Peripheral Neuropathy is a sign of neurotoxicity that may cause therapy withdrawal or dosage adjustment. Therefore, trials for this etiology used adverse event scales to identify neurotoxicity. The most applied instrument to evaluate neurotoxicity was the National Cancer Institute Common Toxicity Criteria, it was used in ten trials.

9   Sample Size

The sample size calculation is a critical step during the design phase of a research study. Underestimation or overestimation of the sample size may occur if it is not well calculated, carrying several methodological issues [50, 51]. For instance, overestimating the sample size can add to the participant unnecessary risks and higher costs to the study. Similarly, underestimating the sample size may generate results with lack of power to demonstrate a clinical effect, thus limiting the generalization of the findings [50, 51]. Therefore, some important parameters should be considered during sample size calculation, such as the study design, the outcome variables, the significance level or alpha value, the power, a measure of variation (Standard Deviation), and the expected effect size (Charan et al.). As mentioned above, understanding the rationale behind the sample size calculation is a fundamental phase for the clinical practice on peripheral neuropathy.

Reporting correctly the methodology employed to calculate the sample size becomes also an important feature in clinical research. In this review of peripheral neuropathy, 45 studies reported how they calculated their sample size. However, only 33 studies reported three of the mentioned parameters (significance level, power and effect size). Adequate report of sample size calculation is of high importance for readers because it allows them to interpret results and bring them to clinical practice on peripheral neuropathy.

Among those parameters, a level of significance or type I error of 0.05 was the most common employed by researchers on peripheral neuropathy. This is expected, since alpha levels below 0.05 are only appropriate in specific settings such as an expensive drug associated with serious side effects. Only two studies on peripheral neuropathy employed a type I error lower than 0.05 [9, 52]. Both of them are phase II multicenter trials using a drug as an intervention. In these specific situations, researchers might want to minimize the type of error to 0.15 [52] and 0.01 [9] to reduce significant risks to the participants. In addition, in this review there was a predominance of a type II error of 80% or more, which means that investigators will accept a risk of at least 20% of do not detect a difference when it actually exists.

The estimation of the difference between groups is often challenging. This estimation should reflect the results of previous investigation on peripheral neuropathy, which in clinical trials is usually reported as pearson correlation (r), mean differences between interventions, or odds ratio (OR). Moreover, in order to predict effect size, a researcher has to make assumptions regarding both the control group response, and the expected effect of the active group. Most studies on peripheral neuropathy did not report the research type from where the authors predicted the treatment response (72%), while 12% were from clinical trials.

Table 10
Statistics

	Total
Alpha level	*100*
0.05	64
Two-sided	15
Not specified	49
0.025	
Two-sided	5
Not specified	1
0.15 two-sided	1
0.01 not specified	1
Not reported	25
Power	*100*
80%	26
83%	1
85%	1
90%	11
95%	1
96%	1
Range 71–96%	1
Not reported	58
Control and active group expected effect	*100*
Observational data	1
Previous clinical trials	12
Not specified	72
Attrition	*100*
50%	2
30%	1
20%	6
10%	4
Not reported	85

An example of a report of sample size on peripheral neuropathy with all the important elements was the study developed by Fonseca et al. [53]. This is a randomized clinical trial testing whether the combination of L-methylfolate, methylcobalamin, and pyridoxal-5′-phosphate improves sensory neuropathy in patients with type 2 diabetes. The authors described the sample size calculation as follows: "the sample size of 216 subjects was calculated based on an analysis of variance model, assuming an SD of 16 for the primary endpoint of change in vibration perception threshold (VPT), with a significance level of 0.05, and assuming a drop-out rate of 20%, such that 90 patients per arm would complete the trial". The cited paragraph provides information of all elements the authors had to take into account in order to calculate their sample size. Thus, sufficient information was provided to replicate the methodology of sample size calculation in another study. Moreover, the alpha (two-sided, 0.05) and power level (80%) reported meet the statistical standards for clinical trials [54]. Further, the effect size was estimated utilizing previous evidence from clinical trials, which minimizes the risk of under or overestimation of the sample size. Table 10 summarizes statistical findings through this review.

10 Randomization

Random allocation qualifies RCTs as the gold standard in medical research. The main goal of randomization is to balance the distribution of known and unknown prognostic factors across groups [55, 56]. Therefore, comparison groups differ only by their intervention, and any observed difference in outcome is likely to result from differences in treatment effect. This way, randomization minimizes the probability of selection and confounding biases [55, 56].

Proper employment and clear reporting of randomization are central for scientific accuracy and credibility. The CONSORT criteria determine that researchers must report the following items in published articles: type of randomization, randomization sequence generation, implementation of randomization, and allocation concealment. We will stress each one of these topics hereon [50].

In the context of Peripheral Neuropathy researchers may use randomization to control for other factors associated with the disease. As we previously discussed, PN is associated with several comorbidities, different levels of disease, affected members and etiologies. It is fundamental to apply a randomization method in order to avoid bias. Of the 100 most cited papers, 28% were of nonrandomized trials.

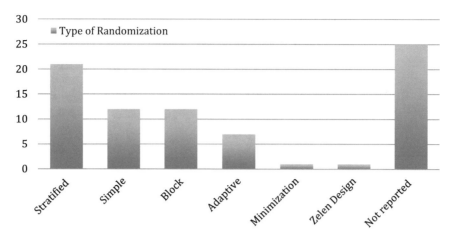

Fig. 7 Types of randomization

10.1 Randomization Sequence Generation

This refers to how the allocation sequence for patients is conducted through the experiment [55]. Common randomization sequence generation is made with the use of a numeric table or computer-generated sequence. For the selected studies of PN, 72% were RCT. However, among the RCTs, 56.9% of the papers did not report sequence generation conduct.

The majority of RCTs used a computer-generated sequence (58%) followed by 16.1% that used a numeric table and approximately 1% that used Interactive Voice Response System. Among the less used strategies, we found a web-based system, heads-and-tails, algorithm, Zelen design, and random numbers from a sealed envelope.

10.2 Type of Randomization

In this systematic review we considered the following types of randomization: simple, stratified, block, adaptive, and minimization. The graph (Fig. 7) illustrates our findings for randomization type between RCTs.

Of the randomized studies, 25 studies did not report the type of randomization used in the trial. This finding may be due to the fact most of the retrieved papers were not published in journals endorsed by CONSORT guidelines.

Stratified randomization is used to control baseline characteristics—covariates—that may influence research results. This method creates separate randomization schedules for selected strata defined by important prognostic factors [55]. For instance, CIPN studies controlled for chemotherapy drugs because neurotoxicity is known to vary among different agents. However, the use of stratified randomization has implications for the statistics analysis. Stratified studies have to address each variable that was used to stratify randomization. Although this adds more complexity to statistical analysis, it is an interesting method for PN, because this is a condition with several possible covariates to be considered [55, 56].

Table 11

Stratification features per PN etiology

Chemotherapy-induced peripheral neuropathy	N	Diabetes peripheral neuropathy—DPN	N
Center	2	Center	2
Age	3	Country	1
Sex	1	Age	2
Chemotherapy agents	5	Sex	3
Chemotherapy treatment status—undergoing; prior or not	4	Median motor nerve conduction velocity	2
Time to PN progression after last treatment	1	HbA1C	3
Chemotherapy dosage	1	Baseline pain intensity	1
Use of opioids or pain medication for PN	2	Use of opioids or oral pain medication	1
Pain risk or ratings	2	Height	1
β2-microglobulin	1	Diabetes type	1
Tumor diameter	1	Michigan diabetic neuropathy score	2
ER/PR status	1	Interventional treatment dosage	2
HER2 status	1	Not reported	2
Auxiliary lymph node status	1		
Surgical procedure	1		
Clinical severity of PN	1		
Number of previous chemotherapy agents	1		

Additionally, stratified randomization must be carefully planned to have enough sample size in each stratum. Balanced groups can be reached when combining stratification with another restriction method such as block or minimization [55, 56]. In our search, only CIPN and DPN used stratified randomization. Table 11 represents the stratified features according to peripheral neuropathy etiology.

Blocked randomization is an alternative when researchers want to ensure same number of subjects across groups [55]. Briefly, this method randomizes several subjects at time in blocks of fixed or permuted sizes. In this way, it is possible to guarantee that comparison groups will have equal sizes throughout the trial. This feature is very important when it comes to carrying out interim analysis or when researchers are working with small sample sizes. On the other hand, this method does not abolish the risk of getting imbalanced groups for important prognostic factors [55, 56]. Blocking can be performed with fixed or permuted sizes. By fixed blocks, it is assumed all have the same number of participants (i.e., blocks of 4), and by permuted the quantity of members in each block may differ. Fixed blocks increase the chance of uncovering participants' allocation concealment. Through our search, only 12 trials adopted this method. Also, some of these studies used block randomization as a restriction for stratified randomization. The majority of papers did not specify the type or number of blocks. Three papers mentioned the use of permuted blocks and three more specified the number of blocks (4; 2, 4 or 6 and 6).

Simple randomization is composed by a single sequence of random assignments [57]. Similarly to a toss of a coin, it gives each subject a 50% chance of receiving active or control treatments. The main drawback of such method is the risk of getting imbalanced groups, or groups with different sample sizes, when applying it to small sample sizes. In contrast, simple random allocation is unpredictable and able to prevent selection bias in relatively big samples. Only 12% out of 100 papers used this methodology.

For researches with Zelen design, randomization sequence generation is not concealed. Participants are informed about the intervention they are allocated to. Afterwards, they may accept or not the treatment allocation and make a decision about consent to enter the study or not. If participants decline to joy the study, other therapies may be offered. Due to this type of study, adherence may increase. However, researchers ask about the capability of this strategy to maintain the advantages of randomization and ethical application. Regardless of using the Zelen design, report about type of randomization, sequence generation and implementation of randomization may add credibility to the study.

Adaptive randomization, minimization method and Zelen design was the less reported strategies through our search. As for adaptive randomization, only seven papers reported the use, as for minimization method one and for zelen one.

10.3 Allocation Concealment

Another key process to achieve appropriate randomization is allocation concealment. Regardless of having a random allocation sequence, prevention of foreknowledge of treatment assignment avoids introduction of selection and confounding biases [58]. Moreover, previous trials have shown that inappropriate allocation concealment undermines the randomization process, because those responsible for enrolment may assign patients for groups based on their beliefs [58, 59]. In this context, sealed envelopes or telephone-based are commonly used to conceal treatment assignment.

Most studies in peripheral neuropathy did not report the employed allocation concealment (76.3% of RCTs). This is a higher frequency than the observed in reports showing an omission rate of 45% in general medical journals. The inadequate report of allocation concealment, as previous commented, have implication on readers' evaluation about introduction of systematic errors.

Alternatively, among those trials that reported concealment methods, approximately 11% used sealed envelopes, whereas less than 1% utilized telephone-based. These methodologies are considered appropriate, however reports have shown that investigators are able to subvert them in some occasions. Therefore, addition strategies to maintain allocation concealment may be applied as well.

10.4 Implementation of Randomization

Besides random sequence generation and proper allocation concealment methods, the implementation must be planned to avoid selection bias [55, 58, 59]. Thus, it is important to ensure that those who generate the allocation sequence and/or conceal the allocation schedule are not involved in enrolment or assignment of patients. From the selected RCTs in this review, 25 studies out of 72 reported the person responsible to generate the randomization list.

11   Blinding

Blinding is a strategy to hide patient's treatment assignment and therefore it is essential to prevent ascertainment and perception bias on clinical research. For trials using self-report data collection for PN, blinding is also necessary to reduce Hawthorn and placebo effect [60, 61]. Awareness of blinding approach assists readers to evaluate how sources of systematic error were evaded. Additionally, approaches to reduce bias suggest more reliable results [60, 61].

A complete report of blinding must contain information about its type, strategy for application, similarity of interventions, the blinded and the assessor investigator. Among our search, only 8% of the publications did not report information regarding the blinding process. Although most of the trials in our evidence reported at least one aspect of blinding, only 8 articles out of 100, provided complete information regarding it.

11.1 Type of Blinding

This review considered three types of blinding: open label, single blinded and double blinded. This classification is based on who is blinded in the study—participants, assessor, physician, and, or statistician [60, 61]. Open label studies do not hide participant's allocation; single blinded have one level of blinding—for instance, participants or physicians; double blinded have two levels of blinding such as participants and assessor. The blinded investigator may

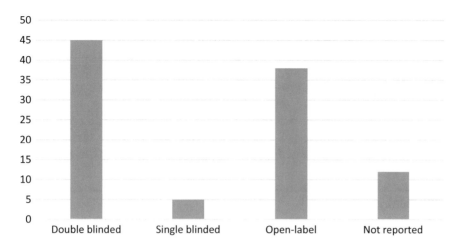

Fig. 8 Blinding level

vary from study to study, the graph (Fig. 8) illustrates the level of blinding in the articles searched.

11.2 Similarity to Intervention

The similarity of interventions has influence on the conservation of blinding. Therefore, a proper manufacturing of drugs and devices is needed [60, 61]. In the context of peripheral neuropathy, when the interventions consisted of drugs, the use of placebo matched in appearance, taste, smell, weight, disposal containers and administration rout secure blinding for participants and study personnel.

Peripheral neuropathy studies using devices face some challenges to protect blinding. The device may be matched in appearance to maintain investigators blinded, however the sensation of the active and sham procedure may allow participants to guess the treatment that they were assigned to, especially if they have previous experience with the intervention. Thus, eligibility criteria may assist to secure blinding by selecting patients who never had treatment before with the device. Another strategy is to use tools to simulate the same sensation that the active treatment provides. For instance, two studies with photo energy used heaters in the shame device to trick participants.

Among the procedure studies with PN, similarity of intervention consisted of similar procedure duration, smell of the ambient, and location of the procedure—these were used by Anastasi and colleagues for acupuncture trial [36]. Comments regarding similarity of intervention for double-blinded studies were present in 28 out of 45 papers. Some authors explained similarity of administration rout and schedule; others highlighted the use of same label for different drugs. The use of a code number to avoid identification of participants' name was also implemented.

11.3 Blinding Assessment

Blinding assessment is recommended to evaluate the risk of bias. It is usually performed by the application of a questionnaire at the end of the study. Study personnel intended to stay blinded (e.g., participants, study personnel) are asked to guess to which treatment they were assigned, similarity of intervention, among other questions. There are several tools to evaluate blinding, however it is not a common practice to report risk of systematic error due to lack of blinding [62]. Through our search, only three research works implemented an evaluation questionnaire at the end of the trial.

12 Recruitment

Participant recruitment in research studies has been one of the major challenges. Investigators should follow appropriate recruitment strategies during each study phase. If this is not the case, low recruitment rates could lead to failure of the research. Therefore, before start the recruitment process, investigators should spend sufficient time planning the recruitment strategies [63, 64].

The first step that studies on peripheral neuropathy should perform is to define the target population in order to enroll them into the protocol study. The target population can be determined based on clear inclusion and exclusion criteria, which allows screening potential participants that will take part of the study. During the recruitment planning investigators should also take into account a common principle called the funnel effect or Lasagna's law [63–65]. This phenomenon refers to an overestimation of the amount of participants willing to participate in a research study. Even more, after enrolling them into the protocol study; investigators may expect a decline in the amount of participants suitable to be randomized. In addition, based on this principle only 10% of the accessible population will be eligible for the study.

For instance, Richter et al. [66] performed a randomized clinical trial that initially included 396 participants. However, a total of 150 participants were screened out, of them 135 participants did not meet the study criteria, 3 participants experienced adverse effects, and 12 participants for other reasons. This process reflects the high impact of the recruitment strategies on the success or failure of a clinical trial.

13 Adherence

Adherence in clinical research refers to the compliance of participants to act in concordance with the study protocol. The poor adherence to the study protocol may have several consequences that affect the regular developing of a clinical trial, such as the exposition of unnecessary risk to the participants, inaccuracy of the results, and longer period of the study [65]. Therefore, poor

Table 12
Methods for improving adherence

	Adherence
Overall (%)	*85%*
Not reported (*n*)	12
Methods to improve adherence (n)	
Run in phase	6
Phone calls	3
Payment for transportation	1
Not reported	90

adherence can impact negatively the performance of the study, leading to stop the trial earlier.

Several factors have been identified as facilitators of low adherence, such as participants, intervention, and the disease. The belief of the participant play an important role to follow the study protocol as well as important is the access of the subjects to the research institution in order to continue the study visits [65]. Secondly, the side effects secondary to the intervention tested by investigators also can decrease the adherence rate affecting the participation during anytime of the study. Contrary to this, some diseases due to the natural course can impose a higher difficulty to follow the study protocol such as HIV, psychiatric conditions, and pediatric populations [63, 65].

In this context, the mean adherence rate of peripheral neuropathy studies was 85% (Table 12), while only in 12 studies the adherence rate was not reported. Similarly, the methods to improve adherence were not properly described in most of the studies (90%). However, only six investigations on peripheral neuropathy reported a run-in phase in order to ensure a participant adherence during the trial. Three studies employed a phone call technique to remind the study visits. Phone calls are useful for trials that have outpatient follow-ups, increasing retention substantially. LeMaster et al. conducted a randomized clinical trial to determine the effect of a lower-extremity exercise and walking intervention program on weight-bearing activity and foot ulcer incident in patients with DM and polyneuropathy [67]. In this trial, the study nurse performed calls every 2 weeks for a minimum of 10 min to motivate participants to follow their study protocol (walking plan). Moreover, the study nurse received 1 week of intensive training in motivational interviewing techniques to assure an effective method to retain participants in the protocol. This study reflects the importance of methods to improve adherence in a research study. The methods to improve adherence increase not only subject retention, also to solve questions or issues from participants related to the study. Table 12 summarizes our findings regarding methods used for adherence.

14   Statistical Analysis

Appropriate statistical analysis is critical for answering the tested research questions. Although there are many ways to analyze data, researchers have to follow basic assumptions when planning their statistical approach (Consult Chap. 1). In addition, researchers have to be acquainted to the main advantages and disadvantages of each analysis method in order to extract the best of their data [68].

As discussed previously, the main outcomes on peripheral neuropathy measure either clinical symptoms, electrophysiological parameters or toxicity. Further, we showed that Numerical Rating Scale and Visual Analogue Scale are the most frequently used assessments of symptoms, mainly pain. In addition, NCI criteria are the most used scale for toxicity, while many electrophysiological measures were employed in the studies.

Study endpoints can be further classified based on their level of measurement (Consult Chap. 1 for more details). Furthermore, the statistical analysis of a study depends strictly on the type of level of measurement considered for the study. Therefore, we will discuss the most frequently used statistical analysis for each type of data in order to guide clinical researchers during the design of a trial.

14.1   Time-to-Event Data

In this review, only one clinical trial used time-to-event data as its primary outcome (Table 13).

14.2   Categorical Data

In this chapter, 13 clinical trials had categorical data as their primary outcome. Chi-square was the most employed statistical test for this type of data. Further, other four trials analyzed categorical endpoints with Fisher's exact test. As discussed on Chap. 1, this test is an alternative to Chi-square when researchers are dealing with small sample sizes, and the main assumption for this test is not fulfilled [68]. Therefore, nine studies in total analyzed their primary outcome using tests that compare observed and expected values in tables of contingence. These tests are relatively simple to perform, and easy to understand.

In addition, two trials used logistic regression models to analyze their primary endpoints. Logistic regression models allow

Table 13
Time to event data

	Frequency	Kaplan-Meier curve
Overall	1	1
Time-to-progression	1	1

Table 14
Statistics approach by outcome

	Frequency	Descriptive	Fisher's exact test	Chi-square	Logistic regression	Cochran-Mantel-Haenszel
Overall	13	2	4	5	2	1
Overall response rate	2	1	1			
NCI toxicity	2		1	1		
NRS (>50%)	2				1	1
Symptoms (mild/moderate/ severe)	2		1	1		
PNP score	1			1		
WHO criteria physical and neurological examination	1		1			
CTCAE	2			2		
Clinical response (improved, unchanged, aggravated, increased or other)	1	1				
Reduced total neuropathy score	1				1	

researchers to adjust the statistical analysis for confounders and effect modifiers (Chap. 1). This approach enabled authors to identify small differences that could have remained undetected in direct comparisons between study groups. Finally, two studies only described frequencies and 95% confidence of intervals, while one employed the Cochran-Mantel-Haenszel model. Table 14 summarizes the statistical analysis by outcome.

14.3 Continuous Data

Studies on PN had continuous data as their primary outcome. Among these, 26 studies studied NRS or VAS with different approaches. Most trials analyzed NRS/VAS means with an analysis of covariance model (ANCOVA) adjusting for baseline values. A significant number of these studies opted for repeated-measures ANOVA model with time as one the independent variables. Seven studies compared means or medians using t-test (five studies), Wilcoxon signed-rank (1) or Mann-Whitney U test (1), depending on the distribution of the data. The option of using a nonparametric test is guided by the normality assumptions explained on Chap. 1. Overall, parametric comparison of means has more power than nonparametric tests. However, in the context of small samples that do not meet these assumptions, the use of nonparametric approaches is valid. Table 15 illustrates our statistical findings for continuous data.

Table 15
Statistics of continuous data

	Frequency	ANCOVA	ANOVA	Student's t test	Linear regression	Mann Whitney U test	Kruskal-Wallis	Wilcoxon	Friedman	Bayesian Markov chain Monte Carlo
Overall	78	13	28	4	4	3	22	1	1	2
NRS	26	10	5		1	1	8	1		1
Electrophisiological measures	20		13				6	1		
Brief Pain Inventory-SF	2	1			1					
VDS (change from baseline)	1									1
SWM	3		1				2			
Functional reach, tandem stance, unipedal stance measures	1		1							
EORTC-QLQ-CIPN20	2		1				1			
SkBF	2			1			1			
CPNE score	1	1								
Capillary blood cell velocity	2					2				
StepWatch-based total daily steps	1			1						
Neuropathic symptom score	1						1			
100 mm PVA scale	1	1								
Pain intensity rating scale	1			1						
Pain relief rating scale	1		1							

Measure							
ADP score	1			1			
Short form of McGill pain questionnaire	2		1		1		
NeuroQoL	1	1					
6-min walk distance	1			1			
Medical Research Council sum score	1						
Clinical total neuropathy score	1			1			
52-item West haven Yale multidimensional pain inventory	1					1	
Two-point discrimination test	1						
Cumulative symptom score	1						1
Cumulative sign score	1						1
Tandem stance, functional reach, unipedal stance	1						
HRV	1						

14.4 Missing Data

The concept of missing data is an important feature to understand in order to handle the data of a research study. If the missing data is not properly handled by the investigator, the analysis of the data will generate inaccurate or biased results. Therefore, investigators should understand the different methods to handle the missing data.

There are three types of missing data mechanisms, missing completely at random (MCAR), missing at random (MAR), and missing not at random (MNAR) [69]. The main idea behind this classification is whether the missing values are related to the study outcome and/or independent variables. Missing completely at random (MCAR) defines the case when missing values are independent of study outcomes and independent variables [69]. In contrast, missing at random (MAR) occurs whenever missing data depends on independent variables but do not relate to the outcome [69]. Finally, in missing not at random (MNAR) the missing data relates to study outcomes [69].

To deal with the aforementioned types of missing data, researchers have different options depending on the context. The most simple but worst methodology consists in analyzing only the subjects who completed the study, which is defined as the Complete Case Analysis (CCA) or Per-protocol analysis. This approach is valid whenever the mechanism of missing data is MCAR. However, this approach leads to decreased study power, since it excludes subjects from the main analysis. Therefore, the most recommended approach by the CONSORT criteria is to use a method of imputation of missing observations and analyze all participants as they were randomized, which is called intention-to-treat (ITT) analysis [50]. This approach allows researchers to preserve the benefit of randomization on statistical inferences, and can be used for MCAR and MAR mechanisms of missing data [69]. Further, there are different methods for imputing missing data and comply with the ITT principle. Those methods range from simple imputation approaches (e.g., last observation carried forward, worst case scenario, mean/median imputation) to sophisticated techniques (e.g., multiple imputation) (consult Chap. 1 for more details). Finally, there is no efficient method to deal with MNAR data, since the missing observations are directly related to the study outcome [69].

Most of the studies on peripheral neuropathy did not report the method of handle missing data (Table 15). Moreover, 28 studies described the type of method used to impute the missing data. Twenty-one performed the last observation carried forward (LOCF), which consists on imputing missing data from the last known outcome value. The LOCF is the most common method of imputation in clinical trials, according to the CONSORT guidelines [50]. However, authors have reported some limitations of this method due to the primary assumption, which consist in the lack of change in the outcome values from

Table 16

Methods to handle missing data

	Frequency
ITT	28
Last observation carried forward	21
Multiple imputation	2
Worst case scenario	1
Not specified	4
Not reported	80

the previous observations. Nevertheless, the Food and Drug Administration (FDA) has described this method as the more conservative alternative to the per-protocol analysis [69]. Conversely, two clinical trials on peripheral neuropathy employed the multiple imputation method, which is a more sophisticated method but more complex approach to data imputation. Finally, one study considered the worst-case scenario method to impute the missing values, which consists on imputing the missing observation according to the worst outcome. This approach is the most conservative method to impute missing data. However, it may lead to robust clinical results in case it demonstrates that missing data did not affect the study outcome. Table 16 summarizes methods to handle missing data.

15 Limitations

Even when taking the necessary time to plan a study with good methodological rational, limitations is part of every research. There are several sources of constraints; it may be due to resources, implementation or lack of planning of a trial. Some aspects can be noticed during study planning and others while the research is running. Regardless of when the authors took awareness of it, it is important to address the limitations of an investigation. Acknowledgment of flaws shows careful and critical thinking about the entire research process to the scientific community. Through our review 55 articles reported limitations and strengths of their studies. In this context, for peripheral neuropathy studies the most reported limitations are related to unblinding, sample size, study duration, self-report of patients for data collection and lack of a control or placebo group.

15.1 Blinding
Unblinding risk or lack of blinding is a source of bias that can be avoided with allocation concealment and proper comparison among similar interventions. Moreover, studies with peripheral neuropathy already have sources of bias such as self-report of outcomes, use of concomitant medication by participants, or comorbidities. Thus, it is highly recommended to address this limitation and the strategies to avoid it.

This was the most reported limitation among PN trial. In this sense, 21 papers reported the possibility of placebo effect in their results. It is noteworthy, that even though the majority of trials in this review had some level of blinding, as discussed on Sect. 11, a small amount of researchers evaluated if it was kept till the end of the study.

15.2 Sample Size
Small sample size was reported as a limitation by 17 articles, which increases the chances of a type II error. This caveat can be addressed during study planning by awareness of the target population, strategies for recruitment and adherence to avoid participant's dropout. It is important to plan how to achieve the necessary sample size and to increase compliance. For instance, a challenge of therapies for PN is intolerable adverse effects that may lead subjects to drop the trial. Therefore, strategies for dose adjustments such as titration and tapering are interesting in this context to maintain sample size.

Additionally, during the study planning some measures can be considerate to reduce the required sample size to achieve sufficient power. The inclusion of a homogeneous narrow population, and robust outcomes assessment increases internal validity and decrease the required sample for the study. However, for PN trials, the use of pain scales as primary outcome may implicate in a larger amount of participants. Therefore, an appropriate study design and strong evidence of the intervention may assist researchers on reducing it.

16 Conclusion

Peripheral neuropathy can be a severe clinical condition, although yet there is no standard highly efficacious treatment regardless of its etiology. Accordingly, there is an unmet clinical need for more research in this field. Nevertheless, it is important to design trials with adequate methodology, as well as proper application, and report of clinical investigation in this field. Learning with the limitations and strengths of previous researches are a good way to move forward. In this manner, this systematic review hopes to have highlighted the main aspects to assist future clinical trials in peripheral neuropathy.

References

1. Vinik A, Park T, Stansberry K (2000) Diabetic neuropathies. Diabetologia 43:957–973

2. Carozzi VA, Canta A, Chiorazzi A (2015) Chemotherapy-induced peripheral neuropathy: what do we know about mechanisms? Neurosci Lett 596:90–107. https://doi.org/10.1016/j.neulet.2014.10.014

3. Driessen CML, de Kleine-Bolt KME, Vingerhoets AJ et al (2012) Assessing the impact of chemotherapy-induced peripheral neurotoxicity on the quality of life of cancer patients: the introduction of a new measure. Support Care Cancer 20:877–881. https://doi.org/10.1007/s00520-011-1336-0

4. Seretny M, Currie GL, Sena ES et al (2014) Incidence, prevalence, and predictors of chemotherapy-induced peripheral neuropathy: a systematic review and meta-analysis. Pain 155:2461–2470. https://doi.org/10.1016/j.pain.2014.09.020

5. Cavaletti G, Marmiroli P (2010) Chemotherapy-induced peripheral neurotoxicity. Nat Rev Neurol 3:535–546. https://doi.org/10.1517/14740338.3.6.535

6. Argyriou AA, Chroni E, Koutras A et al (2006) Preventing paclitaxel-induced peripheral neuropathy: a phase II trial of vitamin E supplementation. J Pain Symptom Manage 32(3):237–244. https://doi.org/10.1016/j.jpainsymman.2006.03.013

7. Kottschade LA, Sloan JA, Mazurczak MA et al (2011) The use of vitamin E for the prevention of chemotherapy-induced peripheral neuropathy: results of a randomized phase III clinical trial. Support Care Cancer 19(11):1769–1777. https://doi.org/10.1007/s00520-010-1018-3

8. Please provide bibliographic details for Ref. 8

9. Offidani M, Corvatta L, Marconi M et al (2004) Common and rare side-effects of low-dose thalidomide in multiple myeloma: focus on the dose-minimizing peripheral neuropathy. Eur J Haematol 72:403–409. https://doi.org/10.1111/j.1600-0609.2004.00238.x

10. Razazian M, Bazyiar M, Moradian N et al (2014) Evaluation of the efficacy and safety of pregabalin, venlafaxine, and carbamazepine in patients with painful diabetic peripheral neuropathy. Neuroscienses (Riyadh) 19(3):192–198

11. Richardson PG, Sonneveld P, Shuster MW et al (2009) Reversibility of symptomatic peripheral neuropathy with bortezomib in the phase III APEX trial in relapsed multiple myeloma: impact of a dose-modification guideline. Br J Haematol 144(6):895–903. https://doi.org/10.1111/j.1365-2141.2008.07573.x

12. Portney LG, Watkins MP (2009) Foundations of clinical research: applications to practice. Pearson/Prentice Hall, Upper Saddle River, NJ

13. Please provide bibliographic details for Ref. 13

14. Vinik AL, Shapiro DY, Rauschkolb C et al (2014) A randomized withdrawal, placebo-controlled study evaluating the efficacy and tolerability of tapentadol extended release in patients with chronic painful diabetic peripheral neuropathy. Diabetes Care 37(8):2302–2309. https://doi.org/10.2337/dc13-2291

15. Raskin P, Huffman C, Toth C et al (2014) Pregabalin in patients with inadequately treated painful diabetic peripheral neuropathy a randomized withdrawal trial. Clin J Pain 30(5):379–390. https://doi.org/10.1097/AJP.0b013e31829ea1a1.

16. Vahdat L, Papadopoulos K, Lange D et al (2001) Reduction of paclitaxel-induced peripheral neuropathy with glutamine. Clin Cancer Res 7(5):1192–1197

17. Stubblefield MD, Vahdat LT, Balmaceda CM et al (2005) Glutainine as a neuroprotective agent in high-dose paclitaxel-induced peripheral neuropathy: a clinical and electrophysiologic study. Clin Oncol 17:271–276. https://doi.org/10.1016/j.clon.2004.11.014

18. Goldblatt EM, Lee W-H (2010) From bench to bedside: the growing use of translational research in cancer medicine. Am J Transl Res 2:1–18

19. Coriat R, Alexandre J, Nicco C et al (2014) Treatment of oxaliplatin-induced peripheral neuropathy by intravenous mangafodipir. J Clin Invest 124:262–272. https://doi.org/10.1172/JCI68730

20. Tamimi NA, Ellis P (2009) Drug development: from concept to marketing. Nephron Clin Pract 113:125–131. https://doi.org/10.1159/000232592

21. Mielke S, Sparreboom A, Steinberg SM et al (2005) Association of Paclitaxel pharmacokinetics with the development of peripheral neuropathy in patients with advanced cancer. Clin Cancer Res 11:4843–4850. https://doi.org/10.1158/1078-0432.CCR-05-0298

22. Verstanppen C, Koeppen S, Heimans J et al (2005) Dose-related induced peripheral neuropathy with unexpected off-therapy worsening. Neurology 64:1076–1077

23. Thisted RA, Klaff L, Schwartz SL et al (2006) Dextromethorphan and quinidine in adult

patients with uncontrolled painful diabetic peripheral neuropathy: a 29-day, multicenter, open-label, dose-escalation study. Clin Ther 28(10):1607–1618

24. Leon AC, Davis LL, Kraemer HC (2012) Role and interpretation of pilot studies in clinical research. J Psychiatr Res 45:626–629. https://doi.org/10.1016/j.jpsychires.2010.10.008. The

25. Paice JA, Ferrans CE, Lashley FR et al (2000) Topical capsaicin in the management of HIV-associated peripheral neuropathy. J Pain Symptom Manage 19(1):45–52

26. Rao RD, Michalak JC, Sloan JA et al (2007) Efficacy of gabapentin in the management of chemotherapy-induced peripheral neuropathy - a phase 3 randomized, double-blind, placebo-controlled, crossover trial (NOOC3). Cancer 110(9):2110–2118

27. Hotta N, Akanuma Y, Kawamori R et al (2006) Long-term clinical effects of epalrestat, an aldose reductase inhibitor, on diabetic peripheralneuropathy - the 3-year, multicenter, comparative aldose reductase inhibitor-diabetes complications trial. Diabetes Care 29(7):1538–1544

28. Strempska B, Bilinska M, Weyde W et al (2013) The effect of high-tone external muscle stimulation on symptoms and electrophysiological parameters of uremic peripheral neuropathy. Clin Nephrol 79(Suppl 1):S24–S27

29. Bestard JA, Toth CC (2011) An open-label comparison of nabilone and gabapentin as adjuvant therapy or monotherapy in the management of neuropathic pain in patients with peripheral neuropathy. Pain Pract 11(4):353–368. https://doi.org/10.1111/j.1533-2500.2010.00427.x.

30. Tuttle KR (2015 Apr) Effect of ruboxistaurin on albuminuria and estimated GFR in people with diabetic peripheral neuropathy: results from a randomized trial. Am J Kidney Dis 65(4):634–636

31. Valensi P (2005 Sep-Oct) A multicenter, double-blind, safety study of QR-333 for the treatment of symptomatic diabeticperipheral neuropathy - a preliminary report. J Diabetes Complications 19(5):247–253

32. Simpson DM, Estanislao L, Brown SJ et al (2008) An open-label pilot study of high-concentration capsaicin patch in painful HIV neuropathy. J Pain Symptom Manage 35(3):299–306

33. Cornblath DR, Hoke A (2006) Recent advances in HIV neuropathy. Curr Opin Neurol 19:446–450

34. Verma S, Estanislao L, Simpson D (2005) HIV-associated neuropathic pain: epidemiol-

ogy, pathophysiology and management. CNS Drugs 19(4):325–334

35. Schifitto G, McDermott MP, McArthur JC, Marder K, Sacktor N, Epstein L et al (2002) Incidence of and risk factors for HIV-associated distal sensory polyneuropathy. Neurology 58:1764–1768

36. Anastasi JK, Capili B, McMahon DJ, Scully C (2013) Acu/moxa for distal sensory peripheral neuropathy in HIV: a randomized control pilot study. J Assoc Nurses AIDS Care 24:268–275. https://doi.org/10.1016/j.jana.2012.09.006

37. Slangen R, Schaper NC, Faber CG et al (2014) Spinal cord stimulation and pain relief in painful diabetic peripheral neuropathy: a prospective two-center randomized controlled trial. Diabetes Care 37:3016–3024. https://doi.org/10.2337/dc14-0684

38. Rosenstock J, Tuchman M, Lamoreaux L, Sharma U (2004) Pregabalin for the treatment of painful diabetic peripheral neuropathy: a double-blind, placebo-controlled trial. Pain 110:628–638 PMID:15288403

39. Raskin P, Huffman C, Toth C et al (2014) Pregabalin in patients with inadequately treated painful diabetic peripheral neuropathy a randomized withdrawal trial. Clin J Pain 30(5):379–390. https://doi.org/10.1097/AJP.0b013e31829ea1a1

40. Richter RW, Portenoy R, Sharma U et al (2005) Relief of painful diabetic peripheral neuropathy with pregabalin: a randomized, placebo-controlled trial. J Pain 6:253–260. https://doi.org/10.1016/j.jpain.2004.12.007

41. Baron R, Brunnmuller U, Brasser M, May M, Binder A (2008) Efficacy and safety of pregabalin in patients with diabetic peripheral neuropathy or postherpetic neuralgia: open-label, non-comparative, flexible-dose study. Eur J Pain 12(7):850–858

42. Arezzo JC, Rosenstock J, LaMoreaux L, Pauer L (2008) Efficacy and safety of pregabalin 600 mg/d for treating painful diabetic peripheral neuropathy: a double-blind placebo-controlled trial. BMC Neurol 8:33. https://doi.org/10.1186/1471-2377-8-33

43. Satoh J, Yagihashi S, Baba M et al (2011) Efficacy and safety of pregabalin for treating neuropathic pain associated with diabetic peripheral neuropathy: a 14 week, randomized, double-blind, placebo-controlled trial. Diabet Med 28(1):109–116. https://doi.org/10.1111/j.1464-5491.2010.03152 PMID: 21166852

44. Smith EML, Pang H, Cirrincione C et al (2013) Effect of duloxetine on pain, function, and quality of life among patients with chemotherapy-induced painful peripheral neuropa-

thy: a randomized clinical trial. JAMA Psychiat 309:1359–1367

45. Walton MK, Iii JHP, Hobart J et al (2015) Clinical outcome assessments: conceptual foundation—report of the ISPOR clinical outcomes assessment – Emerging good practices for outcomes research task force. Value Health 18:741–752

46. Ravina B, Cummings J, Mcdermott M, Poole RM (2012) Selecting outcome measures. In: Clinical trials in neurology. Cambridge University Press, Cambridge, pp 69–77

47. Stephens RJ, Hopwood P, Girling DJ, Machin D (1997) Randomized trials with quality of life endpoints: are doctors' ratings of patients' physical symptoms interchangeable with patients' self-ratings? Qual Life Res 6:225–236. https://doi.org/10.1023/A: 1026458604826

48. Nagano H, Sanai H, Muraoka M, Takagi K (2012) Efficacy of lafutidine, a histamine H2-receptor antagonist, for taxane-induced peripheral neuropathy in patients with gynecological malignancies. Gynecol Oncol 127:172–174. https://doi.org/10.1016/j.ygyno.2012. 06.029

49. Atkinson AJJ, Colburn WA, DeGruttola VG et al (2001) Biomarkers and surrogate endpoints: preferred definitions and conceptual framework. Clin Pharmacol Ther 69:89–95. https://doi.org/10.1067/mcp.2001. 113989

50. Schulz KF, Altman DG, Moher D, Group C (2010) Academia and clinic annals of internal medicine CONSORT 2010 statement: updated guidelines for reporting parallel group randomized trials. Ann Intern Med 152:727–732. https://doi.org/10.7326/ 0003-4819-152-11-201006010-00232

51. Schulz KF, Grimes DA (2005) Sample size calculations in randomised trials: mandatory and mystical. Lancet 365:1348–1353. https://doi.org/10.1016/ S0140-6736(05)61034-3

52. Hotta N, Toyota T, Matsuoka K et al (2001) Clinical efficacy of fidarestat, a novel aldose reductase inhibitor, for diabetic peripheral neuropathy. Diabetes Care 24:1776–1782

53. Fonseca VA, Lavery LA, Thethi TK et al (2013) Metanx in type 2 diabetes with peripheral neuropathy: a randomized trial. Am J Med 126:141–149. https://doi.org/10.1016/j. amjmed.2012.06.022

54. Cohen J (1988) Statistical power analysis for the behavioral sciences, 2nd edn. Academic, New York

55. Schulz KF, Grimes DA (2002) Generation of allocation sequences in randomised trials: chance, not choice. Lancet 359:515–519

56. Vickers AJ (2008) How to randomize. J Soc Integr Oncol 4:194–198

57. Altman DG, Bland JM (1999) Treatment allocation in controlled trials: why randomise? BMJ 318:1209. https://doi.org/10.1136/ bmj.329.7458.168

58. Schulz KF, Grimes DA (2002) Allocation concealment in randomised trials: defending against deciphering. Lancet 359:614–618

59. Schulz KF (1995) Subverting randomization in controlled trials. JAMA 274:1456–1458

60. Schulz KF (2002) The landscape and lexicon of blinding in randomized trials. Ann Intern Med 136:254–259

61. Schulz KF, Grimes DA (2002) Blinding in randomised trials: hiding who got what. Lancet 359:696–700

62. Bang H, Ni L, Davis CE (2004) Assessment of blinding in clinical trials. Control Clin Trials 25:143–156. https://doi.org/10.1016/j. cct.2003.10.016

63. Nicholson LM, Schwirian PM, Groner JA (2015) Recruitment and retention strategies in clinical studies with low-income and minority populations: progress from 2004–2014. Contemp Clin Trials. https://doi. org/10.1016/j.cct.2015.07.008

64. Thoma A, Farrokhyar F, Mcknight L, Bhandari M (2010) How to optimize patient recruitment. Can J Surg 53:205–210

65. Robiner WN (2005) Enhancing adherence in clinical research. Contemp Clin Trials 26:59–77. https://doi.org/10.1016/j.cct.2004.11.015

66. Richter RW, Portenoy R, Sharma U et al (2005) Relief of painful diabetic peripheral neuropathy with pregabalin: a randomized, placebo-controlled trial. J Pain 6:253–260. https://doi. org/10.1016/j.jpain.2004.12.007

67. Lemaster JW, Mueller MJ, Reiber GE et al (2008) Effect of weight-bearing activity on foot ulcer incidence in people with diabetic peripheral neuropathy: feet first randomized controlled trial. Phys Ther 88:1385–1398. https://doi.org/10.2522/ptj.20080019

68. Bailar JC, Mosteller F (1988) Guidelines for reporting in articles for medical journals. Ann Intern Med 108:266–273. https://doi. org/10.7326/0003-4819-108-2-266

69. Dziura JD, Post LA, Zhao Q et al (2013) Strategies for dealing with missing data in clinical trials: from design to analysis. Yale J Biol Med 86:343–358

Chapter 13

Epilepsy

Mirret M. El-Hagrassy, Ana C. Texeira-Santos, and Felipe Fregni

Abstract

The aim of this chapter is to help investigators design interventional clinical trials in epilepsy, by reviewing and analyzing the top cited articles on such trials. A literature search on "epilepsy" was performed using Web of Science, and the top 100 cited articles on interventional clinical trials in epilepsy between 2010 and 2015 were analyzed and used to provide guidance to future investigators. Based on CONSORT guidelines for reporting clinical trials, we discuss relevant findings in detail, showing the methods used to design various types of interventional epilepsy trials. We make recommendations on best practice methods in epilepsy trial design.

Key words Epilepsy, Seizures, Clinical trials in neurology, Clinical trials in epilepsy, Scientific research, CONSORT guidelines

1 Introduction

Epilepsy is one of the most prevalent, disabling, and misunderstood chronic neurologic disorders. It affects an estimated 69 million people worldwide, roughly 90% of them living in "developing" countries [1]. This number may underestimate epilepsy's burden, as mortality rates can be high, particularly in the early course of the disease. For this reason, Ngugi et al. evaluated the (limited) incidence literature via systematic review and meta-analysis, and reported epilepsy's median incidence at 50.4/100,000/year (IQR 33.6–75.6) [2].

Epilepsy may manifest at any age, from newborns to the elderly, and may or may not have a known etiology. It exists when a patient's brain shows a pathological tendency for the recurrence of seizures. Epilepsy may occur as part of an epileptic syndrome, or may follow any of a number of insults, such as strokes, encephalitis, and traumatic brain injury, even after resolution of the acute injury. Epilepsy may progress over time, and patients often have associated comorbidities [3], such as cognitive or psychiatric comorbidities; children may have abnormal development.

Felipe Fregni (ed.), *Clinical Trials in Neurology*, Neuromethods, vol. 138,
https://doi.org/10.1007/978-1-4939-7880-9_13, © Springer Science+Business Media, LLC, part of Springer Nature 2018

For practical purposes, epilepsy is typically defined by the occurrence of two or more unprovoked (or reflex) seizures more than 24 h apart, or only one such seizure but in the presence of an equivalent risk of seizure recurrence (60% or more) over the following 10 years, or in case of an epilepsy syndrome diagnosis [3].

The goal of treatment is freedom from seizures and the absence, or at least minimization, of treatment-related side effects. However, not all patients respond sufficiently to antiepileptic drugs (AEDs); 25–30% of patients will have medically intractable epilepsy [4], with "failure of adequate trials of two tolerated and appropriately chosen and used AED schedules (whether as monotherapies or in combination) to achieve sustained seizure freedom" [5]. Such patients are unlikely to receive much benefit from the addition of more AEDs, and evaluation for epilepsy surgery (in focal epilepsy) becomes the next step. Alternatives to resective surgery (mainly if nonfeasible or too risky) include gamma knife radiosurgery [6], deep-brain and responsive neurostimulation [7, 8], extracranial device implants (e.g., vagal nerve stimulation), or specific diets. Other therapeutics used only experimentally at this time include noninvasive neuromodulation techniques such as repetitive transcranial magnetic stimulation (rTMS), and transcranial direct current stimulation (tDCS) [9].

Epilepsy can have devastating effects on patients' lives and socio-occupational functioning. Some of these effects are seizure-related, e.g., motor vehicle accidents, falls, injuries, and the risk of aspiration or drowning. Some are therapy-related adverse events, whether due to AEDs (e.g., cognitive decline, congenital malformations), surgery (e.g., focal weakness, memory loss) or other interventions. Death may result from complications related to seizures (e.g., accidents, status epilepticus), AEDs, or underlying brain abnormalities [10]. Additionally, an estimated 1.16/1000 patients with epilepsy die annually of SUDEP (sudden unexpected death in epilepsy) [11].

1.1 Challenges of Epilepsy Clinical Trials

Epilepsy clinical trials have numerous challenges. Not all seizures are clinically identifiable, and even clinical seizures may not be recognized or appropriately recorded by patients and/or observers. This becomes even more challenging in very young or cognitively impaired patients. Pediatric epilepsies occur in a changing brain and the varying pathophysiology of epilepsy across the human life span can make it difficult to extrapolate between pediatric and adult studies. Even in adults, the manifestations of epilepsy may evolve over time. Epilepsy populations are overall heterogeneous, and seizure/epilepsy classifications can be variable [12–14].

Data collection may be limited clinically by the need for patient and observer recognition of seizures, or by poor record keeping. Medication dosing, patient compliance, fluctuating drug levels, the occurrence of and tolerance to side effects may cause variables that need to be controlled for. Electroencephalograms (EEGs) often provide valuable data; however, routine EEGs as well as long-term video-EEG (VEEG) monitoring may be limited by artifacts and spatial resolution, and invasive evaluations (e.g., stereo-EEG (SEEG) or subdural electrodes) have the added challenge of being blind to regions outside of the relatively confined electrode locations [15].

Epilepsy surgery and invasive neurostimulation trials are also constrained by suboptimal patterns of patient referral for presurgical workups, even in advanced urban settings; opportunities in developing countries and rural regions drop significantly, and low feasibility for appropriate neuroimaging further compounds the issue.

1.2 Room for Improvement in Epilepsy Clinical Trial Design, and Need for More Trials

The ILAE sub-commission on evidence based guidelines for the treatment of epilepsy had attempted to identify the variables affecting a specific AED's suitability for newly diagnosed or previously untreated epilepsy patients [16]. However, only seizure/epilepsy syndrome-specific efficacy/effectiveness lent itself to evidence-based analysis. The sub-commission found an absence of comprehensive AED adverse effects data, and they reported, among other things, that the existing RCTs had significant methodological problems.

As explained in those guidelines, "efficacy" indicates a medication's ability to produce seizure freedom; "tolerability" indicates the "incidence, severity, and impact" of adverse effects related to the AEDs [17, 18], and "effectiveness" indicates AED efficacy and tolerability, reflected in retention on treatment [17]. Thus, identifying an intervention's efficacy and effectiveness should be the main end goal while designing any interventional clinical trial, with effectiveness encompassing safety/tolerability. When developing a hypothesis for phase I or exploratory studies, it is important to keep such goals in mind for future phases, and to establish appropriate Go/No Go criteria.

1.3 On this Chapter

Clinical trials in epilepsy need to address multiple challenges in order to be clinically useful. An appropriate trial design is crucial for success, and also helps minimize patient risk, as well as financial and effort-related burdens to researchers and institutions.

The purpose of this chapter is to illustrate how to design clinical trials in epilepsy, using the 100 most highly cited interventional trials in epilepsy between 2010 and 2015 by way of example (we also included trials the articles called "observational," but which were actually experimental studies as they involved groups receiving an intervention, e.g., Sergot et al. [19]) (Fig. 1).

2   Methods

On October 12, 2015, we performed a systematic review of the literature, searching all databases in Web of Science for "epilepsy" under Title. We included only clinical trials (as a Document Type) from 2010 to 2015, and then sorted the results with the highest

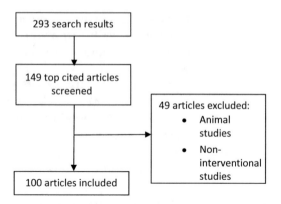

Fig. 1 Schematic illustration of search results and exclusions

cited articles on top. After that, we manually inspected the titles, abstracts, and when necessary, the articles for interventional trials in humans.

2.1 Initial Glance: CONSORT Criteria, Citations, and Positivity for Primary Outcomes

Of the 100 articles, only 28 were in journals that endorsed following CONSORT criteria. Table 1 summarizes the general characteristics of these 100 articles.

In this dataset, only 16 articles had been cited over 30 times (half of them published in 2010); only 4 of those 16 articles had been cited more than 100 times, having been published respectively in Epilepsia, Neurology, NEJM, and JAMA. Epilepsia, which has an impact factor (IF) of 4.424, had published the greatest number of articles in this dataset ($n = 29$), followed by Epilepsy & Behavior (IF 2.225, $n = 14$), Seizure (IF 1.879, $n = 9$), and Neurology (IF 8.092, $n = 7$). Only one article was published in NEJM (IF 56.170), one in JAMA (IF 33.569), and two in Lancet Neurology (IF 23.736).

When designing a clinical trial, it is best to plan ahead for publication (ideally in the case of both positive and negative studies, although the latter can be challenging), to the point of selecting a journal or journals of interest and reviewing their publication requirements in advance. The paragraph above highlights the importance of selecting the appropriate journal for your study; specialized epilepsy journals tend to have a lower IF than neurology journals, which have a lower IF than medicine journals. However, this dataset suggests that more specialized journals are often a better target (possibly because they are more widely read by the target audience).

Overall, the likelihood that your article will get published in a high-end journal *and* become highly cited will depend a great deal on the clinical relevance of the research question, the design of the trial (taking feasibility, funding, and the minimization of biases into account), its implementation and whether your results showed a

Table 1
General characteristics of article in datasets

			Number (Total *N* = 100)
Year	2010		23
	2011		26
	2012		21
	2013		19
	2014		10
	2015		1
Phase	Drug	Pilot	5
		Phase I	6
		Phase I/II	1
		Phase II	9
		Phase II/III	0
		Phase III	29
		Phase IV	2
	Device/diet/procedure	Exploratory	29
		Pivotal	17
		Post-marketing	2
Trial design	RCT (*n* = 57)	2 parallel arms	43
		3 parallel arms	4
		4 or more parallel arms	6
		Factorial	0
		Cross over	4
	Non RCT (*n* = 43)	Single arm	31
		2 parallel arms	10
		3 parallel arms	2
Sample size	0–20		22
	21–50		22
	51–100		18
	101–500		33
	501–1000		3
	1001 + Above		2
	Not applicable		0
Type of intervention	Drug		52
	Device		26
	Diet/procedure		22
Placebo controlled	Yes		20
	No		80
Number of citations (34 articles had <10 citations)	10–30		50
	31–40		7
	41–50		1
	51–60 ·		2
	61–70		0
	71–100		2
	101–200		3
	>200		1

(continued)

Table 1
(continued)

		Number (Total *N* = 100)
Journal impact factor	1–5	83
	5–10	12
	10–20	1
	>20	4
CONSORT criteria	Yes	28
	No	72
Positive for primary outcome	Yes	76
	No	12
	Not applicable	12

positive primary outcome, all of which will, in turn, depend in large part on the design. We did not compare how frequently articles with positive vs. negative primary outcomes were cited because many articles in this dataset did not clearly distinguish between their primary and secondary outcomes.

It is important to stress that strong research begins with a strong research question. At a minimum, a good research question follows the FINER criteria: it should be Feasible, Interesting, Novel, Ethical, and Relevant. Please refer to Chap. 1 for further exploration of the FINER criteria.

3 Trial Design

3.1 What Type of Trial Should You Design?

The gold standard for a clinical trial is the randomized, placebo-control, double-blind trial. However, such a design is not always feasible.

In our dataset, 57 articles were on RCTs (32 on drugs, 25 on devices or procedures), while 43 were on nonrandomized trials (20 on drugs, 23 on devices or procedures). For the purposes of this chapter, articles on open label extensions of RCT trials were analyzed as RCTs. The number of arms during the open label phase was analyzed, e.g., Halasz et al. had three arms during the placebo-control phase, and two arms during the open label phase [20], so it was analyzed as a two arm RCT. Open label phases were not marked as placebo-control, even if the previous phase had been placebo-controlled.

Only 20 articles were on placebo-controlled studies. There were an additional three articles on open label extensions of placebo-controlled studies, and one article on a study with an active control. This may reflect the difficulty of placebo-control in some studies overall, either due to technical or ethical issues. It is not typically feasible to have placebo-control in diet studies, for example, and placebo surgical procedures would be ethically and technically challenging. However, some device trials allow for sham stimulation (e.g., DBS, RNS, and VNS), and subjects can be crossed over from sham to active stimulation.

There is always the ethical challenge of giving placebo even in drug trials, considering the prevalence of trials on patients with refractory epilepsy. In some cases, a placebo is not necessary, e.g., when comparing established interventions in the case of *Ethosuximide, Valproic Acid, and Lamotrigine in Childhood Absence Epilepsy*, the landmark study published in NEJM [21]. Superiority and noninferiority trials in general do not require a placebo arm. However, it is worth noting that there can be versatility in placebo designs, e.g., Uberos et al. reported on an uncontrolled non-blinded single-arm study that had placebo intervals initially [22].

We will further discuss the benefits and challenges of randomization, placebo-control and blinding later in this chapter.

In this section we will discuss clinical trial design as relates to trial phases, number of arms, and study settings. As you will see, the type of intervention will play a significant role in your options for designing a trial.

3.1.1 Relationship Between Phase and Trial Design

For pharmaceutical (drug) studies, phases are divided into the following:

Pilot studies are small preliminary studies "testing the waters" of a new approach or novel therapy prior to developing larger scale Phase I or II studies. Pilot studies can evaluate the feasibility of therapy and its side effects, as well as challenges related to costs, recruitment, randomization, adherence, and possible effect sizes. Pilot studies are conducted on small population samples, and do not evaluate safety, efficacy or effectiveness. They can be used to test or compare new drugs, or established drugs for new indications (e.g., divalproex sodium for bipolar disorder [23]), new ways of dosing or modes of administration (e.g., loading vs. titration of valproate in acute mania [24]), etc. It is important for trials to have well thought out hypotheses with specific and clinically relevant outcomes likely to contribute to the literature. Pilot studies are thus avoided unless necessary (e.g., to reduce the risk of failure of a larger study on a new therapeutic or method).

Our dataset included five pilot studies, two of them RCTs as seen in Table 2 below, both double-blind, placebo-controlled crossover pilot trials [25–29]. For example, in Gonzalez-Heydrich et al. pediatric subjects with Attention Deficit Hyperactivity Disorder (ADHD) and epilepsy were assigned to one out of three OROS-methylphenidate maximum dose groups or to placebo, designed such that if they tolerated the dosing, they would be crossed over to the next-higher dose group; but if one child's epilepsy worsened significantly on active OROS-methylphenidate, then three or more other subjects would be tested at that dose level, and if two subjects got worse, the dose level just below would be the maximum dose for the remainder of the study (however, none of the subjects required these adaptations) [25]. This is one way to test if therapy addressing a comorbid condition (e.g., ADHD) risks worsening a patient's epilepsy.

Phase I clinical trials evaluate the safety of a new therapy, evaluate its pharmacokinetics, determine a safe dosage range, and identify side effects. They are tested in a small group of subjects, typically 20–100 subjects, although the six articles on phase I trials in our dataset had a minimum sample of 10 and median of 19 (see Table 3 above) [22, 30–34]. More objective measures of primary endpoints, e.g., surrogate values, may help support smaller samples.

Phase II clinical trials further evaluate safety and provide preliminary efficacy data, typically with samples of 100–300 subjects. Our dataset included nine articles on phase II trials, although again, sample sizes were relatively small (see Table 3) [35–43]. *Phase I/II* trials share features of both phase I and II, and are seldom chosen; we had one such article in our dataset [44]. We had no phase II/III trials.

Phase III clinical trials test the efficacy of interventions in a larger study population, typically 300–1000s of subjects, comparing the new therapeutic to standard treatment (or in combination with it), or comparing it to placebo. Our dataset included 29 articles on phase III trials, including 23 RCTs in multicenter settings and 6 non-RCTs (see Table 2) [19, 20, 45–71]. Mean and median sample sizes tended to be in lower typical range (see Table 3).

Phase IV clinical trials are conducted after the drug is FDA approved, and they are designed to gather more efficacy data in larger populations, typically 100–10,000s, and to evaluate potential long-term effects. Our sample included two articles on phase IV trials [72, 73], with the smaller sample ($n = 80$) belonging to the antiepileptic drug and osteoporosis prevention trial (ADOPT) [73], which measured bone-marrow density (a surrogate outcome) and had sample size and power calculations from previous studies.

For devices, diet/procedures, phases are divided into the following

Exploratory studies evaluate safety, design/parameter improvement and define the limits of tolerability in a small sample, typically fewer than 30 subjects. They include pilot and feasibility studies (similar to phase I and II studies in drugs). Our dataset included 29 articles on exploratory studies [6, 7, 74–100].

Pivotal studies test safety and effectiveness in a larger population, typically comprising 30–1000 subjects, and are usually multicenter RCTs. Our dataset included articles on 17 pivotal studies, including 15 RCTs [8, 101–116].

Post-marketing studies test intervention effectiveness as well as long-term or rare adverse events, typically on 100–10,000s of sub-

Table 2
Trial design: number of articles by trial phase, study setting, and intervention

Trial design	Study setting	Phases										Total per setting	N = 100
		Drugs								Devices/procedures			
		Pilot	I	I/II	II	II/III	III	IV	Exploratory	Pivotal	Post-marketing		
RCT													
2 parallel arms	Single center	0	1	0	2	0	0	1	5	8	0	17	43
	Multicenter	0	0	0	2	0	13	0	2	7	2	26	
3 parallel arms	Single center	0	0	0	0	0	0	0	0	0	0	0	4
	Multicenter	0	0	0	0	0	4	0	0	0	0	4	
4 or more parallel arms	Single center	0	0	0	0	0	0	0	0	0	0	0	6
	Multicenter	0	0	0	0	0	6	0	0	0	0	6	
Factorial	Single center	0	0	0	0	0	0	0	0	0	0	0	0
	Multicenter	0	0	0	0	0	0	0	0	0	0	0	
Cross over	Single center	2	0	0	1	0	0	0	1	0	0	4	4
	Multicenter	0	0	0	0	0	0	0	0	0	0	0	
Non RCT (quasi-experimental)	Single center	2	4	0	3	0	4	0	16	2	0	31	43
	Multicenter	1	1	1	1	0	2	1	5	0	0	12	

Table 3
Sample size per article by intervention and phase of trial

Sample size by intervention/phase	Drug								Device		Procedure		
	Pilot	I	I/II	II	II/III	III	IV	Exploratory (pilot and feasibility)	Pivotal	Post-market	Exploratory (pilot and feasibility)	Pivotal	Post-market
Mean	16	31	20	68	–	356	166	28	150	2143	35	81	207
Median	12	19	20	45	–	314	166	22	122	2143	30	76	207
Minimum	3	10	20	13	–	30	80	2	30	2143	10	37	207
Maximum	33	86	20	126	–	1698	251	65	358	2143	68	209	207
Number of articles in category	5	6	1	9	0	29	2	18	7	1	11	10	1

jects; they are similar to phase IV drug trials. Our dataset included only two articles on post-marketing studies, one of which was on the International Subarachnoid Aneurysm Trial (ISAT), and which had the largest sample in the dataset ($n = 2143$); Hart et al. was an article on epilepsy after subarachnoid hemorrhage (SAH) treated by coil embolization vs. clip occlusion of ruptured aneurysms [117]. Note that the ISAT's original outcomes related to Rankin scales, not seizures, and this post-marketing study was performed due to the paucity of data investigating epilepsy risk after coil occlusion, and due to the lack of data comparing seizures following coil embolization vs. clip occlusion. They investigated time from SAH to first seizure after discharge (years). The other post-marketing study compared seizure freedom rates (Engel Outcome Scale Class I) following 2.5-cm versus 3.5-cm mesial temporal resection in temporal lobe epilepsy (TLE) in an effort to optimize seizure freedom following surgery [118].

The sample sizes of the device and diet/procedure studies were more typical, compared to the drug trials.

Our dataset had nearly twice as many articles on exploratory studies as pivotal studies, largely on similar interventions, mostly relating to various neuromodulation techniques (whether invasive or noninvasive), diet or behavioral therapies, laser or radio-surgery techniques, surgical techniques (e.g., *Early Surgical Therapy for Drug-Resistant Temporal Lobe Epilepsy: A Randomized Trial*, which had a primary outcome of freedom from disabling seizures at year 2 of follow-up [102]). The sample also included more novel interventions in epilepsy, e.g., Mozart K.448 music [89]. The ratio of exploratory to pivotal studies may relate to the perceived need for more therapeutic options in various epilepsy populations, but some types of intervention rest on a continuum between exploratory and pivotal studies (e.g., if a patient undergoes invasive neurostimulation techniques in the exploratory phase without complications, you should try to keep them for the pivotal phase).

The relatively small numbers of articles on post-marketing studies (two for drug and two for device/procedures) in this sample may hint at the need for more long-term research into interventions once they are approved for clinical use. A chronic disorder with as much heterogeneity as epilepsy, which may evolve over the years, and where the pharmacodynamics and efficacy of treatment may change over time, would likely benefit from such studies.

3.1.2   Trial Design and Trial Arms

Subjects in a parallel group designed trial are randomized to the intervention or control. This is the most commonly used design (probably because it is feasible in most cases, but it can require a larger sample than other designs).

The number of arms required for an RCT is typically fairly straightforward. If you are comparing two interventions or two conditions, then you will need two arms; if comparing three, then you will need three arms, and so on. One of those two or more arms may be for placebo. Non-RCTs may have one or more arms, but most of the articles on non-RCTs in our dataset were on single-arm studies ($n = 30$).

Most of the articles in our dataset overall ($n = 54$) were on trials with two parallel arms (43 on RCTs and 10 on non-RCTs); another six were on trials with three parallel arms (including four articles on RCTs, three of which were on the same trial comparing ethosuximide, valproic acid, and lamotrigine in childhood absence epilepsy). Finally, six were on RCTs with four or more parallel arms; two out of those six articles were on the same study which compared three dose groups of eslicarbazepine with each other and with a placebo group in the blinded phase, and on the open-label extension (OLE) of this study [48, 58]; as we considered OLEs part of the RCT group (but not placebo control), we included it here, and included it as four arms based on the original trial (however, in the OLE, the placebo arm had been added to the 800 mg dose group, and the dosing of all groups had been adjusted freely, so it was no longer truly four arms).

Trinka et al. was on an international superiority trial utilizing a two parallel group design (considered four arms for the purpose of this chapter), where newly diagnosed epilepsy patients were allocated to one of two preferred standard therapies as first line treatment: the VPA stratum (for generalized epilepsy) or the CBZ stratum (for focal epilepsy); patients in these two strata were then randomized to levetiracetam vs. VPA-ER (in the VPA stratum) or vs. CBZ-CR (in the CBZ stratum) [63].

When taking repeated measures (a repeated measure design), crossover trials are an option: patients are assigned to a sequence of (at least two) treatments, which may be in the form of different dosing assignments and/or placebo. Subjects may serve as their own controls. However, selecting an appropriate washout period is important, particularly in the case of drugs with long half-lives, and in stimulation trials. For example, McLachlan et al. tested the effects of chronic bilateral direct hippocampal stimulation on seizure frequency in only two patients with intractable TLE: 3 months (baseline period) after implantation, the stimulator was randomly turned ON or OFF for another 3 months; they then followed each ON interval with a 3 month washout period (of no stimulation), looking for holdover effects [75]. Our dataset included four crossover trials, all RCTs.

We found no factorial trials. In a 2×2 factorial design trial, subjects are randomized to a treatment (A) or placebo to test one hypothesis, then randomized again in each group to another treatment (B) or its corresponding placebo in order to test a second hypothesis. In this way, two different hypotheses can be tested:

the independent effects of the two treatments (A and B) or the interaction between the two treatments. Factorial designs allow the measurement of effects or interactions which may not have been apparent otherwise [119].

3.1.3 *Study Setting*   It is important to have an appropriate setting for your trial. Is your center a good place to recruit the necessary sample? Do you need to collaborate with another center or centers? There are a number of issues to consider: first and foremost, the validity of your research may depend at least in part on your center. Single center studies may have results that are not sufficiently generalizable, limiting their clinical utility—or, a single center might be the best place to recruit a highly specific population, e.g., families with a particular genetic mutation. But if you want your research to have utility to all or most epilepsy patients, then larger samples in multicenter settings are more likely to increase the external validity of the data, and its generalizability to the epilepsy population at large.

Trials in multiple countries and continents are even better, although those will have other challenges regarding variability in data collection and interpretation, operator variability for procedures, etc. When designing a collaborative trial, it is important to answer the following questions: Why is it better to collaborate, why with this person or group, and what is each person's role in the trial?

In some cases, investigator participation is limited to a few individuals with highly specific training, restricting opportunities for collaboration. One thing to note is that many tertiary epilepsy centers have a large catchment area with little competition, sometimes even crossing state boundaries. Populations in such single centers may be sufficiently diverse for your trial's purposes. However, there may be recruitment challenges whenever patients are expected to travel far, especially for multiple visits or frequent follow-ups.

The ability to recruit and retain patients should always be an important part of the study design, and should inform decisions about study duration and visit frequency. In some cases the intervention and data collection may be performed long distance (e.g., it is often possible to mail drugs to patients, and for patients to keep seizure diaries or respond to questionnaires over the phone). On the other hand, some studies will require inpatient admissions, e.g., for VEEG monitoring, which may help with regard to adherence to protocol and obtaining more objective data, but will present the challenge of finding patients who are able and willing to stay.

In our dataset, 48 articles reported on studies in multicenter settings (36 RCTs, 12 non-RCTs), and 52 on studies in single-center settings (21 RCTs, 31 non-RCTs). In rare cases it was unclear if a trial utilized one or more centers; these were counted as single-center trials.

"Multicenter" was defined as two centers and above, but multicenter trials ranged from 2 to 269 centers, across one country (typically the USA) to "23 European countries and Australia" [63]. About 78% of articles on drug RCTs (25/32) reported on multicenter trials, compared to 44% of articles on device/procedure RCTs (11/14). Drug trials were overall more likely to be multicenter than device/procedure trials (62% or 32/52 for drug trials overall vs. 33% or 16/48 for device trials overall; see Table 4). As explained above, the design and feasibility of device/procedure multicenter trials is often more challenging than that of multicenter drug trials. Further breakdown by study phases and arms is given in Table 2.

Not all of the articles on multicenter trials gave their exact locations, but it seems that they tended to be more heavily distributed in developed countries, with many of them taking place only in the USA. It is unclear how much of this may be due to population size, accessibility, institutional support, funding for research and how much might be an artifact of this dataset resulting from an English language-based search.

4   Participants

In this section we focus on participant inclusion and exclusion criteria in our dataset, and discuss some considerations when designing your own trial. It is important to plan for scientific validity as well as recruitment feasibility when deciding on participant inclusion and exclusion criteria. When it is necessary to alter these criteria, delays of weeks or months can occur due to lost recruitment time, time for IRB approval and sometimes time to change recruitment materials, e.g., advertisements and flyers.

4.1   Age

For the purposes of this chapter, trials that allowed the inclusion of pediatric or adult populations but did not actually recruit any were considered pediatric or adult, rather than both (e.g., Kwan et al. allowed for subjects down to 16 years of age, but all partici-

Table 4
Study center setting by intervention and randomization

		Drugs	Devices/procedures	Totals
RCT	Single center	7	14	21
	Multicenter	25	11	36
Non RCT	Single center	13	18	31
	Multicenter	7	5	12
Totals		52	48	100

pants were adults [72]). About a fifth of the articles reported including both pediatric and adult populations; in those articles, most age ranges (when reported) began from adolescents 12–16 years of age (to middle-aged or elderly subjects); less often age ranges began as young as 2 years of age (to 20 or 21 years of age). Overall, 71 articles included adult subjects (ranging from 18-year-olds to the elderly), and 49 included pediatric subjects, while two articles did not report on age groups. Device trials were more than twice as likely to include adult as pediatric populations (see Table 5 and Fig. 2).

Age is an important variable in epilepsy studies due to heterogeneity in the likely underlying pathophysiological conditions, comorbidities, and medication types and effects. Controlling for age group is important for the internal validity of studies evaluating age-related syndromes, e.g., childhood absence epilepsy (CAE). However, when investigating intractable epilepsy as a whole, age inclusivity is better (subgroup analyses based on age can be performed later), as long as ethical considerations are satisfied. While pediatric populations are typically defined as under 18 years of age, it is helpful to specify the ages included in your study, because the cutoff between adolescence and adulthood may vary from one country to another; sometimes language is also used differently, e.g., Herzog et al. referred to their sample from age 13–45 years as "women" [56].

4.2 Sex

Two articles reported on progesterone therapy in women with epilepsy [34, 56], and one on men with epilepsy and hypogonadism [32]. Most of the other articles reported recruiting males and females.

4.3 Pregnancy and Breastfeeding

Over a dozen trials specifically excluded pregnant subjects, about a handful excluded those who were breastfeeding, and a number of trials required adequate contraception (some even describing the type of contraception). When requiring hormonal contraception, it is important to consider drug interactions. Glauser et al., which included CAE patients over 2.5 and under 13 years of age, required that girls either be premenarchal, or agree to abstinence (oral contraception was not allowed, even for acne) [53].

Table 5
Number of articles including adult vs. pediatric populations by type of intervention

	Drugs	Devices	Procedures	Total
Adult	34	23	14	71
Pediatric	29	10	10	49

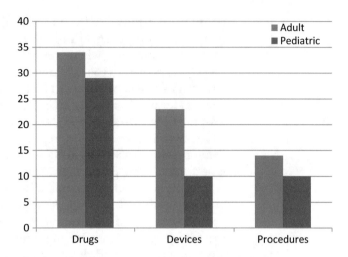

Fig. 2 Articles on type of intervention by age group (excluding two articles which did not report age groups)

4.4 Epilepsy Diagnosis/Classification/Etiology

Nearly half the articles included subjects with focal epilepsy, typically enrolling only those with focal epilepsy (e.g., TLE), although a few studies focused on primary generalized epilepsy syndromes, mainly CAE. Most trials were on intractable epilepsy, but a few were on newly diagnosed epilepsy. Most articles did not specify their classification criteria, but those that did nearly always used ILAE classifications. It may be helpful to specify the classification, as this can be a moving target.

While most subjects had epilepsy, a few trials included healthy controls as well. One article reported only on psychogenic nonepileptic seizures (PNES) [95], but 19 articles specifically excluded subjects with a history of nonepileptic events.

4.5 Who Diagnoses Patients?

IRBs often require documentation of exactly how subjects were diagnosed. Some articles in our dataset specified that subjects were recruited from their epilepsy clinics or centers (thus presumably diagnosed by epileptologists or neurologists), some included subjects diagnosed by pediatric epileptologists or neurologists, and others allowed diagnoses from other practitioners such as neurosurgeons or general practitioners. Most articles did not specify who diagnosed the patients. However, an IRB might request this information.

4.6 Seizure Types/Frequency/Duration

Most trials had some sort of requirement for seizure frequency, whether it was medically intractable epilepsy (which they then often—but not always—defined in terms of seizure frequency and/or failing a specific number of AEDs, often 2 or 3 AEDs), or having a certain number of seizures (e.g., more than one seizure a month, daily seizures) over a certain period of time (e.g., 3 months,

1 year). Some trials required newly diagnosed epilepsy cases (e.g., the KOMET trial [63]) and some required that chronic epilepsy diagnosis was established, e.g., diagnosed 2 years earlier. Motta et al. required seizures in the "entire second half of the menstrual cycle" [34]. Most trials did not require seizures of a particular duration, aside from Holsti et al. which tested rescue intranasal midazolam vs. rectal diazepam [103]. On the other hand, multiple trials excluded subjects with a history (or a recent history) of status epilepticus or seizure clusters.

When designing a clinical trial (especially in the case of RCTs) where the outcomes include seizure frequency, it is important to specify the exact methods of quantifying those seizures, e.g., by diary (kept by patients/caregivers/family/significant other) and VEEG, as well as to prespecify the baseline seizure frequency and follow-up period. It is also important to identify, record, and communicate to patients which seizures would be counted. For example, will you count auras (simple partial seizures), or only seizures with cognitive or motor components? If you will evaluate "disabling seizures," how will you define them? For example, Heck et al. stated that disabling seizures "included simple partial motor, complex partial, and secondarily generalized tonic–clonic" seizures [8], while Rong et al. included "frequent and disabling seizures with falls, injuries, and significant impairment of the quality of life" [100].

In our dataset, most trials did not require a motor component, but some specifically excluded auras, or excluded seizures that suggested a different diagnosis, or where an intervention would not be appropriate (e.g., primary generalized seizures in a trial comparing levetiracetam with carbamazepine-CR [38]).

Again, it is important to consider both scientific validity as well as recruitment challenges in order to avoid having to make alterations to the protocol if possible. Yuen et al. relaxed the inclusion criteria from at least four seizures per month to two seizures per month [27].

4.7 Diagnostic Findings: Neurophysiology and Imaging

While we assume that all subjects had an EEG at some point, many articles did not specifically report this. Trinka et al. did not consider EEG or imaging compulsory in cases with a clear clinical diagnosis [63]. However, many trials included only those with diagnostic findings consistent with epilepsy, or epilepsy origin (e.g., bilateral temporal), and primary generalized epilepsy trials in particular tended to specify exact EEG findings. Other articles (e.g., on surgical procedures and devices) required appropriate presurgical workup, and/or excluded cases that were candidates for resective surgery.

4.8 Drugs, Devices and Previous Surgery

Many trials identified the number of AEDs failed, the number that could be used at the time of the trial, duration on AEDs or on stable AED doses, etc. Some trials excluded subjects who had failed similar medications, or those on medications with potential interactions.

Some trials excluded subjects based on surgical history (e.g., failed SEGA surgery [30]), or with device implants (e.g., VNS; note, Heck et al. required VNS to be turned off during the baseline period, and to have the VNS pulse generator (not the leads) removed before RNS implantation [8]).

4.9 Hypersensitivity Reactions

Some trials excluded those who had hypersensitivity reactions to similar drugs (e.g., eslicarbazepine trials excluded subjects with known hypersensitivity to carbamazepine or oxcarbazepine), or sulfa (e.g., in Lu et al.'s zonisamide trial [42]). Baulac et al. excluded subjects with the HLA-B*1502 allele [50].

4.10 Lab Findings

Some trials specified that AED levels (or at least one of them) were stable/in therapeutic range prior to the trial. Some excluded subjects with abnormal lab findings, e.g., elevated liver enzymes, low absolute neutrophil counts or platelets, and hyperlipidemia. This was especially true for enzyme-inducing AED trials, as well as diet trials. Herzog et al. and Motta et al. included low BAT and progesterone level requirements, respectively [32, 34]. Bochyńska et al. (a trial on vitamin B supplementation) excluded "vegetarianism or a history of a genetic defect of homocysteine, cobalamin, or folate metabolism" [43].

4.11 Caregivers

Epilepsy is rather unique in that family members and caregivers are often critical to history taking, even for adult patients. Many of the articles required that caregivers or family members be willing to consent to study procedures, and/or keep seizure diaries or records.

4.12 Compliance

Some trials included only subjects who would come for baseline evaluations and/or follow-ups. Some excluded subjects or caregivers with a history of poor compliance, or those who did not seem sufficiently motivated; such designs risk excluding subjects less likely to benefit from the intervention, adding bias to the study.

Recruitment may be a challenge if such criteria are too restrictive, and in RCTs excluding subjects who did not complete sufficient baseline or follow-up records, or who did not come for measurements or follow-ups, etc. may compromise randomization in the case of ITT analysis. While this is less of an issue for non-RCTs, excluding patients for not following protocol may risk biasing results anyway, and may lead to missing important information relating to the sustainability of an intervention's use.

For example, if an intervention requires more frequent follow-ups or measurements (e.g., AED levels and weight), more precise diaries, traveling, or if patients dropped out early due to side effects, etc., then it is likely that patients in the real world will also be less compliant, and it would be better to account for such challenges when translating the intervention into clinical practice. There are additional seasonal and geographical considerations, e.g., compliance with follow-ups may decrease during bad winters, the school year, or holiday seasons.

Unsurprisingly, some studies specifically excluded those unable to receive the intervention, such as being unable to swallow in case of drug or diet trials. Diet trials were not performed in patients too young for solid intake.

4.13 Comorbid Neurologic, Medical, Metabolic, Psychiatric, or Visual Disorders

Many articles in this dataset specifically excluded subjects with comorbid neurologic and/or medical disorders (particularly neurodegenerative disorders, brain tumors, and cardiovascular disorders). Many excluded those with inborn errors of metabolism, and those with psychiatric disorders (e.g., mood disorders), including a history of substance abuse. Some specified nonuse of benzodiazepines except as epilepsy treatment. Of course, trials investigating comorbid disorders such as ADHD and depression included populations with those disorders. Some trials (e.g., on vigabatrin or radiosurgery for mTLE) excluded those with visual field defects or ophthalmic disorders.

4.14 Intellectual Impairment and Communication

Some articles reported excluding subjects with intellectual impairment, with some specifying certain cutoff scores on the IQ test or the Mini-Mental Status Exam. Other articles simply specified that subjects had to be able to give informed consent, follow study procedures, and answer questions. Depending on the intended population, it can be helpful to prespecify whom it is you plan to include. For example, if you intend to include patients who developed cognitive problems following epilepsy, but to exclude those with developmental delays, then it helps to clarify this distinction.

Some trials only included English-speaking subjects. Depending on the geographic location and the region's ethnic populace, language restrictions may limit recruitment and/or bias the sample. This is especially important for immigrant countries such as the USA, Canada, and Australia, as well as many regions in the Eurozone. Where possible, it is usually best to acquire translator services (as well as translated consent forms, and other documents), and to specify how and when they are to be used (e.g., what if someone speaks simple English, but falters on more complex topics?) However, in trials evaluating very specific populations, a particular ethnic group, or where there are no validated translations of important questionnaires, it may be necessary to set language or ethnic restrictions to maintain the internal validity of the trial.

4.15 Completing a Previous Phase of the Trial (Or Completing a Run-In Phase)

Unsurprisingly, some trials enrolled only subjects who had completed a previous phase of the trial (e.g., only those completing the blinded phase could enter into OLE), and consented to enroll in the following phase. When designing an open-label phase of a trial, consider whether there is a scientifically valid reason to only include subjects who completed the first phase (e.g., the purpose of the OLE is to follow up those who received an intervention long term, and/or to compare them with subjects who had previously received placebo and would now receive the intervention). Otherwise, consider enrolling additional subjects to increase your sample size. Keep in mind that samples are often smaller in the OLE rather than the blinded phase due to dropouts.

However, some subjects will enroll only in open-label studies to avoid receiving placebo, and because insurance will not cover the active intervention. When this is a concern, it may be better for the external validity of the study to restrict open-label enrollment to subjects who participated in the blinded phase. This may be particularly important for efficacy trials on novel interventions.

Another issue to consider is *when* subjects would be allowed to enroll in the OLE (or new phase) following the previous phase. For example, in drug trials you may want to enroll them right away in order to continue following up the patients who already received active drugs; in other cases you may need to consider carryover effects from the intervention (e.g., neurostimulation). Depending on your protocol, you may need to delay enrollment for a few weeks or months, and may consider getting a new baseline.

4.16 Participation in Another Clinical Trial

Finally, some articles specified that subjects could not participate in another clinical trial during the study period, or up to a certain period of time before the study began (e.g., within 12 weeks of selection). This may be necessary when lasting effects of a previous interventional trial may interfere with the new study.

4.17 A Final Word on Subject Selection

When designing your clinical trial, it is important to consider how your target population compares to populations in the existing literature and also, if it does differ significantly, whether this difference is associated with a better response and safety (in your trial). If your target population has not been previously studied and you have a good hypothesis as to why it might differ from previously studied populations, then you may have identified a defect in the literature lending itself to a valid need for your trial, and potentially to funding opportunities.

5 Trial Duration

When designing a trial, it is important to determine the duration of the trial as well as its phases (e.g., baseline, intervention, follow-up), and the time points (e.g., number of visits). A trial that is too short to properly apply the therapy or measure outcomes may not have internal validity, and a trial that is too long may suffer from dropouts or low recruitment, beside the fact that long trials require more resources.

In this section, we will discuss the total trial durations, follow-up periods and time points in our dataset.

5.1 Trial Duration

Trial durations (i.e., total study durations including baseline, intervention, and follow-up periods) when specified, ranged from 1 day to 9 years in our dataset. This large range seemed to relate to the variety of interventions and phases in our dataset. We divided trial durations into four categories ranging from under 6 months to over 2 years for more clarity.

The total duration was not clear in six articles, so we did not include those in this analysis. Additionally, some articles reported trial duration as greater than a given number, rather than giving the exact duration; for the purposes of this analysis, we considered the cutoff number as the total duration.

The 94 articles we analyzed reported the following trial durations:

- Under 6 months ($n = 33$, 35%);
- From 6 months to under 12 months ($n = 13$, 14%);
- From 1 year to under 2 years ($n = 20$, 21%);
- 2 years or more ($n = 28$, 30%).

Table 6 compares the number of articles according to the four total trial duration categories, by study phase and type of intervention. Drug studies overall had durations under 12 months in 30/50 (60%) of articles, while device/procedure trials had longer durations, lasting 1 year or longer in 28/44 (64%) of articles. It is notable that most of the device/procedure trials were exploratory, yet 14/26 (54%) of those exploratory trials lasted a year or longer. By comparison, more than half the articles on drug trials were on phase III trials (27/50, 54%), and over a third of those lasted under 6 months (10/27, 37%). Overall, 21/50 (42%) of the articles on drug trials reported durations of less than 6 months; this included one phase IV trial (the other phase IV trial lasted over 2 years). Meanwhile, 3/6 (50%) of the phase I articles reported on trials lasting 2 years or longer. The comparison between trial duration according to phases can be observed in Fig. 3.

It is unsurprising that drug trials have shorter durations than invasive trials such as those on invasive brain stimulation and surgery, in which participants are more dependent on providers and a longer follow-up period is necessary to assess intervention effects. For example, in surgical trials, participants typically had regular follow-ups every 2–3 months postoperatively for at least 1 year. Invasive brain stimulation involves several phases described in other sections of this chapter, and subjects are often enrolled in open-label extensions (OLEs) indefinitely. As a result, trial durations were longer in those cases, and adherence to duration period was also better than that for drug trials (see Sect. 12 below).

Table 6
Number of articles according to the total duration categories, compared by type of intervention and study phase

Total duration/phase	Drug							Procedure and device			
	Pilot	Phase I	Phase I/II	Phase II	Phase III	Phase IV	Total	Exploratory	Pivotal	Post-market	Total
<6 months	3	2	1	4	10	1	21	8	4	0	12
≥6 to <12 months	2	1	0	1	5	0	9	4	0	0	4
≥1 to <2 years	0	0	0	1	5	0	6	6	7	1	14
≥2 years	0	3	0	3	7	1	14	8	5	1	14

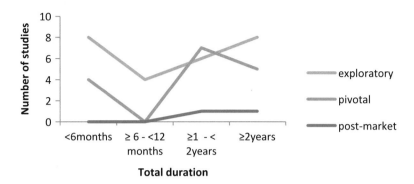

Fig. 3 Durations in drug and exploratory/pivotal trial articles according to their phases

Taking into consideration the 48 articles classified as device and procedure, 10/13 (77%) of the articles on device trials reported durations of 2 years or longer, while most procedure trial articles reported durations less than 2 years (17/21, 81%, over half of which were under 6 months' duration).

5.1.1 Trial Periods

Some studies (n = 31 articles) had a prospective baseline period (typically ranging from 2 to 12 weeks) to determine seizure frequency and other outcomes. Most studies (n = 76 articles) had a follow-up period that allowed the verification of offline or long-term intervention effects. The other 24 articles were unclear about follow-up (n = 7) and had no follow-up (n = 17). It is noteworthy that some studies (e.g., Chaytor et al.) performed the follow-up interview by telephone [107], an option that may potentially improve feasibility and adherence. Twenty-seven articles reported on long follow-ups of 2 or more years, while 49 articles reported follow-ups of less than 2 years. The longest follow-ups were in device (DBS, RNS, VNS) and procedure trials (especially surgery, radiosurgery, and diet trials), where participants may have been more dependent on providers in general. Of the seven articles classified as unclear, four had long-term OLE periods but did not specify the duration.

The articles not reporting follow-up periods were on the following types of trials: drug (n = 11); neurostimulation (n = 3); diet (n = 2); and behavioral intervention trials (n = 1). We failed to find a relationship between presence of follow-up or phase of the study and type of outcome. In general, studies without follow-up periods were short trials with small sample sizes and a total duration of fewer than 3 months, published mostly in journals with impact factors ranging from 1 to 3 (n = 10).

5.1.2 Time Points

About half the articles did not clearly report the trial time points (n = 49). Studies each had 7 visits on average, with a range of 1–47 visits. To illustrate, two studies had only one time point because they aimed to verify the pharmacokinetics following a single dose of a medicine ($n = 2$), while four articles (on drug, surgery, or ketogenic diet trials) required more than 17 visits; it is unclear why those particular studies had more visits than average.

Thus, when designing a clinical trial, the duration, follow-ups and time points may be influenced more by the intervention and outcomes of interest than the phase. It is important to make sure that you have enough time to apply the therapy and measure what needs to be measured, but ideally no longer than that to avoid burdening subjects and research staff.

6 Type of Intervention

We divided types of intervention into three categories, as seen above: drug, device, and procedure. For the purposes of this chapter, we considered as "device" the following: DBS, RNS, VNS, TMS, and procedures involving laser, gamma knife, radiosurgery or endovascular coils. We considered the following as "procedures": surgical procedures not involving "devices," behavioral therapy, and diets. Please refer to Tables 3, 4, and 5 for breakdowns of type of intervention by trial design, study setting, sample size and age groups as well as Tables 7 and 8 for breakdowns by trial duration.

In our dataset, 51 articles were on drug trials, 26 were on devices, and 23 were on procedures.

6.1 Drug Trials

Most of the drug trials were on AEDs, with levetiracetam being the most commonly studied drug ($n = 9$ articles), tested in various populations and study designs.

Asking new questions for frequently used drugs may result in relevant information. For example, levetiracetam is largely considered a safe drug that can be used in focal and primary generalized epilepsies with little risk of drug interactions, but Freitas-Lima et al., noting that there had been no formal investigation into the

Table 7

Number of articles according to the total duration categories, compared by device/procedure classification

	Device	Procedure
<6 months	4	8
≥6 to <12 months	2	2
≥1 to <2 years	7	7
≥2 years	10	4
Unclear	3	1

Table 8
Follow-up duration per number of articles

Follow-up duration	Number of articles
<1 month	2
1 to <6 months	23
≥6 to <12 months	7
≥1 to <2 years	17
≥2 years	27
No follow-up	17
Unclear duration	7

effects of enzyme-inducing AEDs (EIAEDs) on levetiracetam pharmacodynamics in humans, used a prospective single dose design to compare 15 adult patients on EIAEDs with a matched control group of 15 patients not on EIAEDs; they showed an increase of levetiracetam clearance in subjects taking EIAEDs (about 25% difference) [69].

Another five articles explored zonisamide in pediatric and adult populations either as adjunctive therapy [42, 60, 64], or as monotherapy [39], or (in a noninferiority trial) as compared to controlled-release carbamazepine in newly diagnosed focal epilepsy [50]. Other AEDs explored included ezogabine [46, 47] and eslicarbazepine [20, 48, 58] in focal epilepsy, brivaracetam (in focal and/or generalized epilepsies) [61, 62, 67], bumetanide (in TLE) [28], pregabalin (on sleep) [40], diazepam (autoinjector in acute repetitive seizures) [71, 103], sulthiame (compared to levetiracetam in benign epilepsy with centrotemporal spikes (German HEAD Study)) [37], perampanel [52], while two studies evaluated vigabatrin (one was an open-label three parallel group multicenter trial quantifying visual field impairment, while the United Kingdom Infantile Spasms Study (UKISS) compared the effects of hormonal treatments with vigabatrin) [19, 49]. One historical-controlled study evaluated conversion to lacosamide monotherapy in focal epilepsy [59].

There were three articles on the study comparing ethosuximide with valproic acid and lamotrigine in CAE, a double blind RCT across 32 US centers, with two reporting on freedom from treatment failure as a primary outcome [53], and one providing Class I evidence that attentional dysfunction was worse with valproic acid than the other two drugs [55].

One article compared the health-related quality of life of elderly patients on carbamazepine vs. lamotrigine (double-dummy design) [68], and another compared rapid initiation of topiramate vs. phenytoin [65]. One study found that weight gain (an important side effect of valproic acid) could be reversed with intensive behavioral therapy without having to discontinue the medication [31].

Other trials investigated whether using AEDs before seizure onset in infants with TSC could have a protective effect [35], whether bone mass loss could be prevented with supplementation of calcium, vitamin D, and risedronate [73]; some investigated the effects of vitamin B supplementation on homocysteine metabolism [43], the effects of an omega-3 fatty acid on seizure frequency in chronic epilepsy [27], or the effects of epilepsy, oxcarbazepine, and valproic acid on nitric oxide and xanthine oxidase levels or lipid peroxidation [66].

Hormonal therapy was used in seven articles, mostly progesterone and melatonin [22, 34, 51, 56], but also testosterone (plus anastrozole or placebo) for male sexual dysfunction [32], and the UKISS compared prednisolone or tetracosactide depot with vigabatrin in infantile spasms [49].

Finally, other medications used included everolimus or sirolimus for TSC/SEGA [30, 44], OROS methylphenidate for ADHD [25], IVIG (considered a drug here) for intractable childhood epilepsy [36].

A note on placebo trials in pediatric populations: Fattore et al. were able to justify using placebo as the only treatment (for a short course) in patients with childhood and juvenile absence epilepsy as the treatment effect could be quickly demonstrated in absence seizures, and as they were unlikely to directly cause serious harm [57].

To conclude, there are many possibilities when designing a drug trial, depending on your sample, and your intervention can be directed toward seizure control, evaluating long term outcomes, comparing specific outcomes between different drugs or in different syndromes or age groups, evaluating drugs as adjunctive therapy, monotherapy, testing different doses, evaluation of pharmacokinetics, improvement of comorbidities, etc.

However, as discussed above, many cases are medically refractory, and require other therapeutic options.

6.2 Device Trials

Of the device trials in this dataset, five were on DBS, including three with stimulation of the anterior thalamic nuclei (ATN). In the SANTE trial, Fisher et al. report that the ATN DBS baseline period was 1 month after implantation, after which patients (n = 110) were randomized to stimulation ON (5 V) or OFF (0 V), at 90 Is pulses, 145 pulses/s. Following 3 months of blinded treatment, all patients were ON from month 4 to 13 (they allowed limited parameter adjustments); and then they entered the long-term follow-up period where stimulation parameters and AEDs could be adjusted freely [101]. Oh et al. on the other hand, also in stimulating the bilateral ATN, activated the implantable pulse generators only 1 week after implantation in all seven patients, with stimulation parameters of 100–185 Hz, 1.5–3.1 V, pulse duration 90–150 ms, continuous stimulating mode in either one or two

DBS electrode contacts [80]. In Lee et al. the same group reported on long-term outcomes (mean follow-up of 39 months) [84].

Valentín et al. reported on stimulation of the centromedian thalamic nucleus for generalized and frontal epilepsies, the primary outcome being generalized tonic clonic (GTC) or complex partial seizure (CPS) frequency. After implantation they had stimulation OFF for at least 3 months (blinded), then a 3-month blind ON phase, with continuous stimulation at 130 Hz, pulse width 90 Is, and up to 5 V (later adjusted for two patients who did not benefit from this protocol). Patients then had a 6-month unblinded ON phase, and in patients who benefited from stimulation the control period was followed by an unblinded extension phase with stimulation ON [7].

Meanwhile McLachlan et al. modified equipment from their movement disorders DBS program to provide bilateral hippocampal stimulation for intractable TLE in two patients. This was a cross-over study with a 3 month post-implant baseline period; then the stimulator was turned ON or OFF in each patient randomly for 3 month intervals. Their primary outcome was the difference in monthly frequency of partial and secondarily generalized seizures between four 3-month epochs: baseline, ON, washout, and OFF [75].

Responsive cortical stimulation (RNS, three articles), on the other hand, had the advantage of ECoG recordings (Sohal and Sun described its effects on rhythmic activity recorded from the hippocampus and neocortex [93]), although patients could only have one or two localized seizure foci. Morrell et al. reports that the neurostimulator was programmed to sense and record ECoG but not to stimulate for 4 weeks postoperatively; then treatment group patients received programmed stimulation over 20 weeks (blinded), after which all subjects were able to receive responsive stimulation over the open label period [116]. Reporting on final results of the RNS trial, Heck et al. state that on open label period completion, about half the patients had stimulation current amplitude of less than 4 mA, while about a third had 4–7.9 mA, 8.7% had 8–11.9 mA, and only 2.7% had the maximum of 12 mA; 75% of patients had received under 7.3 min of stimulation a day (average 5.9 min/day) [8].

Only two other articles explored intracranial implants in intractable epilepsy, one a first-in-man study on an implanted seizure advisory system [74], and the other a closed loop asynchronous drug delivery system detecting seizure onset and triggering the delivery of a predefined drug dose [82]. Both systems recorded intracranial EEG as well.

There were four articles on VNS trials, studying both pediatric and adult populations [87, 94, 99, 112]; another two articles were on transcutaneous auricular VNS [78, 100], and two more were on trigeminal nerve stimulation [88, 106]. The last two neurostimulation articles were on low frequency rTMS in refractory focal epilepsy [76, 91]. Thus, there are various ways of providing neurostimulation in epilepsy patients, with varying degrees of invasiveness (from intracranial to noninvasive brain stimulation). The variability in procedures and protocols can make comparisons challenging, but indicates a need for further research to optimize neurostimulation in epilepsy subpopulations.

For other device trials, three articles discussed gamma knife radiosurgery in mTLE patients [6, 98, 105], while another investigated cognitive outcomes following MRI-guided stereotactic laser amygdalohippocampotomy [97]. One article was on the ISAT trial, comparing seizure outcomes following coil embolization vs. clip occlusion of ruptured cerebral aneurysms [117]. The final article was on a pivotal trial that tested the Intranasal Midazolam Mucosal Atomization Device against rectal diazepam, comparing times to seizure cessation following the administration of each medication [103].

6.3 Procedure Trials

6.3.1 Surgery

Of the five articles on surgical trials, four were on refractory TLE, relating to early surgical therapy vs. AEDs alone [102], length of resection [108], and evaluating sleep quality in refractory patients who had undergone resective therapy [113]. One exploratory trial examined the efficacy and risk of awake surgery with cortical mapping and resection of epileptogenic zones in or adjacent to eloquent cortex in nonlesional epilepsy cases [96].

6.3.2 Diet

There were nine articles on diet trials in our dataset. These included three on the ketogenic diet (KD) in adults and/or children, with ratios of fat–protein + carbohydrate ranging from 2:1 to 4:1 [77, 86, 114]. There were five articles reporting on Modified Atkins Diet (MAD) trials: one RCT comparing MAD to regular diet [110], one on Danish children (comparing MAD to KD) [109], one on Japanese children [83], and one on Iranian children (diet was culturally adapted); two of the trials were on adults, including one where patients followed with investigators by email only [81, 90].

Almost all the diet trials were exploratory, the rest were pivotal, all but two were non-RCTs, and all but one had samples of less than 100 subjects. Of course, none were blinded or placebo-control, as it is not feasible to blind for these diets.

6.3.3 Behavioral Therapy

There were eight articles on behavioral therapy in our dataset. Two of the articles were on PEARLS depression treatment for adults with epilepsy, where problem-solving therapy enhanced patient skills addressing the types of problems causing and maintaining depression, such as lack of social and physical activation [104, 107].

The COPE intervention (Coping Openly and Personally with Epilepsy) pilot trial, aimed at pediatric patients and caregivers, evaluated cognitive behavioral therapy (CBT) for primary and secondary coping [79]. Another trial compared CBT (modified) to relaxation technique sessions/exercises in patients 60 years old and above, investigating seizure frequency and neuropsychological test scores [111].

One trial evaluated an educational program on self-management in epilepsy [115], and another evaluated psychoeducation group session effects in PNES [95].

Finally, one trial explored Mozart K.448 as potential add-on therapy for children with intractable epilepsy [89], and another targeted AED nonadherence, evaluating a family tailored adherence intervention (AI) to "treatment as usual" [85].

All of these articles on behavioral trials reported positive primary outcomes. Sample sizes were all under 50 (except for PNES) but subjects adhered to the duration of these trials 80–100% of the time in the epilepsy trials (64% in the PNES trial). This suggests that behavioral trials should be further investigated in patients with epilepsy (as well as PNES), as they are well received and can often be helpful for both seizure and neuropsychological outcomes.

7   Outcomes

The predetermination and careful selection of outcomes, particularly primary outcomes, is essential for a good study design. In epilepsy trials, the selection of primary and secondary outcomes depends on multiple factors, including the type of intervention at hand, as well as the study phase.

Due to articles' lack of clarity in defining primary outcomes, or a clear distinction from endpoints and measures, all are referred to as outcomes in this chapter. In this section we discuss only the primary outcomes of the trials in our dataset (as primary and secondary outcomes were often qualitatively similar).

Many trials had more than one primary outcome; we considered those as co-primary outcomes. Additionally, many articles did not clearly specify which outcomes they considered primary, so we made a best guess based on the language and organization of the article. For analysis of the primary outcome data, please refer to Sect. 13 below.

7.1 Types of Outcomes

For the purposes of this chapter, we identified three types of outcomes: clinical, surrogate, and composite (Fig. 4).

Clinical outcomes include any outcomes based on clinical measures, such as seizure frequency, adverse event reporting, and BMI. Functional outcomes, where neuropsychological and other scales assess a patient's function, quality of life, etc., are a subset of clinical outcomes, and we grouped them all together.

Surrogate outcomes include indirect measures of health and function, e.g., serum levels of drugs, hormones, and EEGs.

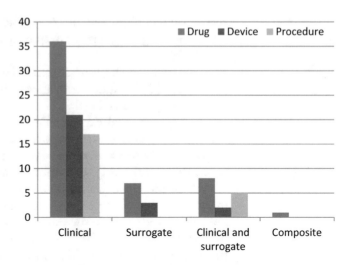

Fig. 4 Type of outcome by type of intervention

Composite outcomes describe a combination of multiple outcomes (such as test scores) into one variable, e.g., a single score.

Co-primary outcomes, multiple primary outcomes of seemingly equal importance to the researchers, were divided for the purposes of this chapter into *co-primary clinical and surrogate outcomes* (a combination of the two types), as well as *co-primary outcomes of the same type (clinical or surrogate)*. There was only one composite outcome in the dataset.

7.1.1   Clinical Outcomes

Most articles (74%) listed clinical-only primary outcomes: 69% for drug trials, 81% for device trials and 77% for procedure trials. About 65% of the clinical-only outcomes were single, while the rest were co-primary clinical outcomes (Fig. 5).

Most of the primary clinical outcomes in all types of trials related to seizure frequency in one form or another. Some examples include seizure frequency reduction as measured by mean seizure frequency over a prespecified period of time (e.g., 1 month, or baseline period) divided by frequency over another period of time (e.g., blinded phase); this could be expressed in numbers, percentages, percentage change, or categorized by percentage reduction (e.g., seizure reduction by <25%, 25–50%, >50%, or 100%). This reduction was often analyzed in the form of responder rates (i.e., 50% or more seizure reduction), and was sometimes analyzed as seizure freedom (i.e., 100% reduction).

Sometimes reductions in particular seizure types/severity were specified, e.g., disabling seizures, or major seizures (described as GTCs or as CPSs with or without secondary generalization). Some articles specified the time of day (e.g., diurnal seizures), the number of seizure days, or the duration of seizures (e.g., in Holsti et al. the primary outcome was time to seizure cessation (i.e., seizure

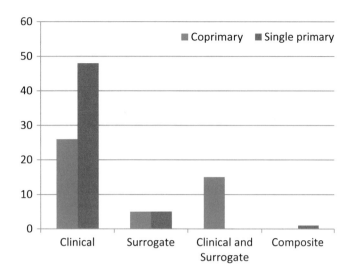

Fig. 5 Number of articles with single primary and co-primary outcomes by outcome type

duration) after intranasal midazolam or rectal diazepam administration in acute seizures [103]).

Some studies used time to seizure as their primary outcome, e.g., time to first seizure after surgery, time to fourth seizure; this was sometimes described as a treatment failure event.

Wechsler et al.'s primary outcome was the percentage of patients receiving lacosamide who met exit criteria by the end of the maintenance phase, and exit criteria included: a twofold or greater increase in either monthly or 2-day partial-onset seizure average frequency compared to baseline, occurrence of a single GTC (if the subject had none 6 months before randomization), worsening of seizure duration/severity/pattern enough to warrant discontinuing the study, or the occurrence of status epilepticus or of new serial or cluster seizures [59].

Other types of clinical outcomes, especially those reflecting safety and tolerability of interventions (particularly in Phase I, II and exploratory trials) included measures of adverse events—or treatment emergent adverse events—over certain periods of time. Clinical outcomes also included neuropsychological test results (Psyche Cattell test, QOLIE-31, HSCL-20 depression score, verbal memory tests, etc.), or clinical measurements such as BMI.

7.1.2 Surrogate Outcomes

Surrogate-only primary outcomes were used in ten of the articles in our dataset (seven on drug, three on device trials); half were co-primary.

Surrogate outcomes included laboratory findings such as plasma levels of AEDs, melatonin, NO, MDA, XO, and urinary levels of 6-sulfatoxy-melatonin; one trial investigated bone mineral density. Surrogate outcomes also included perimetry data (for visual field defects), actigraphy measurements (for sleep–wake patterns), polysomnography, ECoG and EEG findings, and MRI changes (e.g., following radiosurgery). Salam et al. used a combination of surrogate outcomes when testing an implantable closed loop asynchronous drug delivery system—the device detected seizure onset and triggered the drug delivery at a predefined dose, and the drug was measured during injection (to validate doses for seizure suppression) [82].

7.1.3 Clinical and Surrogate Outcomes

A combination of clinical and surrogate primary outcomes (i.e., all co-primary) were reported in 15 articles, 8 on drug, 2 on device, and 5 on procedure studies.

Sometimes the combinations were related to the same outcome, e.g., freedom from both clinical and EEG seizures [57]. Others related to outcomes that were clinical such as seizure frequency or Epworth Sleepiness scale, in addition to separate laboratory values (e.g., liver enzymes and LDL in a trial on MAD [83]), or MRI changes (e.g., SEGA volume [30]), etc.

Combining clinical and surrogate measures can validate the outcomes both ways, by correlating subjective with more objective measures (e.g., decreased seizure frequency with therapeutic drug levels), and also by helping quantify or describe clinical measures (e.g., seizure recordings on VEEG or ECoG in addition to seizure diaries). Particularly in the case of seizure self-reporting, patients should receive clear instructions on what to record, how and when to record it (e.g., they may not think to mention milder seizures or auras).

7.1.4 Composite Outcomes

The UKISS trial used a composite score of VABS subdomains (communication, living skills, socialization, motor function) for adaptive behavior; this was the only trial with a composite outcome in our dataset [49].

Composite outcomes can be more objective and easier to compare between treatment groups. However, researchers must decide if the score would be clinically relevant, and whether it would mask variations in subsets of the score (e.g., total score may be similar in subjects with large variations in different subsets of the score).

7.2 When Choosing a Primary Outcome

Always ask the following questions: what is the purpose of this investigation (e.g., do you want to change clinical practice, or avoid a complication)? Will it be clinically relevant (e.g., will it help you identify a solution to a problem)? How you will identify the outcome (e.g., seizure diaries)? How will you quantify results, and over what period of time (e.g., how long after DBS implantation will you begin counting seizures, considering the implantation itself may lead to seizure reduction in the early period, even without stimulation)?

When designing a clinical trial is important to determine whether the instruments used actually measure what they are intended to measure. For example, in Cook et al., most patients greatly underestimated their seizure frequency (e.g., having 20 or 30 seizures per month more than they estimated at their baselines, captured by iEEG). One patient who overestimated the seizures was later found to have PNES on EMU monitoring [74]. This is not uncommon in EMUs, where patients' accounts of their seizures do not always correlate with VEEG findings.

Thus, while it is best to use objective measurements in clinical trials, when subjective instruments are necessary, having additional measures may help improve data quality and interpretation; it can be helpful to anticipate such measures when choosing primary outcomes.

A final note: sometimes changes in outcomes are necessitated by small samples or difficulty of evaluation, but it is typically best to select outcomes carefully and in a hypothesis-driven manner, and then to be faithful to the results. This may help minimize biases and enhance interpretability of results.

8   Sample Size

8.1   Introduction

Statistical power consideration and analysis is essential for any clinical trial, as power indicates the chance of detecting statistically significant effects that truly exist in the trial. That is, power helps control the probability of making Type I and/or Type II errors [120].

The five statistical elements of concern in power analysis are the sample size (n), the significance criterion (α), sample variance (s^2), effect size (ES), and power ($1 - \beta$) [121, 122].

The general consensus is that in order to test meaningful hypotheses, one has to have at least 80% power for preliminary studies and 90% power for confirmatory studies. It is only when sample size calculations are reported that readers can evaluate whether the sample size and power calculations adequately represent concerns (including clinical, scientific, and statistical concerns).

It is crucial to calculate the sample size prior to the beginning of any interventional clinical trial, particularly if it may change existing therapeutic parameters. Otherwise, the study runs multiple risks to its validity, including risks of Type I and II errors due to underestimation or overestimation of sample sizes, leading to the rejection of potentially beneficial interventions, or leading to excessive financial burdens and potential increase in patient risk (e.g., more patients receiving sham or placebo therapy, or new therapies of unclear safety).

In this chapter we discuss the methods used in our dataset to estimate the minimum sample size required for a fixed power. Articles on open label extensions (OLEs) were included among those without sample size calculations, as were articles on trials where the original sample had a different primary outcome than the article in this dataset.

8.2 Sample Size Calculations

Although the CONSORT 2010 Statement (item 7a)3,4 [120] requires that authors explain how their study's sample size was determined, only 33 articles in our dataset reported sample size calculations (31 RCTs, 2 non-RCTs); many of those were adjusted to allow for dropouts.

However, only 20 articles reported the three parameters required to allow readers to recalculate sample size: alpha level of significance (Type I error), power, and expected effect size. See Table 9 for the number of articles reporting each item.

Alpha level was set at 0.05 in 19 articles, 0.025 in 3, 0.017 in 4, and 0.005 in 1 article; 73 articles did not report alpha level.

The alpha levels set at 0.017 were typically calculated as 0.05 alpha level divided by 3; in two of those articles, a Bonferroni adjustment was reported, e.g., comparing three dose groups of eslicarbazepine with placebo in one [48], and comparing three co-primary outcomes for two intervention groups in another [106].

Brodie et al. and Glauser et al. described the process of sample size calculation well, additionally stating when the interim analysis was planned to be performed (when 50% of subjects reached the primary outcome); they also describe how the sample size was increased to account for two stratification factors as well as a 5% dropout rate [45, 53].

Gonzalez-Heydrich et al. set their p-value at <0.005 (0.05/10), to account for multiple comparisons in their adaptive phase I trial [25].

Power estimations were reported in only 30 articles; about two thirds were 80% and above, while only 9 articles reported on trials powered at 90% and above. Only a quarter of articles reported on effect sizes, and less than a fifth estimated attrition.

Sample size calculations were typically based on the intended primary outcomes. Sometimes they were intended to show noninferiority of one drug versus another (e.g., Borggraefe et al. [37]). Sometimes the assumptions were derived from preliminary trial data [42], or previously published studies, e.g., Herzog et al. and Kwan et al. [32, 67]. Some researchers used historical controls, e.g., Wechsler et al. calculated a sample size intended to provide 90% power to compare the Kaplan–Meier estimate of the patients (percent) exiting by day 112 vs. the 65.3% historical control exit percentage [59].

Sometimes the sample size was not calculated but chosen based on a previous study, e.g., in Cervenka et al. [90]. Some authors explained why sample sizes were not estimated, e.g., Raju et al. explained that they did not do the estimation as this was an exploratory trial, and due to the absence of data on the 2.5:1 ketogenic diet [114].

Table 9
Sample size calculations

Sample size calculations	Total
Alpha level	*100*
0.05	19
One-sided	1
Two-sided	7
Not specified	11
0.025 one-sided	3
0.017	4
0.005	1
Not reported	73
Power	*100*
80% and above	21
90% and above	9
Not reported	70
Sample size estimated	*100*
Yes	33
No or not specified	67
Effect size calculated	*100*
Yes	25
No or not specified	75
Attrition	*100*
5%	3
10%	7
15%	3
Other	5
Not reported	82

8.3 Actual Sample Sizes

Actual sample sizes in our dataset are summarized by number of articles and by intervention in Tables 10 and 11. Please refer to Table 3 for a comparison between sample sizes in our dataset by intervention and phase of trial. As discussed previously, trials in our dataset tended to have relatively small samples for type.

To conclude, sample size estimation should be a part of all clinical trials in order to increase the likelihood of clinically meaningful data and to avoid having an underpowered study.

Table 10
Sample size ranges by number of articles in our dataset

Sample size range	0–20	21–50	51–100	101–500	501–1000	1001 and above
Articles	22	22	18	33	3	2

Table 11
Same size mean, minimum, and maximum by intervention

Sample size	Drug	Device	Procedure
Mean	222	142	64
Minimum	3	2	10
Maximum	1698	2143	209

These estimations are often based on assumptions, may include biases, and may not always be achieved. It is thus important to have a strong hypothesis as to feasibility of recruitment and likely dropout rates, and (as much as possible) to make sure that sufficient funding is available to avoid early termination of the study.

9   Interim Analysis and Stoppage Criteria

Factors determining the end of a trial include reaching trial goals, a scheduled date of closure, or when the trial becomes unviable. A statistical procedure called interim analysis can be performed at any point during a trial to decide on whether to adjust or stop a trial prematurely due to lack of or overwhelming positive effect. The trial can also be interrupted early in the case of life-threatening or severely debilitating illnesses. However, to avoid type I error, the interim analysis should be stated ad hoc and the p-value adjusted. Therefore, the research protocol should state the need for an interim analysis, the number of times it would be performed, and the stopping rules. It is recommended that a person who is not directly involved in the experiment carry out the interim analysis in order to maintain blinding. Interim analyses that do not follow these rules should be avoided as they may bias the results of the trial [123].

In this section, we will discuss premature termination, interim analysis and the stoppage criteria adopted by the epilepsy trials in our dataset. In our sample, seven articles reported stopping trials early, and six performed interim analyses. As shown by Ryvlin et al., this can lead to changing their initial statistical plans for intent-to-treat and per-protocol populations because of the premature study termination. This (PuLsE) trial was ended earlier than originally intended by the sponsor due to a low enrollment rate [112]. Similarly, Engel et al. and Borggraefe et al. had difficulties in recruiting study participants [37, 102]. Cook et al. concluded the trial early due to funding issues [74].

Considering the studies that performed interim analyses, 4/6 of those articles did not clearly explain how the analyses were addressed [19, 52, 53, 56, 101]. Two articles reported using the O'Brien-Fleming method, a widely used prospective statistical strategy for stopping due to overwhelming evidence of efficacy, and adjustment with the Lan-DeMets spending function approach [53].

R. Fisher et al. conducted an interim analysis, and the results of the analysis did not change the sample size or study course [101]. On the other hand, Glauser et al. performed an interim analysis for both efficacy and futility when 50% of the subjects reached the primary outcome). At this point, the sample size was expanded by 5% due to stratification and by another 5% due to dropout [53]. Rektor et al. performed an interim analysis to evaluate the long-term tolerability, safety, and efficacy of adjunctive perampanel in patients with refractory partial-onset seizures 4 years after the beginning of the study [52]. Herzog et al. assessed progesterone treatment of intractable seizures in women with partial epilepsy, and performed an interim analysis for futility and efficacy for each stratum (catamenial and noncatamenial) [56]. The authors stated clear stopping rules for each stratum as follows:

"1. **Overwhelming evidence of progesterone efficacy**

   (a) If planned interim efficacy analysis when 144 women in the catamenial group completed the study showed a statistically significant beneficial effect of progesterone at the $p \leq 0.001$ level.

   (b) If planned interim efficacy analysis when 288 women in the noncatamenial group completed the study showed a statistically significant beneficial effect of progesterone at the $p \leq 0.001$.

2. **Statistical evidence of futility**

   (a) If planned interim futility analysis of the catamenial group at the end of the initial phase of the investigation, i.e., after a total of 150 subjects were recruited, showed a conditional power <50%.

   (b) If planned futility analysis of the noncatamenial group at the end of the initial phase of the investigation, i.e., after a total of 150 subjects were recruited, and again after 1/3 (128) and 42% (162) of the noncatamenial group completed the study showed a conditional power of <50%.

3. **Adverse Events**

   (a) If death possibly related to participation in the investigation occurred

   (b) If the dropout rate during the treatment phase exceeded 20% or twice the dropout rate of the baseline phase."

It can be helpful to identify potential future problems that may lead to concluding the trial earlier than intended, such as funding or recruitment issues that may occur down the line, hopefully allowing those issues to be preemptively addressed. An interim analysis would be helpful to help decide on whether to conclude your trial at the provisional stop date, or if it should be concluded earlier or later than that. However this analysis should be planned ad hoc, anticipating its cost (such as modification of p-value), justifying its use, as well as preplanning the time it should be performed and the stopping rules.

10   Randomization

Randomized clinical trials (RCTs) are considered to have the most powerful experimental design in medical research, one that may lead to causal inference. Randomization is important because it avoids selection bias [124]. Also, randomization balances both unknown and known factors that may affect outcomes between the different trial arms, assuming both sample size and randomization method are suitable [124, 125]. Randomization implies that participants have an equal probability of being allocated into each trial arm during the randomization process, as allocation occurs by chance.

CONSORT 2010 guidelines state that information about the methods used to generate the random allocation sequence, type of randomization, allocation concealment mechanism and implementation of randomization should be described in the body of the main article [130]. In this section, we will discuss the use of these methods in our epilepsy dataset. It is worth noting that 43% of the articles in our dataset ($n = 43$) reported on non-RCTs, mostly single-arm studies ($n = 29$) or open-label extensions of randomized controlled trials ($n = 5$). Moreover, some studies were not randomized because they had healthy subjects in the control group (e.g., Motta et al. [34]) or they had surgical interventions where randomization may be less feasible (e.g., Carrion et al. and Drane et al. [97, 113]). There were 57 articles reporting on RCTs.

10.1   Sequence Generation Method

Using the sequence generation method, subjects are randomly allocated to experimental or control groups, e.g., through a computerized random number generator or random-number table. An interactive voice response system, through a telephone call, can also be utilized to manage the randomization; it is frequently used in multicenter trials. On the other hand, some studies use inappropriate nonrandom methods such as date of appointment, hospital/physician identification number, and even date of birth for group allocation.

Randomization implies that participants have the same probability of receiving any of the available interventions, and to protect

Table 12
Frequency and percentage of RCT articles using each sequence generation method

Sequence generation method	Frequency	Percentage
Telephone-based	1	2
Computer-generated	21	37
Random number tables or randomization algorithm	4	8
Not clear	31	54

randomization benefits, previous allocations should be unknown before a new subject is assigned to a group. When using inappropriate methods such as those mentioned above, researchers may know in advance which treatment arm the next subject will be assigned to, which can introduce selection bias into the study. Fortunately, these pseudo-randomized methodologies were not found in the articles in our dataset.

Although CONSORT guidelines recommend reporting information on sequence generation methods explicitly in articles, this is not a common practice in medical literature. As shown in Table 12, 31/57 articles on RCTs (54%) in our dataset did not clearly report the method used to generate the allocation sequence. Of the articles on RCTs reporting sequence generation methods, 21/26 used a computerized random number generator, 4/26 used random number tables or randomization algorithms, and 1/26 used a telephone-based method.

10.2 Type of Randomization

The most common types of randomization are simple, block or stratified randomization [126].

Simple randomization is the most basic method, in which subjects are randomly assigned to each group without any restrictions. This method works very well in trials with large sample sizes, generating similar numbers of participants in each group. However, for trials with small sample sizes, simple randomization can lead to different numbers of participants in each group.

Alternatively, block randomization, where participants are randomly assigned to blocks of equal numbers, is recommended to avoid imbalance between experimental conditions at any time during the trial. According to the CONSORT statement, block randomization trials should provide information about block generation, block size and whether the block size was fixed or randomly varied.

In stratified randomization, researchers can control for confounding variables, known as prognostic or risk factors (e.g., gender, center, and age). These prognostic factors are baseline variables correlating with outcome variables, thus potentially confounding interpretation. It is possible to control for these factors by using a separate randomization list for each variable, such that confounding variables are equally distributed between the different arms. In the stratification, blocking can be used to guarantee an equal number of participants for treatments in each stratum or the researcher can use simple randomization. Authors should report which factors were controlled for, the cutoff values, and the method used (e.g., simple randomization and block randomization).

Stratified randomization with many prognostic covariates (e.g., more than 4) requires a large sample size. In this case, an alternative stratification technique such as minimization can be used. *Minimization* considers important prognostic factors together, allowing balancing over a larger number of covariates. For the first subject, the allocation is truly randomized; after that, the allocation that minimizes the imbalance between groups with regard to the selected factors is identified. Then this allocation is chosen, or a choice is made at random with heavier weight in favor of this allocation (e.g., a probability of 0.8). Considerable programming and computing resources are required in minimization technique. The CONSORT statement indicates that minimization should be explicitly reported when used, as should the variables added into the scheme.

Other randomization methods include *biased coin randomization*, which randomizes participants with equal probability among different groups if the number of participants in each group is nearly balanced. Otherwise, if the number of participants in one group is higher than a prespecified value, the allocation probability (p) is adjusted such that the group with fewer participants has a higher p.

10.2.1 Methods Used in Our Dataset

In our dataset, only two trials reported using simple randomization. One used computer-generated random numbers with 34 participants allocated to the intervention group and 30 allocated to the control group [95]. The second trial had 50 participants in each group (short versus long resection) by chance [108]. Both trials had approximately balanced experimental and control groups, even though their sample sizes were not large.

The type of randomization employed was not made clear in 26 articles (45.6%) as seen in Fig. 6. As to block randomization, six articles reported using block randomization only, while nine used block randomization divided by strata. Of those 15 articles that reported blocking, 9 used fixed block sizes; the most commonly used block sizes were 4, 6, and 3. Meanwhile, two articles used random permuted blocks, meaning that the blocks had different sizes chosen randomly. Variable block sizes help ensure allocation concealment as they avoid the predictability of treatment allocation for the participants involved at this stage. Otherwise, if experimenters are aware of the block size, they would know the treatment the last person of each block would receive, and no longer be blinded. Four articles did not report on block size.

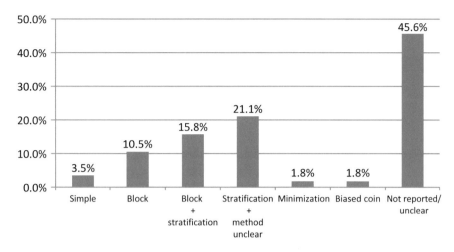

Fig. 6 Type of randomization by percentage of article on RCTs reporting it

Another 12 articles reported the use of stratification but did not specify the type of randomization within each stratum. One article used biased coin randomization with stratification. Commonly used strata included investigational site (in multicenter trials), location and number of seizures, age of participants (in pediatric studies), baseline partial seizures, prior therapeutic surgery for epilepsy, seizures reported in the prior 6 months, type of seizure (e.g., catamenial or noncatamenial), gender, baseline adherence data (to eliminate participant imbalances across groups), concomitant levetiracetam use and the best recommended treatment. Finally, one article reported using minimization method.

10.3 Implementation of Randomization

The implementation of randomization is the description of who generated the allocation sequence, who enrolled participants, and who assigned participants to different interventions. Careful implementation of randomization is important to guarantee allocation concealment, but 35/57 (61%) of the articles on RCTs did not clearly state how it was performed.

As most of our RCTs (48/57 articles, 84%) were in multicenter settings, 18 articles reported that a central site performed sequence generation. Only one article reported that each research unit independently carried out randomization. Another three articles reported that sequence generation was conducted in their trial by an independent team assistant who was not involved in interventions (i.e., a statistician, pharmacist, or trial nurse).

Only three articles reported on who enrolled participants in the study (research assistants in all three cases); one of those articles mentioned that this person was independent from the other steps of the study. We identified only eight articles reporting on who assigned participants to interventions (e.g., secretary, nurse, pharmacist, a person not involved in outcome assessment, or the central pharmacy that prepared the medication).

It is best practice to have an independent investigator enroll participants, another to generate the randomization sequence and/or assign participants to their groups, and not to have these investigators involved in intervention or outcome assessments. The bottom line is to avoid influencing allocation to different groups, and to avoid accidentally (or intentionally) unblinding assessors and participants.

10.4 Allocation Concealment

Allocation concealment means hiding the allocation sequence from the people assigning trial subjects into their intervention groups, until the subjects are actually assigned [127]. This helps reduce selection bias by preventing researchers from influencing which participants are assigned to which group.

While selection bias may occur prior to randomization, allocation concealment may still prevent bias during randomization once patients are enrolled. Allocation concealment is thus critical to randomization. It is separate from random sequence generation, and properly generating a randomization sequence is of little use if proper allocation concealment is not performed [128]. Unlike blinding, allocation concealment is always possible and necessary.

Only 13 articles in our dataset clearly reported allocation concealment, even though they did not always report their methods; it was unclear whether it had taken place in several other articles, but most did not report allocation concealment at all.

The measures used for allocation concealment in our dataset included: (a) interactive voice response systems [50, 63]; (b) randomization by assistants that were not involved in intervention, or by blinded third parties [104, 108]; (c) using sealed envelopes containing the randomization codes (two articles specified that the envelopes were opaque) [110, 114, 118]; (d) using numbered containers to conceal random treatment groups [42] (note, containers should be opaque and appear identical); (e) communicating the patient group to the surgeon via letter only on the day of surgery [108].

11   Blinding

Another important method to minimize the introduction of biases is study blinding (one of the items in the CONSORT 2010 checklist). Blinding refers to keeping those involved in an RCT unaware of the assigned intervention.

Allocation concealment protects the assignment sequence before and up to the point of allocation; it primarily prevents selection biases, and is always possible and necessary.

Blinding, on the other hand, protects the sequence after allocation, such that subjects and/or assessors are not aware of group assignment. Blinding helps prevent ascertainment bias; however, it cannot always be implemented (e.g., in Holsti et al. unblinding to the medications—intranasal midazolam vs. rectal diazepam—was unavoidable [103]); additionally, when the outcome is objective (e.g., mortality rate), blinding may not be necessary.

Lack of blinding can lead to distortion of outcome recognition and/or interpretation (ascertainment bias). Even when blinded, subjects and/or assessors may suspect that a subject received a particular intervention (e.g., due to side effects) or placebo; thus, blinding may not completely erase the risk of ascertainment bias.

11.1 Level of Blinding

The levels of blinding are nonblinded (open or open label), single-blind, double-blind, and triple-blind studies. Unfortunately, the terms are often confused and interpreted differently in the literature. Additionally, the term quadruple-blind is rarely used.

11.1.1 Nonblinded, Open or Open Label

Typically, phase I drug trials and exploratory device or procedure trials are open label, emphasizing safety and gathering preliminary data. Blinded RCTs also often undergo an open label phase, where patients who received placebo are now given the active treatment.

11.1.2 Single-Blind

Only one of the three categories of people involved in the trial (subjects, investigators, and assessors) remains unaware of the group assignment throughout the trial. Typically, it is the subjects that are blinded.

However, sometimes the term is confusingly intended to denote that the subject and investigator but not the assessor are aware of the group assignment (a situation which is more properly termed *dual-blind*).

11.1.3 Double-Blind

This term is rather misleading as it typically denotes that the subjects, investigators, and assessors (who are often also the investigators) are all unaware of the intervention.

11.1.4 Triple-Blind

This usually means that the data analyst is also kept blinded in a double-blind trial. However, it is sometimes used to mean that the subjects, investigators, and assessors (who are separate from the investigators in this case, and may be statisticians, pharmacists, etc.) are all blinded.

11.1.5 Quadruple-Blind

This denotes that subjects, investigators, assessors, and data analysts (four distinct groups in this case) are all blinded to the group assignment. This term is rarely used as triple-blind may indicate these levels of blinding—in fact, it is important to specify who is blinded as terms can vary across studies [129].

11.2 Blinding in Our Dataset

In our dataset, 30 articles reported on double-blind studies, 7 on single-blind studies, while 1 was on a triple-blind study (other studies in our dataset could potentially be considered triple-blind,

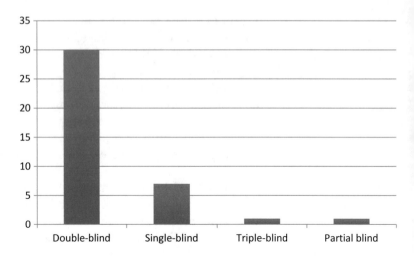

Fig. 7 Article on blinded studies in our dataset

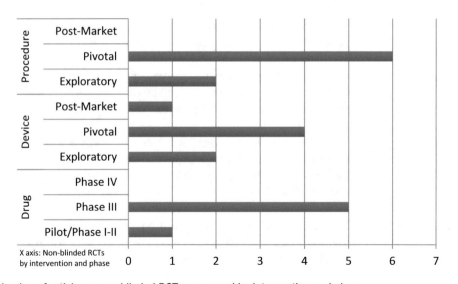

Fig. 8 Number of articles on nonblinded RCTs compared by intervention and phase

but were not given that term), and one was on a partially blinded study (Fig. 7). The two articles that did not report on blinding were assumed to be nonblinded, making a total of 61 articles on nonblinded (open label) studies. These included 21 articles on nonblinded RCTs (including open label extensions of double-blind studies). See Fig. 8 for the distribution of nonblinded RCTs by intervention and phase.

Some relevant examples of different types of blinding and the reasoning behind them are given below.

In Mcdermott et al., as it was not possible to blind patients or investigators to whether or not patients received anteromesial temporal resection in addition to pharmacotherapy, they instead had what they called "partial blinding," where a blinded independent panel monitored both groups' seizure pharmacotherapy, and a blinded committee classified patients' epileptic seizures based on their diaries [102].

Baulac et al. is an example of a triple blind study (patients, investigators, and the sponsor personnel who were administering AEDs, assessing outcomes, and analyzing data, were all blinded), as well as a masking method, where they ensured that patients took the same number of identical capsules by using matching placebo tablets and by overencapsulation [50].

Single blind studies included Wei Sun et al. (others may call it double-blind) where the staff members performing rTMS treatment were unblinded, but the patients and other experimenters [76], including those reading the EEGs [91], were blinded, and Cho et al., where only the physician reading the polysomnogram was blinded [38].

Another single-blind study was reported in Chaytor et al., where blinded interviewers conducted outcome assessments, but subjects, as well as the therapists and psychiatrists (both of whom reviewed PEARLS cases) could not have been blinded [107]. Additionally, in the first article on PEARLS [104], they reported allocation concealment by having assistants (who did not have an intervention role) generate the randomization sequences, enroll and assign patients to each group.

In general, when the mode of therapeutic administration is different (e.g., tablets vs. depot, intranasal vs. rectal), it is more challenging to blind those administering or receiving the medications, although it may still be possible by double-dummy design (giving a placebo of one drug route and an active drug of the other route). Additionally, certain aspects or assessors of the study can be blinded, such as interviewers who are not involved in prescribing or giving the medications, and EEG readers.

When a team approach is necessary to evaluate cases, such as where epilepsy surgical conferences are in order, or when behavioral interventions need to be administered, blinding may be limited by necessity. However, it is best to blind as many participants as possible. This may increase the costs of the trial, but improves the quality of the data and its interpretation.

Even in open label studies, some measures of reducing ascertainment bias can be employed. For example, when testing an implanted seizure advisory system, patients, investigators, and other EEG reviewers did not know the advisory indicators during the data collection phase [74].

Glauser et al. described a double-blind design where subjects and site personnel were blinded to treatment allocation, with the central pharmacy sending subject-specific study kits based on the subject's clinical response, prescribed by the local site investigator [53]. However, subjects reaching treatment failure in the double-blind phase were allowed to enter into the open-label phase, in which case they would be randomized to one of the other two AEDs. Blinding was then maintained by giving subjects the same number of capsules at all steps and weight groups. They describe a double dummy design allowing patients who could not swallow to participate, by instead taking liquid ethosuximide, a capsule containing crushed chewable (lamotrigine) with filler, or (valproic acid) sprinkles with filler. The placebos looked and tasted similar to the active drug.

Other studies were called double-blind, but may be considered triple-blind by other readers. For example, Masur et al. [55], using the same cohort as Glauser et al. [53], sought to measure attention; however, blinded evaluators performed central scoring, and so, it may be called triple-blind rather than double-blind. Another example is reported by Schramm et al. where they blinded the patients, epileptologists and the people performing MRI volumetry, but not the surgeons performing the resections [118].

Of the nonblinded pivotal RCTs, two were diet trials, which could not be blinded to patients or investigators; both used opaque sealed envelopes for allocation concealment [110, 114].

It can be assumed that studies on infants and very young children, when using surrogate outcomes, are less likely to be affected by patient biases. Caregiver or assessor biases may influence clinical outcome reporting, but EEG findings, serum levels of AEDs etc. are unlikely to be affected (note that it may be helpful to standardize timing of measurement, even in objective surrogate outcomes). In adults and older children, however, a careful assessment of potential biases is commendable even when relatively objective measures are used. For example, patients who believe that a particular therapy can reduce their seizures or improve their sleep may feel less stress, may sleep better, may be more compliant with therapy, and this may affect EEG findings.

Blinding may be unethical in some cases, e.g., the open-label PuLsE trial (on VNS), where a long follow-up was needed, and as physicians adjusted AEDs clinically [112].

Finally, several studies mentioned blinded outcome reporting (an important measure to reduce biases in interpreting data) [53, 57, 118], but the majority did not.

In conclusion, as much blinding as feasible and ethical should be the goal when designing clinical trials, particularly phase III and pivotal trials. Blinded outcome assessments should be attempted.

11.3 Similarity of Interventions

One way to ensure proper blinding is to use interventions that are not easily distinguishable from one another.

In over a dozen reports on similarity of intervention in drug trials [32, 37, 42, 46, 47, 50, 57, 59, 60, 65, 67, 68, 73, 85], most articles reported that the active drug and placebo were matched, or were similar or identical in appearance (almost always oral form, though sometimes they specified similar size, shape, color, and/or taste, and in Herzog et al. depot testosterone was combined with either (oral) anastrozone or placebo [32]; in Abou-Khalil et al. the placebo and diazepam autoinjectors were identical) [71]. In titration studies, different doses of the active drug, as well as placebo when present, were also matched. Some studies reported identical packaging, overencapsulation, and several reported a double dummy design. Some reported that the same pharmacist prepared the medications, or that the same company manufactured them [46, 47]. Some of the better measures are already mentioned in the subsection above, e.g., in Glauser et al. [53].

Likewise, devices and procedures tended to be administered in roughly similar ways. For example, articles reported having all subjects undergo actual or sham neurostimulator programming [116], thus spending a similar amount of time with the physician, or adjusting DBS settings such that patients would be unaware of stimulator status [75]; Masur et al. administered the order of neuropsychological testing similarly in each site [55].

Thus, there is typically some way to make interventions at least partly similar across groups, and the greater the similarity, the better the blinding.

11.4 Blinding Assessment

It is important to strive to maintain blinding integrity. Degiorgio et al. mentioned multiple measures to this end, e.g., trigeminal nerve stimulation device programming was performed away from blinded study physicians and staff; subjects were stimulated at only treatment or control settings (so that they would not be able to compare and recognize any differences); they were told that sensory experiences may vary among subjects; and they were instructed to describe their sensations only to the unblinded coordinator (not the blinded staff); finally device settings were covered by opaque adhesive labels [106].

However, even the best blinding efforts can be thwarted. For example, subjects coming into contact with unblinded personnel may potentially pick up clues as to their group assignments, or they may be able to detect whether they received active intervention or not. Researchers may guess at randomization (especially if improperly randomized). Unblinded staff may be pressured to give clues as to the grouping by researchers higher on the totem pole. Adhesive labels can be exchanged.

None of the articles in our dataset reported using strategies to assess for successful blinding. However, there are a number of such strategies. One method is to simply ask subjects and researchers at the end of the trial (before unblinding them) which intervention they thought they had received. If the aggregated correct responses exceed those expected by chance alone, the blinding of the study may have been compromised. Other examples are mentioned in Chap. 1.

It should be noted that the recommendations for this type of testing have been removed from the 2010 CONSORT statement; the reasoning is that participants and healthcare providers typically know if the primary outcome was reached, making it difficult to tell whether correct responses simply reflect correct assumptions about intervention efficacy, or failure of blinding. Now they only recommend that researchers disclose known blinding compromise [130].

It should also be noted that while it can be helpful to directly assess subject and researcher blinding, unblinded parties might still lie if asked; for example, researchers may not want the data to appear compromised, and subjects may not want to disappoint researchers or appear guilty.

12   Adherence

Adherence refers to how well clinical trial participants do what they are asked to do for the trial, an important measure of the trial's effectiveness (efficacy and tolerability).

It is crucial to consider adherence when designing and conducting a clinical trial, as it can be impacted by nearly all other design considerations. Poor adherence to the intervention regimen or to follow-up visits can lead to missing data, underpowered studies and misleading results, such as thinking an intervention is less effective than it really is (when used properly), or not recognizing side effects. Knowing the adherence is thus critical to interpreting trial results.

Researchers should, when necessary and possible, incorporate measures to improve adherence. These strategies should not influence the outcomes themselves, and should be equivalent in each trial arm. In order to incorporate such strategies, researchers must anticipate potential causes for poor adherence in their study, with their specific intervention, patient population, disease severity and study duration. There are also global measures to improve patient adherence, such as having good rapport and communication with researchers, as well as trust in the investigators (perhaps particularly in the case of surgical or invasive procedures).

In epilepsy trials, transportation should be an important consideration, as patients entering clinical trials often have intractable epilepsy and cannot (or should not) drive. Reliance on family members, particularly during business hours, may reduce adherence. Cost and burden of adherence, ease of intervention administration (e.g., drug or diet), side effects and perceived effectiveness are some of the most important factors impacting adherence. Finally, considering the subjectivity of many clinical measures (which depend on patients' and/or caregivers' memories), adherence to drugs, record-keeping, and follow-up appointments can be particularly important. Ideally, adherence strategies would address all relevant factors.

12.1   Adherence to Study

As mentioned in the introduction to this chapter, effectiveness involves both efficacy and tolerability, as reflected by adherence to the drug (or intervention), and it is one of the main questions clinical trials are intended to answer. Only two articles reported adherence in our dataset: Klein et al. measured urine ketone bodies and serum β-hydroxybutyrate levels, and calculated an overall compliance score [77]. They were the first ketogenic diet study to their knowledge to quantitatively measure compliance.

On the other hand, Wagner et al. reported how many patients completed the COPE program, how many sessions were attended, etc. To improve adherence, they included activities to help subjects master the material and interact with the group; participants also received homework assignments and a colorful, developmentally appropriate COPE notebook. Other measures included gas cards, modest monetary compensation, and when children completed their homework, they also received small trinkets and gift cards [79].

In our dataset, less than a tenth of all articles reported measures to either improve or measure adherence or compliance, and some of those were incomplete, e.g., lasting for only part of the trial duration [109].

In drug studies, attempts to improve adherence included pill counts, such as in French et al. and Glauser et al. [47, 53]; the latter also measured the amount of liquid returned, and asked how many doses were missed. Glauser et al. retained 98.5% of patients at the end of the trial, while French et al. retained 73.2% (many of the discontinuations had been due to adverse events or unsatisfactory responses, however).

As with Klein et al. [77], others took more objective measurements, e.g., comparing serum melatonin levels before and after therapy, which also helped confirm compliance [51]. Checklists kept by subjects and audiotapes of intervention sessions helped other studies [31, 104].

Using quantitative means to measure adherence, especially when repeated, may help improve adherence; they may also help achieve meaningful interpretations of study outcomes. A caveat is that repeated measures may become a less accurate reflection of adherence in real-life scenarios if similar measures are not taken; if measured regularly in real life, they may become an extra burden to the patients, providers and to overall healthcare costs.

In device and surgical trials, patients may be more likely to adhere to follow-up visits and record keeping due to the invasiveness of the interventions, the need for implant setting adjustments, or close inspection of adverse events. However, there are other measures that patients may need to adhere to. For example, patients with an RNS implant must collect Neurostimulator data daily, and send this data to the PDMS weekly [131]. Thus, published protocol adherence rates would be helpful in interpreting the data.

Knowing the degree of adherence to study protocol is essential for the PI to interpret the data, and reporting it is an important aspect of the peer review process.

12.2 Adherence to Study Duration

As we are unable to obtain protocol adherence rates for the studies in our dataset, we instead calculated adherence to the study period, by dividing the number of subjects who completed the study (numerator), by (denominator) either the number randomized (in RCTs), or the sample size (in non-RCTs).

It was not possible to calculate this number in 16 articles, where the variables were not clearly reported.

In our subset of 84 articles, the adherence to study duration ranged from 39.1% to 100%, with a mean of 85.5% and a median of 90%. We compared studies that scored <80% adherence to study duration to those that scored 80% and above by type of intervention (see Table 13).

Table 14 shows that device trials fared best, with 73% of the articles reporting duration adherence rates of 80% and above, compared to only 8% with under 80% adherence (the remaining device articles were missing this information). Procedure trials did not do quite as well, with 68% reporting rates of 80% and above, and drug trials fared worst of all, with only 48% of articles on drug trials reporting rates of 80% and above, and 33% reporting rates below 80%.

As mentioned above, patients may be more likely to continue to follow up after invasive procedures. Procedures requiring brain implants/surgery may have better rates of follow-up compared to simpler procedures. For example, adherence in the RNS trials was >90% at 2 years (8117), and in DBS trials rates tended to be higher as well. For example, in Fisher et al., 82.7% of 110 subjects remained in the study at >2 years [101], and in Lee et al., all remained at 2 years [80, 84].

In the meantime, a VNS trial in Budapest had an adherence to study rate of 61.5% at 2 years [94]. In another VNS trial, exact adherence to study at 2 years was unclear (it was not included in the calculations above), but by the time of study termination, only 49.2% of the 122 patients had completed their 1 year follow-up [112].

It is notable that discontinuations or dropouts increased with longer study durations. Adherence to the study period can also vary between subjects receiving one drug vs. another, or between those receiving a drug vs. placebo. The discontinuations are often related to side effects or lack of efficacy, and it is important to report such differences.

As for diet trials, adherence to the durations of the trials ranged from 78.7% to 91.7% in KD, while MAD was more variable—as low as 47% in the Iranian children [92], and up to 95% at 1 month for the adult trial on email management, but then it went down to 64% at 3 months [90]. Thus, there are very specific challenges for diet trials.

Epilepsy patients tended to adhere to behavioral trial durations at high rates (80–100%) as mentioned previously.

Table 13
Number of articles using each sequence randomization method

Simple randomization	Block randomization	Stratification	Stratification + blocking	Stratification without specifying the method used for randomization	Minimization	Biased coin randomization	Urn design	Not reported/ unclear	N/A
2	6	9		12	1	1	0	26	43

Table 14
Adherence to study duration (<80% or more) by intervention (where information available)

	Drug		Device		Procedure	
	<80	≥80	<80	≥80	<80	≥80
Percent adherence to study duration (84 articles total)						
Number of articles	17	25	2	19	6	15
Percent per each intervention	40	60	10	90	29	71
Percent per total of all interventions	20	30	2	23	7	18
Percent per intervention total (e.g., drugs with <80% adherence/all drug articles, including articles with missing or unclear data)	33	48	8	73	27	68

To conclude, it is important when designing a trial to antici-pate lack of adherence, to take measures to prevent it, and to record it when it happens in order to better interpret the data.

13   Statistical Analysis

When designing clinical trials, researchers should first identify their intended outcomes and plan their statistical analysis accordingly, using the methods needed to answer the research question. Otherwise, data interpretation will be compromised.

The CONSORT statement highlights the assumption that data points are independent within standard methods of analysis (usually meaning that there is one observation per participant in a controlled clinical trial). Each participant should be counted once for data analysis, or more complex statistical methods should be used. However, it is a serious error to treat multiple observations from one participant (e.g., when data are measured in different parts of the body) as independent data [132–135]. It is also impor-tant to know when to use the assistance of statisticians and how to communicate with them.

Please refer to Chap. 1 for more details on planning statistical analyses for clinical trials.

13.1   Intention-to-Treat (ITT) Analysis

ITT analysis in an RCT indicates that all participants and groups in the trial are analyzed exactly as they were randomized. This pre-serves the strength of RCTs in cause and effect inference. Yet it is often hard to have a strict ITT analysis due to missing data or lack of adherence to treatment protocol [136].

Of the 57 articles on RCTs in our dataset, 35 reported using ITT analysis, while 22 did not use an ITT or did not mention it. However, only about a third of those 35 articles reported using strict ITT analyses; the rest either reported using a modified ITT, or simply reported ITT but its description stipulated requirements beyond randomization.

Many reported that all efficacy assessments were based on the ITT population, which they considered to be all patients who had received at least one dose of the study drug or intervention; sometimes they included only those who also had at least one post-baseline efficacy assessment, e.g., Ben-Menachem et al. [48]. Sun et al. considered all randomized patients who had received at least three sessions of rTMS to be the ITT population [76]. However, in a strict sense, ITT means that all randomized patients are analyzed, irrespective of whether they received treatment. Yet, in that particular study, all randomized subjects were analyzed; thus, even though their intended analysis was not strictly ITT, the end result was that they performed a strict ITT analysis. Not all studies claiming an ITT analysis but excluding randomized subjects fall into this fortunate category.

Excluding randomized patients compromises the randomization of the trial. For example, if a trial has three arms, and a patient randomized into one arm does not receive treatment and is thus excluded from analysis, then the other two arms would hold greater weight. Excluding patients from analysis because they received only part of the intervention may reduce the likelihood that intervention-related factors be addressed, potentially limiting the intervention's future clinical use.

Additionally, if one or more treatments are completed but there is an absence of data pertaining to treatment effects (e.g., some patients did not complete their seizure diaries), patients should still be analyzed as they were randomized (ITT), ideally with the protocol laying out the plan to deal with missing data (see Sect. 13.3 below). An ITT analysis is the best approach for efficacy measurements.

The most important reason to use an ITT analysis is to ensure that groups will be comparable, thus preserving the main benefit of the randomization. Using a per protocol analysis (i.e., analyzing only subjects that fulfilled the study protocol) often reduces the study's power and leads to biases, which can be more detrimental to its interpretation than missing data in an ITT. A per protocol (PP) analysis also restricts the generalizability of the trial results to ideal patients (which are uncommon in the real world). In this dataset, many studies used both ITT and PP analyses; such an approach is recommended in noninferiority trials [137]. It would also be recommended in subjects who, for example, are unable to receive the intervention (e.g., unable to swallow study pills) after randomization.

It is particularly important to minimize nonadherence to study procedures when running ITT analyses, as estimates of treatment effects (which are often diluted by noncompliance) tend to be conservative. In other words, subtle outcome differences between the intervention groups may be missed, or may not reach statistical significance. In crossover studies, particularly if many patients cross over to the opposite treatment arm, data interpretation may become difficult [138]. The final analysis would best control for such factors.

Interestingly, three articles on non-RCTs also reported using ITT (not necessarily strict), despite all being single-arm (drug and diet) studies.

Thus, it is important to understand the purpose of the ITT analysis, its advantages and disadvantages, and to report on whether it has been followed in the study.

13.2 Outcome Data

It is important to know the outcome of interest and how it is to be analyzed in order to select an appropriate statistical test. In this chapter, we report only on statistical tests used for primary outcomes, as they tended to be similar to secondary outcomes (Fig. 9). We classified those in this dataset as continuous, categorical, and time-to-event outcomes.

In the following subsections, we discuss the statistical tests most commonly used and when to use them.

In our dataset, primary outcomes were analyzed as continuous variables in 74 articles, as categorical variables in 45, and as time-to-event variables in 9 articles. There is overlap because many articles had multiple primary outcomes (or their primary outcomes were not clear and we had to make a best guess), and as similar variables can be treated in various ways (e.g., continuous

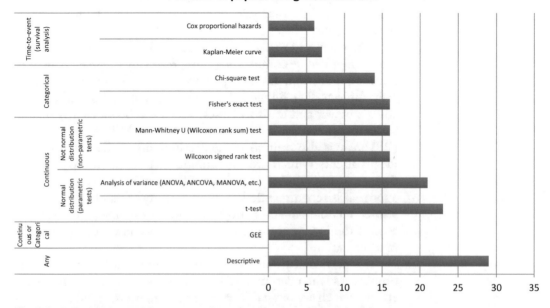

Fig. 9 Statistical tests used five or more times for primary outcomes in dataset

variables can be dichotomized and analyzed as categorical data, e.g., responder rates). As a result, multiple articles had more than one continuous primary outcome data point, or a mixture between continuous, categorical and/or time-to-event data.

About a third of the articles utilized descriptive statistics (mostly for continuous and categorical data, although those were the data types for most articles); in most of those cases, only descriptive analyses were performed (e.g., when sample sizes were small), but sometimes they were performed in conjunction with other statistical tests.

13.2.1 Continuous Primary Outcome Data

Continuous data are those measured on a numerical scale, including percentages and means, e.g., seizure frequency per month, changes in seizure frequency, and QOL measures. In this dataset, normal continuous primary outcome data was mainly analyzed using t-tests and analyses of variance (mostly ANOVA and ANCOVA), while nonparametric tests such as Mann–Whitney U and Wilcoxon signed rank test were used for nonnormal continuous data.

Most of the continuous data related to seizure frequency, or some variation of it, e.g., seizure frequency or its reduction, seizure frequency normalized over 30 days, etc. (as a mean or median). Sometimes it related to neuropsychological scores, or number of adverse events.

How to Use Continuous Data

It is ideal to design a clinical trial with a hypothesis as to how you want the data to behave. For example, if you want to have valid statistical inferences with optimal power, then you should design around data that are likely to be normally distributed without many outliers; in such cases you would be able to analyze the actual data, e.g., average seizure frequency per month, using a t-test; the results of using this normal-based test on normal data would be valid and with optimal power. The more homogenous the sample population, the more likely it is that the data will be normally distributed. For example, if the primary outcome is average seizure frequency per month, and the sample includes only subjects with a relatively narrow average; e.g., 4–6 seizures/month, then the resulting data in the trial are more likely to be normally distributed, and you can also use a confidence interval (CI) for the difference of the means. (Note that means are the most stable measurements of normal data).

On the other hand, if a trial includes patients with both well-controlled and intractable epilepsies, or intractable epilepsies with a wide range of or highly variable average seizure frequencies per month, e.g., 1–50 seizures/month, then the data is likely to have many outliers, or to not be normally distributed; that is, the data is unstable. As it can be quite difficult to design trials around very homogenous populations, many trials are likely to fall into the latter category, and may instead benefit from using ranks; e.g., ranking seizure frequency per month from lowest to highest (with some subjects potentially having the same rank), and using a Wilcoxon (nonparametric) test to compare average ranks. The results will still be valid, minimizing the false positive rates, and the power should still be good. Note, compared to an optimal test (e.g., normal based test on a normal data distribution), the loss of power when using a nonparametric test is equivalent to losing about 5% of the sample size; thus, this loss of power may be negligible, or close to it, with a sufficiently large sample size.

Fortunately, ranks are robust against outliers (e.g., in a sample of 20 subjects, where one subject has a count of 150 seizures compared to 19 other subjects who average a count of 10 seizures each, if you include this outlier the mean seizure count would increase dramatically, while if the subjects are ranked from 1 to 20, this subject would remain at the 20th rank, even if the subject at the 19th rank had 10 seizures.) One drawback is that using a nonparametric test, while valid (thus minimizing the false positive rate), may reduce your ability to clinically interpret the results.

It is also possible to use a parametric test based on the normal distribution even in the case of non-normally distributed data, but only when the sample size is large. Based on the Central Limit Theorem (CLT) [122], the test will still be valid if the sample size is large enough (typically a sample of at least 30) but the CLT does not promise good power. If the sample size is too small, it would always be invalid to use a normal based test if the data is not normally distributed (but a nonparametric test would be valid, with good if not optimal power).

One more option for nonnormal data (when comparing multiple variables) is to use a transformation when possible (e.g., log transformation for right-skewed data) that makes the data normal. Then normal based tests can be used on the transformed data, and these tests will be valid with optimal power—but it may be difficult to come up with clinically interpretable results. Log transformations are often performed on highly skewed data.

Other parametric models can be used. For example, for count data (e.g., number of days a patient is hospitalized for monitoring, or following surgery), you may use a Poisson distribution. Sun et al. compared the primary outcome (seizure frequency, which is count data) and interictal epileptiform discharges between the high- and low-intensity rTMS groups using a GEE model for repeated measures, which was based on a Poisson loglinear distribution [76]. For right-skewed data (e.g., cost of a hospitalization), you can use a lognormal distribution.

13.2.2 Categorical Primary Outcome Data

Categorical outcomes include dichotomous and ordinal outcomes.

Dichotomous outcomes are variables that can belong to only one of two mutually exclusive categories, e.g., yes or no, present or absent, 50% or greater seizure reduction (responder) or not.

The most commonly used categorical data are dichotomous (i.e., binary) outcomes. They often lead to 2-by-2 tables (assuming the exposure is also binary).

When subjects are classified in a category and the categories have a natural order (e.g., mild, moderate, severe), outcome data are of an ordinal type. *Ordinal outcomes* include measurement scales, e.g., some neuropsychological scales, which can have total scores, the levels of which are not equivalent to other total scores.

In our dataset, categorical primary outcome data was typically analyzed by Fisher's exact test, and chi-square test. Categorical data was often in the form of responder rates (50% or greater reduction in seizures, yes or no), or sometimes freedom from seizures at a point in time (or a particular type of seizure, e.g., focal seizures, GTCs, and disabling seizures), as well as Engel class, adverse events (as opposed to total number of subjects with adverse events, which would be continuous data), neuropsychological test results (e.g., in Jóźwiak et al. the Psyche Cattell test was categorized from normal to severe/profound mental retardation based on the score [35]), etc.

How to Use Categorical Data

Some types of data are naturally dichotomous (e.g., mortality, where the subject is either alive or dead), but continuous data is also often dichotomized and then treated as categorical data (e.g., seizure frequency dichotomized into responder rates, or categorized into percentage seizure reduction groups); or data can be treated as both continuous and categorical. Ordinal outcomes with increasing numbers of categories (e.g., presence or absence of certain seizure types), and longer ordinal scales are often analyzed as continuous data, while shorter ones are often combined together into dichotomous data, especially if there is an obvious cutoff. Less obvious cutoffs can lead to bias and it is important to decide on any cutoffs prior to the beginning of the study in order to avoid biases that may maximize the appearance of benefits of an intervention [139].

When using any type of scale as your primary outcome, it is important to know if such scales have been validated (although some trials may aim to validate a new scale). Even when the same rating scale is used in a clinical trial, sometimes modifications are made. Some modifications are necessary, such when a scale is applied to a different culture or translated into a different language, e.g., Elkhayat et al. used versions of the Children's Sleep Habits Questionnaire and the Epworth Sleepiness Scale-pediatric version which had been previously translated and modified to Arabic versions [51].

As to statistical testing, run the exact test (Fisher's exact test) when possible, as it will give you an accurate *p*-value. However, with large sample sizes, it may not be possible to run the exact test, or it may take a very long time. In such cases, the chi-square test can be carried out; it is simple and quick to run, and although the resulting *p*-value will be an approximation, it should be a good approximation of the true *p*-value if the sample size is large enough. The expected count in each cell should be greater than 5 to reasonably approximate normality.

13.2.3 Time-to-Event Primary Outcome Data

Time-to-event outcomes focus on the elapsed time before the events occur.

In our dataset, time-to-event data was typically analyzed by Kaplan–Meier curves and Cox proportional hazards. Some outcome examples are time to fourth seizure, to first complex partial or generalized convulsive seizure, time to first seizure after SAH following hospital discharge, and time to treatment withdrawal.

Freedom from treatment failure was sometimes treated as both time-to-event data (e.g., time to failure, analyzed with Kaplan–Meier curves), and sometimes dichotomized and treated as categorical data (e.g., freedom from treatment failure at a certain point in time, yes or no, analyzed with Fisher's exact test and or chi-square test [45, 53]).

How to Use Time-to-Event Data

Such data include the length of time during which no event was observed, and an indication of whether the period ended at the occurrence of an event, or just the end of the prespecified observation period [140]. Time-to-event data are often called survival data, as the event of interest is often death. However, any future outcome may be included, as seen in the examples above.

Time-to-event data are analyzed differently because the data are often censored, meaning that as we want to measure the time from a certain starting point (e.g., baseline) to some event (e.g., seizure or death), we will not necessarily have complete data (e.g., some patients will not have a seizure, some patients will still be alive). The starting point and the event must be clearly defined in advance, as each observation requires data on the time the subject was observed for, and whether the event occurred.

Typically, Kaplan–Meier curves are used to estimate the survival data, while Cox proportional hazards regression compares survival distributions and allows researchers to adjust for confounders and also to estimate the magnitude of effect of a factor. Thus, all available data are used.

If instead, you choose to select a specific time point and dichotomize the data (e.g., treatment failure, yes or no), you may have the challenges of selecting that time point, and how to deal with censored data before that time point (e.g., treatment failure before time "x"); however, this method will not introduce bias, and is a valid option.

13.2.4 Other Statistical Tests

Generalized estimating equations (GEEs) are utilized to extract the right amount of correlated data by using weighted combinations of observations [141]. As you can see in Fig. 8, GEEs have been used multiple times in our dataset. The ability of GEEs to compare treatment groups while accounting for such variables as time, group-by-time interactions, age, AEDs, etc. can be useful for epilepsy clinical trials. GEEs were mostly used for continuous data, e.g., seizure frequency, or neuropsychological scores [107]. However, GEEs were sometimes used for categorical data, e.g., in Degiorgio et al. responder rates were compared within group and between groups by the GEE logistic model [106].

Some of the articles also reported using a Bonferroni correction for multiple comparisons, e.g., Glauser et al. [53]; some used various forms of regression, e.g., logistic regression comparing country groups for categorical data (proportion of patients achieving seizure freedom for 26 weeks on stable medication dose) [50].

Please note some of the common statistical test terms that are used interchangeably: Wilcoxon test = Wilcoxon signed rank test = Wilcoxon signed rank sum test = Wilcoxon matched pairs test. Mann–Whitney U test = Wilcoxon rank sum test (which is for unmatched data). Cox proportional hazards regression analysis = Cox proportional hazards model. Pearson's chi-squared test = chi square test = X^2. We classified all analyses of variance as ANOVA (whether ANOVA, ANCOVA, MANOVA, MANCOVA, or Kruskal–Wallis analysis of variance).

13.3 Missing Data

In a review on published RCTs in major medical journals, more than 10% of randomized patients may have had missing outcomes in half the trials [142].

Once randomized patients are excluded, analysis is not strictly ITT. The common approach of simply excluding patients without observed outcomes tends to reduce the sample size (and thus power) and may introduce bias, as described above. Yet, in order to include participants with missing outcomes in the analysis, their outcomes must be imputed (meaning that the other information collected will be used to estimate their likely outcomes). This helps in conforming to the ITT analysis, but the strong assumptions this approach requires may be hard to justify.

Sensitivity analyses help to explore the effects of departures from the assumptions used in the main analysis; all randomized patients should be accounted for. The main analysis should include a plausible assumption about missing data, and all observed data that are valid under such assumptions should be analyzed. Under ITT, it is also important to try to follow up on subjects who withdrew from the study; this can be challenging, but may help identify differences between those who withdrew versus those who completed the study [143].

Once all possible data are included, imputing by "the last observation carried forward" (LOCF) method is typically seen as the alternative to handing missing data without omitting participants. For example, if a subject misses a visit, then the seizure frequency from the last visit might be filled in for the missing visit.

This approach is typically seen as conservative, but in some settings can be biased in favor of a new treatment. White et al. gave an example of a trial on a novel drug for Alzheimer's disease, where the analysis was criticized because it was seen to assume that disease progression was halted by loss to follow-up [143, 144]. Instead, they recommended to base primary analyses on the most plausible assumptions, and then to add conservatism by utilizing sensitivity analyses applied to departures from those assumptions.

In our dataset, six articles reported using LOCF, and two reported using multiple imputations for missing data (one was a single arm non-RCT [145]).

Another article reported using two approaches for responder rates and seizure freedom rates, one of which was the conservative approach where all discontinuations were considered treatment failures (irrespective of seizure status), and a nonconservative "LOCF-like" approach where they counted discontinuations as either successes or failures, based on seizure count before dropout [72].

Other articles did not use imputations or LOCF, but only used the "worst case scenario analysis" where subjects with missing data (e.g., postseizure data [46, 48]) were considered nonresponders; or, in the case of Sharma et al., they treated the diet and control groups differently, such that children with missing outcome data were considered to have 0% seizure reduction if in the diet group, and 100% seizure control if in the control group (worst and best case scenario respectively) [110]. Chhuhn et al. classified as nonresponders all subjects lost to follow-up, those with lack of efficacy, with adverse events and with protocol deviations [54].

However, most articles did not report on missing data at all. It is always useful to plan ahead how to deal with missing data and to report on it (at least in supplementary material). While it is mainly important for the ITT analysis, it is also useful for other analyses, and for the overall transparency and reproducibility of your trial. The longer the trial the greater the likelihood for missing data, dropouts, etc., but likewise an intervention's beneficial effects or complications may take weeks, months or longer to be seen (e.g., the beneficial effects of radiosurgery may take more than 6 months to develop [98]). If you take a conservative approach and consider all those with missing data as nonresponders, you might still correctly evaluate the efficacy of an intervention, but if you are looking for small differences between study populations, then a more sensitive approach may work better.

Note that in practice, clinicians often do not have access to all necessary or useful data prior to making clinical decisions. In clinical trials, it is important to keep records of missing data and withdrawals from the study before and after randomization, as well as the reasons behind their absence. Such data can have real implications for the usability and efficacy of interventions, as well as for the design of future studies.

13.4 Nonadherence to Protocol

As discussed above in Sect. 12, the vast majority of articles did not report on protocol adherence, nor did they provide sufficient data allowing readers to make that determination. However, some articles did report on common causes of nonadherence within their particular trial, e.g., randomized participants who did not meet inclusion criteria, did not take the intended treatment, or not all of it, etc.

Some authors reported on PP analyses (where participants inadequately adhering to protocol are excluded) without ITT. However, PP analyses should be labeled as nonrandomized, observational comparisons, and it should be understood that such patient exclusion compromises randomization and leads to a risk of bias in outcomes. The CONSORT checklist requires a clear description of who was included in each analysis [146]. See Sect. 13.1 for more information on ITT and when to use PP analyses.

13.5 Subgroup Analysis

Subgroup analyses should not be confused with sensitivity analyses. In subgroup analyses, there are estimates produced for each subgroup, and statistical comparisons are formally made for the subgroups. Sensitivity analyses (while they may analyze only a subset of the whole) do not try to estimate intervention effects in studies removed from the analysis, and only informally compare different ways of estimating the same thing. Sensitivity analyses ask, to quote from the Cochrane Handbook for Systematic Reviews of Interventions, "Are the findings robust to the decisions made in the process of obtaining them?" [147]

In our dataset, 55 articles reported using post hoc or subgroup analyses. For example, Fisher et al. used both prespecified and post hoc subgroup analyses of subjects with previous VNS or resective surgeries [101]. Glauser et al. used post hoc analyses to compare drug concentrations of treatment failures and successes within treatment groups [21]. Meanwhile, Brodie et al. used prespecified subgroup analyses of the primary outcomes stratified by geographic region (this was a multicenter trial across 71 centers internationally) and baseline seizure frequency; they also calculated numbers needed to treat (NNTs) post hoc [46]. Other subgroup analyses included age, sex, epilepsy type, etiology, duration, etc. Some articles reported using descriptive subgroup analyses, e.g., comparing the response to RNS in MTLE vs. nonMTLE subjects [8].

Subgroup analyses are helpful in identifying differences and confounders, and their selection will vary depending on the type of trial and its aims.

To conclude, proper statistical planning is essential when designing a clinical trial, and it begins with appropriate outcome data selection. Without proper planning, the study has a higher risk of failure or limited clinical utility, and the accompanying waste of money, time and effort.

14   Limitations

CONSORT guidelines recommend that researchers discuss the weaknesses of their studies, such as possible biases or imprecision in the results. Additionally, it is important to discuss possible methods to overcome these limitations.

In our dataset, 29 articles did not report their limitations. Of those articles, 22 (76%) had positive results for their primary outcomes versus 4 (14%) that did not. On the other hand, 52/71 (73%) of the articles reporting trial limitations also had positive results. Therefore, there was no clear association between reporting limitations and trial results.

14.1   Examples of the Main Limitations

14.1.1   Small Sample Size and Lack of Statistical Power

Most of the articles ($n = 44$) reported on the limitation of small sample size and lack of statistical power, particularly due to difficulties during recruitment (Table 15). For example, Engel et al. ran a RCT with a sample size of 38 participants instead of the planned 200 (that had been based on the sample size calculation). The trial was discontinued on recommendation of the data and safety monitoring board (DSMB), due to lack of feasibility in recruitment. This study aimed to verify if surgery soon after the failure of two AED trials was superior to continued medical management in controlling seizures and improving quality of life (QOL). The eligible participants were male or female, 12 years of age or older, with mesial temporal lobe epilepsy and disabling seizures persisting for no more than 2 years after failure of adequate trials of two AEDs. Although resective surgery plus AED treatment resulted in a lower probability of seizures during the second year of follow-up, the small sample does not allow definitive conclusions about group differences [102].

On the other hand, Wagner et al.'s pediatric study did not find a significant improvement on many outcome measures, probably due to the small sample size and weak statistical power. This study aimed to evaluate the effectiveness of the COPE program with regard to its ability to improve attitudes toward illness, parental distress, parents' self-efficacy in managing their children's seizures, and other coping skills. The authors revealed that the enrollment was much lower than expected, not representing the time and effort spent on recruitment. They listed the following difficulties encountered: (1) scheduling conflicts; (2) availability of families and ability to travel to the intervention; (3) mild behavioral and/or emotional concerns; and (4) limited contact with epilepsy research [145].

14.1.2   Selection Bias/ Lack of Generalization or Ecological Validity

Other major limitation of epilepsy trials was selection bias, lack of generalization or ecological validity. For example, Brodie et al. reported the limitation of having highly selected participants [46].

Table 15
Study limitations

Limitation	N
Small sample size and weak statistical power	44
Selection bias/lack of generalization or ecological validity	18
Nonblinded	18
Short or no follow-up	11
Lack of placebo or control group	11
Nonrandomization	8
Sample heterogeneity (especially regarding type of seizure)	7
Subjective reporting	6
Dropout rate and adherence problems	5
Baseline differences between groups	4
Inaccuracy in reporting nocturnal seizure frequency based on recall	2
Dosage characteristics (low, fixed, or flexible)	3
Difficulty in identifying seizure onset/cessation	2
Lack of randomization	2
Retrospective study	2
Difference in intervention administration by parents or different centers	2
Retrospective study	2
Study observational	1
Informative censuring	1
High placebo response	1
Single center	1
Protocol violation	1
Practice effect from repeated neuropsychological assessment	1

Krueger et al. reported that participants could have been unintentionally selected with (an undetermined) favorable advantage to respond to everolimus treatment [44].

14.1.3 Nonblinded

Eighteen articles reported absence of blinding as a limitation. For example, in Engel et al. the main outcomes (i.e., seizures and QOL) were reported by participants who could not be blinded to whether they had received surgery or not. The researchers had tried to minimize biases by including a blind committee to assess whether events (from participant diaries) should be classified as epileptic, and to monitor the appropriateness of pharmacotherapy for each participant [102].

Holsti et al. reported unblinding as a chief limitation of their study. The study aimed to compare intranasal midazolam with rectal diazepam for home treatment of seizures in children with epilepsy. To incorporate blinding, it would have been necessary to use a double-dummy design in which the caretakers would have to give both treatment and placebo to their children during the seizure—one would be intranasal, and the other rectal. The authors did not think caretakers would find this acceptable, so they opted for the nonblind design instead [103].

14.1.4 Short or No Follow-Up and Lack of Placebo/Control Group

Several articles reported short or no follow-ups ($n = 11$) and lack of placebo or control groups ($n = 11$) as limitations.

Wechsler et al. conducted a historical-control study to evaluate the efficacy and safety of conversion to lacosamide monotherapy in adults with focal epilepsy. The choice was based on the Food and Drug Administration (FDA) recommendation that monotherapy trials compare AEDs to historical-control groups instead of low-dose active control groups. One limitation the authors reported was that the study duration was relatively short compared to actual clinical practice. Another limitation was the AED differences between the treated group and the historical control. By the time patients were treated there was a greater variety of AEDs compared to the time of historical-control studies. For instance, the use of carbamazepine was lower in the treated group than in historical-control studies [59].

Ryvlin et al. had a shorter follow-up duration than initially planned, consequent to a low enrollment rate that led to early study termination by the sponsor [112].

In Elkhayat et al. the control group was composed of healthy controls that did not receive melatonin, unlike the intervention group. Therefore, there is no way to know if the observed decrease in seizure severity may have been due to placebo effect [51].

14.1.5 Other Limitations

Other limitations reported include nonrandomized design ($n = 8$) (e.g., Klein et al., Stefan et al. [77, 78]); sample heterogeneity ($n = 7$) (e.g., Bochynska et al., McLaughlin et al. [43, 111]); subjective data provided by self or parental report ($n = 6$) (e.g., Raju et al., Sharma et al. [110, 114]); high dropout rate and low adherence ($n = 3$) (e.g., Lazzari et al., Tonekaboni et al. [73, 92]); potential group differences in demographic/medical variables at baseline ($n = 3$) (e.g., Guerrini et al., Modi et al. [60, 85]); inaccuracy in reporting nocturnal seizures and their frequency based on recall ($n = 2$) (e.g., Carrion et al., Sharma et al. [110, 113]); characteristics of drug dosage—low, fixed, or flexible ($n = 3$) (e.g., Cho et al., Brodie et al. [38, 46]); difficulty in identifying seizure onset/cessation ($n = 2$) (e.g., Biton et al., Holsti et al. [61, 103]); retrospective study ($n = 2$) (e.g., Miranda et al., Carrion et al. [109, 113]); differences in intervention administration by parents or different centers ($n = 2$) (e.g., Holsti et al., McLachlan et al. [103, 111]); observational characteristic of the study ($n = 1$) (Werhahn et al. [70]); informative censuring ($n = 1$) (Ramsay et al. [65]); high placebo response ($n = 1$) (Guerrini et al. [64]); single center study ($n = 1$) (Modi et al. [85]); protocol violation ($n = 1$) (Borggraefe et al. [37]) and practice effect from repeated neuropsychological assessment ($n = 1$) (Oh et al. [80]).

14.1.6 Conclusion of Limitation Section

All trials have limitations. When designing your trial, make sure that the anticipated limitations do not negatively affect your trial's internal validity (a priority), and external validity as well.

An important take home message is the necessity to invest in recruitment strategies. First, researchers should make sure that the inclusion/exclusion criteria are not too restrictive (to ensure the internal and external validity of the trial). Second, it is important to carefully consider the available budget in terms of what is feasible. Third, it is important to monitor enrollment progress and to identify measures to take during the trial if recruitment is low. Simple steps may improve recruitment (e.g., calling participants, identifying favorable appointment times, sending appointment reminders, helping with transport, etc.). Finally, researchers should know their IRB regulations, including the typical turnover times or time to approval, and strategize recruitment efforts accordingly.

15 Conclusions

On reviewing the top 100 cited (and some of the most impactful) articles on interventional clinical trials in epilepsy, we demonstrate some of the challenges arising from investigating this population.

We highlight some important points to remember when designing an interventional clinical trial in epilepsy:

- Your research question is key.
- Careful design is critical to the internal and external validity of your trial.
- Choice of design will depend on the existing literature, your study population, experience, resources, and collaborations. Study setting is also important.
- Your inclusion and exclusion criteria should be optimized in advance, depending on whom you intend to study, with an eye on external validity and recruitment.
- There are many potential interventions, and the field is expanding. Ethical considerations and resources are some of the bigger challenges.

- Choose the best outcome(s) to answer your research question, and decide on the best ways to collect and analyze this data.

- Calculate your sample size in advance, including allowances for potential dropouts.

- In RCTs, randomization should not be undermined.

- Always perform allocation concealment. Blinding should be implemented when and where possible.

- Adherence to the protocol can be a challenge, particularly with longer study durations. However, you can implement measures to maximize adherence.

- Know when to ask for help and/or collaboration.

A final note: when designing a clinical trial, keep in mind that it is *quite possible to obtain statistical significance without clinical relevance*. However, a clinically relevant result that is underpowered may lead to another trial, and the question may get answered in the end. Additionally, when designing your trial, aim to reduce biases as well as the perception of them; so take what measures you can, but understand that in some scenarios it may not be possible to remove the limitations, so at least try to minimize them. We hope this chapter will help guide you in your mission of designing an epilepsy trial.

References

1. Ngugi AK, Bottomley C, Kleinschmidt I, Sander JW, Newton CR (2010) Estimation of the burden of active and life-time epilepsy: a meta-analytic approach. Epilepsia 51(5):883–890

2. Ngugi AK, Kariuki SM, Bottomley C, Kleinschmidt I, Sander JW, Newton CR (2011) Incidence of epilepsy: a systematic review and meta-analysis. Neurology 77(10):1005–1012

3. Fisher RS, Acevedo C, Arzimanoglou A, Bogacz A, Cross JH, Elger CE et al (2014) ILAE official report: a practical clinical definition of epilepsy. Epilepsia 55(4):475–482

4. Brodie MJ, Barry SJE, Bamagous GA, Norrie JD, Kwan P (2012) Patterns of treatment response in newly diagnosed epilepsy. Neurology 78(20):1548–1554

5. Kwan P, Arzimanoglou A, Berg AT, Brodie MJ, Hauser WA, Mathern G et al (2010) Definition of drug resistant epilepsy: consensus proposal by the ad hoc task force of the ILAE commission on therapeutic strategies. Epilepsia 51(6):1069–1077

6. Quigg M, Broshek DK, Barbaro NM, Ward MM, Laxer KD, Yan G et al (2011) Neuropsychological outcomes after Gamma Knife radiosurgery for mesial temporal lobe epilepsy: a prospective multicenter study. Epilepsia 52(5):909–916

7. Valentín A, García Navarrete E, Chelvarajah R, Torres C, Navas M, Vico L et al (2013) Deep brain stimulation of the centromedian thalamic nucleus for the treatment of generalized and frontal epilepsies. Epilepsia 54(10):1823–1833

8. Heck CN, King-Stephens D, Massey AD, Nair DR, Jobst BC, Barkley GL et al (2014) Two-year seizure reduction in adults with medically intractable partial onset epilepsy treated with responsive neurostimulation: final results of the RNS system pivotal trial. Epilepsia 55(3):432–441

9. Alexander R (2013) Epilepsy. Handb Clin Neurol 116:491–497

10. Devinsky O, Spruill T, Thurman D, Friedman D (2016) Recognizing and preventing epilepsy-related mortality: a call for action. Neurology 86(8):779–786. https://doi.org/10.1212/WNL.0000000000002253

11. Thurman DJ, Hesdorffer DC, French JA (2014) Sudden unexpected death in epilepsy:

assessing the public health burden. Epilepsia 55(10):1479–1485

12. Baxendale S, Thompson P (2016) The new approach to epilepsy classification: cognition and behavior in adult epilepsy syndromes. Epilepsy Behav 64(Pt A):253–256. http://www.ncbi.nlm.nih.gov/pubmed/27776297

13. Parisi P, Verrotti A, Paolino M, Castaldo R, Ianniello F, Ferretti A et al (2011) "Electro-clinical syndromes" with onset in paediatric age: the highlights of the clinical-EEG, genetic and therapeutic advances. Ital J Pediatr 37(1):58. http://www.ncbi.nlm.nih.gov/pubmed/22182677

14. Korff CM, Scheffer IE (2013) Epilepsy classification. Curr Opin Neurol 26(2):163–167. http://www.ncbi.nlm.nih.gov/pubmed/23406910

15. Taussig D, Montavont A, Isnard J (2015) Invasive EEG explorations. Neurophysiol Clin 45(1):113–119. http://www.ncbi.nlm.nih.gov/pubmed/25703438

16. Glauser T, Ben-Menachem E, Bourgeois B, Cnaan A, Chadwick D, Guerreiro C et al (2006) ILAE treatment guidelines: evidence-based analysis of antiepileptic drug efficacy and effectiveness as initial monotherapy for epileptic seizures and syndromes. Epilepsia 47(7):1094–1120

17. ILAE (1998) Considerations on designing clinical trials to evaluate the place of new anti-epileptic drugs in the treatment of newly diagnosed and chronic patients with epilepsy. Epilepsia 39(7):799–803

18. Chadwick D (1997) Monotherapy clinical trials of new antiepileptic drugs: design, indications, and controversies. Epilepsia 38(Suppl 9):S16–S20

19. Sergott RC, Bittman RM, Christen EM, Sagar SM (2010) Vigabatrin-induced peripheral visual field defects in patients with refractory partial epilepsy. Epilepsy Res 92(2–3):170–176. https://doi.org/10.1016/j.eplepsyres.2010.09.004

20. Halasz P, Cramer JA, Hodoba D, Czlonkowska A, Guekht A, Maia J et al (2010) Long-term efficacy and safety of esli-carbazepine acetate: results of a 1-year open-label extension study in partial-onset seizures in adults with epilepsy. Epilepsia 51(10):1963–1969

21. Glauser TA, Cnaan A, Shinnar S, Hirtz DG, Dlugos D, Masur D et al (2010) Ethosuximide, valproic acid, and lamotrigine in childhood absence epilepsy. N Engl J Med 362(9):790–799. https://doi.org/10.1056/NEJMoa0902014

22. Uberos J, Augustin-Morales MC, Molina Carballo A, Florido J, Narbona E, Muñoz-Hoyos A (2011) Normalization of the sleep-wake pattern and melatonin and 6-sulphatoxy-melatonin levels after a therapeutic trial with melatonin in children with severe epilepsy. J Pineal Res 50(2):192–196

23. Miller IW, Shea MT, Kazim A, Keller MB (1997) A pilot study of lithium carbonate plus divalproex sodium for the continuation and maintenance treatment of patients with bipolar I disorder. J Clin Psychiatry 58(3):95–99

24. Oluboka OJ, Bird DC, Kutcher S, Kusumakar VA (2002) A pilot study of loading versus titration of valproate in the treatment of acute mania. Bipolar Disord 4:341–345

25. Gonzalez-Heydrich J, Whitney J, Waber D, Forbes P, Hsin O, Faraone SV et al (2010) Adaptive phase I study of OROS methylphenidate treatment of attention deficit hyperactivity disorder with epilepsy. Epilepsy Behav 18(3):229–237. http://www.pubmedcentral.nih.gov/articlerender.fcgi?artid=2902631&tool=pmcentrez&rendertype=abstract

26. Goldberg-Stern H, Oren H, Peled N, Garty B-Z (2012) Effect of melatonin on seizure frequency in intractable epilepsy: a pilot study. J Child Neurol 27(12):1524–1528

27. Yuen AWC, Flugel D, Poepel A, Bell GS, Peacock JL, Sander JW (2012) Non-randomized open trial of eicosapentaenoic acid (EPA), an omega-3 fatty acid, in ten people with chronic epilepsy. Epilepsy Behav 23(3):370–372. https://doi.org/10.1016/j.yebeh.2011.11.030

28. Eftekhari S, Mehvari Habibabadi J, Najafi Ziarani M, Hashemi Fesharaki SS, Gharakhani M, Mostafavi H et al (2013) Bumetanide reduces seizure frequency in patients with temporal lobe epilepsy. Epilepsia 54(1):10–13

29. Clark AM, Kriel RL, Leppik IE, White JR, Henry TR, Brundage RC et al (2013) Intravenous topiramate: safety and pharmacokinetics following a single dose in patients with epilepsy or migraines taking oral topiramate. Epilepsia 54(6):1106–1111

30. Cardamone M, Flanagan D, Mowat D, Kennedy SE, Chopra M, Lawson JA (2014) Mammalian target of rapamycin inhibitors for intractable epilepsy and subependymal giant cell astrocytomas in tuberous sclerosis complex. J Pediatr 164(5):1195–1200. https://doi.org/10.1016/j.jpeds.2013.12.053

31. Kanemura H, Sano F, Maeda YI, Sugita K, Aihara M (2012) Valproate sodium enhances body weight gain in patients with childhood

epilepsy: a pathogenic mechanisms and open-label clinical trial of behavior therapy. Seizure 21(7):496–500. https://doi.org/10.1016/j.seizure.2012.05.001

32. Herzog AG, Farina EL, Drislane FW, Schomer DL, Smithson SD, Fowler KM et al (2010) A comparison of anastrozole and testosterone versus placebo and testosterone for treatment of sexual dysfunction in men with epilepsy and hypogonadism. Epilepsy Behav 17(2):264–271. https://doi.org/10.1016/j.yebeh.2009.12.003

33. Santos K, Palmini A, Radziuk AL, Rotert R, Bastos F, Booij L et al (2013) The impact of methylphenidate on seizure frequency and severity in children with attention-deficit-hyperactivity disorder and difficult-to-treat epilepsies. Dev Med Child Neurol 55(7):654–660

34. Motta E, Golba A, Ostrowska Z, Steposz A, Huc M, Kotas-Rusnak J et al (2013) Progesterone therapy in women with epilepsy. Pharmacol Rep 65(1):89–98

35. Jóźwiak S, Kotulska K, Domańska-Pakieła D, Łojszczyk B, Syczewska M, Chmielewski D, Dunin-Wąsowicz D, Kmieć T, Szymkiewicz-Dangel J, Kornacka M, Kawalec W, Kuczyński D, Julita Borkowska MR-L (2011) Antiepileptic treatment before the onset of seizures reduces epilepsy severity and risk of mental retardation in infants with tuberous sclerosis complex. Eur J Paediatr Neurol 15(5):424–431

36. Mikati MA, Kurdi R, El-Khoury Z, Rahi A, Raad W (2010) Intravenous immunoglobulin therapy in intractable childhood epilepsy: open-label study and review of the literature. Epilepsy Behav 17(1):90–94. http://www.ncbi.nlm.nih.gov/pubmed/20004620

37. Borggraefe I, Bonfert M, Bast T, Neubauer BA, Schotten KJ, Maßmann K et al (2013) Levetiracetam vs. sulthiame in benign epilepsy with centrotemporal spikes in childhood: a double-blinded, randomized, controlled trial (German HEAD Study). Eur J Paediatr Neurol 17(5):507–514

38. Cho YW, Kim DH, Motamedi GK (2011) The effect of levetiracetam monotherapy on subjective sleep quality and objective sleep parameters in patients with epilepsy: compared with the effect of carbamazepine-CR monotherapy. Seizure 20(4):336–339. https://doi.org/10.1016/j.seizure.2011.01.006

39. Eun SH, Kim HD, Eun BL, Lee IK, Chung HJ, Kim JS et al (2011) Comparative trial of low- and high-dose zonisamide as monotherapy for childhood epilepsy. Seizure 20(7): 558–563. https://doi.org/10.1016/j.seizure.2011.04.005

40. Bazil CW, Dave J, Cole J, Stalvey J, Drake E (2012) Pregabalin increases slow-wave sleep and may improve attention in patients with partial epilepsy and insomnia. Epilepsy Behav 23(4):422–425

41. Klein P, Herr D, Pearl PL, Natale J, Levine Z, Nogay C et al (2012) Results of phase 2 safety and feasibility study of treatment with levetiracetam for prevention of posttraumatic epilepsy. Arch Neurol 69(10):1290–1295. http://www.ncbi.nlm.nih.gov/pubmed/22777131

42. Lu Y, Xiao Z, Yu W, Xiao F, Xiao Z, Hu Y et al (2011) Efficacy and safety of adjunctive zonisamide in adult patients with refractory partial-onset epilepsy: a randomized, double-blind, placebo-controlled trial. Clin Drug Investig 31(4):221–229

43. Bochyńska A, Lipczyńska-ŁOjkowska W, Gugała-Iwaniuk M, Lechowicz W, Restel M, Graban A et al (2012) The effect of vitamin B supplementation on homocysteine metabolism and clinical state of patients with chronic epilepsy treated with carbamazepine and valproic acid. Seizure 21(4):276–281

44. Krueger DA, Wilfong AA, Holland-Bouley K, Anderson AE, Agricola K, Tudor C et al (2013) Everolimus treatment of refractory epilepsy in tuberous sclerosis complex. Ann Neurol 74(5):679–687

45. Brodie MJ (2011) Ethosuximide, valproic acid, and lamotrigine in childhood absence epilepsy. Yearb Neurol Neurosurg 2011:35–36

46. Brodie MJ, Lerche H, Gil-Nagel A, Elger C, Hall S, Shin P et al (2010) Efficacy and safety of adjunctive ezogabine (retigabine) in refractory partial epilepsy. Neurology 75(20):1817–1824

47. French JA, Abou-Khalil BW, Leroy RF, Yacubian EM, Shin P, Hall S, Mansbach H, Nohria V (2011) Randomized, double-blind, placebo-controlled trial of ezogabine (retigabine) in partial epilepsy. Neurology 76(18):1555–1563

48. Ben-Menachem E, Gabbai AA, Hufnagel A, Maia J, Almeida L, Soares-da-Silva P (2010) Eslicarbazepine acetate as adjunctive therapy in adult patients with partial epilepsy. Epilepsy Res 89(2–3):278–285. https://doi.org/10.1016/j.eplepsyres.2010.01.014

49. Darke K, Edwards SW, Hancock E, Johnson AL, Kennedy CR, Lux AL et al (2010) Developmental and epilepsy outcomes at age 4 years in the UKISS trial comparing hormonal treatments to vigabatrin for infantile

spasms: a multi-centre randomised trial. Arch Dis Child 95(5):382–386. https://doi.org/10.1136/adc.2009.160606

50. Baulac M, Brodie MJ, Patten A, Segieth J, Giorgi L (2012) Efficacy and tolerability of zonisamide versus controlled-release carbamazepine for newly diagnosed partial epilepsy: a phase 3, randomised, double-blind, non-inferiority trial. Lancet Neurol 11(7):579–588. https://doi.org/10.1016/S1474-4422(12)70105-9

51. Elkhayat HA, Hassanein SM, Tomoum HY, Abd-Elhamid IA, Asaad T, Elwakkad AS (2010) Melatonin and sleep-related problems in children with intractable epilepsy. Pediatr Neurol 42(4):249–254. http://linkinghub.elsevier.com/retrieve/pii/S0887899409005281

52. Rektor I, Krauss GL, Bar M, Biton V, Klapper JA, Vaiciene-Magistris N et al (2012) Perampanel study 207: long-term open-label evaluation in patients with epilepsy. Acta Neurol Scand 126(4):263–269

53. Glauser TA, Cnaan A, Shinnar S, Hirtz DG, Dlugos D, Masur D et al (2013) Ethosuximide, valproic acid, and lamotrigine in childhood absence epilepsy: initial monotherapy outcomes at 12 months. Epilepsia 54(1):141–155

54. Chhun S, Troude P, Villeneuve N, Soufflet C, Napuri S, Motte J et al (2011) A prospective open-labeled trial with levetiracetam in pediatric epilepsy syndromes: continuous spikes and waves during sleep is definitely a target. Seizure 20(4):320–325. https://doi.org/10.1016/j.seizure.2010.12.017

55. Masur D, Shinnar S, Cnaan A, Shinnar RC, Clark P, Wang J et al (2013) Pretreatment cognitive deficits and treatment effects on attention in childhood absence epilepsy. Neurology 81(18):1572–1580

56. Herzog AG, Fowler KM, Smithson SD, Kalayjian LA, Heck CN, Sperling MR et al (2012) Progesterone vs placebo therapy for women with epilepsy: a randomized clinical trial. Neurology 78(24):1959–1966. http://www.ncbi.nlm.nih.gov/pubmed/22649214

57. Fattore C, Boniver C, Capovilla G, Cerminara C, Citterio A, Coppola G et al (2011) A multicenter, randomized, placebo-controlled trial of levetiracetam in children and adolescents with newly diagnosed absence epilepsy. Epilepsia 52(4):802–809

58. Hufnagel A, Ben-Menachem E, Gabbai AA, Falcão A, Almeida L, Soares-da-Silva P (2013) Long-term safety and efficacy of eslicarbazepine acetate as adjunctive therapy in the treatment of partial-onset seizures in adults with epilepsy: results of a 1-year open-label extension study.

Epilepsy Res 103(2–3):262–269. https://doi.org/10.1016/j.eplepsyres.2012.07.014

59. Wechsler RT, Li G, French J, O'Brien TJ, D'Cruz O, Williams P et al (2014) Conversion to lacosamide monotherapy in the treatment of focal epilepsy: results from a historical-controlled, multicenter, double-blind study. Epilepsia 55(7):1088–1098

60. Guerrini R, Rosati A, Bradshaw K, Giorgi L (2014) Adjunctive zonisamide therapy in the long-term treatment of children with partial epilepsy: results of an open-label extension study of a phase III, randomized, double-blind, placebo-controlled trial. Epilepsia 55(4):568–578

61. Biton V, Berkovic SF, Abou-Khalil B, Sperling MR, Johnson ME, Lu S (2014) Brivaracetam as adjunctive treatment for uncontrolled partial epilepsy in adults: a phase III randomized, double-blind, placebo-controlled trial. Epilepsia 55(1):57–66

62. Ryvlin P, Werhahn KJ, Blaszczyk B, Johnson ME, Lu S (2014) Adjunctive brivaracetam in adults with uncontrolled focal epilepsy: results from a double-blind, randomized, placebo-controlled trial. Epilepsia 55(1):47–56

63. Trinka E, Marson AG, Van Paesschen W, Kalviainen R, Marovac J, Duncan B et al (2013) KOMET: an unblinded, randomised, two parallel-group, stratified trial comparing the effectiveness of levetiracetam with controlled-release carbamazepine and extended-release sodium valproate as monotherapy in patients with newly diagnosed epilepsy. J Neurol Neurosurg Psychiatry 84:1138–1147. https://doi.org/10.1136/jnnp-2011-300376

64. Guerrini R, Rosati A, Segieth J, Pellacani S, Bradshaw K, Giorgi L (2013) A randomized phase III trial of adjunctive zonisamide in pediatric patients with partial epilepsy. Epilepsia 54(8):1473–1480

65. Ramsay E, Faught E, Krumholz A, Naritoku D, Privitera M, Schwarzman L et al (2010) Efficacy, tolerability, and safety of rapid initiation of topiramate versus phenytoin in patients with new-onset epilepsy: a randomized double-blind clinical trial. Epilepsia 51(10):1970–1977

66. Arhan E, Serdaroglu A, Ozturk B, Ozturk HS, Ozcelik A, Kurt N et al (2011) Effects of epilepsy and antiepileptic drugs on nitric oxide, lipid peroxidation and xanthine oxidase system in children with idiopathic epilepsy. Seizure 20(2):138–142. https://doi.org/10.1016/j.seizure.2010.11.003

67. Kwan P, Trinka E, Van Paesschen W, Rektor I, Johnson ME, Lu S (2014) Adjunctive brivar-

acetam for uncontrolled focal and generalized epilepsies: results of a phase III, double-blind, randomized, placebo-controlled, flexible-dose trial. Epilepsia 55(1):38–46

68. Saetre E, Abdelnoor M, Perucca E, Taubøll E, Isojärvi J, Gjerstad L (2010) Antiepileptic drugs and quality of life in the elderly: results from a randomized double-blind trial of carbamazepine and lamotrigine in patients with onset of epilepsy in old age. Epilepsy Behav 17(3):395–401. https://doi.org/10.1016/j.yebeh.2009.12.026

69. Freitas-Lima P, Alexandre V, Pereira LRL, Feletti F, Perucca E, Sakamoto AC (2011) Influence of enzyme inducing antiepileptic drugs on the pharmacokinetics of levetiracetam in patients with epilepsy. Epilepsy Res 94(1–2):117–120. https://doi.org/10.1016/j.eplepsyres.2011.01.007

70. Werhahn KJ, Klimpe S, Balkaya S, Trinka E, Krämer G (2011) The safety and efficacy of add-on levetiracetam in elderly patients with focal epilepsy: a one-year observational study. Seizure 20(4):305–311. http://linkinghub.elsevier.com/retrieve/pii/S1059131111000057

71. Abou-Khalil B, Wheless J, Rogin J, Wolter KD, Pixton GC, Shukla RB et al (2013) A double-blind, randomized, placebo-controlled trial of a diazepam auto-injector administered by caregivers to patients with epilepsy who require intermittent intervention for acute repetitive seizures. Epilepsia 54(11):1968–1976

72. Kwan P, Lim SH, Chinvarun Y, Cabral-Lim L, Aziz ZA, Lo YK et al (2010) Efficacy and safety of levetiracetam as adjunctive therapy in adult patients with uncontrolled partial epilepsy: the Asia SKATE II study. Epilepsy Behav 18(1–2):100–105. https://doi.org/10.1016/j.yebeh.2010.03.016

73. Lazzari AA, Dussault PM, Thakore-James M, Gagnon D, Baker E, Davis SA et al (2013) Prevention of bone loss and vertebral fractures in patients with chronic epilepsy - antiepileptic drug and osteoporosis prevention trial. Epilepsia 54(11):1997–2004

74. Cook MJ, O'Brien TJ, Berkovic SF, Murphy M, Morokoff A, Fabinyi G et al (2013) Prediction of seizure likelihood with a long-term, implanted seizure advisory system in patients with drug-resistant epilepsy: a first-in-man study. Lancet Neurol 12(6):563–571

75. McLachlan RS, Pigott S, Tellez-Zenteno JF, Wiebe S, Parrent A (2010) Bilateral hippocampal stimulation for intractable temporal lobe epilepsy: impact on seizures and memory. Epilepsia 51(2):304–307

76. Sun W, Mao W, Meng X, Wang D, Qiao L, Tao W et al (2012) Low-frequency repetitive transcranial magnetic stimulation for the treatment of refractory partial epilepsy: a controlled clinical study. Epilepsia 53(10):1782–1789

77. Klein P, Janousek J, Barber A, Weissberger R (2010) Ketogenic diet treatment in adults with refractory epilepsy. Epilepsy Behav 19(4):575–579. https://doi.org/10.1016/j.yebeh.2010.09.016

78. Stefan H, Kreiselmeyer G, Kerling F, Kurzbuch K, Rauch C, Heers M et al (2012) Transcutaneous vagus nerve stimulation (t-VNS) in pharmacoresistant epilepsies: a proof of concept trial. Epilepsia 53(7):115–118

79. Wagner JL, Smith G, Ferguson P, van Bakergem K, Hrisko S (2010) Pilot study of an integrated cognitive-behavioral and self-management intervention for youth with epilepsy and caregivers: Coping Openly and Personally with Epilepsy (COPE). Epilepsy Behav 18(3):280–285. https://doi.org/10.1016/j.yebeh.2010.04.019

80. Oh YS, Kim HJ, Lee KJ, Kim YI, Lim SC, Shon YM (2012) Cognitive improvement after long-term electrical stimulation of bilateral anterior thalamic nucleus in refractory epilepsy patients. Seizure 21(3):183–187. https://doi.org/10.1016/j.seizure.2011.12.003

81. Smith M, Politzer N, MacGarvie D, McAndrews MP, Del Campo M (2011) Efficacy and tolerability of the Modified Atkins Diet in adults with pharmacoresistant epilepsy: a prospective observational study. Epilepsia 52(4):775–780

82. Salam MT, Mirzaei M, Ly MS, Nguyen DK, Sawan M (2012) An implantable closedloop asynchronous drug delivery system for the treatment of refractory epilepsy. IEEE Trans Neural Syst Rehabil Eng 20(4):432–442. https://doi.org/10.1109/TNSRE.2012.2189020

83. Kumada T, Miyajima T, Oda N, Shimomura H, Saito K, Fujii T (2012) Efficacy and tolerability of modified Atkins diet in Japanese children with medication-resistant epilepsy. Brain Dev 34(1):32–38. https://doi.org/10.1016/j.braindev.2010.12.010

84. Lee KJ, Shon YM, Cho CB (2012) Long-term outcome of anterior thalamic nucleus stimulation for intractable epilepsy. Stereotact Funct Neurosurg 90(6):379–385

85. Modi AC, Guilfoyle SM, Rausch J (2013) Preliminary feasibility, acceptability, and efficacy of an innovative adherence intervention for children with newly diagnosed epilepsy. J Pediatr Psychol 38(6):605–616

86. Thammongkol S, Vears DF, Bicknell-Royle J, Nation J, Draffin K, Stewart KG et al (2012) Efficacy of the ketogenic diet: which epilepsies respond? Epilepsia 53(3):55–59

87. Klinkenberg S, Aalbers MW, Vles JSH, Cornips EMJ, Rijkers K, Leenen L et al (2012) Vagus nerve stimulation in children with intractable epilepsy: a randomized controlled trial. Dev Med Child Neurol 54(9):855–861

88. Pop J, Murray D, Markovic D, DeGiorgio CM (2011) Acute and long-term safety of external trigeminal nerve stimulation for drug-resistant epilepsy. Epilepsy Behav 22(3):574–576. https://doi.org/10.1016/j.yebeh.2011.06.024

89. Lin LC, Lee WT, Wang CH, Chen HL, Wu HC, Tsai CL et al (2011) Mozart K.448 acts as a potential add-on therapy in children with refractory epilepsy. Epilepsy Behav 20(3):490–493. https://doi.org/10.1016/j.yebeh.2010.12.044

90. Cervenka MC, Terao NN, Bosarge JL, Henry BJ, Klees AA, Morrison PF et al (2012) E-mail management of the modified Atkins diet for adults with epilepsy is feasible and effective. Epilepsia 53(4):728–732

91. Sun W, Fu W, Mao W, Wang D, Wang Y (2011) Low-frequency repetitive transcranial magnetic stimulation for the treatment of refractory partial epilepsy. Clin EEG Neurosci 42(1):40–44. https://doi.org/10.1177/155005941104200109

92. Tonekaboni SH, Mostaghimi P, Mirmiran P, Abbaskhanian A, Abdollah Gorji F, Ghofrani M et al (2010) Efficacy of the Atkins diet as therapy for intractable epilepsy in children. Arch Iran Med 13(6):492–497. http://www.ncbi.nlm.nih.gov/pubmed/21039004

93. Sohal VS, Sun FT (2011) Responsive neurostimulation suppresses synchronized cortical rhythms in patients with epilepsy. Neurosurg Clin N Am 22(4):481–488. https://doi.org/10.1016/j.nec.2011.07.007

94. Müller K, Fabó D, Entz L, Kelemen A, Halász P, Rásonyi G et al (2010) Outcome of vagus nerve stimulation for epilepsy in Budapest. Epilepsia 51:98–101. http://www.ncbi.nlm.nih.gov/pubmed/20618411

95. Chen DK, Maheshwari A, Franks R, Trolley GC, Robinson JS, Hrachovy RA (2014) Brief group psychoeducation for psychogenic non-epileptic seizures: a neurologist-initiated program in an epilepsy center. Epilepsia 55(1):156–166

96. Kim YH, Kim CH, Kim JS, Lee SK, Chung CK (2011) Resection frequency map after awake resective surgery for non-lesional neo-cortical epilepsy involving eloquent areas. Acta Neurochir 153(9):1739–1749

97. Drane DL, Loring DW, Voets NL, Price M, Ojemann JG, Willie JT et al (2015) Better object recognition and naming outcome with MRI-guided stereotactic laser amygdalohippocampotomy for temporal lobe epilepsy. Epilepsia 56(1):101–113

98. Hensley-Judge H, Quigg M, Barbaro NM, Newman SA, Ward MM, Chang EF et al (2013) Visual field defects after radiosurgery for mesial temporal lobe epilepsy. Epilepsia 54(8):1376–1380

99. Klinkenberg S, Van Den Bosch CNCJ, Majoie HJM, Aalbers MW, Leenen L, Hendriksen J et al (2013) Behavioural and cognitive effects during vagus nerve stimulation in children with intractable epilepsy—a randomized controlled trial. Eur J Paediatr Neurol 17(1):82–90

100. Rong PJ, Liu AH, Zhang JG, Wang YP, Yang AC, Li L et al (2014) An alternative therapy for drug-resistant epilepsy: transcutaneous auricular vagus nerve stimulation. Chin Med J (Engl) 127(2):300–304

101. Fisher R, Salanova V, Witt T, Worth R, Henry T, Gross R et al (2010) Electrical stimulation of the anterior nucleus of thalamus for treatment of refractory epilepsy. Epilepsia 51(5):899–908

102. Engel J, McDermott MP, Wiebe S, Langfitt JT, Stern JM, Dewar S et al (2012) Early surgical therapy for drug-resistant temporal lobe epilepsy: a randomized trial. JAMA 307(9):922–930

103. Holsti M, Dudley N, Schunk J, Adelgais K, Greenberg R, Olsen C et al (2010) Intranasal midazolam vs rectal diazepam for the home treatment of acute seizures in pediatric patients with epilepsy. Arch Pediatr Adolesc Med 164(8):747–753

104. Ciechanowski P, Chaytor N, Miller J, Fraser R, Russo J, Unutzer J et al (2010) PEARLS depression treatment for individuals with epilepsy: a randomized controlled trial. Epilepsy Behav 19(3):225–231. https://doi.org/10.1016/j.yebeh.2010.06.003

105. Chang EF et al (2010) Predictors of efficacy after stereotactic radiosurgery for medial temporal lobe epilepsy. Neurology 74(2):165–172

106. Degiorgio CM, Soss J, Cook IA, Christopher M (2013) Randomized controlled trial of trigeminal nerve stimulation for drug-resistant epilepsy. Neurology 80(9):786–791

107. Chaytor N, Ciechanowski P, Miller JW, Fraser R, Russo J, Unutzer J et al (2011) Long-term outcomes from the PEARLS

randomized trial for the treatment of depression in patients with epilepsy. Epilepsy Behav 20(3):545–549. https://doi.org/10.1016/j.yebeh.2011.01.017

108. Helmstaedter C, Roeske S, Kaaden S, Elger CE, Schramm J (2011) Hippocampal resection length and memory outcome in selective epilepsy surgery. J Neurol Neurosurg Psychiatry 82(12):1375–1381

109. Miranda MJ, Mortensen M, Povlsen JH, Nielsen H, Beniczky S (2011) Danish study of a Modified Atkins diet for medically intractable epilepsy in children: can we achieve the same results as with the classical ketogenic diet? Seizure 20(2):151–155

110. Sharma S, Sankhyan N, Gulati S, Agarwala A (2013) Use of the modified Atkins diet for treatment of refractory childhood epilepsy: a randomized controlled trial. Epilepsia 54(3):481–486

111. McLaughlin DP, McFarland K (2011) A randomized trial of a group based cognitive behavior therapy program for older adults with epilepsy: the impact on seizure frequency, depression and psychosocial well-being. J Behav Med 34(3):201–207

112. Ryvlin P, Gilliam FG, Nguyen DK, Colicchio G, Iudice A, Tinuper P et al (2014) The long-term effect of vagus nerve stimulation on quality of life in patients with pharmacoresistant focal epilepsy: the PuLsE (Open Prospective Randomized Long-term Effectiveness) trial. Epilepsia 55(6):893–900

113. Carrion MJM, Nunes ML, Martinez JVL, Portuguez MW, da Costa JC (2010) Evaluation of sleep quality in patients with refractory seizures who undergo epilepsy surgery. Epilepsy Behav 17(1):120–123. https://doi.org/10.1016/j.yebeh.2009.11.008

114. Raju KNV, Gulati S, Kabra M, Agarwala A, Sharma S, Pandey RM et al (2011) Efficacy of 4:1 (classic) versus 2.5:1 ketogenic ratio diets in refractory epilepsy in young children: a randomized open labeled study. Epilepsy Res 96(1–2):96–100

115. Aliasgharpour M, Dehgahn Nayeri N, Yadegary MA, Haghani H (2013) Effects of an educational program on self-management in patients with epilepsy. Seizure 22(1):48–52. https://doi.org/10.1016/j.seizure.2012.10.005

116. Morrell MJ, RNS System in Epilepsy Study Group (2011) Responsive cortical stimulation for the treatment of medically intractable partial epilepsy. Neurology 77(13):1295–1304. http://www.ncbi.nlm.nih.gov/pubmed/21917777

117. Hart Y, Sneade M, Birks J, Rischmiller J, Kerr R, Molyneux A (2011) Epilepsy after subarachnoid hemorrhage: the frequency of seizures after clip occlusion or coil embolization of a ruptured cerebral aneurysm: results from the International Subarachnoid Aneurysm Trial. J Neurosurg 115(6):1159–1168. http://www.ncbi.nlm.nih.gov/pubmed/21819189

118. Schramm J, Lehmann TN, Zentner J, Mueller CA, Scorzin J, Fimmers R et al (2011) Randomized controlled trial of 2.5-cm versus 3.5-cm mesial temporal resection in temporal lobe epilepsy - Part 1: Intent-to-treat analysis. Acta Neurochir 153(2):209–219

119. Cornu C, Kassai B, Fisch R, Chiron C, Alberti C, Guerrini R et al (2013) Experimental designs for small randomised clinical trials: an algorithm for choice. Orphanet J Rare Dis 8(Icd):1

120. Borkowf CB, Lee Johnson L, Albert PS (2012) Chapter 22 - Power and sample size calculations. In: Principles and practice of clinical research, 3rd edn, pp 271–283

121. Portney LG, Watkins MP. Foundations of clinical research: applications to practice. 892 p. Accessed 13 Mar 2017. http://www.fadavis.com/product/physical-therapy-foundations-clinical-research-portney-3

122. Portney LG, Watkins MP (2009) Foundations of clinical research: applications to practice. Prentice Hall, Upper Saddle River, 892 p

123. Sankoh AJ (1999) Interim analyses: an update of an FDA reviewer's experience and perspective. Drug Inf J 33(1):165–176

124. Kunz R, Vist GE, Oxman AD (2007) Randomisation to protect against selection bias in healthcare trials. Cochrane Database Syst Rev (4):MR000012

125. Zheng L, Zelen M (2008) Multi-center clinical trials: randomization and ancillary statistics. Ann Appl Stat 2(2):582–600

126. Kim J, Shin W (2014) How to Do Random Allocation (Randomization). Clinics in Orthopedic Surgery. 6(1):103–109. https://doi.org/10.4055/cios.2014.6.1.103.

127. Consort - Glossary. http://www.consort-statement.org/resources/glossary. Accessed 13 Mar 2017

128. Schulz KF, Grimes DA (2002) Allocation concealment in randomised trials: defending against deciphering. Lancet 359(9306):614–618

129. Schulz KF, Grimes DA (2002) Blinding in randomised trials: hiding who got what. Lancet 359(9307):696–700

130. Schulz KF, Altman DG, Moher D, Barbour V, Berlin JA, Boutron I et al (2010) CONSORT 2010 statement: updated guidelines for reporting parallel group randomised

trials (Chinese version). J Chinese Integr Med 8(7):604–612

131. NeuroPace® RNS® System Brief Statement. http://www.neuropace.com/wp-content/uploads/2015/11/Brief_Statement.pdf. Accessed 4 Dec 2017

132. Freeman G, Hjortdahl P (1997) What future for continuity of care in general practice? BMJ 314(7098):1870–1870. http://www.bmj.com/content/314/7098/1870.abstract

133. Greenland S, Bolton S, Freeman G, Hjortdahl P, Bolton S (1998) Independence and statistical inference in clinical trial designs: a tutorial review. J Clin Pharmacol 38(5):408–412. http://www.bmj.com/content/314/7098/1870.abstract

134. Greenland S (2000) Principles of multilevel modelling. Int J Epidemiol 29(1):158–167

135. Pildal J, Chan A-W, Hróbjartsson A, Forfang E, Altman DG, Gøtzsche PC (2005) Comparison of descriptions of allocation concealment in trial protocols and the published reports: cohort study. BMJ 330(April):1049

136. Moher D, Hopewell S, Schulz KF, Montori V, Gøtzsche PC, Devereaux PJ et al (2012) CONSORT 2010 explanation and elaboration: updated guidelines for reporting parallel group randomised trials. Int J Surg 10(1):28–55

137. Shah PB (2011) Intention-to-treat and per-protocol analysis. CMAJ 183(6):696; author reply 696. http://www.ncbi.nlm.nih.gov/pubmed/21464181

138. Gupta S (2011) Intention-to-treat concept: a review. Perspect Clin Res 2(3):109. http://www.picronline.org/text.asp?2011/2/3/109/83221

139. 9.2.4 Effect measures for ordinal outcomes and scales. http://handbook.cochrane.org/chapter_9/9_2_4_effect_measures_for_

ordinal_outcomes_and_measurement.htm. Accessed 13 Mar 2017

140. 9.2.6 Effect measures for time-to-event (survival) outcomes. http://handbook.cochrane.org/chapter_9/9_2_6_effect_measures_for_time_to_event_survival_outcomes.htm. Accessed 13 Mar 2017

141. Hanley JA, Negassa A, Edwardes MD, Forrester JE (2003) Statistical analysis of correlated data using generalized estimating equations: an orientation. Am J Epidemiol 157(4):364–375. http://www.ncbi.nlm.nih.gov/pubmed/12578807

142. Wood AM, White IR, Thompson SG (2004) Are missing outcome data adequately handled? A review of published randomized controlled trials in major medical journals. Clin Trials 1:368–376

143. White IR, Horton NJ, Carpenter J, Pocock SJ (2011) Strategy for intention to treat analysis in randomised trials with missing outcome data. BMJ 342:1–4

144. Doody RS, Gavrilova SI, Sano M, Thomas RG, Aisen PS, Bachurin SO et al (2008) Effect of dimebon on cognition, activities of daily living, behaviour, and global function in patients with mild-to-moderate Alzheimer's disease: a randomised, double-blind, placebo-controlled study. Lancet 372(9634):207–215

145. Wagner JL, Smith G, Ferguson P, Van Bakergem K, Hrisko S (2011) Feasibility of a pediatric cognitive-behavioral self-management intervention: Coping Openly and Personally with Epilepsy (COPE). Seizure 20(6):462–467. https://doi.org/10.1016/j.yebeh.2010.04.019

146. Consort (2014) CONSORT 2010 checklist. pp 11–2. http://www.consort-statement.org/consort-statement/checklist

147. 9.7 Sensitivity analyses. http://handbook.cochrane.org/chapter_9/9_7_sensitivity_analyses.htm. Accessed 13 Mar 2017

Chapter 14

Future of Clinical Trials in Neurology

Beatriz Teixeira Costa, Isadora Santos Ferreira, and Felipe Fregni

Abstract

This chapter discusses the main challenges of the 12 conditions reviewed in this book according to the 100 most cited clinical trials in each of these conditions, with the aim of providing future perspectives for the development of high-quality clinical trials in neurology. Therefore, the limitations of the studies are discussed, possible solutions as well as the areas in which research is still needed. For this purpose, relevant practices in these clinical trials are analyzed and suggestions are made for the improvement on the design and conduction of further clinical studies in the field.

Key words Limitations, Future perspectives, Study design, Review, Clinical trial

1 Introduction

Developing a feasible and high-quality trial requires several skills as well as specific expertise, including critical thinking and scientific knowledge. After reviewing innumerous clinical trials of different conditions, it is reasonable to state that conducting research is a both challenging and demanding task. However, although investigators may face limitations and unanticipated drawbacks, it is undeniable how important and fundamental the emergence of new clinical trials is, regardless of the neurological condition. By following guidelines and addressing all the elements previously mentioned in all the chapters, researchers are able to improve scientific evidence, consequently transforming clinical practice.

Overall, all the chapters reported and discussed the main challenges and limitations involved with each neurological condition addressed in this book. Among them, the most common limitation was the lack of reporting. Several authors did not describe with all the necessary details or even failed to describe completely fundamental aspects which influence the interpretation of the final results.

Another limitation highly mentioned in several studies was the small sample size, which could be a result of different factors.

Felipe Fregni (ed.), *Clinical Trials in Neurology*, Neuromethods, vol. 138,
https://doi.org/10.1007/978-1-4939-7880-9_14, © Springer Science+Business Media, LLC, part of Springer Nature 2018

One of them is the difficulty in recruiting patients. A considerable amount of studies related to different conditions mention that recruitment is a main concern as, even though patients often have the condition, they do not always meet all the inclusion or exclusion criteria. This leads to the issue of either investigators thinking in new ways to enhance recruitment, including multicenter trials, or in novel methods such as development of markers that would make small sample size studies valid. Also, due to the complexity of neurological conditions, there usually is a high dropout rate which end up affecting the final sample. Hence, knowing the characteristics of the target population as well as properly defining the inclusion and exclusion criteria, not too restringing or too broad, must be one of the primary aspects to consider when starting a trial.

In spite of the different challenges involved with any trial, the perspectives for future studies in the neurological field are still very positive. The will to improve health care in conjunction with discovering alternative therapies which may be more effective keep encouraging researchers to overcome these challenges and develop novel scientific evidence. Whether it is a phase I, II, or III trial, investigators are increasingly taking a step forward into improving patient's quality of life. Indeed, more studies are still necessary; however, clinical research seems to be advancing in a promising direction. Below we discuss the main areas for future development in this field for each condition analyzed in this book.

2 Stroke

The most mentioned limitation in stroke trials was a small sample size. This was possibly related to difficulties in recruiting patients who had a stroke, as they are usually seniors, with multiple comorbidities and unstable health condition. Also, inclusion criteria sometimes restrict the available sample. The investigator needs then to carefully consider when designing the eligibility criteria for a given trial (see Chap. 2 for an extended discussion on this topic). Indeed, recruitment is one of the biggest challenges faced by investigators conducting stroke trials, which frequently restricts the advance of clinical research in this field. Also, as the stroke population is highly heterogeneous, a limited external validity is another issue these trials must address in the future; however, for early trials, external validity should not be the main goal. Interestingly, the studies of this review mostly included male patients, which shows the need for using broader inclusion criteria and balance the randomization process in upcoming trials.

Another important challenge mentioned in multiple stroke trials was the limitation of assessment instruments, as several of them do not address all the necessary information for a proper evaluation of clinical outcomes. For instance, few scales mostly assess motor function aspects, while other instruments focus on speech recovery or different neurological characteristics. Therefore, in order to correctly measure study endpoints and consequently generate useful conclusions, it is fundamental that investigators become familiarized with the existing types of scales and choose the one that best fits the study design.

Contemplating future perspectives, the development of stroke clinical trials, specifically focused on innovative therapeutic methods, is imperative for advancing the field and improving patients' quality of life. Indeed, future studies must be well planned in order to prevent major issues regarding recruitment, participants' adherence, and study's external validity. Furthermore, we encourage clinical investigators to identify which features in stroke health care is a priority to be explored, with the goal of creating promising research questions, thus positively impacting the field.

3   Disorders of Consciousness

In clinical trials involved with disorders of consciousness some important aspects are noteworthy to be discussed. In addition to the lack of reporting, the small sample size was a concerning limitation, followed by a heterogeneity in sample distribution. The number of trials which reported either blinding issues or a lack of follow-up to assess the long-term effect of the treatment although was small and not highly frequent; these are important considerations to be addressed. The assessment of long-term effects as well as assuring adequate blinding procedures for the assessors is crucial to ensure patient's safety and minimize the risk of bias, respectively. Also, as subjects' awareness and vigilance are compromised to various extents in DOC trials; family members, caregivers and/or the assessors are usually blinded to the intervention, which results in ethical and methodological concerns. According to this review, 1-, 3-, and 6-month follow-ups are recommended as the ideal time points.

Hence, for future studies in this field, as most trials were parallel, phase II, single-center drug trials, it is important to develop new phase III trials as well as multicenter trials in order to test the alternative drugs in a broader and more heterogeneous population. In fact, there is an unmet clinical need to confirming results of phase II to change clinical practice in DOC care.

4   Traumatic Brain Injury

Research studies conducted in traumatic brain injury (TBI) have shown promising results for advancing the field, although investigators still need to overcome numerous limitations. From this review, it became clear that a lack of blinding and randomization was frequent in several clinical trials in TBI, although most of them corresponded to phase II and pivotal trials. This significant limitation shows that a large number of these trials fail to control for confounders and are vulnerable to different types of biases, such as measurement and performance bias. Also, the majority of TBI studies did not work with any covariate adjustment, which could lead to unreliable results. The lack of reporting missing data imputation methods was another limitation in TBI trials, which shows that research methodology guidelines need to be better explored and complied in future trials. In fact, as compared to other neurological areas, TBI trials still show less robustness in their methodology. This could be seen at the same time as a limitation and an opportunity to change this field.

In order to successfully advance knowledge on TBI and improve medical care in this field, innovative research projects are being developed. For instance, the National Institute of Neurological Disorders and Stroke (NINDS) supports preclinical and clinical research on TBI, also in collaboration with international institutes, with the main goal of making progress on treatment and care in this field. Additionally, for this purpose, it is evident that future clinical trials need not only to address the main limitations of previous studies, but also to go beyond the already explored diagnostic/therapeutic methods and bring innovation to research studies in TBI. Also, deeply exploring different populations with TBI, such as pediatric and senior patients, is another possible area for researchers aiming to generate novel initiatives in the field.

5   Parkinson's Disease

Alike stroke trials and the ones related to disorders of consciousness, the most frequent reported limitation was the sample size, followed by the duration of follow-ups. However, differently from the trials related to disorders of consciousness which did not have enough follow-ups, clinical trials of Parkinson's disease reported extensive follow-up visits. These extensive follow-ups may lead to potential limitations, especially attrition bias. As a result, systematic differences between the groups from a study can be noticed. Other common limitations were also reported, such as blinding and restrictive inclusion/exclusion criteria.

Besides clinical outcomes, future studies in this field should focus on the search for biomarkers, especially those from neuroimaging, as surrogate endpoints are often easier to be investigated and they require less resources and time investment. In addition, incorporating innovative designs such as adaptive and delayed-start design may generate benefits toward research in this field. Finally, the use of new outcome-measuring tools such as Internet-based applications should be encouraged as this may facilitate adherence and the recruitment of patients.

6 Dystonia

As noticed in other neurological disorders, clinical investigators in dystonia often fail to report the limitations of clinical trials, as well as the main strategies applied to overcome them. The lack of statistical power due to small sample size was one of the main issues mentioned in the manuscripts, which may be justified by the difficulty for recruiting patients with different levels of disease severity and given the prevalence of dystonia is lower than other disorders such as stroke. Also, ethical aspects related to the use of placebo in clinical research may often compromise the development of large trials in this field. Another limitation reported in a few studies was the variability within patients with dystonia as they have different clinical presentations, which may directly affect study internal validity if randomization and allocation concealment is not performed.

Sample size calculation and study design are indeed critical determinants of a successful clinical trial. By overcoming these main limitations, researchers will certainly develop successful trials for improving clinical practice and ensuring that patients will have access to the best available health care. Although the research field in dystonia has advanced over the past 30 years, there are still multiple aspects of disease management to explore. In order to accomplish that, different organizations, such as the Dystonia Medical Research Foundation, are applying significant efforts on cultivating clinical trials, which will enhance effectiveness of dystonia diagnostic and therapeutic tools [1].

7 Meningitis

Meningitis is still a major concern of health care authorities due to its life-threatening characteristics, especially bacterial meningitis. Thus, the development of innovative clinical trials is fundamental for enhancing this field. In this chapter, it could be noticed that a meaningful limitation of most research studies in meningitis was the lack of blinding methods and allocation concealment. Although these components are not indispensable in clinical trials, this limitation may have added biases to the results of some of these trials. Hence, an important task for future investigators in meningitis is to pursue better methods of blinding, randomization and allocation concealment while designing new studies. With these components ensured in the trials, fairly comparisons of experimental groups can be performed, and reliable conclusion may be done.

As observed in different neurological diseases, small sample size was also an issue mentioned in most meningitis manuscripts. The main issue with a sample size smaller than expected is the lack of statistical power and consequently an increase in the probability of type II error. An ideal strategy to address this issue in future trials is to properly know the target population, as well as the accessible and study population. By assuring this, researchers are able to create useful approaches for recruiting enough participants and enhancing patients' adherence.

8 Multiple Sclerosis

The most common limitations such as a small sample size and lack of reporting were also noticed in multiple sclerosis trials. However, some particular challenges were also noticed. Besides referring to a small sample size, several studies reported an attrition rate higher than expected in conjunction with a low recruitment yield. In addition, many outcomes were based on self-reported quality of life scales, which imposes a limitation about the variability of the results.

In contrast with the conditions that report high heterogeneity of the population, multiple sclerosis trials described the population as being often too homogenous, thus not representing the diversity of the general population. Although a more homogeneous population is expected in phase II trials, this may lead to a lack of external validity in phase III trials. Here, the investigator needs to consider how different may be the results in another sample of multiple sclerosis with other characteristics.

Therefore, future studies should aim in finding a balance between the characteristics of the final sample, consequently applying efficient inclusion/exclusion criteria in order to truly test the effects of the chosen intervention, not harming the external validity of the trial. Also, by improving self-reported scales, investigators would be able to gather more information and better analyze chosen outcomes.

9 Alzheimer's Disease

In regard to Alzheimer's disease trials, the main limitations were related to the difficulty and subjectivity of the diagnosis as well as the assessment and applicability of cognitive tests. As the symptoms related with Alzheimer's disease involve cognitive impairment, memory deficits mood swings and symptoms which might be similar to other conditions, providing an effective diagnosis and testing alternative interventions is definitely not an easy task.

Therefore, in order to ensure patient's safety, it is fundamental that investigators choose wisely how to measure outcomes and whether the intervention being tested is actually safe. One suggestion is the use of biomarkers both as diagnostic and surrogate markers. The current markers have not shown great correlation with clinical outcome; thus, this is a critical area of development to make Alzheimer's trials more feasible and efficient.

Additionally, once the data on safety and efficacy is gathered, the development of multicenter trials is suggested as to recruit demographically diverse individuals and enhance the generalizability of the results.

10   Tumors of the Central Nervous System

Unlike other conditions, clinical trials of brain tumors reported the lack of previous studies which the results could be compared to as one of the main limitations. Consequently, it is difficult for researchers to develop new trials, especially those in phase III and IV. In addition, these clinical trials also reported the limitations that were previously mentioned in other neurological conditions, such as small sample size, a short follow up period, as well as a heterogeneity among groups, which particularly in brain tumors trials, may lead to confounders and possible bias.

Another important limitation worth mentioning was the heterogeneity of recurrent brain tumors, especially meningiomas and glioblastomas. This aspect has a fundamental importance when it comes to conducting a trial and defining the initial steps because of how this may affect the patient's response to the intervention.

Therefore, future investigators should aim on the development of new clinical trials which are able to investigate alternative therapies, taking into account the variability of patient's responses to each drug and how brain tumors behave. Consequently, future researchers may need to take one step back to further understand the differences across patients with these tumors as to use this knowledge in designing more efficient clinical trials and hopefully obtain the approval of drugs that have a significant impact in this field.

11   Spinal Cord Injury

Besides the commonly reported limitations such as difficulties in recruiting patients, the lack of reporting vital information and heterogeneity of the population, spinal cord injury clinical trials also reported a short duration of the intervention. This aspect was mainly highlighted by four trials, in which they mentioned that longer trials with also a longer intervention period would provide stronger results. As a result, it is possible to question whether those trials have truly investigated the effects of the therapy being tested.

In addition to those limitations, it was noticed in spinal cord injury clinical trials some particular challenges. For instance, a large number of dropouts, a population to which the results could not be generalized in the future and large differences between control and intervention group were too high. The last challenge, specifically, is extremely concerning as the results obtained in this trial might be due to the discrepancy between interventional and control group and not because of the intervention itself, thus resulting in type I and/or II error.

Hence, future trials must consider all these limitations by overcoming them and consequently improving the way that research is being conducted in this field. Although there are several challenges involving spinal cord injury, there seems to have an increase in trials published in this field as well as the impact of these publications.

12   Peripheral Neuropathy

In spite of the challenges already mentioned above such as a small sample size and the lack of blinding, there is one important limitation related not only to the development of peripheral neuropathy trials, but to the lack of standard established treatments for this condition regardless of the etiology.

As a result, clinicians and researchers are constantly trying to relief patient's symptoms and develop an effective therapy. Given this stage of clinical development for this condition, it is critical to design clinical trials that are also mechanistic as to understand further the mechanisms of peripheral neuropathy as well as potential treatments. Indeed, when there is a lack of mechanistic data on the disease and the intervention the chances of success are smaller [2].

A field that is being widely explored and should be the aim of future discussions is neuromodulation therapy, particularly the combination of this intervention with behavioral therapies as well as therapies with visual-augmented reality through the contribution of technology. Since patients often do not respond to medications in the same way and peripheral approaches are not always effective enough, investigators are looking for different alternatives such as the use of electrical stimulation which have shown positive results in other conditions.

Therefore, although there are still several aspects to be elucidated in terms of the neurophysiologic elements involving this condition as well as the effects of these new therapies, researchers should not only design strong clinical trials but also strong mechanistic trials as to find better and novel treatments for this condition.

13   Final remarks

The art of clinical research is understanding the current level of science and finding the best tools to answer important clinical questions. The pathway to adding knowledge to this field is not an easy one but is certainly rewarding. The new investigator needs to understand that clinical research is not only about randomizing two groups of patients to receive drug A and B. There is much more to that. A clinical trial should be informative regardless of its results (positive or negative). When a negative clinical trial is not published because results are "not interesting," this means that the clinical trial was not properly designed and indeed it resulted in a waste of time and resources. Finally, a clinical trial is a costly venture and therefore appropriate training and time during the design phases are necessary in order to maximize their gains. We hope this book, besides providing a basic guide to the investigator, also provides important insights that are specific to trials in neurological disorders. We look forward to reading future contributions from the readers of this book.

References

1. Dystonia Medical Research Foundation (n.d) https://www.dystonia-foundation.org/research. Accessed 8 Nov 2017
2. Edelmuth RC, Nitsche MA, Battistella L, Fregni F (2010) Why do some promising brain-stimulation devices fail the next steps of clinical development? Expert Rev Med Devices 7(1):67–97. https://doi.org/10.1586/erd.09.64. Review

Appendix

Chapter 2

1. Broderick et al (2013) Endovascular therapy after intravenous t-PA versus t-PA alone for stroke

2. Saver et al (2012) Solitaire flow restoration device versus the Merci Retriever in patients with acute ischaemic stroke (SWIFT): a randomised, parallel-group, non-inferiority trial

3. Wallentin et al (2010) Efficacy and safety of dabigatran compared with warfarin at different levels of international normalised ratio control for stroke prevention in atrial fibrillation: an analysis of the RE-LY trial

4. Kidwell et al (2013) A trial of imaging selection and endovascular treatment for ischemic stroke

5. Ciccone et al (2013) Endovascular treatment for acute ischemic stroke

6. Nogueira et al (2012) Trevo versus Merci retrievers for thrombectomy revascularisation of large vessel occlusions in acute ischaemic stroke (TREVO 2) a randomised trial

7. Lo et al (2010) Robot-assisted therapy for long-term upper-limb impairment after stroke

8. Rothwell et al (2010) Effects of β blockers and calcium-channel blockers on within-individual variability in blood pressure and risk of stroke

9. Lee et al (2010) A long-term follow-up study of intravenous autologous mesenchymal stem cell transplantation in patients with ischemic stroke

10. Furlan et al (2012) Closure or medical therapy for cryptogenic stroke with patent foramen ovale

11. Chollet et al (2011) Fluoxetine for motor recovery after acute ischaemic stroke (FLAME): a randomised placebo-controlled trial

12. Halliday et al (2010) 10-year stroke prevention after successful carotid endarterectomy for asymptomatic stenosis (ACST-1): a multicentre randomised trial

13. Shinohara et al (2010) Cilostazol for prevention of secondary stroke (CSPS2): an aspirin-controlled, double-blind, randomised non-inferiority trial

Felipe Fregni (ed.), *Clinical Trials in Neurology*, Neuromethods, vol. 138, https://doi.org/10.1007/978-1-4939-7880-9, © Springer Science+Business Media, LLC, part of Springer Nature 2018

14. Baker et al (2010) Using transcranial direct-current stimulation to treat stroke patients with aphasia

15. Fox et al (2011) Prevention of stroke and systemic embolism with rivaroxaban compared with warfarin in patients with non-valvular atrial fibrillation and moderate renal impairment

16. Diener et al (2010) Dabigatran compared with warfarin in patients with atrial fibrillation and previous transient ischaemic attack or stroke: a subgroup analysis of the RE-LY trial

17. Powers et al (2011) Extracranial-intracranial bypass surgery for stroke prevention in hemodynamic cerebral ischemia: the carotid occlusion surgery study randomized trial

18. Carroll et al (2013) Closure of patent foramen ovale versus medical therapy after cryptogenic stroke

19. Sandset et al (2011) The angiotensin-receptor blocker candesartan for treatment of acute stroke (SCAST): a randomised, placebo-controlled, double-blind trial

20. Saposnik et al (2010) Effectiveness of virtual reality using Wii gaming technology in stroke rehabilitation: a pilot randomized clinical trial and proof of principle

21. Lindenberg et al (2010) Bihemispheric brain stimulation facilitates motor recovery in chronic stroke patients

22. Duncan et al (2011) Body-weight-supported treadmill rehabilitation after stroke

23. Lip et al (2010) Identifying patients at high risk for stroke despite anticoagulation: a comparison of contemporary stroke risk stratification schemes in an anticoagulated atrial fibrillation cohort

24. Parsons et al (2012) A randomized trial of tenecteplase versus alteplase for acute ischemic stroke

25. Wang et al (2013) Clopidogrel with aspirin in acute minor stroke or transient ischemic attack

26. Honmou et al (2011) Intravenous administration of auto serum-expanded autologous mesenchymal stem cells in stroke

27. Weitz et al (2010) Randomised, parallel-group, multicentre, multinational phase 2 study comparing edoxaban, an oral factor Xa inhibitor, with warfarin for stroke prevention in patients with atrial fibrillation

28. Reddy et al (2012) Percutaneous left atrial appendage closure for stroke prophylaxis in patients with atrial fibrillation 2.3-year follow-up of the PROTECT AF (Watchman left atrial appendage system for embolic protection in patients with atrial fibrillation) trial

29. Abou-Chebl et al (2010) Conscious sedation versus general anesthesia during endovascular therapy for acute anterior circulation stroke preliminary results from a retrospective, multicenter study

30. Piccini et al (2013) Renal dysfunction as a predictor of stroke and systemic embolism in patients with nonvalvular atrial fibrillation validation of the R(2)CHADS(2) index in the ROCKET AF (Rivaroxaban Once-daily, oral, direct factor Xa inhibition Compared with vitamin K antagonism for prevention of stroke and Embolism Trial in Atrial Fibrillation) and ATRIA (AnTicoagulation and Risk factors In Atrial fibrillation) study cohorts

31. Berkhemer et al (2015) A randomized trial of intraarterial treatment for acute ischemic stroke

32. Hemmen et al (2010) Intravenous thrombolysis plus hypothermia for acute treatment of ischemic stroke (ICTuS-L) final results

33. Oldgren et al (2011) Risks for stroke, bleeding, and death in patients with atrial fibrillation receiving dabigatran or warfarin in relation to the CHADS (2) score: a subgroup analysis of the RE-LY trial

34. Reilly et al (2014) The effect of dabigatran plasma concentrations and patient characteristics on the frequency of ischemic stroke and major bleeding in atrial fibrillation patients: the RE-LY trial (Randomized Evaluation of Long-Term Anticoagulation Therapy)

35. Hijazi et al (2012) Cardiac biomarkers are associated with an increased risk of stroke and death in patients with atrial fibrillation: a randomized evaluation of long-term anticoagulation therapy (RE-LY) substudy

36. Hankey et al (2012) Rivaroxaban compared with warfarin in patients with atrial fibrillation and previous stroke or transient ischaemic attack: a subgroup analysis of ROCKET AF

37. Parsons et al (2010) Pretreatment diffusion- and perfusion-MR lesion volumes have a crucial influence on clinical response to stroke thrombolysis

38. Viles-Gonzalez et al (2012) The clinical impact of incomplete left atrial appendage closure with the Watchman device in patients with atrial fibrillation

39. Sanna et al (2014) Cryptogenic stroke and underlying atrial fibrillation

40. Michielsen et al (2011) Motor recovery and cortical reorganization after mirror therapy in chronic stroke patients: a phase II randomized controlled trial

41. Everett et al (2010) Rosuvastatin in the prevention of stroke among men and women with elevated levels of C-reactive protein justification for the use of statins in prevention: an intervention trial evaluating rosuvastatin (JUPITER)

42. Wolf et al (2010) The EXCITE stroke trial comparing early and delayed constraint-induced movement therapy

43. Cumming et al (2010) Very early mobilization after stroke fast-tracks return to walking further results from the phase II AVERT randomized controlled trial

44. Lopes et al (2012) Efficacy and safety of apixaban compared with warfarin according to patient risk of stroke and of bleeding in atrial fibrillation: a secondary analysis of a randomized controlled trial

45. Easton et al (2012) Apixaban compared with warfarin in patients with atrial fibrillation and previous stroke or transient ischaemic attack: a subgroup analysis of the ARISTOTLE trial

46. Jorge et al (2010) Escitalopram and enhancement of cognitive recovery following stroke

47. Middleton et al (2011) Implementation of evidence-based treatment protocols to manage fever, hyperglycaemia, and swallowing dysfunction in acute stroke (QASC): a cluster randomized controlled trial

48. Liebeskind et al (2011) Collaterals dramatically alter stroke risk in intracranial atherosclerosis

49. Khedr et al (2010) Long-term effect of repetitive transcranial magnetic stimulation on motor function recovery after acute ischemic stroke

50. Weiduschat et al (2011) Effects of repetitive transcranial magnetic stimulation in aphasic stroke: a randomized controlled pilot study

51. Bayard et al (2010) PLAATO (Percutaneous Left Atrial Appendage Transcatheter Occlusion) for prevention of cardioembolic stroke in non-anticoagulation eligible atrial fibrillation patients: results from the European PLAATO study

52. Goyal et al (2015) Randomized assessment of rapid endovascular treatment of ischemic stroke

53. Ang et al (2011) A large clinical study on the ability of stroke patients to use an EEG-based motor imagery brain-computer interface

54. Campbell et al (2015) Endovascular therapy for ischemic stroke with perfusion-imaging selection

55. Birkenmeier et al (2010) Translating animal doses of task-specific training to people with chronic stroke in 1-hour therapy sessions: a proof-of-concept study

56. Ietswaart et al (2011) Mental practice with motor imagery in stroke recovery: randomized controlled trial of efficacy

57. Patel et al (2013) Outcomes of discontinuing rivaroxaban compared with warfarin in patients with nonvalvular atrial fibrillation analysis from the ROCKET AF trial (Rivaroxaban Once-Daily, Oral, Direct Factor Xa Inhibition Compared With Vitamin K Antagonism for Prevention of Stroke and Embolism Trial in Atrial Fibrillation)

58. Hill et al (2012) Safety and efficacy of NA-1 in patients with iatrogenic stroke after endovascular aneurysm repair (ENACT): a phase 2, randomised, double-blind, placebo-controlled trial

59. Dávalos et al (2012) Citicoline in the treatment of acute ischaemic stroke: an international, randomised, multicentre, placebo-controlled study (ICTUS trial)

60. Barwooda et al (2011) Improved language performance subsequent to low-frequency rTMS in patients with chronic non-fluent aphasia post-stroke

61. Wallentin et al (2013) Efficacy and safety of apixaban compared with warfarin at different levels of predicted international normalized ratio control for stroke prevention in atrial fibrillation

62. You et al (2011) Cathodal transcranial direct current stimulation of the right Wernicke's area improves comprehension in subacute stroke patients

63. Haley et al (2010) Phase IIB/III trial of tenecteplase in acute ischemic stroke results of a prematurely terminated randomized clinical trial

64. Ware et al (2012) Stroke with Transfusions Changing to Hydroxyurea (SWiTCH)

65. Robinson et al (2010) Effects of antihypertensive treatment after acute stroke in the Continue or Stop Post-Stroke Antihypertensives Collaborative Study (COSSACS): a prospective, randomized, open, blinded-endpoint trial

66. Ackerley et al (2010) Combining theta burst stimulation with training after subcortical stroke

67. Campbell et al (2010) Regional very low cerebral blood volume predicts hemorrhagic transformation better than diffusion-weighted imaging volume and thresholded apparent diffusion coefficient in acute ischemic stroke

68. Pereira et al (2013) Prospective, multicenter, single-arm study of mechanical thrombectomy using solitaire flow restoration in acute ischemic stroke

69. Murguialday et al (2013) Brain-machine interface in chronic stroke rehabilitation: a controlled study

70. Wu et al (2011) Randomized trial of distributed constraint-induced therapy versus bilateral arm training for the rehabilitation of upper-limb motor control and function after stroke

71. Emara et al (2010) Repetitive transcranial magnetic stimulation at 1Hz and 5Hz produces sustained improvement in motor function and disability after ischaemic stroke

72. Dengler et al (2010) Early treatment with aspirin plus extended-release dipyridamole for transient ischaemic attack or ischaemic stroke within 24 h of symptom onset (EARLY trial): a randomised, open-label, blinded-endpoint trial

73. James et al (2012) Ticagrelor versus clopidogrel in patients with acute coronary syndromes and a history of stroke or transient ischemic attack

74. Goyal et al (2010) Effect of baseline CT scan appearance and time to recanalization on clinical outcomes in endovascular thrombectomy of acute ischemic strokes

75. Diringer et al (2009) Thromboembolic events with recombinant activated factor VII in spontaneous intracerebral hemorrhage: results from the Factor Seven for Acute Hemorrhagic Stroke (FAST) trial

76. Olsson et al (2010) Safety and tolerability of an immediate-release formulation of the oral direct thrombin inhibitor AZD0837 in the prevention of stroke and systemic embolism in patients with atrial fibrillation

77. Schäbitz et al (2010) AXIS a trial of intravenous granulocyte colony-stimulating factor in acute ischemic stroke

78. Jüttler et al (2014) Hemicraniectomy in older patients with extensive middle-cerebral-artery stroke

79. Hesse et al (2011) Combined transcranial direct current stimulation and robot-assisted arm training in subacute stroke patients: an exploratory, randomized multicenter trial

80. Avenanti et al (2012) Low-frequency rTMS promotes use-dependent motor plasticity in chronic stroke: a randomized trial

81. Battistella et al (2011) Safety of autologous bone marrow mononuclear cell transplantation in patients with nonacute ischemic stroke

82. Meng et al (2012) Upper limb ischemic preconditioning prevents recurrent stroke in intracranial arterial stenosis

83. Ryuji et al (2010) Botulinum toxin type A in post-stroke lower limb spasticity: a multicenter, double-blind, placebo-controlled trial

84. Demaerschalk et al (2010) Stroke team remote evaluation using a digital observation camera in Arizona: the initial Mayo Clinic experience trial

85. Sun et al (2010) Combined botulinum toxin type A with modified constraint-induced movement therapy for chronic stroke patients with upper extremity spasticity: a randomized controlled study

86. Khatri et al (2014) Time to angiographic reperfusion and clinical outcome after acute ischaemic stroke: an analysis of data from the Interventional Management of Stroke (IMS III) phase 3 trial

87. Hori et al (2013) Dabigatran versus warfarin effects on ischemic and hemorrhagic strokes and bleeding in Asians and non-Asians with atrial fibrillation

88. Globas et al (2012) Chronic stroke survivors benefit from high-intensity aerobic treadmill exercise: a randomized control trial

89. Parra et al (2011) Early treatment of obstructive apnoea and stroke outcome: a randomized controlled trial

90. Turton et al (2010) A single blinded randomized controlled pilot trial of prism adaptation for improving self-care in stroke patients with neglect

91. Diener et al (2012) Apixaban versus aspirin in patients with atrial fibrillation and previous stroke or transient ischaemic attack: a predefined subgroup analysis from AVERROES, a randomized trial

92. Skali et al (2011) Stroke in patients with type 2 diabetes mellitus, chronic kidney disease, and anemia treated with Darbepoetin Alfa: the trial to reduce cardiovascular events with Aranesp therapy (TREAT) experience

93. Chang et al (2010) Long-term effects of rTMS on motor recovery in patients after subacute stroke

94. Sabut et al (2010) Restoration of gait and motor recovery by functional electrical stimulation therapy in persons with stroke

95. Moniche et al (2012) Intra-arterial bone marrow mononuclear cells in ischemic stroke: a pilot clinical trial

96. Kim et al (2010) Effect of transcranial direct current stimulation on motor recovery in patients with subacute stroke

97. Fridriksson et al (2010) Preservation and modulation of specific left hemisphere regions is vital for treated recovery from anomia in stroke

98. Lin et al (2010) Minimal detectable change and clinically important difference of the stroke impact scale in stroke patients

99. Shaw et al (2010) BoTULS: a multicentre randomised controlled trial to evaluate the clinical effectiveness and cost-effectiveness of treating upper limb spasticity due to stroke with botulinum toxin type A

100. Mirelman et al (2010) Effects of virtual reality training on gait biomechanics of individuals post-stroke

Chapter 3

1. Megha et al (2013) Effect of frequency of multimodal coma stimulation on the consciousness levels of traumatic brain injury comatose patients

2. Whyte et al (2013) Medical complications during inpatient rehabilitation among patients with traumatic disorders of consciousness

3. Lopez-de-Sa et al (2012) Hypothermia in comatose survivors from out-of-hospital cardiac arrest: pilot trial comparing 2 levels of target temperature

4. Shakeri et al (2013) Effect of progesterone administration on prognosis of patients with diffuse axonal injury due to severe head trauma

5. Thonnard et al (2013) Effect of zolpidem in chronic disorders of consciousness: a prospective open-label study

6. Thibaut et al (2014) tDCS in patients with disorders of consciousness: sham-controlled randomized double-blind study

7. Snyman et al (2010) Zolpidem for persistent vegetative state—a placebo-controlled trial in pediatrics

8. Clifton et al (2011) Very early hypothermia induction in patients with severe brain injury (the National Acute Brain Injury Study: Hypothermia II): a randomised trial

9. Page et al (2013) Effect of intravenous haloperidol on the duration of delirium and coma in critically ill patients (Hope-ICU): a randomised, double-blind, placebo-controlled trial

10. Giacino et al (2012) Placebo-controlled trial of amantadine for severe traumatic brain injury

11. Whyte et al (2009) Incidence of clinically significant responses to zolpidem among patients with disorders of consciousness

12. Shi et al (2013) Vagus nerve stimulation to augment recovery from severe traumatic brain injury impeding consciousness: a prospective pilot clinical trial

13. Machado et al (2014) Zolpidem arousing effect in persistent vegetative state patients: autonomic, EEG and behavioral assessment

14. Mwanga-Amumpaire et al (2012) Effect of vitamin A adjunct therapy for cerebral malaria in children admitted to Mulago hospital: a randomized controlled trial

15. Cruz et al (2004) Successful use of the new high-dose mannitol treatment in patients with Glasgow Coma Scale scores of 3 and bilateral abnormal pupillary widening: a randomized trial

16. Skolnick et al (2014) A clinical trial of progesterone for severe traumatic brain injury

17. Rockswold et al (2013) A prospective, randomized Phase II clinical trial to evaluate the effect of combined hyperbaric and normobaric hyperoxia on cerebral metabolism, intracranial pressure, oxygen toxicity, and clinical outcome in severe traumatic brain injury

18. Cottenceau et al (2011) Comparison of effects of equiosmolar doses of mannitol and hypertonic saline on cerebral blood flow and metabolism in traumatic brain injury

19. Cooper et al (2011) Decompressive craniectomy in diffuse traumatic brain injury

20. Firsching et al (2012) Early survival of comatose patients after severe traumatic brain injury with the dual cannabinoid CB1/CB2 receptor agonist KN38-7271: a randomized, double-blind, placebo-controlled phase II trial

21. Taftachi et al (2012) Lipid emulsion improves Glasgow coma scale and decreases blood glucose level in the setting of acute non-local anesthetic drug poisoning—a randomized controlled trial

22. Bulger et al (2010) Out of hospital hypertonic resuscitation following severe traumatic brain injury: a randomized controlled trial

23. Eide et al (2010) A randomized and blinded single-center trial comparing the effect of intracranial pressure and intracranial pressure wave amplitude-guided intensive care management on early clinical state and 12-month outcome in patients with aneurysmal subarachnoid hemorrhage

24. Zafonte et al (2012) Effect of citicoline on functional and cognitive status among patients with traumatic brain injury: Citicoline Brain Injury Treatment Trial (COBRIT)

25. Shaikh et al (2011) Effect of dexamethasone on brain oedema following acute ischemic stroke

26. Maude et al (2011) Timing of enteral feeding in cerebral malaria in resource-poor settings: a randomized trial

27. Mourvillier et al (2013) Induced hypothermia in severe bacterial meningitis: a randomized clinical trial

28. Olivecrona et al (2012) Prostacyclin treatment and clinical outcome in severe traumatic brain injury patients managed with an ICP-targeted therapy: a prospective study

29. Wright et al (2014) Very early administration of progesterone for acute traumatic brain injury

30. Bernard et al (2010) Prehospital rapid sequence intubation improves functional outcome for patients with severe traumatic brain injury: a randomized controlled trial

31. Moein et al (2013) The effect of Boswellia Serrata on neurorecovery following diffuse axonal injury

32. Morrison et al (2011) The Toronto prehospital hypertonic resuscitation—head injury and multiorgan dysfunction trial: feasibility study of a randomized controlled trial

33. Ruijter et al (2014) Treatment of electroencephalographic status epilepticus after cardiopulmonary resuscitation (TELSTAR): study protocol for a randomized controlled trial

34. Adelson et al (2005) Phase II clinical trial of moderate hypothermia after severe traumatic brain injury in children

35. Adelson et al (2005) Phase II clinical trial of moderate hypothermia after severe traumatic brain injury in children

36. Jiang et al (2005) Efficacy of standard trauma craniectomy for refractory intracranial hypertension with severe traumatic brain injury: a multicenter, prospective, randomized controlled study

37. Moein et al (2006) Effect of methylphenidate on ICU and hospital length of stay in patients with severe and moderate traumatic brain injury

38. Jiang et al (2006) Effect of long-term mild hypothermia or short-term mild hypothermia on outcome of patients with severe traumatic brain injury

39. He et al (2005) Acupuncture treatment for 15 cases of post-traumatic coma

40. Marmarou et al (2005) A single dose, three-arm, placebo-controlled, phase I study of the bradykinin B2 receptor antagonist Anatibant (LF16-0687Ms) in patients with severe traumatic brain injury

41. Karma et al (2006) Effect of stimulation in coma

42. Etemadrezaie et al (2006) The effect of fresh frozen plasma in severe closed head injury

43. Ghori et al (2007) Effect of midazolam versus propofol sedation on markers of neurological injury and outcome after isolated severe head injury: a pilot study

44. Idris et al (2007) Prognostic study of using different monitoring modalities in treating severe traumatic brain injury

45. Namutangula et al (2007) Mannitol as adjunct therapy for childhood cerebral malaria in Uganda: a randomized clinical trial

46. Xiao et al (2007) Improved outcomes from the administration of progesterone for patients with acute severe traumatic brain injury: a randomized controlled trial

47. Aquilani et al (2008) Branched-chain amino acids may improve recovery from a vegetative or minimally conscious state in patients with traumatic brain injury: a pilot study

48. McMahon et al (2005) Effects of amantadine in children with impaired consciousness caused by acquired brain injury: a pilot study

49. Abbasi et al (2009) Effect of a regular family visiting program as an affective, auditory, and tactile stimulation on the consciousness level of comatose patients with a head injury

50. Yan et al (2010) Cerebral oxygen metabolism and neuroelectrophysiology in a clinical study of severe brain injury and mild hypothermia

51. Clifton et al (2009) Multicenter trial of early hypothermia in severe brain injury

52. SAFE Study Investigators et al (2007) Saline or albumin for fluid resuscitation in patients with traumatic brain injury

53. Aceng et al (2005) Rectal artemether versus intravenous quinine for the treatment of cerebral malaria in children in Uganda: randomised clinical trial

54. Cruz et al (2004) Successful use of the new high-dose mannitol treatment in patients with Glasgow Coma Scale scores of 3 and bilateral abnormal pupillary widening: a randomized trial

55. Yurkewicz et al (2005) The effect of the selective NMDA receptor antagonist Traxoprodil in the treatment of traumatic brain injury

56. Cooper et al (2004) Prehospital hypertonic saline resuscitation of patients with hypotension and severe traumatic brain injury

57. Das et al (2003) Pentoxifylline adjunct improves prognosis of human cerebral malaria in adults

58. Zhi et al (2003) Study on therapeutic mechanism and clinical effect of mild hypothermia in patients with severe head injury

59. Giacino and Whyte (2003) Amantadine to improve neurorecovery in traumatic brain injury-associated diffuse axonal injury: a pilot double-blind randomized trial

60. Bernard et al (2002) Treatment of comatose survivors of out-of-hospital cardiac arrest with induced hypothermia

61. Chen et al (2002) Curative effect of wilsonii injecta on severe head injury

62. Ren et al (2001) Glasgow Coma Scale, brain electric activity mapping and Glasgow Outcome Scale after hyperbaric oxygen treatment of severe brain injury

63. Marshall et al (1998) A multicenter trial on the efficacy of using tirilazad mesylate in cases of head injury

64. Barennes et al (1998) An open randomized clinical study of intrarectal versus infused Quinimax for the treatment of childhood cerebral malaria in Niger

65. Zuccarello et al (1999) Early surgical treatment for supratentorial intracerebral hemorrhage

66. Young et al (1996) Effects of pegorgotein on neurologic outcome of patients with severe head injury

67. Morris et al (1999) Failure of the competitive N-methyl-D-aspartate antagonist Selfotel (CGS 19755) in the treatment of severe head injury: results of two Phase III clinical trials

68. Taylor et al (1999) Prospective, randomized, controlled trial to determine the effect of early enhanced enteral nutrition on clinical outcome in mechanically ventilated patients suffering head injury

69. Marion et al (1997) Treatment of traumatic brain injury with moderate hypothermia

70. Cincotta et al (2015) No effects of 20 Hz-rTMS of the primary motor cortex in vegetative state: a randomised, sham-controlled study

Chapter 4

1. Adamides et al (2009) Focal cerebral oxygenation and neurological outcome with or without brain tissue oxygen-guided therapy in patients with traumatic brain injury

2. Whyte et al (2008) The effects of bromocriptine on attention deficits after traumatic brain injury: a placebo-controlled pilot study

3. Wright et al (2007) ProTECT: a randomized clinical trial of progesterone for acute traumatic brain injury

4. Bernard et al (2010) Prehospital rapid sequence intubation improves functional outcome for patients with severe traumatic brain injury: a randomized controlled trial

5. Rivera et al (2008) Problem-solving training for family caregivers of persons with traumatic brain injuries: a randomized controlled trial

6. Vanderploeg et al (2008) Rehabilitation of traumatic brain injury in active duty military personnel and veterans: defense and veterans brain injury center randomized controlled trial of two rehabilitation approaches

7. Sarajuuri et al (2005) Outcome of a comprehensive neurorehabilitation program for patients with traumatic brain injury

8. Sarajuuri et al (2005) Outcome of a comprehensive neurorehabilitation program for patients with traumatic brain injury

9. High et al (2006) Early versus later admission to postacute rehabilitation: impact on functional outcome after traumatic brain injury

10. Bell et al (2005) The effect of a scheduled telephone intervention on outcome after moderate to severe traumatic brain injury: a randomized trial

11. Cicerone et al (2008) A randomized controlled trial of holistic neuropsychologic rehabilitation after traumatic brain injury

12. Dahlberg et al (2007) Treatment efficacy of social communication skills training after traumatic brain injury: a randomized treatment and deferred treatment controlled trial

13. Ashman et al (2009) A randomized controlled trial of sertraline for the treatment of depression in persons with traumatic brain injury

14. Ouellet et al (2007) Efficacy of cognitive-behavioral therapy for insomnia associated with traumatic brain injury: a single-case experimental design

15. Bhambhani et al (2005) Effects of circuit training on body composition and peak cardiorespiratory responses in patients with moderate to severe traumatic brain injury

16. Tiersky et al (2005) A trial of neuropsychologic rehabilitation in mild-spectrum traumatic brain injury

17. Braga et al (2005) Direct clinician-delivered versus indirect family-supported rehabilitation of children with traumatic brain injury: a randomized controlled trial

18. Kraus et al (2005) Effects of the dopaminergic agent and NMDA receptor antagonist amantadine on cognitive

function, cerebral glucose metabolism and D2 receptor availability in chronic traumatic brain injury: a study using positron emission tomography (PET)

19. Zhu et al (2007) Does intensive rehabilitation improve the functional outcome of patients with traumatic brain injury (TBI)? A randomized controlled trial

20. Wilson et al (2005) A randomized control trial to evaluate a paging system for people with traumatic brain injury

21. Cheng et al (2006) Management of impaired self-awareness in persons with traumatic brain injury

22. Bourgeois et al (2007) The effects of cognitive teletherapy on reported everyday memory behaviors of persons with chronic traumatic brain injury

23. Silver et al (2008) Long-term effects of rivastigmine capsules in patients with traumatic brain injury

24. Moein et al (2006) Effect of methylphenidate on ICU and hospital length of stay in patients with severe and moderate traumatic brain injury

25. Kim et al (2006) Effects of single-dose methylphenidate on cognitive performance in patients with traumatic brain injury: a double-blind placebo-controlled study

26. McPherson et al (2009) Pilot study of self-regulation informed goal setting in people with traumatic brain injury

27. Katz-Leurer et al (2009) The effects of a 'home-based' task-oriented exercise programme on motor and balance performance in children with spastic cerebral palsy and severe traumatic brain injury

28. Vespa et al (2012) Tight glycemic control increases metabolic distress in traumatic brain injury: a randomized controlled within-subjects trial

29. Xiao et al (2008) Improved outcomes from the administration of progesterone for patients with acute severe traumatic brain injury: a randomized controlled trial

30. Qiu et al (2009) Effects of unilateral decompressive craniectomy on patients with unilateral acute post-traumatic brain swelling after severe traumatic brain injury

31. Zygun et al (2009) The effect of red blood cell transfusion on cerebral oxygenation and metabolism after severe traumatic brain injury

32. Bayir et al (2009) Therapeutic hypothermia preserves antioxidant defenses after severe traumatic brain injury in infants and children

33. Zhang et al (2008) A combined procedure to deliver autologous mesenchymal stromal cells to patients with traumatic brain injury

34. Lee et al (2005) Comparing effects of methylphenidate, sertraline and placebo on neuropsychiatric sequelae in patients with traumatic brain injury

35. Ponsford et al (2008) Gender differences in outcome in patients with hypotension and severe traumatic brain injury

36. Acosta-Escribano et al (2010) Gastric versus transpyloric feeding in severe traumatic brain injury: a prospective, randomized trial

37. Tisdall et al (2008) Increase in cerebral aerobic metabolism by normobaric hyperoxia after traumatic brain injury

38. Wade et al (2008) Preliminary efficacy of a web-based family problem-solving treatment program for adolescents with traumatic brain injury

39. Wade et al (2010) A randomized trial of teen online problem solving for improving executive function deficits following pediatric traumatic brain injury

40. Harris et al (2009) Discrete cerebral hypothermia in the management of traumatic brain injury: a randomized controlled trial Clinical article

41. Bulger et al (2010) Out-of-hospital hypertonic resuscitation following severe traumatic brain injury: a randomized controlled trial

42. Zafonte et al (2012) Effect of citicoline on functional and cognitive status among patients with traumatic brain injury citicoline brain injury treatment trial (COBRIT)

43. Robertson et al (2014) Effect of erythropoietin and transfusion threshold on neurological recovery after traumatic brain injury a randomized clinical trial

44. Jiang et al (2006) Effect of long-term mild hypothermia or short-term mild hypothermia on outcome of patients with severe traumatic brain injury

45. Patrick et al (2006) Dopamine agonist therapy in low-response children following traumatic brain injury

46. Wright et al (2005) Steady-state serum concentrations of progesterone following continuous intravenous infusion in patients with acute moderate to severe traumatic brain injury

47. Castriotta et al (2009) Treatment of sleep disorders after traumatic brain injury

48. Qiu et al (2007) Effects of therapeutic mild hypothermia on patients with severe traumatic brain injury after craniotomy

49. Cooper et al (2008) Early decompressive craniectomy for patients with severe traumatic brain injury and refractory intracranial hypertension - a pilot randomized trial

50. Wade et al (2006) Putting the pieces together - preliminary efficacy of a family problem-solving intervention for children with traumatic brain injury

51. Bombardier et al (2009) The efficacy of a scheduled telephone intervention for ameliorating depressive symptoms during the first year after traumatic brain injury

52. Brown et al (2005) Body weight-supported treadmill training versus conventional gait training for people with chronic traumatic brain injury

53. Beers et al (2005) Neurobehavioral effects of amantadine after pediatric traumatic brain injury - a preliminary report

54. Liu et al (2006) Effects of selective brain cooling in patients with severe traumatic brain injury: a preliminary study

55. Willmott et al (2009) Efficacy of methylphenidate in the rehabilitation of attention following traumatic brain injury: a randomised, crossover, double blind, placebo controlled inpatient trial

56. Rockswold et al (2010) A prospective, randomized clinical trial to compare the effect of hyperbaric to normobaric hyperoxia on cerebral metabolism, intracranial pressure, and oxygen toxicity in severe traumatic brain injury

57. Morgalla et al (2008) Do long-term results justify decompressive craniectomy after severe traumatic brain injury

58. Jiang et al (2006) Efficacy of standard trauma craniectomy for refractory intracranial hypertension with severe traumatic brain injury: a multicenter, prospective, randomized controlled study

59. Ji-Yao Jian et al (2005) Efficacy of standard trauma craniectomy for refractory intracranial hypertension with severe traumatic brain injury: a multicenter, prospective, randomized controlled study

60. Buttram et al (2007) Multiplex assessment of cytokine and chemokine levels in cerebrospinal fluid following severe pediatric traumatic brain injury: effects of moderate hypothermia

61. Yurkewicz et al (2005) The effect of the selective NMDA receptor antagonist traxprodil in the treatment of traumatic brain injury

62. High et al (2010) Effect of growth hormone replacement therapy on cognition after traumatic brain injury

63. Mazzeo et al (2009) Safety and tolerability of cyclosporin A in severe traumatic brain injury patients: results from a prospective randomized trial

64. Andelic et al (2012) Does an early onset and continuous chain of rehabilitation improve the long-term functional outcome of patients with severe traumatic brain injury

65. Mazzeo et al (2008) Brain metabolic and hemodynamic effects of cyclosporin A after human severe traumatic brain injury: a microdialysis study

66. Empey et al (2006) Cyclosporin A disposition following acute traumatic brain injury

67. Tapia-Perez et al (2008) Effect of rosuvastatin on amnesia and disorientation after traumatic brain injury

68. Marmarou et al (2005) A single dose, three-arm, placebo-controlled, phase I study of the bradykinin B-2 receptor antagonist anatibant (LF16-0687Ms) in patients with severe traumatic brain injury

69. Mazzeo et al (2006) Severe human traumatic brain injury, but not cyclosporin A treatment, depresses activated T lymphocytes early after injury

70. Baker et al (2009) Resuscitation with hypertonic saline-dextran reduces serum biomarker levels and correlates with outcome in severe traumatic brain injury patients

71. Olivecrona et al (2009) Prostacyclin treatment in severe traumatic brain injury: a microdialysis and outcome study

72. Cooper et al (2013) Albumin resuscitation for traumatic brain injury: is intracranial hypertension the cause of increased mortality

73. Wolf G et al (2012) The effect of hyperbaric oxygen on symptoms after mild traumatic brain injury

74. Ghaffar et al (2006) Randomized treatment trial in mild traumatic brain injury

75. Adelson et al (2013) Comparison of hypothermia and normothermia after severe traumatic brain injury in children (Cool Kids): a phase 3, randomised controlled trial

76. Temkin et al (2007) Magnesium sulfate for neuroprotection after traumatic brain injury: a randomised controlled trial

77. Maas et al (2006) Efficacy and safety of dexanabinol in severe traumatic brain injury: results of a phase III randomised, placebo-controlled, clinical trial

78. Andrews et al (2011) European society of intensive care medicine study of therapeutic hypothermia (32–35°C) for intracranial pressure reduction after traumatic brain injury (the Eurotherm3235Trial)

79. Schiff et al (2007) Behavioural improvements with thalamic stimulation after severe traumatic brain injury

80. Cooper et al (2011) Decompressive craniectomy in diffuse traumatic brain injury

81. Chesnut et al (2012) A trial of intracranial-pressure monitoring in traumatic brain injury

82. Giacino et al (2012) Placebo-controlled trial of amantadine for severe traumatic brain injury

83. Puccio et al (2009) Induced normothermia attenuates intracranial hypertension and reduces fever burden after severe traumatic brain injury

84. Bilotta et al (2008) Intensive insulin therapy after severe traumatic brain injury: a randomized clinical trial

85. Silver et al (2006) Effects of rivastigmine on cognitive function in patients with traumatic brain injury

86. Kaiser et al (2010) Modafinil ameliorates excessive daytime sleepiness after traumatic brain injury

87. Hewitt et al (2006) Theory driven rehabilitation of executive functioning: improving planning skills in people with traumatic brain injury through the use of an autobiographical episodic memory cueing procedure

88. Bornhofen et al (2008) Treating deficits in emotion perception following traumatic brain injury

89. Couillet et al (2010) Rehabilitation of divided attention after severe traumatic brain injury: a randomised trial

90. O'Brien et al (2007) An investigation of the differential effect of self-generation to improve learning and memory in multiple sclerosis and traumatic brain injury

91. Galbiati et al (2009) Attention remediation following traumatic brain injury in childhood and adolescence

92. Cox et al (2011) Autologous bone marrow mononuclear cell therapy for severe traumatic brain injury in children

93. Adelson et al (2005) Phase II clinical trial of moderate hypothermia after severe traumatic brain injury in children

94. Hutchison et al (2008) Hypothermia therapy after traumatic brain injury in children

95. Yang et al (2009) Intensive insulin therapy on infection rate, days in NICU, in-hospital mortality and neurological outcome in severe traumatic brain injury patients: a randomized controlled trial

96. Hoffer et al (2013) Amelioration of acute sequelae of blast induced mild traumatic brain injury by N-acetyl cysteine: a double-blind, placebo controlled study

97. Jha et al (2008) A randomized trial of modafinil for the treatment of fatigue and excessive daytime sleepiness in individuals with chronic traumatic brain injury

98. Sakellaris et al (2006) Prevention of complications related to traumatic brain injury in children and adolescents with creatine administration: an open label randomized pilot study

99. Coester et al (2010) Intensive insulin therapy in severe traumatic brain injury: a randomized trial

100. Thaut et al (2009) Neurologic music therapy improves executive function and emotional adjustment in traumatic brain injury rehabilitation

Chapter 5

1. Follet et al (2010) Pallidal versus subthalamic deep-brain stimulation for Parkinson's disease

2. Marks et al (2010) Gene delivery of AAV2-neurturin for Parkinson's disease: a double blind, randomised, controlled trial

3. LeWitt et al (2011) AAV2 - GAD gene therapy for advanced Parkinson's disease: a double blind, sham surgery controlled, randomised trial

4. Williams et al (2010) Deep brain stimulation plus best medical therapy versus best medical therapy alone for advanced parkinson's disease (PD SURG trial) randomised, open-label trial

5. Li F et al (2012) Tai Chi and postural stability in patients with Parkinson's disease

6. Schuepbach et al (2013) Neurostimulation for Parkinson's disease with early motor complications

7. Barone et al (2010) Pramipexole for the treatment of depressive symptoms in patients with Parkinon's disease: a randomised, double-blind, placebo-controlled trial

8. Moro et al (2010) Long-term results of multicenter study and subthalamic and pallidum stimulation in Parkinson's disease

9. Trendwalder et al (2010) Rotigotine effects on early morning motor function and sleep in Parkinson's disease: a double-blind, randomized, placebo-controlled study

10. Murat et al (2010) Memantine for patients with Parkinson's disease dementia or dementia with Lewy bodies: a randomised, double-blind, placebo-controlled trial

11. Snow et al (2010) A double-blind, placebo-controlled study to assess the mitochondria-targeted antioxidant MitoQ as a disease-modifying therapy in Parkinson's disease

12. Odekerken et al (2013) Subthalamic nucleus versus globus pallidus bilateral deep brain stimulation for advanced Parkinon's disease (NSTAPS study): a randomised controlled trial

13. Okun et al (2012) Subthalamic deep brain stimulation with a constant current device in Parkinson's disease: an open label randomised controlled trial

14. Hauser et al (2011) Preladenant in patients with Parkinson's disease and motor fluctuations: a phase 2, double blind, randomised trial

15. Rascol et al (2011) A double-blind, dealyed start trial of rasagine in Parkinson's disease (the ADAGIO study) pre-specified and posthoc analyses of the need for additional therapies, changes in UPDRS scores and non motor outcomes

16. Mizuno et al (2010) Clinical efficacy of Istradefylline (KW-6002) in Parkinson's disease: a randomised, controlled study

17. Wolf et al (2010) Long term antidiskinetic efficacy of amantadine in Parkinson's disease

18. Benninger et al (2010) Transcranial direct current stimulation for the treatment of Parkinson's disease

19. Ma et al (2010) Dopamine cell implantation in Parkinson's disease: long-term clinical and F-18 F DOPA PET outcomes

20. Allen et al (2010) The effects of an exercise program on fall risk factors in people with Parkinson's disease: a randomised controlled trial

21. Ebersbach et al (2010) Comparing exercise in Parkinson's disease - the Berlin LSVT® BIG Study

22. Dobkin et al (2011) Cognitive - behavorial therapy for depression in Parkinson's disease: a randomised, controlled trial

23. Moreau et al (2012) Methylphenidate for gait hypokinesia and freezing in patients with Parkinson's disease undergoing subthalamic stimulation: a multicenter, parallel, randomized, placebo controlled trial

24. Gross et al (2011) Intrastriatal transplantation of microcarrier-bound human retinal pigment epithelial cells versus sham surgery in patients with advanced Parkinson's disease: a double-blind, randomised, controlled trial

25. Smania et al (2010) Effects of balance training on postural inestability in patients with idiopathic Parkinson's disease

26. Munneke et al (2010) Efficacy of community based physiotherapy networks for patients with Parkinson's disease: a cluster-randomised trial

27. Olanow et al (2014) Continuous intrajejunal infusion of levodopa-carbidopa intestinal gel for patients with advanced Parkinson's disease: a randomised controlled double-blind, double dummy study

28. Hackney et al (2010) Effects of dance on gait and balance in Parkinson's disease: a comparasion of partnered and non partnered dance movement

29. Palfi et al (2014) Long term safety and tolerability of ProSavin a lentiviral vector based gene therapy for Parkinson's disease: a dose escalation, open label phase1/2 trial

30. Hauser et al (2010) Randomized, double blind, multicenter evaluation of pramipexole extended release once daily in early Parkinson's disease

31. Tickle-Degnen et al (2010) Self-management rehabilitation and health-related quality of life in Parkinson's disease: a randomized controlled trial

32. Hauser et al (2011) Crossover comparison of IPX066 and a standard levodopa formulation in advanced Parkinson's disease

33. Pal et al (2010) The impact of left prefrontal repetitive transcranial magnetic stimulation on depression in Parkinson's disease: a randomized, double-blind, placebo-controlled study

34. Venkataramana et al (2010) Open-labeled study of unilateral autologous bone-marrow-derived mesenchymal stem cell transplantation in Parkinson's disease

35. Muramatsu et al (2010) A phase I study of aromatic L-amino acid decarboxylase gene therapy for Parkinson's disease

36. Meltzer et al (2010) Pimavanserin, a serotonin(2A) receptor inverse agonist, for the treatment of Parkinson's disease psychosis

37. Mirelman et al (2011) Virtual reality for gait training: can it induce motor learning to enhance complex walking and reduce fall risk in patients with Parkinson's disease

38. Rochester et al (2010) Evidence for motor learning in Parkinson's disease: acquisition, automaticity and retention of cued gait performance after training with external rhythmical cues

39. Aviles-Olmos et al (2013) Exenatide and the treatment of patients with Parkinson's disease

40. Schneider et al (2010) GM1 ganglioside in Parkinson's disease: results of a five year open study

41. Shiner et al (2012) Dopamine and performance in a reinforcement learning task: evidence from Parkinson's disease

42. Rosa et al (2010) Time dependent subthalamic local field potential changes after DBS surgery in Parkinson's disease

43. Esposito et al (2013) Rhythm-specific modulation of the sensorimotor network in drug-naive patients with Parkinson's disease by levodopa

44. Mittermeyer et al (2012) Long-term evaluation of a phase 1 study of AADC gene therapy for Parkinson's disease

45. Pompeu et al (2012) Effect of Nintendo Wii (TM)-based motor and cognitive training on activities of daily living in patients with Parkinson's disease: a randomised clinical trial

46. Corvol et al (2011) The COMT Val158Met polymorphism affects the response to entacapone in Parkinson's disease: a randomized crossover clinical trial

47. Delaveau et al (2010) Dopaminergic modulation of the default mode network in Parkinson's disease

48. Sawada et al (2010) Amantadine for dyskinesias in Parkinson's disease: a randomized controlled trial

49. Mizuno et al (2013) Adenosine A(2A) receptor antagonist istradefylline reduces daily OFF time in Parkinson's disease

50. Goodwin et al (2011) An exercise intervention to prevent falls in people with Parkinson's disease: a pragmatic randomised controlled trial

51. Eggert et al (2010) Safety and efficacy of perampanel in advanced Parkinson's disease: a randomized, placebo-controlled study

52. Watts et al (2010) Onset of dyskinesia with adjunct ropinirole prolonged-release or additional levodopa in early Parkinson's disease

53. Bartel et al (2010) [11C]-PK11195 PET: quantification of neuroinflammation and a monitor of anti-inflammatory treatment in Parkinson's disease

54. Witt et al (2013) Relation of lead trajectory and electrode position to neuropsychological outcomes of subthalamic neurostimulation in Parkinson's disease: results from a randomized trial

55. Dubois et al (2012) Donepezil in Parkinson's disease dementia: a randomized, double-blind efficacy and safety study

56. Lees et al (2012) Perampanel, an AMPA antagonist, found to have no benefit in reducing "off" time in Parkinson's disease

57. Pourcher et al (2012) Istradefylline for Parkinson's disease patients experiencing motor fluctuations: results of the KW-6002-US-018 study

58. Prats et al (2011) Blind randomized controlled study of the efficacy of cognitive training in Parkinson's disease

59. Dorsey et al (2010) Increasing access to specialty care: a pilot, randomized controlled trial of telemedicine for Parkinson's disease

60. Factor et al (2010) A long-term study of istradefylline in subjects with fluctuating Parkinson's disease

61. Puente et al (2010) Eighteen month study of continuous intraduodenal levodopa infusion in patients with advanced Parkinson's disease: impact on control of fluctuations and quality of life

62. Hauser et al (2013) Extended-release carbidopa-levodopa (IPX066) compared with immediate-release carbidopa-levodopa in patients with Parkinson's disease and motor fluctuations: a phase 3 randomised, double-blind trial

63. Hanagasi et al (2011) The effects of rasagiline on cognitive deficits in Parkinson's disease patients without dementia: a randomized, double-blind, placebo-controlled, multicenter study

64. Saša et al (2010) Slow (1 Hz) repetitive transcranial magnetic stimulation (rTMS) induces a sustained change in cortical excitability in patients with Parkinson's disease

65. Peron et al (2010) Subthalamic nucleus stimulation affects fear and sadness recognition in Parkinson's disease

66. Devos et al (2014) Targeting chelatable iron as a therapeutic modality in Parkinson's disease

67. Weiss et al (2013) Nigral stimulation for resistant axial motor impairment in Parkinson's disease? A randomized controlled trial

68. Mueller et al (2011) Elevation of total homocysteine levels in patients with Parkinson's disease treated with duodenal levodopa/carbidopa gel

69. Stocchi et al (2011) PREPARED: comparison of prolonged and immediate release ropinirole in advanced Parkinson's disease

70. Guidubaldi et al (2011) Botulinum toxin A versus B in sialorrhea: a prospective, randomized, double-blind, crossover

pilot study in patients with amyotrophic lateral sclerosis or Parkinson's disease

71. Constantinescu et al (2011) Treating disordered speech and voice in Parkinson's disease online: a randomized controlled non-inferiority trial

72. Fernandez et al (2013) Levodopa–carbidopa intestinal gel in advanced Parkinson's disease open-label study: interim results

73. Schneider et al (2013) A randomized, controlled, delayed start trial of GM1 ganglioside in treated Parkinson's disease patients

74. Deuschl et al (2012) Stimulation of the subthalamic nucleus at an earlier disease stage of Parkinson's disease: concept and standards of the EARLYSTIM-study

75. Stocchi et al (2012) A randomized, double-blind, placebo-controlled trial of safinamide as add-on therapy in early Parkinson's disease patients

76. Simuni et al (2010) Tolerability of isradipine in early Parkinson's disease: a pilot dose escalation study

77. Conte et al (2010) Subthalamic nucleus stimulation and somatosensory temporal discrimination in Parkinson's disease

78. Gierthmühlen et al (2010) Influence of deep brain stimulation and levodopa on sensory signs in Parkinson's disease

79. Kwak et al (2010) Effect of dopaminergic medications on the time course of explicit motor sequence learning in Parkinson's disease

80. Cummings et al (2014) Pimavanserin for patients with Parkinson's disease psychosis: a randomised, placebo-controlled phase 3 trial

81. Rogers et al (2011) Postural preparation prior to stepping in patients with Parkinson's disease

82. Daniels et al (2010) Risk factors for executive dysfunction after subthalamic nucleus stimulation in Parkinson's disease

83. Schapira et al (2013) Pramipexole in patients with early Parkinson's disease (PROUD): a randomised delayed-start trial

84. Schapira et al (2013) Long-term efficacy and safety of safinamide as add-on therapy in early Parkinson's disease

85. Gonzalez-Garcia et al (2011) Effects of rTMS on Parkinson's disease: a longitudinal fMRI study

86. Dellapina et al (2011) Apomorphine effect on pain threshold in Parkinson's disease: a clinical and positron emission tomography study

87. Pelosin et al (2010) Action observation improves freezing of gait in patients with Parkinson's disease

88. Cho et al (2012) Effectiveness of acupuncture and bee venom acupuncture in idiopathic Parkinson's disease

89. Vivas et al (2011) Aquatic therapy versus conventional land-based therapy for Parkinson's disease: an open-label pilot study

90. Eggers et al (2010) Theta burst stimulation over the primary motor cortex does not induce cortical plasticity in Parkinson's disease

91. A'Campo et al (2010) The benefits of a standardized patient education program for patients with Parkinson's disease and their caregivers

92. Corcos et al (2013) A two-year randomized controlled trial of progressive resistance exercise for Parkinson's disease

93. Moreau et al (2012) Memantine for axial signs in Parkinson's disease: a randomised, double-blind, placebo-controlled pilot study

94. Woltz et al (2010) Levetiracetam for levodopa-induced dyskinesia in Parkinson's disease: a randomized, double-blind, placebo-controlled trial

95. Menza et al (2010) Treatment of insomnia in Parkinson's disease: a controlled trial of eszopiclone and placebo

96. Tyne et al (2010) Modafinil for Parkinson's disease fatigue

97. Kishore et al (2012) Early, severe and bilateral loss of LTP and LTD-like plasticity in motor cortex (M1) in de novo Parkinson's disease

98. Wuellner et al (2010) Transdermal rotigotine for the perioperative management of Parkinson's disease

99. Lim et al (2010) Does cueing training improve physical activity in patients with Parkinson's disease

100. Costa et al (2010) Effects of deep brain stimulation of the peduncolopontine area on working memory tasks in patients with Parkinson's disease

Chapter 6

1. Vidailhet et al (2005) Bilateral deep-brain stimulation of the globus pallidus in primary generalized dystonia

2. Kupsch et al (2006) Pallidal deep-brain stimulation in primary generalized or segmental dystonia

3. Vidailhet et al (2007) Bilateral, pallidal, deep-brain stimulation in primary generalised dystonia: a prospective 3 year follow-up study

4. Bittar et al (2005) Deep brain stimulation for generalised dystonia and spasmodic torticollis

5. Benecke et al (2005) A new botulinum toxin type A free of complexing proteins for treatment of cervical dystonia

6. Vidailhet et al (2009) Bilateral pallidal deep brain stimulation for the treatment of patients with dystonia-choreoathetosis cerebral palsy: a prospective pilot study

7. Hung et al (2007) Long-term outcome of bilateral pallidal deep brain stimulation

8. Kiss et al (2007) The Canadian multicentre study of deep brain stimulation for cervical dystonia

9. Starr et al (2005) Spontaneous pallidal neuronal activity in human dystonia: comparison with Parkinson's disease and normal macaque

10. Comella et al (2005) Comparison of botulinum toxin serotypes A and B for the treatment of cervical dystonia

11. Truong et al (2005) Efficacy and safety of botulinum type A toxin (Dysport) in cervical dystonia: results of the first US randomized, double-blind, placebo-controlled study

12. Mueller et al (2007) Pallidal deep brain stimulation improves quality of life in segmental and generalized dystonia: results from a prospective, randomized sham-controlled trial

13. Ostrem et al (2011) Subthalamic nucleus deep brain stimulation in primary cervical dystonia

14. Pappert et al (2009) Botulinum toxin type B vs. type A in toxin-naive patients with cervical dystonia: randomized, double-blind, noninferiority trial

15. Volkmann et al (2012) Pallidal deep brain stimulation in patients with primary generalised or segmental dystonia: 5-year follow-up of a randomised trial

16. Pillon et al (2006) Preservation of cognitive function in dystonia treated by pallidal stimulation

17. Truong et al (2010) Long-term efficacy and safety of botulinum toxin type A (Dysport) in cervical dystonia

18. Rijn et al (2009) Intrathecal baclofen for dystonia of complex regional pain syndrome

19. Pretto et al (2008) A prospective blinded evaluation of deep brain stimulation for the treatment of secondary dystonia and primary torticollis syndromes

20. Bihari et al (2005) Safety, effectiveness, and duration of effect of BOTOX after switching from Dysport for blepharospasm, cervical dystonia, and hemifacial spasm

21. Diamond et al (2006) Globus pallidus deep brain stimulation in dystonia

22. Moro et al (2009) Pallidal stimulation in cervical dystonia: clinical implications of acute changes in stimulation parameters

23. Sanger et al (2007) Prospective open-label clinical trial of trihexyphenidyl in children with secondary dystonia due to cerebral palsy

24. Houeto et al (2007) Acute deep-brain stimulation of the internal and external globus pallidus in primary dystonia

25. Borich et al (2009) Lasting effects of repeated rTMS application in focal hand dystonia

26. Blood et al (2006) White matter abnormalities in dystonia normalize after botulinum toxin treatment

27. Grips et al (2007) Patterns of reoccurrence of segmental dystonia after discontinuation of deep brain stimulation

28. Tassorelli et al (2006) Botulinum toxin and neuromotor rehabilitation: an integrated approach to idiopathic cervical dystonia

29. Comella et al (2011) Efficacy and safety of incobotulinumtoxinA (NT 201, XEOMIN (R), botulinum neurotoxin type A, without accessory proteins) in patients with cervical dystonia

30. Valldeoriola et al (2009) Efficacy and safety of pallidal stimulation in primary dystonia: results of the Spanish multicentric study

31. Bonanni et al (2007) Botulinum toxin treatment of lateral axial dystonia in parkinsonism

32. Factor et al (2005) Efficacy and safety of repeated doses of botulinum toxin type B in type A resistant and responsive cervical dystonia

33. Tinazzi et al (2005) TENS for the treatment of writer's cramp dystonia: a randomized, placebo-controlled study

34. Skogseid et al (2012) Good long-term efficacy of pallidal stimulation in cervical dystonia: a prospective, observer-blinded study

35. Sanger et al (2007) Botulinum toxin type B improves the speed of reaching in children with cerebral palsy and arm dystonia: an open-label, dose-escalation pilot study

36. Hamani et al (2008) Location of active contacts in patients with primary dystonia treated with globus pallidus deep brain stimulation

37. Volkmann et al (2014) Pallidal neurostimulation in patients with medication-refractory cervical dystonia: a randomised, sham-controlled trial

38. Benninger et al (2011) Transcranial direct current stimulation for the treatment of focal hand dystonia

39. Grabli et al (2009) Interruption of deep brain stimulation of the globus pallidus in primary generalized dystonia

40. Cersosimo et al (2009) Micro lesion effect of the globus pallidus internus and outcome with deep brain stimulation in patients with Parkinson disease and dystonia

41. Byl et al (2009) Focal hand dystonia: effectiveness of a home program of fitness and learning-based sensorimotor and memory training

42. Magariños-Ascone et al (2008) Deep brain stimulation in the globus pallidus to treat dystonia: electrophysiological characteristics and 2 years' follow-up in 10 patients

43. Kojovic et al (2011) Botulinum toxin injections reduce associative plasticity in patients with primary dystonia

44. Rice et al (2009) Pilot study on trihexyphenidyl in the treatment of dystonia in children with cerebral palsy

45. Hanajima et al (2008) Difference in intracortical inhibition of the motor cortex between cortical myoclonus and focal hand dystonia

46. Gregori et al (2008) Fast voluntary neck movements in patients with cervical dystonia: a kinematic study before and after therapy with botulinum toxin type A

47. Munts et al (2009) Intrathecal glycine for pain and dystonia in complex regional pain syndrome

48. Buttkus F (2011) Single-session tDCS-supported retraining does not improve fine motor control in musician's dystonia

49. Schjerling et al (2013) A randomized double-blind crossover trial comparing subthalamic and pallidal deep brain stimulation for dystonia. Clinical article

50. Furuya et al (2013) Finger-specific loss of independent control of movements in musicians with focal dystonia

51. Avanzino et al (2013) Temporal expectation in focal hand dystonia

52. Ondo et al (2005) A pilot study of botulinum toxin A for headache in cervical dystonia

53. Haring et al (2007) An open trial of levetiracetam for segmental and generalized dystonia

54. Quagliato et al (2010) A prospective, randomized, double-blind study comparing the efficacy and safety of type A botulinum toxins botox and prosigne in the treatment of cervical dystonia

55. Thomas et al (2006) Frozen versus fresh reconstituted botox laryngeal dystonia

56. Pelosin et al (2009) Cervical dystonia affects aimed movements of nondystonic segments

57. Kimberley et al (2013) Multiple sessions of low-frequency repetitive transcranial magnetic stimulation in focal hand dystonia: clinical and physiological effects

58. Morgante et al (2011) Abnormal tactile temporal discrimination in psychogenic dystonia

59. Lew et al (2010) Rimabotulinumtoxin B effects on pain associated with cervical dystonia: results of placebo and comparator-controlled studies

60. Benecke et al (2009) Xeomin (R) in the treatment of cervical dystonia

61. Guettard et al (2009) Management of spasticity and dystonia in children with acquired brain injury with rehabilitation and botulinum toxin A

62. Vercueil et al (2006) Effects of pulse width variations in pallidal stimulation for primary generalized dystonia

63. Queiroz et al (2012) Physical therapy program for cervical dystonia: a study of 20 cases

64. Opavský et al (2011) Sensorimotor network in cervical dystonia and the effect of botulinum toxin treatment: a functional MRI study

65. McKenzie et al (2009) Differences in physical characteristics and response to rehabilitation for patients with hand dystonia: musicians' cramp compared to writers' cramp

66. Dowson et al (2008) Clinical profile of botulinum toxin A in patients with chronic headaches and cervical dystonia - a prospective, open-label, longitudinal study conducted in a naturalistic clinical practice setting

67. Dressler et al (2013) Long-term efficacy and safety of incobotulinumtoxinA injections in patients with cervical dystonia

68. Charles et al (2012) Efficacy, tolerability, and immunogenicity of onabotulinumtoxinA in a randomized, double-blind, placebo-controlled trial for cervical dystonia

69. Dresel et al (2011) Botulinum toxin modulates basal ganglia but not deficient somatosensory activation in orofacial dystonia

70. Scontrini et al (2011) Somatosensory temporal discrimination tested in patients receiving botulinum toxin injection for cervical dystonia

71. Chan et al (2010) A randomised controlled study of risperidone and olanzapine for schizophrenic patients with neuroleptic-induced acute dystonia or parkinsonism

72. Lo et al (2005) Identification and treatment of cervical and oromandibular dystonia in acutely brain-injured patients

73. Koch et al (2014) Effects of two weeks of cerebellar theta burst stimulation in cervical dystonia patients

74. Tisch et al (2008) Cortical evoked potentials from pallidal stimulation in patients with primary generalized dystonia

75. Allam et al (2007) Sustained attention in cranial dystonia patients treated with botulinum toxin

76. Tarsy et al (2006) An open-label trial of levetiracetam for treatment of cervical dystonia

77. Evidente et al (2013) A randomized, double-blind study of repeated incobotulinumtoxin A (Xeomin (R)) in cervical dystonia

78. Winner et al (2012) Concurrent onabotulinumtoxinA treatment of cervical dystonia and concomitant migraine

79. Huang et al (2012) Modulation of the disturbed motor network in dystonia by multisession suppression of premotor cortex

80. Kim et al (2011) Treatment of secondary dystonia with a combined stereotactic procedure: long-term surgical outcomes

81. Opladen et al (2010) Phenylalanine loading in pediatric patients with dopa-responsive dystonia: revised test protocol and pediatric cutoff values

82. Young et al (2014) Cathodal transcranial direct current stimulation in children with dystonia: a sham-controlled study

83. Pelosin et al (2013) KinesioTaping reduces pain and modulates sensory function in patients with focal dystonia: a randomized crossover pilot study

84. Bonouvrie et al (2011) Effects of intrathecal baclofen on daily care in children with secondary generalized dystonia: a pilot study

85. Chinnapongse et al (2010) Open-label, sequential dose-escalation, safety, and tolerability study of rimabotulinumtoxinb in subjects with cervical dystonia

86. Ostrem et al (2014) Effect of frequency on subthalamic nucleus deep brain stimulation in primary dystonia

87. Dressler et al (2014) The Dystonia Discomfort Scale (DDS): a novel instrument to monitor the temporal profile of botulinum toxin therapy in cervical dystonia

88. Van den Dool et al (2013) Cervical dystonia: effectiveness of a standardized physical therapy program; study design and protocol of a single blind randomized controlled trial

89. Coleman et al (2012) Immunoresistance in cervical dystonia patients after treatment with abobotulinumtoxinA

90. Evidente et al (2014) IncobotulinumtoxinA (Xeomin (R)) injected for blepharospasm or cervical dystonia according to patient needs is well tolerated

91. Hauser et al (2013) AbobotulinumtoxinA (Dysport) dosing in cervical dystonia: an exploratory analysis of two large open-label extension studies

92. Starr et al (2014) Interventional MRI-guided deep brain stimulation in pediatric dystonia: first experience with the Clear Point system

93. Boyce et al (2013) Active exercise for individuals with cervical dystonia: a pilot randomized controlled trial

94. Levin et al (2014) Onset latency of segmental dystonia after deep brain stimulation cessation: a randomized, double-blind crossover trial

95. Rieu et al (2014) Motor cortex stimulation does not improve dystonia secondary to a focal basal ganglia lesion

96. Mordin et al (2014) Factors affecting the health-related quality of life of patients with cervical dystonia and impact of treatment with abobotulinumtoxinA (Dysport): results from a randomised, double-blind, placebo-controlled study

97. Van der Plas et al (2013) An explanatory study evaluating the muscle relaxant effects of intramuscular magnesium sulphate for dystonia in complex regional pain syndrome

98. Morgan et al (2008) A single-blind trial of bilateral globus pallidus internus deep brain stimulation in medically refractory cervical dystonia

99. Yun et al (2015) Dysport and botox at a ratio of 2.5:1 units in cervical dystonia: a double-blind, randomized study

100. O'Reilly et al (2012) A comparison of facial muscle squeezing versus non-facial muscle squeezing on the efficacy of botulinumtoxin-A injections for the treatment of facial dystonia

Chapter 7

1. Kanra et al (1995) Beneficial effects of dexamethasone in children with pneumococcal meningitis

2. Klugman et al (1995) Bacterial activity against cephalosporin-resistant streptococcus pneumoniae in cerebrospinal fluid in children with acute bacterial meningitis

3. Menichetti et al (1995) High-dose fluconazole therapy for cryptococcal meningitis in patients with AIDS

4. Kilpi et al (1995) Oral glycerol and intravenous dexamethasone in preventing neurologic and audiologic sequelae of childhood bacterial meningitis

5. Klugman et al (1995) Randomized comparison of meropenem with cefotaxime for treatment of bacterial meningitis

6. Schmutzhard et al (1995) A randomised comparison of meropenem with cefotaxime or ceftriaxone for the treatment of bacterial meningitis in adults

7. Wald et al (1995) Dexamethasone therapy for children with bacterial meningitis

8. Sharkey et al (1996) Amphotericin B lipid complex compared with amphotericin B in the treatment of cryptococcal meningitis in patients with AIDS

9. Leenders et al (1997) Liposomal amphotericin B (AmBisome) compared with amphotericin B both followed by oral fluconazole in the treatment of AIDS-associated cryptococcal meningitis

10. Horst et al (1997) Treatment of cryptococcal meningitis associated with the acquired immunodeficiency syndrome

11. Schoeman et al (1997) Effect of corticosteroids on intracranial pressure, computed tomographic findings, and clinical outcome in young children with tuberculous meningitis

12. Chamberlain et al (1998) Carcinoma meningitis secondary to non-small cell lung cancer

13. Mayanja-Kizza et al (1998) Combination therapy with fluconazole and flucytosine for cryptococcal meningitis in Ugandan patients with AIDS

14. Glantz et al (1999) Randomized trial of a slow-release versus a standard formulation of cytarabine for the intrathecal treatment of lymphomatous meningitis

15. Glantz et al (1999) A randomized controlled trial comparing intrathecal sustained-release cytarabine to intrathecal methotrexate in patients with neoplastic meningitis from solid tumors

16. Saag et al (1999) A comparison of Itraconazole versus Fluconazole as maintenance therapy for AIDS-associated Cryptococcal Meningitis

17. Schoeman et al (2000) Adjusnctive thalidomide therapy of childhood tuberculous meningitis

18. Chotmongkol et al (2000) Corticosteroid treatment of eosinophilic meningitis

19. Roine et al (2000) Randomized trial of four vs. seven days of ceftriaxone treatment for bacterial meningitis in children with rapid initial recovery

20. Schoeman et al (2001) The effect of adjuvant steroid treatment on serial cerebrospinal fluid changes in tuberculous meningitis

21. Chamberlain et al (2002) A phase II trial of intra-cerebrospinal fluid alpha interferon in the treatment of neoplastic meningitis

22. Duke et al (2002) Management of meningitis in children with oral fluid restriction or intravenous fluid at maintenance volumes: a randomised trial

23. Gans et al (2002) Dexamethasone in adults with bacterial meningitis

24. Gijwani et al (2002) Dexamethasone therapy for bacterial meningitis in adults: a double-blind placebo control study

25. Molineux et al (2002) Dexamethasone treatment in childood bacterial meningitis in Malawi: a randomised controlled trial

26. Newton et al (2002) A randomized, double-blind, placebo-controlled trial of acetazolamide for the treatment of elevated intracranial pressure in Cryptococcal Meningitis

27. Saez-Llorens et al (2002) Quinolone treatment for pediatric bacterial meningitis: a comparative study of trovafloxacin and ceftriaxone with or without vancomycin

28. Singhi et al (2002) Seven days vs. 10 days ceftriaxone therapy in bacterial meningitis

29. Sungkanuparph et al (2002) Efavirez-based regimen as treatment of advanced AIDS with cryptococcal meningitis

30. Jaeckle et al (2002) An open-label trial of sustained-release cytarabine for the intrathecal treatment of solid tumor neoplastic meningitis

31. Blaney et al (2003) Phase I clinical trial of intrathecal topotecan in patients with neoplastic meningitis

32. Abzug et al (2003) Double blind placebo-controlled tril of pleconaril in infants with enterovirus meningitis

33. Buke et al (2003) Does dexamethasone affect ceftriazone penetration into cerebrospinal fluid in adult bacterial meningitis

34. Cole et al (2003) Quality-of-life survival comparison of sustained-release cytosine al abinoside versus intrathecal methotrexate for treatment of solid tumor neoplastic meningitis

35. Kapoor et al (2003) Spinal epidermoid tumors: novel approach to aseptic meningitis

36. Molyneux et al (2003) The effect of HIV infection on paediatric bacterial meningitis in Blantyre, Malawi

37. Bomgaars et al (2004) Phase I trial of intrathecal liposomal cytarabine in children with neoplastic meningitis

38. Brouwer et al (2004) Combination antifungal therapies for HIV-associated cryptococcal meningitis - a randomised trial

39. Pappas et al (2004) Recombinant interferon-gama1b as adjunctive therapy for AIDS-related acute cryptococcal meningitis

40. Schoeman et al (2004) Adjunctive thalidomide therapy for childhood tuberculous meningitis: results of a randomized study

41. Thwaites et al (2004) Dexamethasone for the treatment of tuberculous meningitis in adolescents and adults

42. Chotmongkol et al (2005) Initial treatment of cryptococcl meningitis in AIDS

43. Nathan et al (2005) Ceftriaxone as effective as long-lasting chloramphenicol in short-course treatment of meningococcal meningitis during epidemics - a randomized non-inferiority trial

44. Simmons et al (2005) The clinical benefit of adjunctive dexamethasone in tuberculous meningitis is not associated with measurable attenuation of peripheral or local immune responses

45. Gururangan et al (2006) Phase I trial of intrathecal spartaject busulfan in children with neoplastic meningitis - a pediatric brain tumor consortium study

46. Chamberlain et al (2006) Phase II trial of intracerebrospinal fluid etoposide in the treatment of neoplastic meningitis

47. Tansuphaswadikul et al (2006) Comparison of one week with two week regimens of Amphotericin B both followed by fluconazole in the treatment of Cryptococcal Meningitis among AIDS patients

48. Weisfelt et al (2006) Dexamethasone and long-term outcome in adults with bacterial meningitis

49. Ellis et al (2007) Cefepime cerebrospinal fluid concentrations in neonatal bacterial meningitis

50. Phuphanich et al (2007) A pharmacokinetic study of intra-CSF administered encapsulated cytarabine for the treatment of neoplastic meningitis in patiets with leukemia, lymphoma or solid tumors

51. Brouwer et al (2007) Oral versus intravenous Flucytosine in patients with Human Immunodeficiency Virus-associated Cryptococcal Meningitis

52. Jitpimolmard et al (2007) Albendazole therapy for eosinophilic meningitis caused by Angiostrongylus cantonensis

53. Peltola et al (2007) Adjuvant glycerol and/or dexamethasone to improve the outcomes of childhood bacterial meningitis - a prospective, randomized, double-blind, placebo-controlled trial

54. Sankar et al (2007) Role of dexamethasone and oral glycerol in reducing hearing and neurological sequelae in children with bacterial meningitis

55. Scarborough et al (2007) Corticosteroids for bacterial meningitis in adults in sub-saharan Africa

56. Thi Hoang Mai et al (2007) Dexamethasone in vietnamese adolescents and adults with bacterial meningitis

57. Bernardi et al (2008) Phase I clinical trial of intrathecal gemcitabine in patients with neoplastic meningitis

58. Bicanic et al (2008) High-dose amphotericin B with flucytosine for the treatment of cryptococcal meningitis in HIV-infected patients

59. Milefchik et al (2008) Fluconazole alone or combined with flucytosine for the treatment of AIDS-associated cryptococcal meningitis

60. Singhi et al (2008) Increase in serum osmolarity is possible mechanism for the beneficial effects of glycerol in childhood bacterial meningitis

61. Pappas et al (2009) A phase II randomized trial of Amphotericin B alone or combined with Fluconazole in the treatment of HIV-associated cryptococcal meningitis

62. Chotmongkol et al (2009) Comparison of prednisolone plus albendazole with prednisolone alone for treatment of patients with eosinophilic meningitis

63. Green et al (2009) Dexamethasone, cerebrospinal fluid matrix metalloproteinase concentrations and clinical outcomes in Tuberculous Meningitis

64. Gujjar et al (2009) HHH regime for arteritis secondary to TB meningitis: a prospective randomized study

65. Malhotra et al (2009) Corticosteroids (dexamethasone versus intravenous methylprednisolone) in patients with tuberculous meningitis

66. Techapornroong et al (2009) Alternate-day verus once-daily admnistration of amphotericin B in the treatment of cryptococcal meningitis: a randomized controlled trial

67. Jadhav et al (2010) Lyposomal amphotericin B for the treatment of cryptococcal meningitis in HIV/AIDS patients in India - a multicentric, randomized controlled trial

68. Spina et al (2010) Phase 2 study of intrathecal, long-acting liposomal cytarabine in the prophylaxis of lymphomatous meningitis in Human Immunodeficiency Virus-related Non-Hodgkin lymphoma

69. Hamill et al (2010) Comparison of 2 doses of liposomol Amphotericin B and conventional Amphotericin B deoxycholate for treatment of AIDS-associated acute Cryptococcal Meningitis: a randomized, double-blind clinical trial of efficacy and safety

70. Makadzange et al (2010) Early versus delayed initiation of ART for concurrent HIV infection and cryptococcal meningitis in subsaharan Africa

71. Manosuthi et al (2010) Monitoring and impact of fluconazole serum and cerebrospinal fluid concentration in HIV-associated cryptococcal menigitis-infected patients

72. Misra et al (2010) Role of aspirin in tuberculous meningitis - a randomized open label placebo controlled trial

73. Nussbaum et al (2010) Combination Flucytosine and high-dose Fluconazole compared with Fluconazole monotherapy for the treatment of Cryptococcal meningitis - a randomized trial in Malawi

74. Peltola et al (2010) Hearing impairment in childhood bacterial meningitis is little relieved by Dexamethasone of Glycerol

75. Thwaites et al (2011) Randomized pharmokinetic and pharmacodynamic comparison of fluoroquinolones for tuberculous meningitis

76. Ajdukiewicz et al (2011) Glycerol adjuvant therapy in adults with bacterial meningitis in a high HIV seroprevalence setting in Malawi - a double-blind, randomised controlled trial

77. Heemskerk et al (2011) Intensified treatment with high dose Rifampicin and Levofloxacin compared to standard treatment for adult patients with Tuberculous Meningitis (TBM-IT): protocol for a randomized controlled trial

78. Molyneux et al (2011) 5 versus 10 days of treatment with ceftriaxone for bacterial meningitis in children - a double-blind randomized equivalence study

79. Pelkonen et al (2011) Slow initial beta-lactam infusion and oral parecetamol to treat childhood bacterial meningitis - a randomised, controlled trial

80. Schoeman et al (2011) The role of aspirin in childhood tuberculous meningitis

81. Torok et al (2011) Timing of initiation of antiretroviral therapy in HIV-associated tuberculous meningitis

82. Torok et al (2011) Dexamethasone and long-term outcome of Tuberculous meningitis in Vietnamese adults and adolescents

83. Jackson et al (2012) A phase II randomized controlled trial adding oral flucytosine to high-dose fluconazole, with short-course Amphotericin B, for Cryptococcal Meningitis

84. Loyse et al (2012) Comparison of the early fungicidal activity of high-dose fluconazole, voriconazole, and flucytosine as second-line drugs given in combination with Amphotericin B for the treatment of HIV-associated Cryptococcal Menigitis

85. Aurelius et al (2012) Long-term Valacyclovir suppressive treatment after herpes simplex virus type 2 meningitis - a double-blind, randomized controlled trial

86. Fritz et al (2012) Dexamethasone and long-term survival in bacterial meningitis

87. Glantz et al (2012) Route of intracerebrospinal fluid chemotherapy administration and efficacy of therapy in neoplastic meningitis

88. Jarvis et al (2012) Adjunctive interferon-[gamma] immunotherapy for the treatment of HIV-associated cryptococcal meningitis

89. Ruslami et al (2013) Intensified regimen containing rifampicin and moxifloxacin for tuberculous meningitis - an open-label, randomised controlled phase 2 trial

90. Bisson et al (2013) Early versus delayed initiation Antiretroviral Therapy and cerebrospianl fluid clearance in adults with HIV and Cryptococcal Meningitis

91. Day et al (2013) Combination antifungal therapy for cryptococcal meningitis

92. Mathur et al (2013) Role of dexamethasone in neonatal meningitis: a randomized controlled trial

93. Molyneux et al (2013) Glycerol and acetaminophen as adjuvant therapy did not affect the outcome of bacterial meningitis in malawian children

94. Mourvillier et al (2013) Induced hypothermia in severe bacterial menigitis

95. Boulware et al (2014) Timing of antiretroviral therapy after diagnosis of Cryptococcal Meningitis

96. Day et al (2014) CryptoDex - a randomised, double-blind, placebo-controlled phase III trial of adjunctive dexamethasone in HIV-infected adults with cryptococcal meningitis - study protocol

97. Roine et al (2104) Factors affecting time to death from start of treatment among children succumbing to bacterial meningitis

98. Rolfes et al (2014) The effect of therapeutic lumbar punctures on acute mortality from cryptococcal meningitis

99. Shah et al (2014) High versus low dose steroids in children with tuberculous meningitis

100. Mfinanga et al (2015) Cryptococcal meningitis screening and community-based early adherence support in people with advanced HIV infection starting antiretroviral therapy in Tanzania and Zambia: an open-label, randomized controlled trial

Chapter 8

1. Newcombe J et al (1991) Histopathology of multiple sclerosis lesions detected by magnetic resonance imaging in unfixed postmortem central nervous system tissue. Brain 114(Pt 2):1013–1023

2. Filippi M, Rocca MA (2011) MR imaging of multiple sclerosis. Radiology 259(3):659–681

3. Rocca MA, Messina R, Filippi M (2013) Multiple sclerosis imaging: recent advances. J Neurol 260(3):929–935

4. Pohl D et al (2007) Paediatric multiple sclerosis and acute disseminated encephalomyelitis in Germany: results of a nationwide survey. Eur J Pediatr 166(5):405–412

5. Mackenzie IS et al (2014) Incidence and prevalence of multiple sclerosis in the UK 1990–2010: a descriptive study in

the General Practice Research Database. J Neurol Neurosurg Psychiatry 85(1):76–84

6. Noonan CW et al (2010) The prevalence of multiple sclerosis in 3 US communities. Prev Chronic Dis 7(1):A12

7. Kinsinger SW, Lattie E, Mohr DC (2010) Relationship between depression, fatigue, subjective cognitive impairment, and objective neuropsychological functioning in patients with multiple sclerosis. Neuropsychology 24(5):573–580

8. Noseworthy JH et al (2000) Multiple sclerosis. N Engl J Med 343(13):938–952

9. Montalban X (2011) Review of methodological issues of clinical trials in multiple sclerosis. J Neurol Sci 311(Suppl 1):S35–S42

10. Cohen JA et al (2010) Oral fingolimod or intramuscular interferon for relapsing multiple sclerosis. N Engl J Med 362(5):402–415

11. Calabresi PA et al (2014) Safety and efficacy of fingolimod in patients with relapsing-remitting multiple sclerosis (FREEDOMS II): a double-blind, randomised, placebo-controlled, phase 3 trial. Lancet Neurol 13(6):545–556

12. Cohen JA et al (2012) Alemtuzumab versus interferon beta 1a as first-line treatment for patients with relapsing-remitting multiple sclerosis: a randomised controlled phase 3 trial. Lancet 380(9856):1819–1828

13. Coles AJ et al (2012) Alemtuzumab for patients with relapsing multiple sclerosis after disease-modifying therapy: a randomised controlled phase 3 trial. Lancet 380(9856):1829–1839

14. Confavreux C et al (2012) Long-term follow-up of a phase 2 study of oral teriflunomide in relapsing multiple sclerosis: safety and efficacy results up to 8.5 years. Mult Scler 18(9):1278–1289

15. Confavreux C et al (2014) Oral teriflunomide for patients with relapsing multiple sclerosis (TOWER): a randomised, double-blind, placebo-controlled, phase 3 trial. Lancet Neurol 13(3):247–256

16. Fox RJ et al (2012) Placebo-controlled phase 3 study of oral BG-12 or glatiramer in multiple sclerosis. N Engl J Med 367(12):1087–1097

17. Freedman MS et al (2012) Teriflunomide added to interferon-beta in relapsing multiple sclerosis: a randomized phase II trial. Neurology 78(23):1877–1885

18. Ghezzi A et al (2010) Safety and efficacy of natalizumab in children with multiple sclerosis. Neurology 75(10):912–917

19. Gold R et al (2014) Assessment of cardiac safety during fingolimod treatment initiation in a real-world relapsing multiple sclerosis population: a phase 3b, open-label study. J Neurol 261(2):267–276

20. Gold R et al (2012) Placebo-controlled phase 3 study of oral BG-12 for relapsing multiple sclerosis. N Engl J Med 367(12):1098–1107

21. Kappos L et al (2012) Effect of BG-12 on contrast-enhanced lesions in patients with relapsing—remitting multiple sclerosis: subgroup analyses from the phase 2b study. Mult Scler 18(3):314–321

22. Kappos L et al (2010) A placebo-controlled trial of oral fingolimod in relapsing multiple sclerosis. N Engl J Med 362(5):387–401

23. Khatri B et al (2011) Comparison of fingolimod with interferon beta-1a in relapsing-remitting multiple sclerosis: a randomised extension of the TRANSFORMS study. Lancet Neurol 10(6):520–529

24. MacManus DG et al (2011) BG-12 reduces evolution of new enhancing lesions to T1-hypointense lesions in patients with multiple sclerosis. J Neurol 258(3):449–456

25. Mehling M et al (2011) Antigen-specific adaptive immune responses in fingolimod-treated multiple sclerosis patients. Ann Neurol 69(2):408–413

26. Mellergard J et al (2010) Natalizumab treatment in multiple sclerosis: marked decline of chemokines and cytokines in cerebrospinal fluid. Mult Scler 16(2):208–217

27. Phillips JT et al (2011) Sustained improvement in Expanded Disability Status Scale as a new efficacy measure of neurological change in multiple sclerosis: treatment effects with natalizumab in patients with relapsing multiple sclerosis. Mult Scler 17(8):970–979

28. O'Connor P et al (2011) Randomized trial of oral teriflunomide for relapsing multiple sclerosis. N Engl J Med 365(14):1293–1303

29. Vermersch P et al (2014) Teriflunomide versus subcutaneous interferon beta-1a in patients with relapsing multiple sclerosis: a randomised, controlled phase 3 trial. Mult Scler 20(6):705–716

30. Soilu-Hanninen M et al (2012) A randomised, double blind, placebo controlled trial with vitamin D3 as an add on

treatment to interferon beta-1b in patients with multiple sclerosis. J Neurol Neurosurg Psychiatry 83(5):565–571

31. Zamboni P et al (2012) Venous angioplasty in patients with multiple sclerosis: results of a pilot study. Eur J Vasc Endovasc Surg 43(1):116–122

32. Khan S et al (2011) Long-term effect on quality of life of repeat detrusor injections of botulinum neurotoxin-A for detrusor overactivity in patients with multiple sclerosis. J Urol 185(4):1344–1349

33. Putzki N et al (2010) Efficacy of natalizumab in second line therapy of relapsing-remitting multiple sclerosis: results from a multi-center study in German speaking countries. Eur J Neurol 17(1):31–37

34. Calabresi PA et al (2014) Pegylated interferon beta-1a for relapsing-remitting multiple sclerosis (ADVANCE): a randomised, phase 3, double-blind study. Lancet Neurol 13(7):657–665

35. Bermel RA et al (2010) Intramuscular interferon beta-1a therapy in patients with relapsing-remitting multiple sclerosis: a 15-year follow-up study. Mult Scler 16(5):588–596

36. Goodin DS et al (2012) Relationship between early clinical characteristics and long term disability outcomes: 16 year cohort study (follow-up) of the pivotal interferon beta-1b trial in multiple sclerosis. J Neurol Neurosurg Psychiatry 83(3):282–287

37. Dorr J et al (2012) Efficacy of vitamin D supplementation in multiple sclerosis (EVIDIMS Trial): study protocol for a randomized controlled trial. Trials 13:15

38. Kimball S et al (2011) Cholecalciferol plus calcium suppresses abnormal PBMC reactivity in patients with multiple sclerosis. J Clin Endocrinol Metab 96(9):2826–2834

39. Mosayebi G et al (2011) Therapeutic effect of vitamin D3 in multiple sclerosis patients. Immunol Invest 40(6):627–639

40. Smolders J et al (2011) Efficacy of vitamin D3 as add-on therapy in patients with relapsing-remitting multiple sclerosis receiving subcutaneous interferon beta-1a: a Phase II, multi-center, double-blind, randomized, placebo-controlled trial. J Neurol Sci 311(1-2):44–49

41. Stein MS et al (2011) A randomized trial of high-dose vitamin D2 in relapsing-remitting multiple sclerosis. Neurology 77(17):1611–1618

42. Serpell MG, Notcutt W, Collin C (2013) Sativex long-term use: an open-label trial in patients with spasticity due to multiple sclerosis. J Neurol 260(1):285–295

43. Kavia RB et al (2010) Randomized controlled trial of Sativex to treat detrusor overactivity in multiple sclerosis. Mult Scler 16(11):1349–1359

44. Collin C et al (2010) A double-blind, randomized, placebo-controlled, parallel-group study of Sativex, in subjects with symptoms of spasticity due to multiple sclerosis. Neurol Res 32(5):451–459

45. Zajicek JP et al (2012) Multiple sclerosis and extract of cannabis: results of the MUSEC trial. J Neurol Neurosurg Psychiatry 83(11):1125–1132

46. Novotna A et al (2011) A randomized, double-blind, placebo-controlled, parallel-group, enriched-design study of nabiximols* (Sativex((R))), as add-on therapy, in subjects with refractory spasticity caused by multiple sclerosis. Eur J Neurol 18(9):1122–1131

47. Notcutt W et al (2012) A placebo-controlled, parallel-group, randomized withdrawal study of subjects with symptoms of spasticity due to multiple sclerosis who are receiving long-term Sativex(R) (nabiximols). Mult Scler 18(2):219–228

48. Langford RM et al (2013) A double-blind, randomized, placebo-controlled, parallel-group study of THC/CBD oromucosal spray in combination with the existing treatment regimen, in the relief of central neuropathic pain in patients with multiple sclerosis. J Neurol 260(4):984–997

49. Broekmans T et al (2011) Effects of long-term resistance training and simultaneous electro-stimulation on muscle strength and functional mobility in multiple sclerosis. Mult Scler 17(4):468–477

50. Cakt BD et al (2010) Cycling progressive resistance training for people with multiple sclerosis: a randomized controlled study. Am J Phys Med Rehabil 89(6):446–457

51. Collett J et al (2011) Exercise for multiple sclerosis: a single-blind randomized trial comparing three exercise intensities. Mult Scler 17(5):594–603

52. Conklyn D et al (2010) A home-based walking program using rhythmic auditory stimulation improves gait performance in patients with multiple sclerosis: a pilot study. Neurorehabil Neural Repair 24(9):835–842

53. Dodd KJ et al (2011) Progressive resistance training did not improve walking but can improve muscle performance, quality of life and fatigue in adults with multiple sclerosis: a randomized controlled trial. Mult Scler 17(11):1362–1374

54. Fimland MS et al (2010) Enhanced neural drive after maximal strength training in multiple sclerosis patients. Eur J Appl Physiol 110(2):435–443

55. Hebert JR et al (2011) Effects of vestibular rehabilitation on multiple sclerosis-related fatigue and upright postural control: a randomized controlled trial. Phys Ther 91(8):1166–1183

56. Prosperini L et al (2013) Home-based balance training using the Wii balance board: a randomized, crossover pilot study in multiple sclerosis. Neurorehabil Neural Repair 27(6):516–525

57. Prosperini L et al (2010) Visuo-proprioceptive training reduces risk of falls in patients with multiple sclerosis. Mult Scler 16(4):491–499

58. Motl RW et al (2011) Internet intervention for increasing physical activity in persons with multiple sclerosis. Mult Scler 17(1):116–128

59. Nilsagard YE, Forsberg AS, von Koch L (2013) Balance exercise for persons with multiple sclerosis using Wii games: a randomised, controlled multi-centre study. Mult Scler 19(2):209–216

60. Mori F et al (2010) Effects of anodal transcranial direct current stimulation on chronic neuropathic pain in patients with multiple sclerosis. J Pain 11(5):436–442

61. Jurynczyk M et al (2010) Immune regulation of multiple sclerosis by transdermally applied myelin peptides. Ann Neurol 68(5):593–601

62. Dalgas U et al (2010) Muscle fiber size increases following resistance training in multiple sclerosis. Mult Scler 16(11):1367–1376

63. Burton JM et al (2010) A phase I/II dose-escalation trial of vitamin D3 and calcium in multiple sclerosis. Neurology 74(23):1852–1859

64. Filippi M et al (2012) Multiple sclerosis: effects of cognitive rehabilitation on structural and functional MR imaging measures--an explorative study. Radiology 262(3):932–940

65. Starck M et al (2010) Acquired pendular nystagmus in multiple sclerosis: an examiner-blind cross-over treatment study of memantine and gabapentin. J Neurol 257(3):322–327

66. Rice CM et al (2010) Safety and feasibility of autologous bone marrow cellular therapy in relapsing-progressive multiple sclerosis. Clin Pharmacol Ther 87(6):679–685

67. Velikonja O et al (2010) Influence of sports climbing and yoga on spasticity, cognitive function, mood and fatigue in patients with multiple sclerosis. Clin Neurol Neurosurg 112(7):597–601

68. Mattioli F et al (2010) Efficacy and specificity of intensive cognitive rehabilitation of attention and executive functions in multiple sclerosis. J Neurol Sci 288(1-2):101–105

69. Naismith RT et al (2010) Rituximab add-on therapy for breakthrough relapsing multiple sclerosis: a 52-week phase II trial. Neurology 74(23):1860–1867

70. Jensen MP et al (2011) Effects of self-hypnosis training and cognitive restructuring on daily pain intensity and catastrophizing in individuals with multiple sclerosis and chronic pain. Int J Clin Exp Hypn 59(1):45–63

71. Barkhof F et al (2010) Ibudilast in relapsing-remitting multiple sclerosis: a neuroprotectant? Neurology 74(13):1033–1040

72. Bielekova B et al (2011) Intrathecal effects of daclizumab treatment of multiple sclerosis. Neurology 77(21):1877–1886

73. Chataway J et al (2014) Effect of high-dose simvastatin on brain atrophy and disability in secondary progressive multiple sclerosis (MS-STAT): a randomised, placebo-controlled, phase 2 trial. Lancet 383(9936):2213-2221

74. Comi G et al (2010) Oral laquinimod in patients with relapsing-remitting multiple sclerosis: 36-week double-blind active extension of the multi-centre, randomized, double-blind, parallel-group placebo-controlled study. Mult Scler 16(11):1360–1366

75. Comi G et al (2011) Phase III dose-comparison study of glatiramer acetate for multiple sclerosis. Ann Neurol 69(1):75–82

76. Comi G et al (2012) Comparison of two dosing frequencies of subcutaneous interferon beta-1a in patients with a first clinical demyelinating event suggestive of multiple sclerosis (REFLEX): a phase 3 randomised controlled trial. Lancet Neurol 11(1):33–41

77. Comi G et al (2012) Placebo-controlled trial of oral laquinimod for multiple sclerosis. N Engl J Med 366(11):1000–1009

78. Connick P et al (2011) The mesenchymal stem cells in multiple sclerosis (MSCIMS) trial protocol and baseline cohort characteristics: an open-label pre-test: post-test study with blinded outcome assessments. Trials 12:62

79. Cortesi M, Cattaneo D, Jonsdottir J (2011) Effect of kinesio taping on standing balance in subjects with multiple sclerosis: a pilot study\m{1}. NeuroRehabilitation 28(4):365–372

80. Cree BA, Kornyeyeva E, Goodin DS (2010) Pilot trial of low-dose naltrexone and quality of life in multiple sclerosis. Ann Neurol 68(2):145–150

81. de Seze M et al (2011) Transcutaneous posterior tibial nerve stimulation for treatment of the overactive bladder syndrome in multiple sclerosis: results of a multicenter prospective study. Neurourol Urodyn 30(3):306–311

82. Devonshire V et al (2012) Relapse and disability outcomes in patients with multiple sclerosis treated with fingolimod: subgroup analyses of the double-blind, randomised, placebo-controlled FREEDOMS study. Lancet Neurol 11(5):420–428

83. Edan G et al (2011) Mitoxantrone prior to interferon beta-1b in aggressive relapsing multiple sclerosis: a 3-year randomised trial. J Neurol Neurosurg Psychiatry 82(12):1344–1350

84. Finlayson M et al (2011) Randomized trial of a teleconference-delivered fatigue management program for people with multiple sclerosis. Mult Scler 17(9):1130–1140

85. Giovannoni G et al (2010) A placebo-controlled trial of oral cladribine for relapsing multiple sclerosis. N Engl J Med 362(5):416–426

86. Gold R et al (2013) Daclizumab high-yield process in relapsing-remitting multiple sclerosis (SELECT): a randomised, double-blind, placebo-controlled trial. Lancet 381(9884):2167–2175

87. Gonsette RE et al (2010) Boosting endogenous neuroprotection in multiple sclerosis: the ASsociation of Inosine and Interferon beta in relapsing-remitting Multiple Sclerosis (ASIIMS) trial. Mult Scler 16(4):455–462

88. Goodman AD et al (2010) A phase 3 trial of extended release oral dalfampridine in multiple sclerosis. Ann Neurol 68(4):494–502

89. Hayes HA, Gappmaier E, LaStayo PC (2011) Effects of high-intensity resistance training on strength, mobility, balance, and fatigue in individuals with multiple sclerosis: a randomized controlled trial. J Neurol Phys Ther 35(1):2–10

90. Hu X et al (2012) A novel PEGylated interferon beta-1a for multiple sclerosis: safety, pharmacology, and biology. J Clin Pharmacol 52(6):798–808

91. Kampman MT et al (2012) Effect of vitamin D3 supplementation on relapses, disease progression, and measures of function in persons with multiple sclerosis: exploratory outcomes from a double-blind randomised controlled trial. Mult Scler 18(8):1144–1151

92. Kapoor R et al (2010) Lamotrigine for neuroprotection in secondary progressive multiple sclerosis: a randomised, double-blind, placebo-controlled, parallel-group trial. Lancet Neurol 9(7):681–688

93. Kappos L et al (2011) Ocrelizumab in relapsing-remitting multiple sclerosis: a phase 2, randomised, placebo-controlled, multicentre trial. Lancet 378(9805):1779–1787

94. Khan O et al (2013) Three times weekly glatiramer acetate in relapsing-remitting multiple sclerosis. Ann Neurol 73(6):705–713

95. Kinkel RP et al (2012) Association between immediate initiation of intramuscular interferon beta-1a at the time of a clinically isolated syndrome and long-term outcomes: a 10-year follow-up of the Controlled High-Risk Avonex Multiple Sclerosis Prevention Study in Ongoing Neurological Surveillance. Arch Neurol 69(2):183–190

96. Krupp LB et al (2011) Multicenter randomized clinical trial of donepezil for memory impairment in multiple sclerosis. Neurology 76(17):1500–1507

97. Lanzillo R et al (2010) Atorvastatin combined to interferon to verify the efficacy (ACTIVE) in relapsing-remitting active multiple sclerosis patients: a longitudinal controlled trial of combination therapy. Mult Scler 16(4):450–454

98. Lovera JF et al (2010) Memantine for cognitive impairment in multiple sclerosis: a randomized placebo-controlled trial. Mult Scler 16(6):715–723

99. Lublin FD et al (2013) Randomized study combining interferon and glatiramer acetate in multiple sclerosis. Ann Neurol 73(3):327–340

100. Miller DH et al (2012) Firategrast for relapsing remitting multiple sclerosis: a phase 2, randomised, double-blind, placebo-controlled trial. Lancet Neurol 11(2):131–139

101. Moller F et al (2011) HAGIL (Hamburg Vigil Study): a randomized placebo-controlled double-blind study with modafinil for treatment of fatigue in patients with multiple sclerosis. Mult Scler 17(8):1002–1009

102. Pilutti LA et al (2011) Effects of 12 weeks of supported treadmill training on functional ability and quality of life in progressive multiple sclerosis: a pilot study. Arch Phys Med Rehabil 92(1):31–36

103. Radue EW et al (2012) Impact of fingolimod therapy on magnetic resonance imaging outcomes in patients with multiple sclerosis. Arch Neurol 69(10):1259–1269

104. Ravnborg M et al (2010) Methylprednisolone in combination with interferon beta-1a for relapsing-remitting multiple sclerosis (MECOMBIN study): a multicentre, double-blind, randomised, placebo-controlled, parallel-group trial. Lancet Neurol 9(7):672–680

105. Selmaj K et al (2013) Siponimod for patients with relapsing-remitting multiple sclerosis (BOLD): an adaptive, dose-ranging, randomised, phase 2 study. Lancet Neurol 12(8):756–767

106. Sorensen PS et al (2011) Simvastatin as add-on therapy to interferon beta-1a for relapsing-remitting multiple sclerosis (SIMCOMBIN study): a placebo-controlled randomised phase 4 trial. Lancet Neurol 10(8):691–701

107. Togha M et al (2010) Simvastatin treatment in patients with relapsing-remitting multiple sclerosis receiving interferon beta 1a: a double-blind randomized controlled trial. Mult Scler 16(7):848–854

108. Torkildsen O et al (2012) Omega-3 fatty acid treatment in multiple sclerosis (OFAMS Study): a randomized, double-blind, placebo-controlled trial. Arch Neurol 69(8):1044–1051

109. Wynn D et al (2010) Daclizumab in active relapsing multiple sclerosis (CHOICE study): a phase 2, randomised, double-blind, placebo-controlled, add-on trial with interferon beta. Lancet Neurol 9(4):381–390

Chapter 9

1. Rinne et al (2010) 11-C-PiB PET assessment of change in fibrillar amyloid-beta load in patients with Alzheimer's disease treated with bapineuzumab: a phase 2, double-blind, placebo-controlled, ascending-dose study

2. Craft et al (2012) Intranasal insulin therapy for Alzheimer disease and amnestic mild cognitive impairment: a pilot clinical trial

3. Quinn et al (2010) Docosahexaenoic acid supplementation and cognitive decline in Alzheimer disease: a randomized trial

4. Laxton et al (2010) A phase I trial of deep brain stimulation of memory circuits in Alzheimer's disease

5. Feldman et al (2010) Randomized controlled trial of atorvastatin in mild to moderate Alzheimer disease LEADe

6. Salloway et al (2014) Two phase 3 trials of bapineuzumab in mild-to-moderate Alzheimer's disease

7. Faux et al (2010) PBT2 rapidly improves cognition in Alzheimer's disease: additional phase II analyses

8. Doody et al (2014) Phase 3 trials of solanezumab for mild-to-moderate Alzheimer's disease

9. Howard et al (2012) Donepezil and memantine for moderate-to-severe Alzheimer's disease

10. Gold et al (2010) Rosiglitazone monotherapy in mild-to-moderate Alzheimer's disease: results from a randomized, double-blind, placebo-controlled phase III study

11. Sato et al (2011) Efficacy of PPAR-gamma agonist pioglitazone in mild Alzheimer disease

12. Doody et al (2013) A phase 3 trial of semagacestat for treatment of Alzheimer's disease

13. Ostrowitzki et al (2012) Mechanism of amyloid removal in patients with Alzheimer disease treated with gantenerumab

14. Siemers et al (2010) Safety and changes in plasma and cerebrospinal fluid amyloid beta after a single administration of an amyloid beta monoclonal antibody in subjects with Alzheimer disease

15. Blennow et al (2012) Effect of immunotherapy with bapineuzumab on cerebrospinal fluid biomarker levels in patients with mild to moderate Alzheimer disease

16. Winblad et al (2012) Safety, tolerability, and antibody response of active ABeta immunotherapy with CAD106 in patients with Alzheimer's disease: randomised, double-blind, placebo-controlled, first-in-human study

17. Scheltens et al (2010) Efficacy of a medical food in mild Alzheimer's disease: a randomized, controlled trial

18. Sano et al (2011) A randomized, double-blind, placebo-controlled trial of simvastatin to treat Alzheimer disease

19. Rosenberg et al (2010) Sertraline for the treatment of depression in Alzheimer disease

20. Farlow et al (2012) Safety and biomarker effects of Solanezumab in patients with Alzheimer's disease

21. Vellas et al (2012) Long-term use of standardised Ginkgo biloba extract for the prevention of Alzheimer's disease (GuidAge): a randomised placebo-controlled trial

22. Clare et al (2010) Goal-oriented cognitive rehabilitation for people with early-stage Alzheimer Disease: a single-blind randomized controlled trial of clinical efficacy

23. Dodel et al (2013) Intravenous immunoglobulin for treatment of mild-to-moderate Alzheimer's disease: a phase 2,

randomised, double-blind, placebo-controlled, dose-finding trial

24. Galasko et al (2012) Antioxidants for Alzheimer disease: a randomized clinical trial with cerebrospinal fluid biomarker measures

25. Salloway et al (2011) A phase 2 randomized trial of ELND005, scyllo-inositol, in mild to moderate Alzheimer disease

26. Virgen et al (2011) Cognitive effects of atypical antipsychotic medications in patients with Alzheimer's disease: outcomes from CATIE-AD

27. Geldmacher et al (2011) A randomized pilot clinical trial of the safety of pioglitazone in treatment of patients with Alzheimer disease

28. Scheltens et al (2012) Efficacy of souvenaid in mild Alzheimer's disease: results from a randomized, controlled trial

29. Black et al (2010) A single ascending dose study of bapineuzumab in patients with Alzheimer disease

30. Coric et al (2012) Safety and tolerability of the gamma-secretase inhibitor avagacestat in a phase 2 study of mild to moderate Alzheimer disease

31. Rafii et al (2011) A phase II trial of huperzine A in mild to moderate Alzheimer disease

32. Farlow et al (2010) Effectiveness and tolerability of high-dose (23 mg/d) versus standard-dose (10 mg/d) donepezil in moderate to severe Alzheimer's disease: a 24-week, randomized, double-blind study

33. Weintraub et al (2010) Sertraline for the treatment of depression in Alzheimer disease: week-24 outcomes

34. Douaud et al (2013) Preventing Alzheimer's disease-related gray matter atrophy by B-vitamin treatment

35. Coteli et al (2011) Improved language performance in Alzheimer disease following brain stimulation

36. Tariot et al (2011) Chronic divalproex sodium to attenuate agitation and clinical progression of Alzheimer disease

37. Dysken et al (2014) Effect of vitamin E and memantine on functional decline in Alzheimer disease: the TEAM-AD VA cooperative randomized trial

38. Harrington et al (2011) Rosiglitazone does not improve cognition or global function when used as adjunctive therapy to AChE inhibitors in mild-to-moderate Alzheimer's disease: two phase 3 studies

39. Baker et al (2010) Aerobic exercise improves cognition for older adults with glucose intolerance, a risk factor for Alzheimer's disease

40. Simmons-Stern et al (2010) Music as a memory enhancer in patients with Alzheimer's disease

41. Sabbagh et al (2011) PF-04494700, an oral inhibitor of receptor for advanced glycation end products (RAGE), in Alzheimer disease

42. Porsteinsson et al (2014) Effect of citalopram on agitation in Alzheimer disease: the CitAD randomized clinical trial

43. del Ser et al (2013) Treatment of Alzheimer's disease with the GSK-3 inhibitor tideglusib: a pilot study

44. Cornelli et al (2010) Treatment of Alzheimer's disease with a cholinesterase inhibitor combined with antioxidants

45. Modrego et al (2010) Memantine versus donepezil in mild to moderate Alzheimer's disease: a randomized trial with magnetic resonance spectroscopy

46. Akhondzadeh et al (2010) A 22-week, multicenter, randomized, double-blind controlled trial of Crocus sativus in the treatment of mild-to-moderate Alzheimer's disease

47. McCurry et al (2011) Increasing walking and bright light exposure to improve sleep in community-dwelling persons with Alzheimer's disease: results of a randomized, controlled trial

48. Alvarez et al (2011) Efficacy and safety of Cerebrolysin in moderate to moderately severe Alzheimer's disease: results of a randomized, double-blind, controlled trial investigating three dosages of Cerebrolysin

49. Ahmed et al (2012) Effects of low versus high frequencies of repetitive transcranial magnetic stimulation on cognitive function and cortical excitability in Alzheimer's dementia

50. Fox et al (2012) Efficacy of memantine for agitation in Alzheimer's dementia: a randomised double-blind placebo controlled trial

51. Fleisher et al (2011) Chronic divalproex sodium use and brain atrophy in Alzheimer disease

52. Maher-Edwards et al (2011) SB-742457 and donepezil in Alzheimer disease: a randomized, placebo-controlled study

53. Cummings et al (2012) Randomized, double-blind, parallel-group, 48-week study for efficacy and safety of a higher-dose rivastigmine patch (15 vs. 10 cm(2)) in Alzheimer's disease

54. Akhondzadeh et al (2010) Saffron in the treatment of patients with mild to moderate Alzheimer's disease: a 16-week, randomized and placebo-controlled trial

55. Venturelli et al (2011) Six-month walking program changes cognitive and ADL performance in patients with Alzheimer

56. Stein et al (2011) A randomized controlled trial of high-dose vitamin D2 followed by intranasal insulin in Alzheimer's disease

57. Nunes et al (2013) Microdose lithium treatment stabilized cognitive impairment in patients with Alzheimer's disease

58. Padalam et al (2010) Methylphenidate for apathy and functional status in dementia of the Alzheimer type

59. Devanand et al (2012) Relapse risk after discontinuation of risperidone in Alzheimer's disease

60. Wahlberg et al (2012) Targeted delivery of nerve growth factor via encapsulated cell biodelivery in Alzheimer disease: a technology platform for restorative neurosurgery

61. Leoutsakos et al (2012) Effects of non-steroidal anti-inflammatory drug treatments on cognitive decline vary by phase of pre-clinical Alzheimer disease: findings from the randomized controlled Alzheimer's disease anti-inflammatory prevention trial

62. Okahara et al (2010) Effects of Yokukansan on behavioral and psychological symptoms of dementia in regular treatment for Alzheimer's disease

63. Penner et al (2010) Increased glutamate in the hippocampus after galantamine treatment for Alzheimer disease

64. Vreugdenhil et al (2012) A community-based exercise programme to improve functional ability in people with Alzheimer's disease: a randomized controlled trial

65. Burns et al (2011) A double-blind placebo-controlled randomized trial of Melissa officinalis oil and donepezil for the treatment of agitation in Alzheimer's disease

66. Buschert et al (2011) Effects of a newly developed cognitive intervention in amnestic mild cognitive impairment and mild Alzheimer's disease: a pilot study

67. Richard et al (2010) Vascular care in patients with Alzheimer disease with cerebrovascular lesions slows progression of white matter lesions on MRI: the evaluation of vascular care in Alzheimer's disease (EVA) study

68. Padala et al (2012) The effect of HMG-CoA reductase inhibitors on cognition in patients with Alzheimer's dementia: a prospective withdrawal and rechallenge pilot study

69. Yagueez et al (2011) The effects on cognitive functions of a movement-based intervention in patients with Alzheimer's type dementia: a pilot study

70. Wharton et al (2011) Short-term hormone therapy with transdermal estradiol improves cognition for postmenopausal women with Alzheimer's disease: results of a randomized controlled trial

71. Cumbo et al (2010) Levetiracetam, lamotrigine, and phenobarbital in patients with epileptic seizures and Alzheimer's disease

72. Farlow et al (2010) A 25-week, open-label trial investigating rivastigmine transdermal patches with concomitant memantine in mild-to-moderate Alzheimer's disease: a post hoc analysis

73. Pitkala et al (2013) Effects of the Finnish Alzheimer Disease Exercise Trial (FINALEX): a randomized controlled trial

74. Hayashi et al (2010) Treatment of behavioral and psychological symptoms of Alzheimer-type dementia with Yokukansan in clinical practice

75. Waldorff et al (2012) Efficacy of psychosocial intervention in patients with mild Alzheimer's disease: the multicentre, rater blinded, randomised Danish Alzheimer Intervention Study (DAISY)

76. Nathan et al (2013) The safety, tolerability, pharmacokinetics and cognitive effects of GSK239512, a selective histamine H-3 receptor antagonist in patients with mild to moderate Alzheimer's disease: a preliminary investigation

77. Kennelly et al (2011) Demonstration of safety in Alzheimer's patients for intervention with an anti-hypertensive drug Nilvadipine: results from a 6-week open label study

78. Serrano-Pozo et al (2010) Effects of simvastatin on cholesterol metabolism and Alzheimer disease biomarkers

79. Nourhashemi et al (2010) Effectiveness of a specific care plan in patients with Alzheimer's disease: cluster randomised trial (PLASA study)

80. Valen-Sendstad et al (2010) Effects of hormone therapy on depressive symptoms and cognitive functions in women with Alzheimer disease: a 12 month randomized, double-blind, placebo-controlled study of low-dose estradiol and norethisterone

81. Ihl et al (2012) Efficacy and tolerability of a once daily formulation of Ginkgo biloba extract EGb 761 (R) in Alzheimer's disease and vascular dementia: results from a randomised controlled trial

82. Alvarez et al (2011) Combination treatment in Alzheimer's disease: results of a randomized, controlled trial with cerebrolysin and donepezil

83. Shinto et al (2014) A randomized placebo-controlled pilot trial of omega-3 fatty acids and alpha lipoic acid in Alzheimer's disease

84. Devanand et al (2011) A 6-month, randomized, double-blind, placebo-controlled pilot discontinuation trial following response to haloperidol treatment of psychosis and agitation in Alzheimer's disease

85. Kamphuis et al (2011) Efficacy of a medical food on cognition in Alzheimer's disease: results from secondary analyses of a randomized, controlled trial

86. Lorenzi et al (2011) Effect of memantine on resting state default mode network activity in Alzheimer's disease

87. Lancioni et al (2010) Technology-aided verbal instructions to help persons with mild or moderate Alzheimer's disease perform daily activities

88. Landen et al (2013) Safety and pharmacology of a single intravenous dose of ponezumab in subjects with mild-to-moderate Alzheimer disease: a phase I, randomized, placebo-controlled, double-blind, dose-escalation study

89. Kume et al (2012) Effects of telmisartan on cognition and regional cerebral blood flow in hypertensive patients with Alzheimer's disease

90. Froelich et al (2011) Effects of AZD3480 on cognition in patients with mild-to-moderate Alzheimer's disease: a phase IIb dose-finding study

91. Winblad et al (2010) Phenserine efficacy in Alzheimer's disease

92. Kurz et al (2012) CORDIAL: cognitive rehabilitation and cognitive-behavioral treatment for early dementia in Alzheimer disease

93. Uenaka et al (2012) Comparison of pharmacokinetics, pharmacodynamics, safety, and tolerability of the amyloid beta monoclonal antibody solanezumab in Japanese and white patients with mild to moderate Alzheimer disease

94. Roach et al (2011) A randomized controlled trial of an activity specific exercise program for individuals with Alzheimer disease in long-term care settings

95. Farlow et al (2011) Safety and tolerability of donepezil 23 mg in moderate to severe Alzheimer's disease

96. Niu et al (2010) Cognitive stimulation therapy in the treatment of neuropsychiatric symptoms in Alzheimer's disease: a randomized controlled trial

97. Rabey et al (2013) Repetitive transcranial magnetic stimulation combined with cognitive training is a safe and effective

modality for the treatment of Alzheimer's disease: a random-ized, double-blind study

98. Claxton et al (2013) Sex and ApoE genotype differences in treatment response to two doses of intranasal insulin in adults with mild cognitive impairment or Alzheimer's disease

99. Sakurai et al (2013) Effects of cilostazol on cognition and regional cerebral blood flow in patients with Alzheimer's disease and cerebrovascular disease: a pilot study

100. Munro (2012) Cognitive outcomes after sertaline treatment in patients with depression of Alzheimer disease

Chapter 10

1. Wen et al (2008) Phase II study of imatinib mesylate for recurrent meningiomas

2. Norden et al (2010) Phase II trials of erlotinib or gefitinib in patients with recurrent meningioma

3. Astner et al (2008) Effect of 11C-methionine-positron emission tomography on gross tumor volume delineation in stereotactic radiotherapy of skull base meningiomas

4. Chamberlain MC, Glantz MJ (2008) Interferon-afor recurrent World Health Organization grade 1 intracranial meningiomas

5. Bartolomei et al (2009) Peptide receptor radionuclide therapy with Y-90-DOTATOC in recurrent meningioma

6. Colombo et al (2009) Cybernife radiosurgery for benign meningiomas: short-term results in 199 patients

7. Combs et al (2010) Carbon ion radiation therapy for high-risk meningiomas

8. Raizer et al (2010) A phase I trial of erlotinib in patients with nonprogressive glioblastoma multiforme postradiation therapy, and recurrent malignant gliomas and meningiomas

9. Jonhson et al (2010) Phase II study of subcutaneous octreotide in adults with recurrent or progressive meningioma and meningeal hemangiopericytoma

10. Reardon et al (2012) Phase II study of Gleevec® plus hydroxyurea (HU) in adults with progressive or recurrent meningioma

11. Minniti et al (2011) Fractionated stereotactic conformal radiotherapy for large benign skull base meningiomas

12. Douglas et al (2009) Stereotactic radiosurgery for radiation-induced meningiomas

13. Combs et al (2010) Treatment of patients with atypical meningiomas Simpson grade 4 and 5 with a carbon ion boost in combination with postoperative photon radiotherapy: the MARCIE trial

14. Kreissl et al (2012) Combination of peptide receptor radionuclide therapy with fractionated external beam radiotherapy for treatment of advanced symptomatic meningioma

15. Sluzewski et al (2013) Embolization of meningiomas: comparison of safety between calibrated microspheres and polyvinyl-alcohol particles as embolic agents

16. Trippa et al (2009) Hypofractionated stereotactic radiotherapy for intracranial meningiomas: preliminary results of a feasible trial

17. Raizer et al (2014) A phase II trial of PTK787/ZK 222584 in recurrent or progressive radiation and surgery refractory meningiomas

18. Tang et al (2011) The effect of hyperbaric oxygen on clinical outcome of patients after resection of meningiomas with conspicuous peritumoral brain edema

19. Kawaji et al (2013) Evaluation of tumor blood flow after feeder embolization in meningiomas by arterial spin-labeling perfusion magnetic resonance imaging

20. Simo et al (2014) Recurrent high-grade meningioma: a phase II trial with somatostatin analogue therapy

21. Stupp et al (2009) Effects of radiotherapy with concomitant and adjuvant temozolomide versus radiotherapy alone on survival in glioblastoma in a randomised phase III study: 5-year analysis of the EORTC-NCIC trial

22. Friedman et al (2009) Bevacizumab alone and in combination with irinotecan in recurrent glioblastoma

23. Kreisl et al (2009) Phase II trial of single-agent bevacizumab followed by bevacizumab plus irinotecan at tumor progression in recurrent glioblastoma

24. Stummer et al (2008) Extent of resection and survival on glioblastoma multiforme-identification of and adiustment for bias

25. Reardon et al (2008) Randomized phase II study of cilengitide, an integrin-targeting arginine-glycine-aspartic acid peptide, in recurrent glioblastoma multiforme

26. Van Den Bent et al (2009) Randomized phase II trial of erlotinib versus temozolomide or carmustine in recurrent glioblastoma: EORTC Brain Tumor Group Study 26034

27. Batchelor et al (2010) Phase II study of cediranib, an oral panvascular endothelial growth factor receptor tyrosine kinase inhibitor, in patients with recurrent glioblastoma

28. Cloughesy et al (2008) Antitumor activity of rapamycin in a phase I trial for patients with recurrent PTEN-deficient glioblastoma

29. Sampson et al (2010) Immunologic escape after prolonged progression-free survival with epidermal growth factor receptor variant III peptide vaccination in patients with newly diagnosed glioblastoma

30. Gorlia et al (2008) Nomograms for predicting survival of patients with newly diagnosed glioblastoma: prognostic factor analysis of EORTC and NCIC trial 26981-22981/CE.3

31. Strupp et al (2010) Phase I/IIa study of cilengitide and temozolomide with concomitant radiotherapy followed by cilengitide and temozolomide maintenance therapy in patients with newly diagnosed glioblastoma

32. Gilbert et al (2014) A randomized trial of bevacizumab for newly diagnosed glioblastoma

33. Lai et al (2011) Phase II study of bevacizumab plus temozolomide during and after radiation therapy for patients with newly diagnosed glioblastoma multiforme

34. Prados et al (2009) Phase II study of erlotinib plus temozolomide during and after radiation therapy in patients with newly diagnosed glioblastoma multiforme or gliosarcoma

35. Chinot et al (2014) Bevacizumab plus radiotherapy-temozolomide for newly diagnosed glioblastoma

36. Malmstrom et al (2012) Temozolomide versus standard 6-week radiotherapy versus hypofractionated radiotherapy in patients older than 60 years with glioblastoma: the Nordic randomised, phase 3 trial

37. Malmström et al (2012) Temozolomide versus standard 6-week radiotherapy versus hypofractionated radiotherapy in patients older than 60 years with glioblastoma: the Nordic randomised, phase 3 trial

38. Maier-Hauff et al (2010) Efficacy and safety of intratumoral thermotherapy using magnetic iron-oxide nanoparticles combined with external beam radiotherapy on patients with recurrent glioblastoma multiforme

39. Wick et al (2010) Phase III study of enzastaurin compared with lomustine in the treatment of recurrent intracranial glioblastoma

40. Galanis et al (2009) Phase II trial of vorinostat in recurrent glioblastoma multiforme: a North Central Cancer Treatment Group Study

41. Wheeler et al (2008) Vaccination elicits correlated immune and clinical responses in glioblastoma multiforme patients

42. Brown et al (2008) Phase I/II trial of erlotinib and temozolomide with radiation therapy in the treatment of newly diagnosed glioblastoma multiforme: North Central Cancer Treatment Group Study N0177

43. Lai et al (2008) Phase II pilot study of bevacizumab in combination with temozolomide and regional radiation therapy for up-front treatment of patients with newly diagnosed glioblastoma multiforme: interim analysis of safety and tolerability

44. Pichlmeier et al (2008) Resection and survival in glioblastoma multiforme: an RTOG recursive partitioning analysis of ALA study patients

45. Kunwar et al (2010) Phase III randomized trial of CED of IL13-PE38QQR vs. Gliadel wafers for recurrent glioblastoma

46. Prins et al (2011) Gene expression profile correlates with T-cell infiltration and relative survival in glioblastoma patients vaccinated with dendritic cell immunotherapy

47. Raizer et al (2010) A phase II trial of erlotinib in patients with recurrent malignant gliomas and nonprogressive glioblastoma multiforme postradiation therapy

48. Reardon et al (2010) Phase 2 trial of erlotinib plus sirolimus in adults with recurrent glioblastoma

49. Sampson et al (2009) An epidermal growth factor receptor variant III-targeted vaccine is safe and immunogenic in patients with glioblastoma multiforme

50. Izumoto et al (2008) Phase II clinical trial of Wilms tumor 1 peptide vaccination for patients with recurrent glioblastoma multiforme

51. Iwamoto et al (2010) Phase II trial of pazopanib (GW786034), an oral multi-targeted angiogenesis inhibitor, for adults with recurrent glioblastoma (North American Brain Tumor Consortium Study 06-02)

52. Hasselbalch et al (2010) Cetuximab, bevacizumab, and irinotecan for patients with primary glioblastoma and progression after radiation therapy and temozolomide: a phase II trial

53. Gilbert et al (2013) Dose-dense temozolomide for newly diagnosed glioblastoma: a randomized phase III clinical trial

54. Sorensen et al (2011) Increased survival of glioblastoma patients who respond to antiangiogenic therapy with elevated blood perfusion

55. Vredenburgh et al (2011) The addition of bevacizumab to standard radiation therapy and temozolomide followed by bevacizumab, temozolomide, and irinotecan for newly diagnosed glioblastoma

56. Thiessen et al (2010) A phase I/II trial of W572016 (lapatinib) in recurrent glioblastoma multiforme: clinical outcomes, pharmacokinetics and molecular correlation

57. Peereboom et al (2010) Phase II trial of erlotinib with temozolomide and radiation in patients with newly diagnosed glioblastoma multiforme

58. Wen et al (2011) A phase II study evaluating the efficacy and safety of AMG 102 (rilotumumab) in patients with recurrent glioblastoma

59. Prins et al (2008) Cytomegalovirus immunity after vaccination with autologous glioblastoma lysate

60. Reardon et al (2009) Multicentre phase II studies evaluating imatinib plus hydroxyurea in patients with progressive glioblastoma

61. Drappatz et al (2010) Phase I study of vandetanib with radiotherapy and temozolomide for newly diagnosed glioblastoma

62. Fadul et al (2011) Immune response in patients with newly diagnosed glioblastoma multiforme treated with intranodal autologous tumor lysate-dendritic cell vaccination after radiation chemotherapy

63. Phuphanich et al (2013) Phase I trial of a multi-epitope-pulsed dendritic cell vaccine for patients with newly diagnosed glioblastoma

64. Reardon et al (2012) Phase II study of carboplatin, irinotecan, and bevacizumab for bevacizumab naive, recurrent glioblastoma

65. Desjardins et al (2011) Bevacizumab and daily temozolomide for recurrent glioblastoma

66. Yung et al (2010) Safety and efficacy of erlotinib in first-relapse glioblastoma: a phase II open label study

67. Butowski et al (2009) A phase II clinical trial of poly-ICLC with radiation for adult patients with newly diagnosed supratentorial glioblastoma: a North American Brain Tumor Consortium (NABTC01-05)

68. Razis et al (2009) Phase II study of neoadjuvant imatinib in glioblastoma: evaluation of clinical and molecular effects of the treatment

69. Lee et al (2009) Association of c-11-methionine pet uptake with site of failure after concurrent temozolomide and radiation for primary glioblastoma multiforme

70. Nabors et al (2012) A safety run-in and randomized phase 2 study of cilengitide combined with chemoradiation for newly diagnosed glioblastoma (NABTT 0306)

71. Brandes et al (2009) Phase I/II study on concomitant and adjuvant temozolomide (TMZ) and radiotherapy (RT) with PTK787/ZK222584 (PTK/ZK) in newly diagnosed glioblastoma

72. Pope et al (2011) Patterns of progression in patients with recurrent glioblastoma treated with bevacizumab

73. Kong et al (2010) Phase II trial of low-dose continuous (metronomic) treatment of temozolomide for recurrent glioblastoma

74. Brandes et al (2009) Fotemustine as second-line treatment for recurrent or progressive glioblastoma after concomitant and/or adjuvant temozolomide: a phase II trial of Gruppo Italiano Cooperativo di Neuro-Oncologia (GICNO)

75. Ardon et al (2012) Integration of autologous dendritic cell-based immunotherapy in the standard of care treatment for patients with newly diagnosed glioblastoma: results of the HGG-2006 phase I/II trial

76. Minniti et al (2008) Hypofractionated radiotherapy followed by adjuvant chemotherapy with temozolomide in elderly patients with glioblastoma

77. Henrikssona et al (2008) Boron neutron capture therapy (BNCT) for glioblastoma multiforme: a phase II study evaluating a prolonged high-dose of boronophenylalanine (BPA)

78. Minniti et al (2011) Phase II study of short-course radiotherapy plus concomitant and adjuvant temozolomide in elderly patients with glioblastoma

79. Carpentier et al (2010) Intracerebral administration of CpG oligonucleotide for patients with recurrent glioblastoma: a phase II study

80. Tsien et al (2009) Phase I three-dimensional conformal radiation dose escalation study in newly diagnosed glioblastoma: radiation therapy oncology group trial 98-03

81. Kesari et al (2008) Phase II study of temozolomide, thalidomide, and celecoxib for newly diagnosed glioblastoma in adults

82. Kawabata et al (2009) Boron neutron capture therapy for newly diagnosed glioblastoma

83. Scoccianti et al (2008) Second-line chemotherapy with fotemustine in temozolomide-pretreated patients with relapsing glioblastoma: a single institution experience

84. Puduvalli et al (2008) Phase II trial of irinotecan and thalidomide in adults with recurrent glioblastoma multiforme

85. Laprie et al (2008) Proton magnetic resonance spectroscopic imaging in newly diagnosed glioblastoma: predictive value for

the site of postradiotherapy relapse in a prospective longitudinal study

86. Stupp et al (2014) Cilengitide combined with standard treatment for patients with newly diagnosed glioblastoma with methylated MGMT promoter (CENTRIC EORTC 26071-22072 study): a multicentre, randomised, open-label, phase 3 trial

87. Fabrini et al (2009) A multi-institutional phase II study on second-line Fotemustine chemotherapy in recurrent glioblastoma

88. Taal et al (2014) Single-agent bevacizumab or lomustine versus a combination of bevacizumab plus lomustine in patients with recurrent glioblastoma (BELOB trial): a randomised controlled phase 2 trial

89. Stragliotto et al (2013) Effects of valganciclovir as an add-on therapy in patients with cytomegalovirus-positive glioblastoma: a randomized, double-blind, hypothesis-generating study

90. Geletneky et al (2012) Phase I/IIa study of intratumoral/intracerebral or intravenous/intracerebral administration of Parvovirus H-1 (ParvOryx) in patients with progressive primary or recurrent glioblastoma multiforme: ParvOryx01 protocol

Chapter 11

1. Feron F et al (2005) Autologous olfactory ensheathing cell transplantation in human spinal cord injury

2. Fregni et al (2006) A sham-controlled, phase II trial of transcranial direct current stimulation for the treatment of central pain in traumatic spinal cord injury

3. Carlos Lima MD et al (2005) Olfactory mucosa autografts in human spinal cord injury: a pilot clinical study

4. Yoon SH et al (2007) Complete spinal cord injury treatment using autologous bone marrow cell transplantation and bone marrow stimulation with granulocyte macrophage-colony stimulating factor: phase I/II clinical trial

5. Siddall PJ et al (2006) Pregabalin in central neuropathic pain associated with spinal cord injury - a placebo-controlled trial

6. Thomas SL, Gorassini MA (2005) Increases in corticospinal tract function by treadmill training after incomplete spinal cord injury

7. Knoller N et al (2005) Clinical experience using incubated autologous macrophages as a treatment for complete spinal cord injury: phase I study results

8. Dobkin B et al (2007) The evolution of walking-related outcomes over the first 12 weeks of rehabilitation for incomplete traumatic spinal cord injury: the multicenter randomized Spinal Cord Injury Locomotor Trial

9. Winchester P et al (2005) Changes in supraspinal activation patterns following robotic locomotor therapy in motor-incomplete spinal cord injury

10. Pal R et al (2009) Ex vivo-expanded autologous bone marrow-derived mesenchymal stromal cells in human spinal cord injury/paraplegia: a pilot clinical study

11. Lima et al (2010) Olfactory mucosal autografts and rehabilitation for chronic traumatic spinal cord injury

12. Edelle et al (2011) Influence of a locomotor training approach on walking speed and distance in people with chronic spinal cord injury: a randomized clinical trial

13. Fehlings et al (2011) Phase I/IIa clinical trial of a recombinant Rho protein antagonist in acute spinal cord injury

14. Soler MD et al (2010) Effectiveness of transcranial direct current stimulation and visual illusion on neuropathic pain in spinal cord injury

15. Casha et al (2012) Results of a phase II placebo-controlled randomized trial of minocycline in acute spinal cord injury

16. Harkema et al (2012) Balance and ambulation improvements in individuals with chronic incomplete spinal cord injury using locomotor training-based rehabilitation

17. Karamouzian et al (2012) Clinical safety and primary efficacy of bone marrow mesenchymal cell transplantation in subacute spinal cord injured patients

18. Trumbower et al (2012) Exposure to acute intermittent hypoxia augments somatic motor function in humans with incomplete spinal cord injury

19. Cardenas et al (2013) A randomized trial of pregabalin in patients with neuropathic pain due to spinal cord injury

20. Lammertse DP et al (2012) Autologous incubated macrophage therapy in acute, complete spinal cord injury: results of the phase 2 randomized controlled multicenter trial

21. Vranken et al (2011) Duloxetine in patients with central neuropathic pain caused by spinal cord injury or stroke: a randomized, double-blind, placebo-controlled trial

22. Mulroy et al (2011) Strengthening and optimal movements for painful shoulders (STOMPS) in chronic spinal cord injury: a randomized controlled trial

23. Alexeeva et al (2011) Comparison of training methods to improve walking in persons with chronic spinal cord injury: a randomized clinical trial

24. Hayes et al (2014) Daily intermittent hypoxia enhances walking after chronic spinal cord injury: a randomized trial

25. Kumru et al (2010) Reduction of spasticity with repetitive transcranial magnetic stimulation in patients with spinal cord injury

26. Tabakow et al (2013) Transplantation of autologous olfactory ensheathing cells in complete human spinal cord injury

27. Heutink et al (2012) The CONECSI trial: results of a randomized controlled trial of a multidisciplinary cognitive behavioral program for coping with chronic neuropathic pain after spinal cord injury

28. Kuppuswamy et al (2011) Action of 5 Hz repetitive transcranial magnetic stimulation on sensory, motor and autonomic function in human spinal cord injury

29. Houldin A et al (2011) Locomotor adaptations and aftereffects to resistance during walking in individuals with spinal cord injury

30. Cardenas et al (2011) Intermittent catheterization with a hydrophilic-coated catheter delays urinary tract infections in acute spinal cord injury: a prospective, randomized, multicenter trial

31. Jan et al (2010) Effect of wheelchair tilt-in-space and recline angles on skin perfusion over the ischial tuberosity in people with spinal cord injury

32. Wu et al (2012) Robotic resistance treadmill training improves locomotor function in human spinal cord injury: a pilot study

33. Bubbear et al (2011) Early treatment with zoledronic acid prevents bone loss at the hip following acute spinal cord injury

34. Groah et al (2010) Intensive electrical stimulation attenuates femoral bone loss in acute spinal cord injury

35. Rintala et al (2010) Effect of dronabinol on central neuropathic pain after spinal cord injury

36. Ottomanelli et al (2012) Effectiveness of supported employment for veterans with spinal cord injuries: results from a randomized multisite study

37. Norrbrink et al (2011) Acupuncture and massage therapy for neuropathic pain following spinal cord injury: an exploratory study

38. Kapadia et al (2011) Functional electrical stimulation therapy for grasping in traumatic incomplete spinal cord injury: randomized control trial

39. Pooyania et al (2010) A randomized, double-blinded, cross-over pilot study assessing the effect of nabilone on spasticity in persons with spinal cord injury

40. Benito-Penalva et al (2012) Gait training in human spinal cord injury using electromechanical systems: effect of device type and patient characteristics

41. Zariffa, J et al (2012) Feasibility and efficacy of upper limb robotic rehabilitation in a subacute cervical spinal cord injury population

42. Johnston et al (2011) Muscle changes following cycling and/or electrical stimulation in pediatric spinal cord injury

43. Bauman et al (2011) A small-scale clinical trial to determine the safety and efficacy of testosterone replacement therapy in hypogonadal men with spinal cord injury

44. Coggrave et al (2010) The need for manual evacuation and oral laxatives in the management of neurogenic bowel dysfunction after spinal cord injury: a randomized controlled trial of a stepwise protocol

45. Lauer et al (2011) Effects of cycling and/or electrical stimulation on bone mineral density in children with spinal cord injury

46. Chen et al (2011) Therapeutic effects of detrusor botulinum toxin A injection on neurogenic detrusor overactivity in patients with different levels of spinal cord injury and types of detrusor sphincter dyssynergia

47. Alexander et al (2011) Sildenafil in women with sexual arousal disorder following spinal cord injury

48. Giangregorio et al (2012) A randomized trial of functional electrical stimulation for walking in incomplete spinal cord injury: effects on body composition

49. Domurath et al (2011) Clinical evaluation of a newly developed catheter (SpeediCath Compact Male) in men with spinal cord injury: residual urine and user evaluation

50. Tan et al (2011) Efficacy of cranial electrotherapy stimulation for neuropathic pain following spinal cord injury: a multi-site randomized controlled trial with a secondary 6-month open-label phase

51. Roth et al (2010) Expiratory muscle training in spinal cord injury: a randomized controlled trial

52. Wrigley et al (2013) Longstanding neuropathic pain after spinal cord injury is refractory to transcranial direct current stimulation: a randomized controlled trial

53. Yang et al (2012) Efficacy and safety of lithium carbonate treatment of chronic spinal cord injuries: a double-blind, randomized, placebo-controlled clinical trial

54. Adams et al (2011) Comparison of the effects of body-weight-supported treadmill training and tilt-table standing on spasticity in individuals with chronic spinal cord injury

55. Boswell-ruys et al (2010) Training unsupported sitting in people with chronic spinal cord injuries: a randomized controlled trial

56. Takahashi et al (2012) Neuroprotective therapy using granulocyte colony-stimulating factor for acute spinal cord injury: a phase I/IIa clinical trial

57. Arienti et al (2011) Osteopathic manipulative treatment is effective on pain control associated to spinal cord injury

58. Knikou et al (2014) Locomotor training improves premotoneuronal control after chronic spinal cord injury

59. Celik et al (2013) The effect of low-frequency TENS in the treatment of neuropathic pain in patients with spinal cord injury

60. Jan, Y-K et al (2013) Effect of durations of wheelchair tilt-in-space and recline on skin perfusion over the ischial tuberosity in people with spinal cord injury

61. Silver, J et al (2012) Barriers for individuals with spinal cord injury returning to the community: a preliminary classification

62. Wu J et al (2012) Clinical observation of fetal olfactory ensheathing glia transplantation (OEGT) in patients with complete chronic spinal cord injury

63. Bassett, RL et al (2011) Risky business: the effects of an individualized health information intervention on health risk perceptions and leisure time physical activity among people with spinal cord injury

64. Gonzalez Sarasua, J. et al (2011) Treatment of pressure ulcers with autologous bone marrow nuclear cells in patients with spinal cord injury

65. Wong YW et al (2011) A three-month, open-label, single-arm trial evaluating the safety and pharmacokinetics of oral lithium in patients with chronic spinal cord injury

66. Harvey LA et al (2010) Electrical stimulation plus progressive resistance training for leg strength in spinal cord injury: a randomized controlled trial

67. Moldenhauer S et al (2010) Mobilization of CD133(+) CD34(−) cells in healthy individuals following whole-body acupuncture for spinal cord injuries

68. Hofstoetter US et al (2014) Modification of spasticity by transcutaneous spinal cord stimulation in individuals with incomplete spinal cord injury

69. Jette F et al (2013) Effect of single-session repetitive transcranial magnetic stimulation applied over the hand versus leg motor area on pain after spinal cord injury

70. Triolo RJ et al (2013) Effects of stimulating hip and trunk muscles on seated stability, posture, and reach after spinal cord injury

71. Hasnan N et al (2013) Exercise responses during functional electrical stimulation cycling in individuals with spinal cord injury

72. Chang Y-J et al (2013) Effects of continuous passive motion on reversing the adapted spinal circuit in humans with chronic spinal cord injury

73. Nussbaum EL et al (2013) Ultraviolet-C irradiation in the management of pressure ulcers in people with spinal cord injury: a randomized, placebo-controlled trial

74. Salinas FA et al (2012) Efficacy of early treatment with carbamazepine in prevention of neuropathic pain in patients with spinal cord injury

75. Khorrami MH et al (2010) Sildenafil efficacy in erectile dysfunction secondary to spinal cord injury depends on the level of cord injuries

76. Furmaniuk L et al (2010) Influence of long-term wheelchair rugby training on the functional abilities of persons with tetraplegia over a two-year period post-spinal cord injury

77. Rosety-Rodriguez M et al (2014) Low-grade systemic inflammation and leptin levels were improved by arm cranking exercise in adults with chronic spinal cord injury

78. Cardenas DD et al (2014) Two phase 3, multicenter, randomized, placebo-controlled clinical trials of fampridine-SR for treatment of spasticity in chronic spinal cord injury

79. McBain RA et al (2013) Abdominal muscle training can enhance cough after spinal cord injury

80. Kressler J et al (2013) Metabolic responses to 4 different body weight-supported locomotor training approaches in persons with incomplete spinal cord injury

81. Jan Y-K et al (2013) Wheelchair tilt-in-space and recline does not reduce sacral skin Perfusion as changing from the upright to the tilted and reclined position in people with spinal cord injury

82. Jia C et al (2013) Detrusor botulinum toxin A injection significantly decreased urinary tract infection in patients with traumatic spinal cord injury

83. Arija-Blazquez A et al (2013) Time-course response in serum markers of bone turnover to a single-bout of electrical stimulation in patients with recent spinal cord injury

84. Wadsworth BM et al (2013) Abdominal binder improves lung volumes and voice in people with tetraplegic spinal cord injury

85. Lee Y-H et al (2012) The effect of semiconditional dorsal penile nerve electrical stimulation on capacity and compliance of the bladder with deformity in spinal cord injury patients: a pilot study

86. Li Q et al (2011) High-dose ambroxol reduces pulmonary complications in patients with acute cervical spinal cord injury after surgery

87. Yamanaka M et al (2010) Impaired immune response to voluntary arm-crank ergometer exercise in patients with cervical spinal cord injury

88. Yilmaz B et al (2014) The effect of repetitive transcranial magnetic stimulation on refractory neuropathic pain in spinal cord injury

89. Yoon EJ et al (2014) Transcranial direct current stimulation to lessen neuropathic pain after spinal cord injury: a mechanistic PET study

90. Fleerkotte BM et al (2014) The effect of impedance-controlled robotic gait training on walking ability and quality in individuals with chronic incomplete spinal cord injury: an explorative study

91. Apostolidis A et al (2012) An exploratory, placebo-controlled, dose-response study of the efficacy and safety of onabotulinumtoxinA in spinal cord injury patients with urinary incontinence due to neurogenic detrusor overactivity

92. Astorino TA et al (2013) Effect of chronic activity-based therapy on bone mineral density and bone turnover in persons with spinal cord injury

93. Ordonez FJ et al (2013) Arm-cranking exercise reduced oxidative damage in adults with chronic spinal cord injury

94. Rice LA et al (2013) Impact of the clinical practice guideline for preservation of upper limb function on transfer skills of persons with acute spinal cord injury

95. Bakkum AJ et al (2013) The effects of hybrid cycle training in inactive people with long-term spinal cord injury: design of a multicenter randomized controlled trial

96. Fornusek C et al (2013) Pilot study of the effect of low-cadence functional electrical stimulation cycling after spinal cord injury on thigh girth and strength

97. Thompson CK et al (2012) Divergent modulation of clinical measures of volitional and reflexive motor behaviors following serotonergic medications in human incomplete spinal cord injury

98. Lavado EL et al (2012) Effectiveness of aerobic physical training for treatment of chronic asymptomatic bacteriuria in subjects with spinal cord injury: a randomized controlled trial

99. Hoekstra F et al (2013) Effect of robotic gait training on cardiorespiratory system in incomplete spinal cord injury

100. Richardson EJ et al (2012) Effects of nicotine on spinal cord injury pain vary among subtypes of pain and smoking status: results from a randomized, controlled experiment

101. Dorstyn et al (2012) Effectiveness of telephone counseling in managing psychological outcomes after spinal cord injury: a preliminary study

102. Lovell D et al (2012) The aerobic performance of trained and untrained handcyclists with spinal cord injury

103. Julia et al (2011) Benefit of triple-strap abdominal binder on voluntary cough in patients with spinal cord injury

104. Theiss et al (2011) Riluzole decreases flexion withdrawal reflex but not voluntary ankle torque in human chronic spinal cord injury

105. Grijalva et al (2010) High doses of 4-aminopyridine improve functionality in chronic complete spinal cord injury patients with MRI evidence of cord continuity

106. Da Slva et al (2010) Is sperm cryopreservation an option for fertility preservation in patients with spinal cord injury-induced anejaculation?

107. Leech et al (2014) Effects of serotonergic medications on locomotor performance in humans with incomplete spinal cord injury

108. Müller-Putz GR et al (2014) Motor imagery-induced EEG patterns in individuals with spinal cord injury and their impact on brain-computer interface accuracy

109. El-kheir WA et al (2014) Autologous bone marrow-derived cell therapy combined with physical therapy induces functional improvement in chronic spinal cord injury patients

110. Inada T et al (2014) Multicenter prospective nonrandomized controlled clinical trial to prove neurotherapeutic effects of granulocyte colony-stimulating factor for acute spinal cord injury

Chapter 12

1. Rosenstock et al (2004) Pregabalin for the treatment of painful diabetic peripheral neuropathy: a double-blind, placebo-controlled trial

2. Richardson et al (2006) A phase 2 study of bortezomib in relapsed, refractory myeloma

3. Richardson et al (2006) Frequency, characteristics, and reversibility of peripheral neuropathy during treatment of advanced multiple myeloma with bortezomib

4. Richardson et al (2006) Bortezomib-induced peripheral neuropathy in multiple myeloma: a comparison between previously treated and untreated patients

5. Richter RW et al (2005) Relief of painful diabetic peripheral neuropathy with pregabalin: a randomized, placebo-controlled trial

6. Richardson PG et al (2009) Reversibility of symptomatic peripheral neuropathy with bortezomib in the phase III APEX trial in relapsed multiple myeloma: impact of a dose-modification guideline

7. Hotta et al (2001) Clinical efficacy of fidarestat, a novel aldose reductase inhibitor, for diabetic peripheralneuropathy - a 52-week multicenter placebo-controlled double-blind parallel group study

8. Hotta et al (2006) Long-term clinical effects of epalrestat, an aldose reductase inhibitor, on diabetic peripheralneuropathy - the 3-year, multicenter, comparative aldose reductase inhibitor-diabetes complications trial

9. Rao et al (2007) Efficacy of gabapentin in the management of chemotherapy-induced peripheral neuropathy - a phase 3 randomized, double-blind, placebo-controlled, crossover trial (NOOC3)

10. Smith et al (2013) Effect of duloxetine on pain, function, and quality of life among patients with chemotherapy-induced painful peripheral neuropathy: a randomized clinical trial

11. Hammack et al (2002) Phase III evaluation of nortriptyline for alleviation of symptoms of cis-platinum-inducedperipheral neuropathy

12. Broyl et al (2010) Mechanisms of peripheral neuropathy associated with bortezomib and vincristine in patients with newly diagnosed multiple myeloma: a prospective analysis of data from the HOVON-65/GMMG-HD4 trial

13. Vinik et al (2005) Treatment of symptomatic diabetic peripheral neuropathy with the protein kinase C beta-inhibitor ruboxistaurin mesylate during a 1-year, randomized, placebo-controlled, double-blind clinical trial

14. Schwartz et al (2011) Safety and efficacy of tapentadol ER in patients with painful diabetic peripheral neuropathy: results of a randomized-withdrawal, placebo-controlled trial

15. Vahdat et al (2001) Reduction of paclitaxel-induced peripheral neuropathy with glutamine

16. Rao et al (2008) Efficacy of lamotrigine in the management of chemotherapy-induced peripheral neuropathy - a phase 3 randomized, double-blind, placebo-controlled trial, N01C3

17. Paice et al (2000) Topical capsaicin in the management of HIV-associated peripheral neuropathy

18. Leonard DR et al (2004) Restoration of sensation, reduced pain, and improved balance in subjects with diabeticperipheral neuropathy

19. Arezzo et al (2008) Efficacy and safety of pregabalin 600 mg/d for treating painful diabetic peripheral neuropathy: a double-blind placebo-controlled trial

20. Verstappen et al (2005) Dose-related vincristine-induced peripheral neuropathy with unexpected off-therapy worsening

21. Mielke et al (2005) Association of paclitaxel pharmacokinetics with the development of peripheral neuropathy in patients with advanced câncer

22. Richardson et al (2001) A focused exercise regimen improves clinical measures of balance in patients with peripheralneuropathy

23. Rauck et al (2007) Lacosamide in painful diabetic peripheral neuropathy - a phase 2 double-blind placebo-controlled study

24. Argyriou et al (2006) A randomized controlled trial evaluating the efficacy and safety of vitamin E supplementation for protection against cisplatin-induced peripheral neuropathy: final results

25. Barton et al (2011) A double-blind, placebo-controlled trial of a topical treatment for chemotherapy-inducedperipheral neuropathy: NCCTG trial N06CA

26. Casellini et al (2007) A 6-month, randomized, double-masked, placebo-controlled study evaluating the effects of the protein kinase C-beta inhibitor ruboxistaurin on skin microvascular blood flow and other measures of diabetic peripheral neuropathy

27. Davis et al (2011) A randomized, double-blinded, placebo-controlled phase II trial of recombinant human leukemia inhibitory factor (rhuLIF, emfilermin, AN1424) to prevent chemotherapy induced peripheralneuropathy

28. Freeman et al (2007) Randomized study of tramadol/acetaminophen versus placebo in painful diabetic peripheralneuropathy

29. Haak et al (2000) Effects of alpha-lipoic acid on microcirculation in patients with peripheral diabetic neuropathy

30. Haak et al (2000) Effects of alpha-lipoic acid on microcirculation in patients with peripheral diabetic neuropathy

31. Satoh et al (2011) Efficacy and safety of pregabalin for treating neuropathic pain associated with diabeticperipheral neuropathy: a 14 week, randomized, double-blind, placebo-controlled trial

32. Stubblefield et al (2005) Glutainine as a neuroprotective agent in high-dose paclitaxel-induced peripheral neuropathy: a clinical and electrophysiologic study

33. Argyriou et al (2006) Preventing paclitaxel-induced peripheral neuropathy: a phase II trial of vitamin E supplementation

34. Valensi et al (2005) A multicenter, double-blind, safety study of QR-333 for the treatment of symptomatic diabeticperipheral neuropathy - a preliminary report

35. Corso et al (2010) Bortezomib-induced peripheral neuropathy in multiple myeloma: a comparison between previously treated and untreated patients

36. LeMaster et al (2008) Effect of weight-bearing activity on foot ulcer incidence in people with diabetic peripheral neuropathy: feet first randomized controlled trial

37. Offidani et al (2004) Common and rare side-effects of low-dose thalidomide in multiple myeloma: focus on the dose-minimizing peripheral neuropathy

38. Kottschade et al (2011) The use of vitamin E for the prevention of chemotherapy-induced peripheral neuropathy: results of a randomized phase III clinical trial

39. Kadiroglu et al (2008) The effect of venlafaxine HCl on painful peripheral diabetic neuropathy in patients with type 2 diabetes mellitus

40. Cavallo et al (2009) Rituximab in cryoglobulinemic peripheral neuropathy

41. Kluding et al (2012) The effect of exercise on neuropathic symptoms, nerve function, and cutaneous innervation in people with diabetic peripheral neuropathy

42. Weintraub et al (2004) Pulsed magnetic field therapy in refractory neuropathic pain secondary to peripheralneuropathy: electrodiagnostic parameters - pilot study

43. Kruse et al (2010) Fall and balance outcomes after an intervention to promote leg strength, balance, and walking in people with diabetic peripheral neuropathy: "feet first" randomized controlled trial

44. Smith et al (2010) Pilot trial of a patient-specific cutaneous electrostimulation device (MC5-A Calmare (R)) for chemotherapy-induced peripheral neuropathy

45. Amin et al (2003) A pilot study of the beneficial effects of amantadine in the treatment of painful diabeticperipheral neuropathy

46. Baron et al (2008) Efficacy and safety of pregabalin in patients with diabetic peripheral neuropathy or postherpetic neuralgia: open-label, non-comparative, flexible-dose study

47. Schroeder et al (2012) Acupuncture for chemotherapy-induced peripheral neuropathy (CIPN): a pilot study using neurography

48. Donald et al (2011) Evaluation of acupuncture in the management of chemotherapy-induced peripheral neuropathy

49. Lavery et al (2008) Does anodyne light therapy improve peripheral neuropathy in diabetes? A double-blind, sham-controlled, randomized trial to evaluate monochromatic infrared photoenergy

50. Clifft et al (2005) The effect of monochromatic infrared energy on sensation in patients with diabetic peripheral neuropathy - a double-blind, placebo-controlled study

51. Donofrio et al (2005) Safety and effectiveness of topiramate for the management of painful diabetic peripheral neuropathy in an open-label extension study

52. Schroeder et al (2007) Acupuncture treatment improves nerve conduction in peripheral neuropathy

53. Li et al (2010) Long term Tai Chi exercise improves physical performance among people with peripheral neuropathy

54. Shimozuma et al (2012) Taxane-induced peripheral neuropathy and health-related quality of life in postoperative breast cancer patients undergoing adjuvant chemotherapy: N-SAS BC 02, a randomized clinical trial

55. Ghoreishi et al (2012) Omega-3 fatty acids are protective against paclitaxel-induced peripheral neuropathy: a randomized double-blind placebo controlled trial

56. Thisted et al (2006) Dextromethorphan and quinidine in adult patients with uncontrolled painful diabetic peripheral neuropathy: a 29-day, multicenter, open-label, dose-escalation study

57. Bestard et al (2011) An open-label comparison of nabilone and gabapentin as adjuvant therapy or monotherapy in the management of neuropathic pain in patients with peripheral neuropathy

58. Tong et al (2010) Fifteen-day acupuncture treatment relieves diabetic peripheral neuropathy

59. Sandercock et al (2009) Gabapentin extended release for the treatment of painful diabetic peripheral neuropathy: efficacy and tolerability in a double-blind, randomized, controlled clinical trial

60. Hotta et al (2008) Stratified analyses for selecting appropriate target patients with diabetic peripheral neuropathy for long-term treatment with an aldose reductase inhibitor

61. Brooks et al (2008) Endothelial and neural regulation of skin microvascular blood flow in patients with diabeticperipheral neuropathy: effect of treatment with the isoform-specific protein kinase C beta inhibitor, ruboxistaurin

62. Fonseca et al (2013) Metanx in type 2 diabetes with peripheral neuropathy: a randomized trial

63. Kawai et al (2010) Effects of epalrestat, an aldose reductase inhibitor, on diabetic peripheral neuropathy in patients with type 2 diabetes, in relation to suppression of N-epsilon-carboxymethyl lysine

64. Gewandter et al (2014) A phase III randomized, placebo-controlled study of topical amitriptyline and ketamine for chemotherapy-induced peripheral neuropathy (CIPN): a University of Rochester CCOP study of 462 cancer survivors

65. Chron et al (2006) Short-term oral isotretinoin therapy does not cause clinical or subclinical peripheral neuropathy

66. Biyik et al (2013) Gabapentin versus pregabalin in improving sleep quality and depression in hemodialysis patients with

peripheral neuropathy: a randomized prospective crossover trial

67. Sandercock et al (2012) A gastroretentive gabapentin formulation for the treatment of painful diabetic peripheral neuropathy: efficacy and tolerability in a double-blind, randomized, controlled clinical trial

68. Teixeira E (2010) The effect of mindfulness meditation on painful diabetic peripheral neuropathy in adults older than 50 years

69. Swislocki et al (2010) A randomized clinical trial of the effectiveness of photon stimulation on pain, sensation, and quality of life in patients with diabetic peripheral neuropathy

70. Leal et al (2014) North Central Cancer Treatment Group/ Alliance Trial N08CA-the use of glutathione for prevention of paclitaxel/carboplatin-induced peripheral neuropathy

71. Ermis et al (2010) Gabapentin therapy improves heart rate variability in diabetic patients with peripheralneuropathy

72. Kanai et al (2009) The analgesic effect of a metered-dose 8% lidocaine pump spray in posttraumatic peripheral neuropathy: a pilot study

73. Coriat et al (2014) Treatment of oxaliplatin-induced peripheral neuropathy by intravenous mangafodipir

74. Vinik et al (2014) A randomized withdrawal, placebo-controlled study evaluating the efficacy and tolerability of tapentadol extended release in patients with chronic painful diabetic peripheral neuropathy

75. Rauck et al (2013) A randomized, controlled trial of gabapentin enacarbil in subjects with neuropathic pain associated with diabetic peripheral neuropathy

76. Mueller et al (2013) Weight-bearing versus nonweight-bearing exercise for persons with diabetes and peripheral neuropathy: a randomized controlled trial

77. Klassen et al (2008) High-tone external muscle stimulation in end-stage renal disease: effects an symptomatic diabetic and uremic peripheral neuropathy

78. Shibuya et al (2014) Safety and efficacy of intravenous ultra-high dose methylcobalamin treatment for peripheral neuropathy: a phase I/II open label clinical trial

79. Kessler   et al (2013) Whole body vibration therapy for painful diabetic peripheral neuropathy: a pilot study

80. Afonseca et al (2013) Vitamin E for prevention of oxaliplatin-induced peripheral neuropathy: a pilot randomized clinical trial

81. Atalay et al (2013) Cross-over, open-label trial of the effects of gabapentin versus pregabalin on painful peripheral neuropathy and health-related quality of life in haemodialysis patients

82. Tuveson et al (2011) Ondansetron, a 5HT3-antagonist, does not alter dynamic mechanical allodynia or spontaneous ongoing pain in peripheral neuropathy

83. Campone et al (2013) A double-blind, randomized phase II study to evaluate the safety and efficacy of acetyl-L-carnitine in the prevention of sagopilone-induced peripheral neuropathy

84. Dixit et al (2014) Effect of aerobic exercise on peripheral nerve functions of population with diabetic peripheralneuropathy in type 2 diabetes: a single blind, parallel group randomized controlled trial

85. Burke et al (2006) The effect of monochromatic infrared energy on sensation in subjects with diabetic peripheral neuropathy: a double-blind, placebo-controlled study - Response to Clifft et al

86. Coyne et al (2013) A trial of Scrambler therapy in the treatment of cancer pain syndromes and chronic chemotherapy-induced peripheral neuropathy

87. Bao et al (2014) A pilot study of acupuncture in treating bortezomib-induced peripheral neuropathy in patients with multiple myeloma

88. Razazian et al (2014) Evaluation of the efficacy and safety of pregabalin, venlafaxine, and carbamazepine in patients with painful diabetic peripheral neuropathy

89. Otis et al (2014) A randomized controlled pilot study of a cognitive-behavioral therapy approach for painful diabetic peripheral neuropathy

90. Syngle et al (2014) Minocycline improves peripheral and autonomic neuropathy in type 2 diabetes: MIND study

91. Raskin et al (2014) Pregabalin in patients with inadequately treated painful diabetic peripheral neuropathy: a randomized withdrawal trial

92. Anjayani et al (2014) Sensory improvement of leprosy peripheral neuropathy in patients treated with perineural injection of platelet-rich plasma

93. Anastasi et al (2013) Acu/Moxa for distal sensory peripheral neuropathy in HIV: a randomized control pilot study

94. York et al (2009) Motor learning of a gait pattern to reduce forefoot plantar pressures in individuals with diabetic peripheral neuropathy

95. Pachman et al (2015) Pilot evaluation of Scrambler therapy for the treatment of chemotherapy-induced peripheral neuropathy

96. Tuttle et al (2015) Effect of ruboxistaurin on albuminuria and estimated GFR in people with diabetic peripheral neuropathy: results from a randomized trial

97. Slangen et al (2014) Spinal cord stimulation and pain relief in painful diabetic peripheral neuropathy: a prospective two-center randomized controlled trial

98. Zhang et al (2014) Investigating the role of backward walking therapy in alleviating plantar pressure of patients with diabetic peripheral neuropathy

99. Strempska et al (2013) The effect of high-tone external muscle stimulation on symptoms and electrophysiological parameters of uremic peripheral neuropathy

100. Nagano et al (2012) Efficacy of lafutidine, a histamine H2-receptor antagonist, for taxane-induced peripheralneuropathy in patients with gynecological malignancies

Chapter 13

1. Fisher R et al (2010) Electrical stimulation of the anterior nucleus of thalamus for treatment of refractory epilepsy

2. Morrell MJ (2011) Responsive cortical stimulation for the treatment of medically intractable partial epilepsy

3. Glauser TA et al (2010) Ethosuximide, valproic acid, and lamotrigine in childhood absence epilepsy

4. Engel Jr J et al (2012) Early surgical therapy for drug-resistant temporal lobe epilepsy: a randomized trial

5. Brodie MJ et al (2010) Efficacy and safety of adjunctive ezogabine (retigabine) in refractory partial epilepsy

6. French JA et al (2011) Randomized, double-blind, placebo-controlled trial of ezogabine (retigabine) in partial epilepsy

7. Ben-Menachem E et al (2010) Eslicarbazepine acetate as adjunctive therapy in adult patients with partial epilepsy

8. Krueger DA et al (2013) Everolimus treatment of refractory epilepsy in tuberous sclerosis complex

9. Cook MJ et al (2013) Prediction of seizure likelihood with a long-term, implanted seizure advisory system in patients with drug-resistant epilepsy: a first-in-man study

10. Jozwiak S et al (2011) Antiepileptic treatment before the onset of seizures reduces epilepsy severity and risk of mental retardation in infants with tuberous sclerosis complex

11. Halasz P et al (2010) Long-term efficacy and safety of esli-carbazepine acetate: results of a 1-year open-label extension study in partial-onset seizures in adults with epilepsy

12. Heck CN et al (2014) Two-year seizure reduction in adults with medically intractable partial onset epilepsy treated with responsive neurostimulation: final results of the RNS System Pivotal trial

13. Darke K et al (2010) Developmental and epilepsy outcomes at age 4 years in the UKISS trial comparing hormonal treatments to vigabatrin for infantile spasms: a multi-centre randomised trial

14. Holsti M et al (2010) Intranasal midazolam vs. rectal diazepam for the home treatment of acute seizures in pediatric patients with epilepsy

15. McLachlan RS et al (2010) Bilateral hippocampal stimulation for intractable temporal lobe epilepsy: Impact on seizures and memory

16. Baulac M et al (2012) Efficacy and tolerability of zonisamide versus controlled-release carbamazepine for newly diagnosed partial epilepsy: a phase 3, randomised, double-blind, non-inferiority trial

17. Elkhayat HA et al (2010) Melatonin and sleep-related problems in children with intractable epilepsy

18. Sun W et al (2012) Low-frequency repetitive transcranial magnetic stimulation for the treatment of refractory partial epilepsy: a controlled clinical study

19. Klein P et al (2010) Ketogenic diet treatment in adults with refractory epilepsy

20. Ciechanowski P et al (2010) PEARLS depression treatment for individuals with epilepsy: a randomized controlled trial

21. Stefan H et al (2012) Transcutaneous vagus nerve stimulation (t-VNS) in pharmacoresistant epilepsies: a proof of concept trial

22. Chang EF et al (2010) Predictors of efficacy after stereotactic radiosurgery for medial temporal lobe epilepsy

23. DeGiorgio CM et al (2011) Randomized controlled trial of trigeminal nerve stimulation for drug-resistant epilepsy

24. Rektor I et al (2012) Perampanel Study 207: long-term open-label evaluation in patients with epilepsy

25. Glauser TA et al (2013) Ethosuximide, valproic acid, and lamotrigine in childhood absence epilepsy: initial monotherapy outcomes at 12 months

26. Wagner JL et al (2010) Pilot study of an integrated cognitive-behavioral and self-management intervention for youth with

epilepsy and caregivers: coping openly and personally with epilepsy (COPE)

27. Oh Y-S et al (2012) Cognitive improvement after long-term electrical stimulation of bilateral anterior thalamic nucleus in refractory epilepsy patients

28. Smith M et al (2011) Efficacy and tolerability of the Modified Atkins Diet in adults with pharmacoresistant epilepsy: a prospective observational study

29. Gonzalez-Heydrich J et al (2010) Adaptive phase I study of OROS methylphenidate treatment of attention deficit hyperactivity disorder with epilepsy

30. Chhun S et al (2011) A prospective open-labeled trial with levetiracetam in pediatric epilepsy syndromes: continuous spikes and waves during sleep is definitely a target

31. Chaytor N et al (2011) Long-term outcomes from the PEARLS randomized trial for the treatment of depression in patients with epilepsy

32. Masur D et al (2013) Pretreatment cognitive deficits and treatment effects on attention in childhood absence epilepsy

33. Mikati MA et al (2010) Intravenous immunoglobulin therapy in intractable childhood epilepsy: open-label study and review of the literature

34. Helmstaedter C et al (2011) Hippocampal resection length and memory outcome in selective epilepsy surgery

35. Uberos J et al (2011) Normalization of the sleep-wake pattern and melatonin and 6-sulphatoxy-melatonin levels after a therapeutic trial with melatonin in children with severe epilepsy

36. Herzog AG et al (2012) Progesterone vs. placebo therapy for women with epilepsy: a randomized clinical trial

37. Quigg M et al (2011) Neuropsychological outcomes after gamma knife radiosurgery for mesial temporal lobe epilepsy: a prospective multicenter study

38. Fattore C et al (2011) A multicenter, randomized, placebo-controlled trial of levetiracetam in children and adolescents with newly diagnosed absence epilepsy

39. Miranda MJ et al (2011) Danish study of a Modified Atkins diet for medically intractable epilepsy in children: can we achieve the same results as with the classical ketogenic diet?

40. Hart Y et al (2011) Epilepsy after subarachnoid hemorrhage: the frequency of seizures after clip occlusion or coil embolization of a ruptured cerebral aneurysm: results from the International Subarachnoid Aneurysm Trial Clinical article

41. Schramm J et al (2011) Randomized controlled trial of 2.5-cm versus 3.5-cm mesial temporal resection in temporal lobe epilepsy-Part 1: Intent-to-treat analysis

42. Cardamone M et al (2014) Mammalian target of rapamycin inhibitors for intractable epilepsy and subependymal giant cell astrocytomas in tuberous sclerosis complex

43. Sharma S et al (2013) Use of the modified Atkins diet for treatment of refractory childhood epilepsy: a randomized controlled trial

44. Hufnagel A et al (2013) Long-term safety and efficacy of eslicarbazepine acetate as adjunctive therapy in the treatment of partial-onset seizures in adults with epilepsy: results of a 1-year open-label extension study

45. Kanemura H et al (2012) Valproate sodium enhances body weight gain in patients with childhood epilepsy: a pathogenic mechanisms and open-label clinical trial of behavior therapy

46. Salam MT et al (2012) An implantable closedloop asynchronous drug delivery system for the treatment of refractory epilepsy

47. Kumada T et al (2012) Efficacy and tolerability of modified Atkins diet in Japanese children with medication-resistant epilepsy

48. McLaughlin DP et al (2011) A randomized trial of a group based cognitive behavior therapy program for older adults with epilepsy: the impact on seizure frequency, depression and psychosocial well-being

49. Borggraefe I et al (2013) Levetiracetam vs. sulthiame in benign epilepsy with centrotemporal spikes in childhood: a double-blinded, randomized, controlled trial (German HEAD Study)

50. Lee KJ et al (2012) Long-term outcome of anterior thalamic nucleus stimulation for intractable epilepsy

51. Cho YW et al (2011) The effect of levetiracetam monotherapy on subjective sleep quality and objective sleep parameters in patients with epilepsy: compared with the effect of carbamazepine-CR monotherapy

52. Wechsler RT et al (2014) Conversion to lacosamide monotherapy in the treatment of focal epilepsy: results from a historical-controlled, multicenter, double-blind study

53. Ryvlin P et al (2014) The long-term effect of vagus nerve stimulation on quality of life in patients with pharmacoresistant focal epilepsy: the PuLsE (Open Prospective Randomized Long-term Effectiveness) trial

54. Guerrini R et al (2014) Adjunctive zonisamide therapy in the long-term treatment of children with partial epilepsy: results of an open-label extension study of a phase III, randomized, double-blind, placebo-controlled trial

55. Modi AC et al (2013) Preliminary feasibility, acceptability, and efficacy of an innovative adherence intervention for children with newly diagnosed epilepsy

56. Thammongkol et al (2012) Efficacy of the ketogenic diet: which epilepsies respond?

57. Valentin et al (2013) Deep brain stimulation of the centromedian thalamic nucleus for the treatment of generalized and frontal epilepsies

58. Klinkenberg et al (2012) Vagus nerve stimulation in children with intractable epilepsy: a randomized controlled trial

59. Pop et al (2011) Acute and long-term safety of external trigeminal nerve stimulation for drug-resistant epilepsy

60. Eun et al (2011) Comparative trial of low- and high-dose zonisamide as monotherapy for childhood epilepsy

61. Lin et al (2011) Mozart K.448 acts as a potential add-on therapy in children with refractory epilepsy

62. Biton et al (2014) Brivaracetam as adjunctive treatment for uncontrolled partial epilepsy in adults: a phase III randomized, double-blind, placebo-controlled trial

63. Bazil et al (2012) Pregabalin increases slow-wave sleep and may improve attention in patients with partial epilepsy and insomnia

64. Kwan et al (2010) Efficacy and safety of levetiracetam as adjunctive therapy in adult patients with uncontrolled partial epilepsy: the Asia SKATE II Study

65. Ryvlin et al (2014) Adjunctive brivaracetam in adults with uncontrolled focal epilepsy: results from a double-blind, randomized, placebo-controlled trial

66. Trinka et al (2012) KOMET: an unblinded, randomised, two parallel-group, stratified trial comparing the effectiveness of levetiracetam with controlled-release carbamazepine and extended-release sodium valproate as monotherapy in patients with newly diagnosed epilepsy

67. Guerrini et al (2013) A randomized phase III trial of adjunctive zonisamide in pediatric patients with partial epilepsy

68. Goldberg-Stern et al (2012) Effect of melatonin on seizure frequency in intractable epilepsy: a pilot study

69. Cervenka et al (2012) E-mail management of the Modified Atkins Diet for adults with epilepsy is feasible and effective

70. Sun et al (2011) Low-frequency repetitive transcranial magnetic stimulation for the treatment of refractory partial epilepsy

71. Tonekaboni et al (2010) Efficacy of the Atkins Diet as therapy for intractable epilepsy in children

72. Herzog et al (2010) A comparison of anastrozole and testosterone versus placebo and testosterone for treatment of sexual dysfunction in men with epilepsy and hypogonadism

73. Machline et al (2010) Evaluation of sleep quality in patients with refractory seizures who undergo epilepsy surgery

74. Santos et al (2012) The impact of methylphenidate on seizure frequency and severity in children with attention-deficit-hyperactivity disorder and difficult-to-treat epilepsies

75. Sohal et al (2011) Responsive neurostimulation suppresses synchronized cortical rhythms in patients with epilepsy

76. Ramsay et al (2010) Efficacy, tolerability, and safety of rapid initiation of topiramate versus phenytoin in patients with new-onset epilepsy: a randomized double-blind clinical trial

77. Muller et al (2010) Outcome of vagus nerve stimulation for epilepsy in Budapest

78. Lazzari et al (2013) Prevention of bone loss and vertebral fractures in patients with chronic epilepsy-Antiepileptic drug and osteoporosis prevention trial

79. Arhan et al (2011) Effects of epilepsy and antiepileptic drugs on nitric oxide, lipid peroxidation and xanthine oxidase system in children with idiopathic epilepsy

80. Chen et al (2014) Brief group psychoeducation for psychogenic nonepileptic seizures: a neurologist-initiated program in an epilepsy center

81. Kwan et al (2014) Adjunctive brivaracetam for uncontrolled focal and generalized epilepsies: results of a phase III, double-blind, randomized, placebo-controlled, flexible-dose trial

82. Klein et al (2012) Results of phase 2 safety and feasibility study of treatment with levetiracetam for prevention of post-traumatic epilepsy

83. Yuen et al (2012) Non-randomized open trial of eicosapentaenoic acid (EPA), an omega-3 fatty acid, in ten people with chronic epilepsy

84. Lu et al (2011) Efficacy and safety of adjunctive zonisamide in adult patients with refractory partial-onset epilepsy: a randomized, double-blind, placebo-controlled trial

85. Eftekhari et al (2013) Bumetanide reduces seizure frequency in patients with temporal lobe epilepsy

86. Bochynska et al (2012) The effect of vitamin B supplementation on homocysteine metabolism and clinical state of patients with chronic epilepsy treated with carbamazepine and valproic acid

87. Kim et al (2011) Resection frequency map after awake resective surgery for non-lesional neocortical epilepsy involving eloquent areas

88. Sergott et al (2010) Vigabatrin-induced peripheral visual field defects in patients with refractory partial epilepsy

89. Saetre et al (2010) Antiepileptic drugs and quality of life in the elderly: results from a randomized double-blind trial of carbamazepine and lamotrigine in patients with onset of epilepsy in old age

90. Drane et al (2015) Better object recognition and naming outcome with MRI-guided stereotactic laser amygdalohippocampotomy for temporal lobe epilepsy

91. Hensley-Judge et al (2013) Visual field defects after radiosurgery for mesial temporal lobe epilepsy

92. Clark et al (2013) Intravenous topiramate: safety and pharmacokinetics following a single dose in patients with epilepsy or migraines taking oral topiramate

93. Klinkenberg et al (2013) Behavioural and cognitive effects during vagus nerve stimulation in children with intractable epilepsy - a randomized controlled trial

94. Motta et al (2013) Progesterone therapy in women with epilepsy

95. Raju et al (2011) Efficacy of 4:1 (classic) versus 2.5:1 ketogenic ratio diets in refractory epilepsy in young children: a randomized open labeled study

96. Freitas-Lima et al (2011) Influence of enzyme inducing antiepileptic drugs on the pharmacokinetics of levetiracetam in patients with epilepsy

97. Peijing et al (2014) An alternative therapy for drug-resistant epilepsy: transcutaneous auricular vagus nerve stimulation

98. Werhahn KJ et al (2011) The safety and efficacy of add-on levetiracetam in elderly patients with focal epilepsy: a one-year observational study

99. Abou-Khalil B et al (2013) A double-blind, randomized, placebo-controlled trial of a diazepam auto-injector administered by caregivers to patients with epilepsy who require intermittent intervention for acute repetitive seizures

100. Aliasgharpour M et al (2013) Effects of an educational program on self-management in patients with epilepsy

INDEX

Printed in the United States
By Bookmasters